Lecture Notes in Mechanical Engineering

Lecture Notes in Mechanical Engineering (LNME) publishes the latest developments in Mechanical Engineering—quickly, informally and with high quality. Original research reported in proceedings and post-proceedings represents the core of LNME. Volumes published in LNME embrace all aspects, subfields and new challenges of mechanical engineering. Topics in the series include:

- Engineering Design
- Machinery and Machine Elements
- Mechanical Structures and Stress Analysis
- Automotive Engineering
- Engine Technology
- Aerospace Technology and Astronautics
- Nanotechnology and Microengineering
- Control, Robotics, Mechatronics
- MEMS
- Theoretical and Applied Mechanics
- Dynamical Systems, Control
- Fluid Mechanics
- Engineering Thermodynamics, Heat and Mass Transfer
- Manufacturing
- Precision Engineering, Instrumentation, Measurement
- Materials Engineering
- Tribology and Surface Technology

To submit a proposal or request further information, please contact the Springer Editor of your location:

China: Ms. Ella Zhang at ella.zhang@springer.com
India: Priya Vyas at priya.vyas@springer.com
Rest of Asia, Australia, New Zealand: Swati Meherishi at swati.meherishi@springer.com
All other countries: Dr. Leontina Di Cecco at Leontina.dicecco@springer.com

To submit a proposal for a monograph, please check our Springer Tracts in Mechanical Engineering at http://www.springer.com/series/11693 or contact Leontina.dicecco@springer.com

Indexed by SCOPUS. All books published in the series are submitted for consideration in Web of Science.

More information about this series at http://www.springer.com/series/11236

Rajeev Agrawal · Jinesh Kumar Jain ·
Vinod Singh Yadav · Vijaya Kumar Manupati ·
Leonilde Varela
Editors

Recent Advances in Industrial Production

Select Proceedings of ICEM 2020

Springer

Editors
Rajeev Agrawal
Department of Mechanical Engineering
Malaviya National Institute of Technology
Jaipur, Rajasthan, India

Jinesh Kumar Jain
Department of Mechanical Engineering
Malaviya National Institute of Technology
Jaipur, Rajasthan, India

Vinod Singh Yadav
Department of Mechanical Engineering
National Institute of Technology
Uttarakhand
Srinagar, Uttarakhand, India

Vijaya Kumar Manupati
Department of Mechanical Engineering
National Institute of Technology Warangal
Warangal, Telangana, India

Leonilde Varela
Department of Production and Systems
University of Minho
Braga, Portugal

ISSN 2195-4356 ISSN 2195-4364 (electronic)
Lecture Notes in Mechanical Engineering
ISBN 978-981-16-5280-6 ISBN 978-981-16-5281-3 (eBook)
https://doi.org/10.1007/978-981-16-5281-3

This Springer imprint is published by the registered company Springer Nature Singapore Pte Ltd.
The registered company address is: 152 Beach Road, #21-01/04 Gateway East, Singapore 189721, Singapore

Contents

Synthesis and Investigation of Mechanical Behavior of Aluminum Oxide/Silicon Carbide Filled Bi-directional Woven E-glass Fiber Reinforcement Epoxy Polymer Composites 1
Rahul Sharma and Rahul Sen

A Contemporary Review of Pushing/Pulling Strength at Different Handle Heights 13
Chinmay Pancholi, Rahul Jain, K. B. Rana, and M. L. Meena

Development of an Optimal PID Controller for the 4-DOF Manipulator Using Genetic Algorithm 23
Shyam Prasad Kodali, Ravi Kumar Mandava, and Boggarapu Nageswara Rao

Assessing the Carbon Foot Print of an Ayurveda Medical Institute: A Case of National Institute of Ayurveda, Jaipur, India 33
Gaurav Gaurav, Tejas Kumar, Chandni Khandelwal, Alok Bihari Singh, M. L. Meena, Sundeep Kumar, and G. S. Dangayach

The Linkages Between Spare-Parts Management and Maintenance Management in Army Supply Chain of Vehicles 45
Chander Sheikhar and Rajesh Matai

An Integrated ISM-AHP Computing Framework for Evaluating Supply Chain Competitiveness 61
Ajay Verma and Nisha Singhal

Experiencing Life Cycle Assessment in Indian Additive Manufacturing Industries: Needs, Challenges and Solutions 67
Alok Yadav, Anbesh Jamwal, Rajeev Agrawal, and Sundeep Kumar

Analysis of Barriers in Sustainable Supply Chain Management for Indian Automobile Industries ... 79
Anbesh Jamwal, Akshay Patidar, Rajeev Agrawal, Monica Sharma, and Vijaya Kumar Manupati

Optimization of Injection Timing for a C.I. Engine Fuelled with Gomutra Emulsified Diesel ... 91
Amit Jhalani, Dilip Sharma, Digambar Singh, and Pushpendra Kumar Sharma

Extraction of 3D Solid Model of Decaying Tooth from 2D DICOM Images ... 101
Vaishnavi V. Gejji, Ravi Yerigeri, and C. M. Choudhari

Analyzing the Drivers for Lean and Green Manufacturing Using ISM Approach ... 111
Sarita Prasad, A. Neelakanteswara Rao, and Krishnanand Lanka

Quantifiable Contribution of Sustainable Manufacturing Enablers in Indian SMEs ... 123
Deepak Sharma, Pravin Kumar, and Rajesh Kr Singh

Circular Economy and Sustainable Manufacturing: A Bibliometric Based Review ... 137
Kiran Gundu, Anbesh Jamwal, Alok Yadav, Rajeev Agrawal, Jinesh Kumar Jain, and Sundeep Kumar

Identification of Challenges & Practices of Sustainability in Indian Apparel and Textile Industries ... 149
Amit Vishwakarma, M. L. Meena, G. S. Dangayach, and Sumit Gupta

Challenges of Adoption of Blockchain Technology in Supply Chain: An Overview ... 157
Dnyaneshwar Jivanrao Ghode, Rakesh Jain, and Gunjan Soni

A Bibliometric Analysis of Sustainable Supply Chain Management: Research Implications and Future Perspectives ... 167
Akshay Patidar, Monica Sharma, Rajeev Agrawal, Kuldip Singh Sangwan, and Anbesh Jamwal

Effect of Handle Orientation on Two-Handed Push Strength in Unorganized Sector Workers ... 189
Rahul Jain, K. B. Rana, Vikky Kumar, and M. L. Meena

Weight Optimization of Gears in the Transmission of an All-Terrain Vehicle ... 197
Shantanu Tiwari, Shikhar Verma, Shashwat Kulshreshtha, Naman Varshney, and Mayank Kushawaha

Ergonomic Interventions in Maintaining Postural Stability in Pregnant Women at Their Workplaces 205
Nikhil Yadav, M. L. Meena, G. S. Dangayach, and Yashvin Gupta

Thermal Analysis of Discontinuity in Deposited Bead 217
Soumen Mandal, Manish Oraon, and Subrata Kumar

A Case Study on Survey Plan for Digital Merchandising System and Consumer Association Management 227
Sundeep Kumar, Vikram Singh Rathore, and Alok Mathur

Medical Applications of Rapid Prototyping Technology 241
Rakesh Chaudhari, Praveen Kumar Loharkar, and Asha Ingle

Prediction of Temperature During Susceptor-Assisted Microwave Heating of Aluminum Using Parametric Simulation 251
Praveen Kumar Loharkar, Asha Ingle, and Himanshu Singh

Tool Path Generation for Layer Specific Infill Density in Fused Filament Fabrication (FFF) 261
Krishnanand, Ankit Nayak, Shivam Soni, and Mohammad Taufik

Cost Minimization in a Scheduling Problem with Unrestricted and Restricted Common Due Date 269
Prasad Bari and Prasad Karande

Prioritization of Sustainability Criteria of Service Only Supply Chain: A Case Study of Indian Hospitals 279
Ramji Nagariya, Divesh Kumar, and Ishwar Kumar

Incremental Sheet Metal Forming: The State of Art and Its Future Prospects 293
Bittu Toppo, Manish Oraon, and Manish Kr. Roy

Lean Implementation Value in Automobile Sector 303
Arshit Kapoor, Krishna Mohan Agarwal, and Aaryan Sheokand

Optimization of Turning Process Parameters Using Entropy-Gra and Dear Methods 315
K. Srinivasulu Reddy, V. Venkata Reddy, and Ravi Kumar Mandava

An Impact of Internet Based Supply Chain Management Using IOT in Current Scenario 325
Sundeep Kumar, Vikram Singh Rathore, and Alok Mathur

An Analytical Study on Big Data Management for Supply Chain Analytics 333
Sundeep Kumar, Vikram Singh Rathore, and Alok Mathur

Die Design and Its Parameters for Grain Refinement of AA6XXX Series Through Equal Channel Angular Pressing 343
Arshit Kapoor, Bhuwan Gupta, Abhishek Singhal, and Krishna Mohan Agarwal

Integrating the Challenges of Cloud Computing in Supply Chain Management 355
Subhodeep Mukherjee, Venkataiah Chittipaka, Manish Mohan Baral, and Sharad Chandra Srivastava

Finite Element Analysis of Infill Density on the Compressive Strength of 3D Printed Parts by Fused Deposition Modelling 365
Anurag Kumar Mishra, Abhishek Kaushal, Rabesh Kumar Singh, and Anuj Kumar Sharma

An Experimental Investigation of Ribbed Solar Air Heater—Free Convection 373
Niraj Kumar, Manoj Kumar Singh, Vinod Singh Yadav, Vineet Singh, and Anurag Maheswari

Insolation Effect on Solar Photovoltaic Performance Parameters 383
Navneet, Neha Khuran, and Smita Pareek

Comparing Theoretical and Practical Aspects of Process Management Practices for Competitive Potential in SMEs 391
Satyajit Mahato, Amit Rai Dixit, and Rajeev Agrawal

Development of a Mathematical Model for the Software Defect Rework Process to Optimize Defect Rework—A Six-Sigma Case Study 403
Satyajit Mahato, Amit Rai Dixit, and Rajeev Agrawal

Theoretical Analysis of 1st Law and 2nd Law Efficiency of a Solar Pump for Geographical Location 28.10′N, 78.23′E 411
Vineet Singh, Vinod Singh Yadav, Vishal Saxena, Niraj Kumar, and Anurag Maheswari

Experimental Analysis of a Thermoelectric Air-Conditioning System with Desiccant Dehumidification 419
Anurag Maheswari, Manoj Kumar Singh, Yogesh K. Prajapati, Niraj Kumar, and Vineet Singh

Fabrication of Isogrids by Conventional and Unconventional Techniques: A Comparative Review Study 429
K. Tripathi, K. Kukreja, and N. Gupta

An Integrated Lean Six Sigma Model for Enhancing the Competitive Advantage of Industries 437
S. K. Tiwari, R. K. Singh, and Sharad Chandra Srivastava

Agri-fresh Supply Chain Management: A Systematic Literature Review ... 449
J. Krishna Manasvi and Rajesh Matai

Methods to Measure Residual Stresses in 3D Printed Objects: A Review ... 459
Devesh, Devender, and N. Gupta

Corrosion Performance in Grain Structure of C22 in Acidic Environment ... 469
Aezeden Mohamed, Kamalakanta Muduli, Devendra K. Yadav, and Pankaj Jena

An Evolutionary Tomographic Reconstruction Procedure for Defect Identification Using Time-of-Flight of Ultrasound ... 477
Shyam Prasad Kodali and Boggarapu Nageswara Rao

Manufacturing Is Not as Usual: Lessons Learnt from COVID-19 Pandemic ... 485
Anbesh Jamwal, Rajeev Agrawal, and Monica Sharma

Machine Learning in CAD/CAM: What We Think We Know So Far and What We Don't ... 495
Smriti Upmanyu, Anil Upmanyu, Anbesh Jamwal, and Rajeev Agrawal

Assesment of Traditional and Hybrid Controller for Controlling Robotic Manipulator System ... 509
Aditi Saxena, Jitendra Kumar, Vinay Kumar Deolia, and Debanik Roy

About the Editors

Dr. Rajeev Agrawal is currently an associate professor at the Department of Mechanical Engineering, Malaviya National Institute of Technology Jaipur, (Rajasthan). He obtained his B.E. (Mechanical) from Govt. Engg. College, Jabalpur (M.P.) and M.E. (Prod. Engg.) and Ph.D. from the MNNIT, Allahabad (U.P.) and BIT, Mesra (Ranchi), respectively. Dr. Rajeev Agrawal is having more than 20 years of professional experience. His areas of the research interests are in soft computing (GA, ANN, Fuzzy) applications to manufacturing, modelling and simulation of manufacturing systems, sustainable manufacturing, lean six sigma, supply chain design and reconfigurable manufacturing system (RMS). He has published more than 55 papers in reputed international journals and conferences. Dr. Agrawal received the Fellow of The Institution of Engineer's (IEI), India, in 2020. He is acting as a reviewer for many peer-reviewed Journals. Currently, he is an editorial board member for the International Journal of Business and Systems Research (IJBSR).

Dr. Jinesh Kumar Jain is currently working as Associate Professor at Malaviya National Institute of Technology Jaipur (MNIT) having research interests in joining and welding, additive manufacturing and bio-fabrication, design and manufacturing technology, operation management, sustainable manufacturing. He completed his Ph.D. and M.Tech. from MNIT and B.E. from MBM Engineering College Jodhpur. He has more than twenty years of teaching, research and industrial experience, published many technical papers in journals and conference proceedings, authored books, chapters in reputed publications and delivered lectures in reputed institutions in India and abroad. Prior to joining MNIT, he has worked in several renowned organizations. Dr. Jain is a life member of many reputed professional bodies and board member in the number of Indian universities. Dr. Jain successfully guided many M.Tech. dissertations and presently guiding many doctoral and masters scholars. He successfully organized many short term courses and faculty development programmes to impart, encourage the young fellows to proceed by sharing knowledge and experiences from eminent researchers.

Dr. Vinod Singh Yadav is currently an assistant professor at the Department of Mechanical Engineering, National Institute of Technology Uttarakhand, Uttarakhand. He obtained his Bachelor of Engineering (Mechanical Engineering) from University of Rajasthan, and M.Tech. (Energy Engineering) and Ph.D. from Malaviya National Institute of Technology, Jaipur. His major interest of research areas are alternate fuels in IC engines, renewable energy technologies, thermal storage techniques and their applications. He has published more than 35 research papers in reputed journals and international conferences. He is the author of 05 text and reference books. He is also reviewer of more than 05 SCI Elsevier journals. He is technical editor of Journal of Automotive Mechanical & Aerospace Engineering Research and Consulting Editor of Hon. Editorial Board Member recognized by Innovative Scientific Research Professional Malaysia.

Dr. Vijaya Kumar Manupati is currently working as an Assistant Professor in the Department of Mechanical Engineering, NIT Warangal. He received his Ph.D. in the Department of Industrial and Systems Engineering from the Indian Institute of Technology Kharagpur. His current research interests include Intelligent manufacturing systems, cyber-physical systems, sustainable supply chain, and health care systems. He has published more than 80 publications which include prestigious journals like International Journal of Production Research, Computers, and Industrial Engineering, International Journal of Advanced Manufacturing Technology, Journal of measurements, International Journal of Computer Integrated Manufacturing, etc. He is acting as a reviewer for more than 30 peer-reviewed journals. Currently, he is acting as an editorial review board member of International Journal of Sustainable Entrepreneurship and corporate social responsibility, International Journal of Web Portals IGI Global publications. He received an Early Carrier Research Grant from Department of Science and Technology (DST) for his research work on Telefacturing Systems. He is a member of the Institute of Industrial and Systems Engineering (IISE), Institute of Engineers (IEI) India, Life member of International Association of Engineers, US, and also acting as a technical committee member of various International conferences.

Dr. Leonilde Varela received her Ph.D. degree in Industrial Engineering and Management from the University of Minho, Portugal, in 2007. She is Assistant Professor at Department of Production and Systems of University of Minho. Her main research interests are in manufacturing management, production planning and control, optimization, artificial intelligence, meta-heuristics, scheduling, web based systems, services and technologies, mainly for supporting engineering and production management, collaborative networks, decision making models, methods and systems, and virtual and distributed enterprises. She has published more than 150 refereed scientific papers in international conferences and in international scientific books and journals, indexed in the Web of Science and/or in the Scopus data bases. She coordinates R&D projects in the area of production and systems engineering, mainly concerning the development of web-based platforms and decision support models, methods and systems. She is a frequent paper reviewer for several journals, such as, Journal of Computer Integrated Manufacturing, Engineering Applications of Artificial Intelligence, IEEE Transactions on Neural Networks and Learning Systems, Journal of Decision Systems, Sustainability, International Journal of Management and Fuzzy Systems, International Journal of Sustainability Management and Information Technologies, International Journal of Intelligent Enterprise, International Journal of Decision Support Systems Technology, Management and Production Engineering Review, Mathematical Problems in Engineering, Journal of Robotics. She is a member of several international networks, such as: Machine Intelligence Research Labs, Scientific Network for Innovation and Research Excellence (MirLabs), Euro Working Group of Decision Support Systems (EWG-DSS), Institute of Electrical and Electronics Engineers (IEEE); System, Man, and Cybernetics Society (IEEE SMC), Industrial Engineering Network (IE Network), and Institute of Industrial and Systems Engineers (IISE).

Synthesis and Investigation of Mechanical Behavior of Aluminum Oxide/Silicon Carbide Filled Bi-directional Woven E-glass Fiber Reinforcement Epoxy Polymer Composites

Rahul Sharma and Rahul Sen

Abstract The research paper describes the development of a new set of polymer composites from Epoxy. The polymer matrix composite (PMC) comprises epoxy resin (LY 556), E-glass fiber as a reinforced material and Alumina and silicon carbide as filler material. The epoxy resin (LY 556) and hardener (HY951) were mixed at 10:1 at room temperature. It is synthesized by different weight percentage of aluminum oxide/silicon carbide (1:1) (0, 2, 4 and 6 wt%). The testing includes the investigations of mechanical properties in tensile and flexural mode. It is observed from experiment work that 4 wt% aluminum oxide-silicon carbide filled glass fiber reinforced epoxy composite sample possess optimum mechanical properties. It comprises yield strength (102.56 MPa), tensile modulus (4167.85 MPa), tensile ultimate strength (170.839 MPa), flexural strength (162.5515 MPa) and flexural modulus (10736.4287 MPa), which is required in all general application areas where the epoxy-based composites are prominently used like—automobile, aircrafts components and sports goods.

Keywords Polymer composite · Epoxy · Alumina · Silicon carbide

1 Introduction

Combine properties of different materials can be achieved by adding two or more materials which are known as composite materials. Composites are implanted by number of continuous and discontinuous phase. Continuous phase is known as matrix where as discontinuous phase is called reinforcement. Reinforcement based composites are harder and stronger than matrix based composites. Reinforcement can either non fibrous or fibrous [1, 2]. The polymer matrix composites (PMC) with wove glass fiber have interesting applications in aerospace, structural and automobile engineering. They have several advantages such as excellent mechanical strength, high

R. Sharma (✉) · R. Sen
Mechanical Engineering Department, Poornima College of Engineering, Jaipur 302022, India

R. Agrawal et al. (eds.), *Recent Advances in Industrial Production*, Lecture Notes in Mechanical Engineering, https://doi.org/10.1007/978-981-16-5281-3_1

modulus and high corrosion resistance and are quite cost-effective [3]. The mechanical properties of PMCs are completely dependent upon the particulate size, filler–matrix interface adhesion and loading characteristics. Strengthening of polymers is generally achieved with the reinforcement of fibers. On the other hand, higher tensile strength and higher impact strength are generally achieved by adding more layers of fiber to the polymer matrix. The glass fibers consist of such properties and that's why these are extensively used as reinforcement material in epoxy matrix-based composites. The mechanical strength of polymer composite can be made better by the addition of silicon-iron oxide particulates along with fibers [4]. Silicon carbide shows favorable chemical and mechanical characteristics at high temperatures in structural applications, such as air craft's, boats and automobiles. SiC is the simple chemical compound of carbon (C) and silicon (Si). Silicon carbide is molten salts up to 800 °C. It has advantages include its high elastic modulus, high hardness, excellent thermal shock resistance, low thermal expansion, low density, high thermal conductivity, high strength and better chemical inertness. Alumina has advantages include dielectric properties, hardness, high strength, wear-resistant, stiffness, alkali attack at elevated temperatures, and resistance to strong acid. Alumina has the most stable phase is hexagonal alpha. For the application of structural, hexagonal alpha phase of alumina interphase at elevated temperatures. Considering all such information, an attempt has been made to develop and synthesis the above-mentioned composites from the materials discussed and their properties investigated by general testing.

1.1 Literature Review

Various research works also been done on such matters. T. Rangaswamy and K. Devendra evaluated epoxy composite for addition and non-addition of filler materials. He used different filling materials like silicon carbide, aluminum oxide and magnesium hydroxide. The addition of filler materials increases overall mechanical properties. The optimum result for all mechanical properties found in 10 wt% SiC samples [5]. Amar Patnaik and Gaurav Agarwal et al. investigated that the overall strength of epoxy-based composite enhanced with the increase in weight percentage of silicon carbide. Although it shows a non-linear increment in all mechanical properties. The tensile and flexural strength increase up to 10 wt%, impact strength and hardness up to 15 wt% and shear strength up to 20 wt% content of silicon carbide [6]. Aluminum oxide and silicon carbide (1:1) particulates were reinforced into the epoxy resin based composites attribute the better mechanical characteristics as compared to polyester resin based composites [7]. Alumina filler GF reinforced epoxy based composites increased the impact energy as well as hardness compared to SiO_2 and TiO_2 modifiers epoxy based composites. Agglomeration is also presented in SEM observation due to the bigger particle size of Al_2O_3 in composites [8]. Sham Prasad et al. investigated the chopped stands E glass fiber composites with filler aluminum oxide. It was observed by the author from the experiment that 4 wt% Al_2O_3 composite possess maximum tensile strength (158.26 MPa), young's modulus (8692.06 MPa) and percentage of elongation (4.53%) among other filler specimens. The epoxy basic composite without filler has a tensile strength (111.21 MPa), Young's modulus (7524.27 MPa) and percentage of elongation (3.51%) [9]. The addition of nano particulars of SiC was improved not only thermal resistance but also mechanical characteristics [10]. Similar erosion wears testing conduction Al_2O_3 filler glass fiber reinforced polyester composite based is examined to lower wear rate as compare to SiC and CBPD modifiers polyester based composites [11]. SaiSravani et al. studied the mechanical response such as the strength of tensile, impact as well as flexural and also the hardness of $CaCO_3$ and Al_2O_3 filled epoxy-based GF reinforced composites. It was examined from the experiment by the author that the flexural and tensile properties of composites with the addition of filler material got decreased as compare to pure epoxy-based composites. So it was evaluated that all filler materials don't show better mechanical properties [12]. Alam and Chowdhury evaluated that mechanical characteristics such as tensile strength (20.353%), impact strength (12.935%) and flexural strength (26.652%) of $CaCO_3$–Al_2O_3–MgO–CuO epoxy based composites are improved as compare with filler $CaCO_3$–Al_2O_3–MgO–TiO_2 epoxy based composites [13]. Al_2O_3/SiC filler GF reinforced epoxy based composites enhanced mechanical characteristics with in 6 vol% than that of the composite made of TiO_2/SiC, only Al_2O_3 and only TiO_2 as filler [14]. So it is clear from the works of literature that the Silicon carbide and aluminum oxide are suitable filler materials, which can be used in developing new epoxy-based composite. Here in the research work both of these are used as filler materials, as discussed in the materials and methodology section (Figs. 1 and 2).

Fig. 1 Epoxy resin (LY 556) and hardener (HY951)

Fig. 2 E-glass fibre

2 Materials and Methodology

Al_2O_3 (200 mesh) and SiC (1200 mesh) (1:1) particulates were reinforced into the epoxy resin (chemical formula shown in Fig. 3) at the varying amount of 0, 2, 4 and 6 wt%. The epoxy was taken into a beaker and the aluminum and silicon carbide were mixed into it in the required ratio, stirred the solution for 5–8 min. Also, curing agent HY951 (Araldite hardener, the chemical formula shown in Fig. 4) is mixed with epoxy composite in 1:10 (as per resin and hardener manufacturer's recommendation). The silicone spray is spread on the inside surface of the bottom mold sheet. Epoxy resin is spread uniformly on the mold sheet using a brush after that E GF is placed on it. A roller is used to remove the air traps present on the mold surface and to remove the excess of resin present on the surface. Mild pressure is applied to the roller while moving it on the surface. The same process is repeated until the eight layers of glass fiber (5 mm thickness) are stacked. The silicone spray is spread on the inside surface of the top mold sheet before placing it on the mold. The pressure is applied by the 3 kg sand lime brick (dimension 230 mm × 110 mm) on the stacked layers. After curing at room temperature in 24 h, the mold is opened and the composite part is taken out.

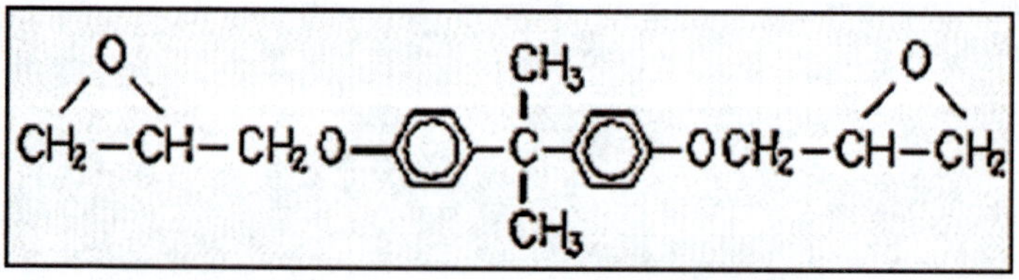

Fig. 3 Formula of epoxy (LY 556) [15]

Fig. 4 Formula of hardener resin (HY951) [16]

$H_2N\text{-}CH_2\text{-}CH_2\text{-}NH\text{-}CH_2\text{-}\ CH_2.NH\text{-}CH_2\text{-}\ CH_2\text{-}NH$

3 Results and Discussions

3.1 Tensile Properties

Mechanical tensile properties of four specimens with variations of filler particles of aluminum oxide and silicon carbide (1:1) into woven E-GF reinforced epoxy composites were analyzed as per ASTM D2344-84 standard (Figs. 5 and 6).

It is well known that the tensile strength property of composites is generally determined by the glass fiber content and the glass fiber strength. The ratio of the

Fig. 5 Specimens after tensile test

Fig. 6 Instron 1195 machine (tensile mode)

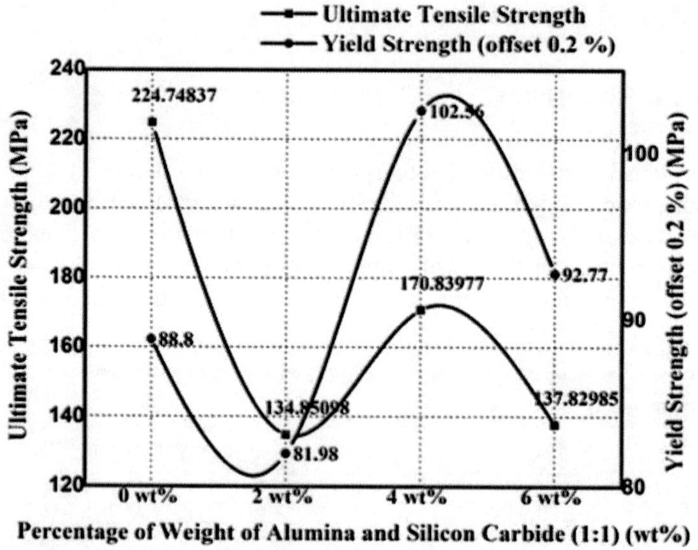

Fig. 7 Ultimate tensile as well as yield strength versus percentage of weight of Al_2O_3/SiC (1:1)

maximum load on a material to the initial cross section of the test specimen is representing the ultimate tensile strength of a material. Figure 7 represents the curve of ultimate tensile strength as well as yield strength versus the percentage of the weight of aluminum oxide and silicon carbide (1:1) particulate with woven E GF-reinforced epoxy composites.

The woven E GF-reinforced epoxy based composite without particulate filler has a maximum ultimate tensile strength of 224.748 MPa due to better bonding strength between the polymer matrix and glass fiber. After that tensile strength suddenly decrease due to alumina/silicon carbide (1:1) particles act as a barrier for transferring stress from one point to another and poor bonding strength between epoxy resin, alumina, silicon carbide and glass fiber. It was evaluated from Fig. 7 that the value of the ultimate tensile strength (134.850–170.839 MPa) is further increases for 2–4 wt% alumina/silicon carbide (1:1) content epoxy composites due to the increase of transfer of stresses from one point to another.

The composite without particulate filler has a maximum tensile modulus of 5253.04 MPa due to better bonding strength between the polymer matrix and glass fiber. It was evaluated from Fig. 8 that the value of the tensile modulus (3812.76–4167.85 MPa) is increased for 2–4 wt% alumina/silicon carbide (1:1) content epoxy composites due to the relatively lower strain rates of epoxy composites.

The woven E GF-reinforced epoxy composite without particulate filler has a maximum tensile elongation at a maximum load of 0.079 mm/mm due to better bonding strength between the polymer matrix and glass fiber. After that tensile strength suddenly decrease due to alumina/silicon carbide (1:1) particles act as a barrier for transferring stress from one point to another and poor bonding strength

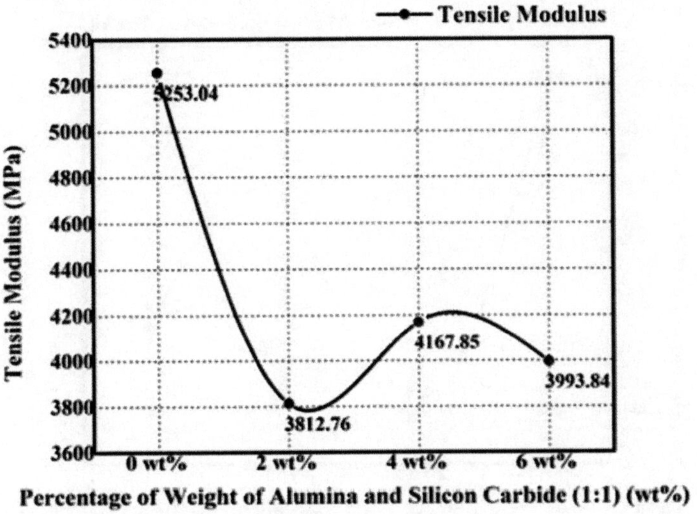

Fig. 8 Tensile modulus versus percentage of weight of Al_2O_3/SiC (1:1)

between epoxy resin, aluminum oxide, silicon carbide and glass fiber. It was evaluated from Fig. 9 that the value of the tensile elongation (0.045–0.0544 mm/mm) is further increases for 2–4 wt% alumina/silicon content epoxy composites due to the increase of transfer of stresses from one point to another.

The woven E GF-reinforced epoxy composite without particulate filler has a maximum barring load of 15.282 kN due to better bonding strength between the

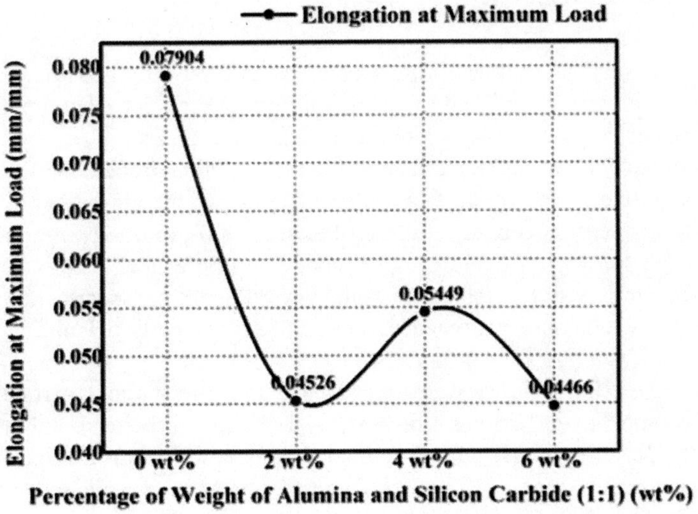

Fig. 9 Elongation at maximum load versus percentage of weight of Al_2O_3/SiC (1:1)

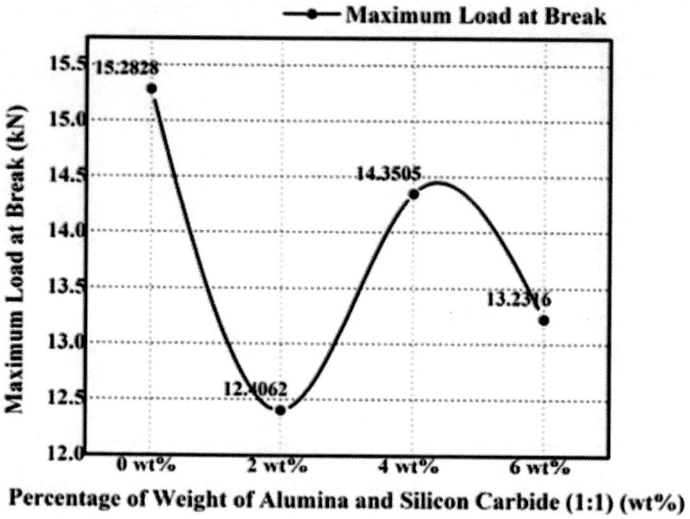

Fig. 10 Maximum loads at break versus percentage of weight of Al_2O_3/SiC (1:1)

polymer matrix and glass fiber. It was evaluated from Fig. 10 that the value of the tensile barring load (12.406–14.350 kN) is further increases for 2–4 wt% alumina/silicon carbide (1:1) content epoxy composites due to the increase of transfer of stresses from one point to another.

3.2 *Flexural Properties*

Mechanical flexural properties of four specimens with variations of filler particles of alumina/silicon carbide (1:1) into woven E-GF reinforced epoxy composites were analyzed as per ASTM D2344-84 standard (Figs. 11 and 12).

It is well known that the flexural strength and modulus properties of composites are generally determined by the glass fiber content and the glass fiber strength. In flexural strength test specimen is placed between two point supports and load is applied at the top layer of mid of specimen. The upper fiber of the specimen is subjected to compressive stress where as the lower fiber of the specimen is subjected to tensile stress. The observation carried out on four woven E-GF reinforced epoxy composites are given in Figs. 13 and 14.

The woven E GF-reinforced epoxy based composite without particulate filler has a maximum flexural strength of 200.027 MPa due to better bonding strength between the polymer matrix and glass fiber. It was evaluated from Fig. 14 that the value of the flexural strength (156.488 MPa to 162.551 MPa) is increased for 2–4 wt% alumina/silicon carbide (1:1) content epoxy composites due to the good compatibility between the polymer matrix and glass fiber.

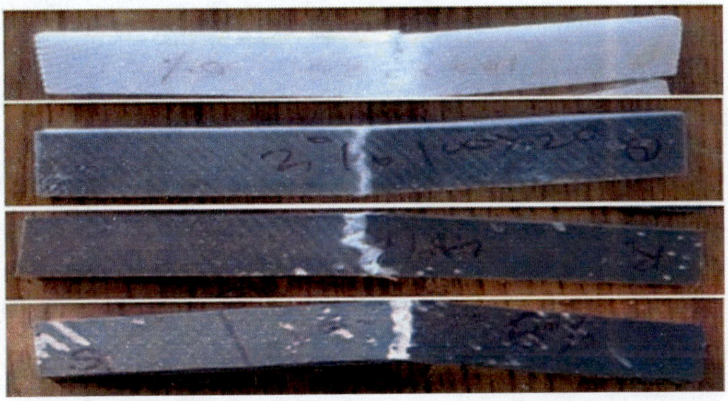

Fig. 11 Specimens after flexural test

Fig. 12 Instron 1195 machine (flexural mode)

The woven E GF-reinforced epoxy based composite without particulate filler has maximum flexural modulus 13944.101 MPa due to better bonding strength between the polymer matrix and glass fiber. It was observed from Fig. 13 that the value of the flexural modulus (9172.60 MPa to 10736.428 MPa) is increased for 2–4 wt% alumina/silicon carbide (1:1) content epoxy composites due to the relatively lower strain rates of epoxy composites.

4 Results and Discussions

Experimental and analytical evaluation of mechanical properties on aluminum oxide and silicon carbide (1:1) filled woven E-GF reinforced epoxy composites have led to the following specific conclusions:

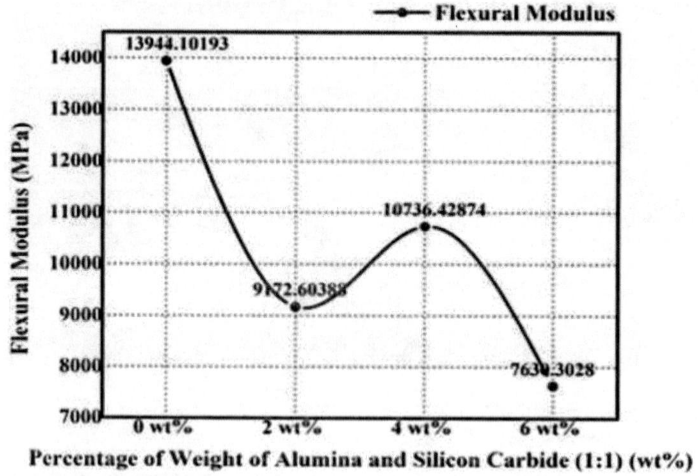

Fig. 13 Flexural modulus versus percentage of weight of Al_2O_3/SiC (1:1)

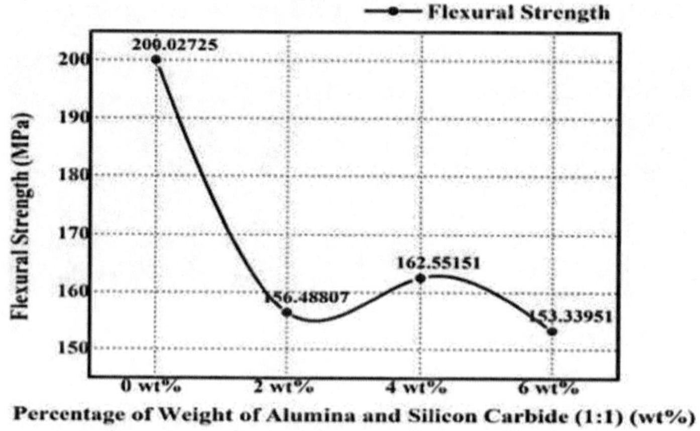

Fig. 14 Flexural strength versus percentage of weight of Al_2O_3/SiC (1:1)

- The woven E GF-reinforced epoxy based composite without particulate filler has maximum values of ultimate strength (224.748 MPa) and yield strength (88.80 MPa).
- The woven E GF-reinforced epoxy based composite without particulate filler has maximum values of flexural strength (200.0273 MPa) and flexural modulus (13,944.1019 MPa).
- Specimen 4 wt% has maximum values of yield strength (102.56 MPa), tensile modulus (4167.85 MPa), elongation at maximum load (0.05449 mm/mm) and maximum load at break (14.3505 kN).

- Specimen 4 wt.% has maximum values of flexural strength (162.5515 MPa) and flexural modulus (10736.4287 MPa).
- The optimum mechanical properties found in the epoxy composite sample with 4 wt.% of silicon carbide and alumina particulates, which comprises mechanical properties in tensile and flexural mode, which is required in all general application areas where the epoxy based composites are prominently used like—automobile components, aircrafts components and sports goods. In future characterisation, inter laminar shear strength and thermal characteristics for developed composite will be checked for different combinations.

References

1. Patnaik A, Satapathy A, Mahapatra SS, Dash RR (2009) Tribo-performance of polyester hybrid composites: damage assessment and parameter optimization using Taguchi design. Mater Des 30(1):57–67. https://doi.org/10.1016/j.matdes.2008.04.057
2. Patnaik A, Kumar P, Biswas S, Kumar M (2012) Investigations on micro-mechanical and thermal characteristics of glass fiber reinforced epoxy based binary composite structure using finite element method. Comput Mater Sci 62:142–151. https://doi.org/10.1016/j.commatsci.2012.05.020
3. Mallick PK (2007) Fiber reinforced composites materials manufacturing and design, 3rd edn. Taylor & Francis Group, United States
4. Rajadurai A (2016) Thermo-mechanical characterization of siliconized E-glass fiber/hematite particles reinforced epoxy resin hybrid composite. Appl Surf Sci 384:99–106. https://doi.org/10.1016/j.apsusc.2016.04.185
5. Devendra K, Rangaswamy T (2012) Determination of mechanical properties of Al_2O_3, Mg $(OH)_2$ and SiC filled E-glass/epoxy composites. Int J Eng Res Appl 2(5):2028–2033
6. Agarwal G, Patnaik A, Sharma RK (2013) Thermo-mechanical properties of silicon carbide filled chopped glass fiber reinforced epoxy composites. Int J Adv Struct Eng 5(1):21. https://doi.org/10.1186/2008-6695-5-21
7. Rajesh S, VijayaRamnath B, Elanchezhian C, Aravind N, Rahul VV, Sathish S (2014) Analysis of mechanical behavior of glass fibre/Al_2O_3-SiC reinforced polymer composites. Procedia Eng 97:598–606. https://doi.org/10.1016/j.proeng.2014.12.288
8. Nayak RK, Dash A, Ray BC (2014) Effect of epoxy modifiers ($Al_2O_3/SiO_2/TiO_2$) on mechanical performance of epoxy/glass fiber hybrid composites. Procedia Mater Sci 6:1359–1364. https://doi.org/10.1016/j.mspro.2014.07.115
9. Ravikumar M, Prasad MS (2014) Fracture toughness and mechanical properties of aluminum oxide filled chopped strand mat e-glass fibre reinforced-epoxy composites. Int J Sci Res Publ 4(7):1–7
10. Kwon DJ, Shin PS, Kim JH, Baek YM, Park HS, DeVries KL, Park JM (2017) Interfacial properties and thermal aging of glass fiber/epoxy composites reinforced with SiC and SiO_2 nanoparticles. Compos B Eng 130:46–53. https://doi.org/10.1016/j.compositesb.2017.07.045
11. Mahapatra SS, Patnaik A (2009) Study on mechanical and erosion wear behavior of hybrid composites using Taguchi experimental design. Mater Des 30(8):2791–2801. https://doi.org/10.1016/j.matdes.2009.01.037
12. SaiSravani K, Reddy BRG, Mohammed R (2017) Effect of $CaCO_3$ and Al_2O_3 fillers on mechanical properties of glass/epoxy composites. Int J Modern Trends Sci Technol 3(6):207–213

13. Alam MS, Chowdhury MA (2020) Characterization of epoxy composites reinforced with $CaCO_3$–Al_2O_3–MgO–TiO_2/CuO filler materials. Alex Eng J. https://doi.org/10.1016/j.aej.2020.07.017
14. Kiran MD, Govindaraju HK, Jayaraju T (2018) Evaluation of mechanical properties of glass fiber reinforced epoxy polymer composites with alumina, titanium dioxide and silicon carbide fillers. Mater Today Proc 5(10):22355–22361. https://doi.org/10.1016/j.matpr.2018.06.602
15. Rao S, Rao RMVGK (2008) Cure studies on bifunctional epoxy matrices using a domestic microwave oven. Polym Testing 27(5):645–652. https://doi.org/10.1016/j.polymertesting.2008.04.005
16. Ratna D, Simon GP (2001) Mechanical characterization and morphology of carboxyl randomized poly (2-ethyl hexyl acrylate) liquid rubber toughened epoxy resins. Polymer 42(18):7739–7747. https://doi.org/10.1016/S0032-3861(01)00278-6

A Contemporary Review of Pushing/Pulling Strength at Different Handle Heights

Chinmay Pancholi, Rahul Jain, K. B. Rana, and M. L. Meena

Abstract A significant amount of research has been done to examine factors affecting pushing and pulling activities such as cart weight/load, exerted forces, handling tasks (push and pull) and more. However, very little has been done to determine the height of the handle to work efficiently. Ergonomics principles are the best technique for implementing in any industry to evaluate and control the risks arising due to various work activities. The purpose of this paper was to undertake a detailed analysis of the handle height used in the cart so that there is no or less musculoskeletal strain resulting from tasks of pushing and pulling. 34 studies are classified for specific application sectors for finding an effective way to work at different handle heights. The selected articles were grouped for publication trend, factors, field wise journals and anthropometric data. The current research gives researchers knowledge about the current status and most effective handle height used by researchers for future work.

Keywords Handle height · Musculoskeletal disease · Factors affecting pushing and pulling forces

1 Introduction

Even after rapid technological advances, the role of human physical fitness in operating equipment or moving objects continues. From farming to building, factories are filled with jobs involving manual tasks such as pushing, dragging, and raising. Workers who constantly push, pull and lift objects may be at risk of developing injuries from sprains and strains to chronic back pain as part of their everyday work activities. Manual jobs, such as moving up and down, trigger workers in ways that can be painful and harmful to reorient their bodies. Overexertion, an accident that is

C. Pancholi · R. Jain (✉) · K. B. Rana
Department of Mechanical Engineering, University Departments, Rajasthan Technical University, Kota 324010, India

M. L. Meena
Department of Mechanical Engineering, MNIT, Jaipur, India

R. Agrawal et al. (eds.), *Recent Advances in Industrial Production*, Lecture Notes in Mechanical Engineering, https://doi.org/10.1007/978-981-16-5281-3_2

all too common, may be the main consequence of inappropriate pushing and pulling. Over-exertion lawsuits with an estimated cost of $8147 were 22% of all employee compensation recorded cases in 2012 in Ohio [1]. For all employees in manual occupations, appropriate ergonomic and biomechanical knowledge is crucial.

Employees may use secure and defensive movements to complete push-and-pull activities, from warehouse operations to farming and forestry. The ergonomic norms and principles are structured to provide an orientation towards the operator's mental and emotional needs. The design is essentially a compromise between operators' biological needs, as specified by the requirements for ergonomics and the facility's physical specifications. The layout is accomplished mainly by taking into account the reciprocal effects of anthropometry and the location of the equipment elements on the position, control, reach, vision, clearance and interference with the elements of the equipment by the body segments. The posture requirements at job or work performance are determined by all of these design factors [2–6].

Prior studies [3–6] have mentioned a significant impact on the push and pull forces of the handle's vertical height against which one pushes and pulls. Lee [7] discovered a dynamic biomechanical model that push and pull height significantly affected the predicted dynamic compression forces at the lower back [8].

Concerning musculoskeletal complaints, the handle's height and the direction of the forces applied (push or pull) appear to be important risk factors for physical movement [9]. Although the etiology cause of lower back issues is still uncertain, a contributing factor may be compressive and shear forces on low back structures. When pushing and pulling, it is also interesting to consider the lower back's compressive and shear forces.

Pushing is characterized by exerting force on someone to drive them away from oneself, whereas pulling is characterized by exerting force on someone or something to move towards oneself. Often, depending on the hands' vertical height during the push/pull process, the direction of the exerted force is not strictly horizontal and probably includes a vertical component. Push/Pull forces are determined by (i) the main force required to initiate the object's motion, (ii) the continuous force-the lower force needed to continue the motion-and (iii) the preventing force needed to stop the object's motion. Most of the ergonomics literature published discusses initial and continuous forces to push and pull [10].

Over-exertion incidents accounted for 24.4% of all recorded workplace injuries, with a cumulative expense of $15.08 billion, the 2016 Liberty Mutual Occupational Safety Index concluded. Repeated movements, repeated push/pull and work in stressful positions contribute to thousands of workplace injuries every year [1]. Services must be skilled and trained to push and pull objects to avoid injury correctly. Injuries linked to the practice of driving and pulling include tear and stretches; slides, rides, and tumbles; pain in the spinal area; and injury to elbows, hands and wrists.

Proper driving and pulling behavior are a must in the workplace to keep employees satisfied and protected. Best practices must be in place, whether it is production or farming, to provide employees with the means to remain secure and protected in the workplace. To the authors' knowledge, there is no comprehensive examination of literature that outlines the current evidence on cart height handling during push/pull activities linked to musculoskeletal disorders. A systematic analysis by Garg et al. [10] discussed lesser about the handle's height. The purpose of this analysis is to systemize the findings of the research papers which have examined how to manage height in pushing and pulling tasks. Other factors other than handle height during push and pull are also known. Also, workspace or handle design suggestions can be generated by synthesizing the findings.

2 Materials and Methods

Different databases have been used to classify the origins for this literature review (Fig. 1). Initially, Google Scholar was used to taking an initial sample of what kind of papers were available. Different journals that provide essential information on the handle height of a cart used during pushing and pulling were selected. The following methodology filters were applied: Firstly, the relevant articles from different databases using the following terms (pushing and pulling force, handle height, musculoskeletal diseases and evaluation of handle height) were searched. Secondly, the relevancy of the articles was checked for final selection. Finally, the selected items were explored for the subsequent dimensions: Year of publication, Journal title, Field wise journal, Handle height used, Factors affecting the handle height, participants category, and anthropometric data used. After collecting the data

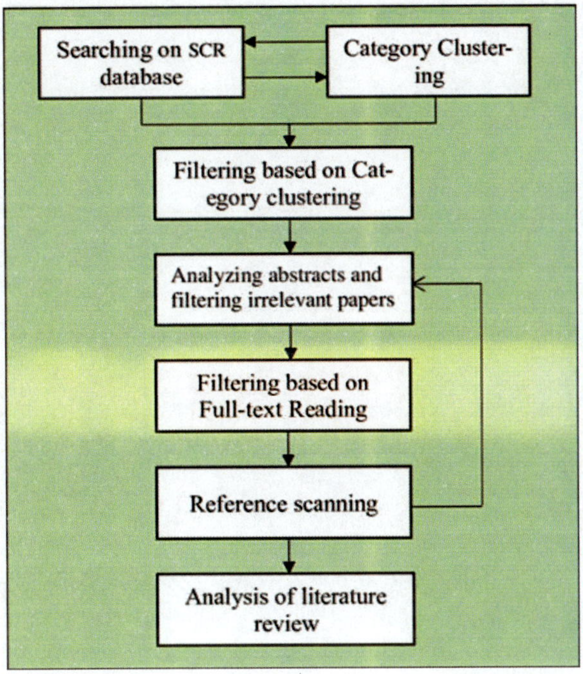

Fig. 1 Flowchart of the literature review

for the above aspects, the data was interpreted statistically, concluding remarks and future research avenues were elaborated.

3 Observations and Recommendations

After applying search finally, 34 articles were selected for further analysis in the current research. This section represents statistical representation of handle height used in the cart while pushing and pulling by the following perspective: (1) trend of publication, (2) field wise journal and (3) handle height used.

3.1 *Year Wise Publication Trend*

The selected papers are treated promptly to model the development by altering the distribution of the number of studies during the 1974–2020 period. It is clear from Fig. 2 that the publishing of papers in this zone is on the rise, which indicates that the need for better handle height has a high priority after 2002 for the selected risk assessment literature.

3.2 *Field Wise Journal*

Figure 3 shows the list of various journals that have published the selected studies in the current research. Most (26%) of the articles are published in "Ergonomics".

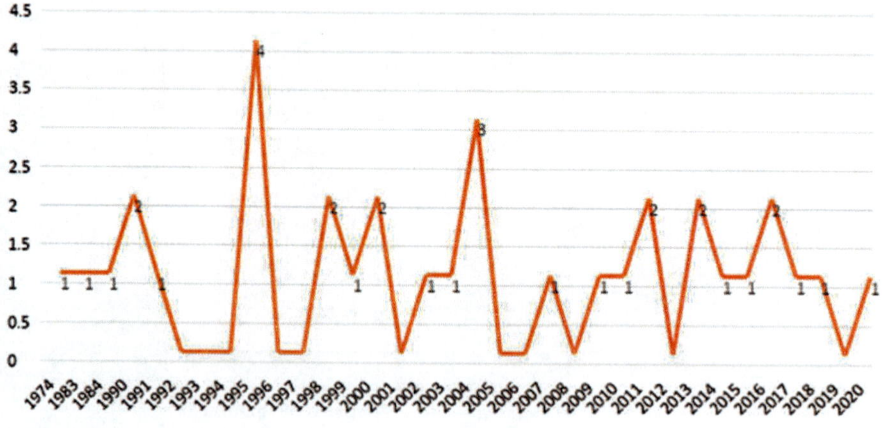

Fig. 2 Year wise publications

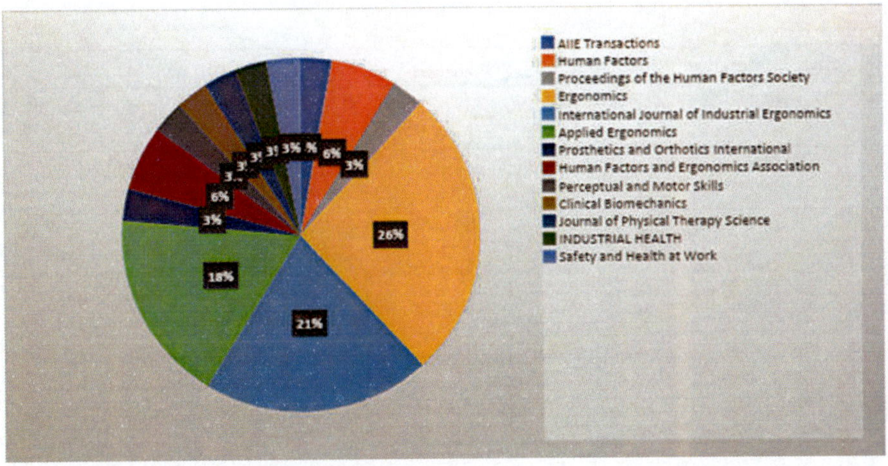

Fig. 3 Journal wise distribution

After this 20% materials are published in the "International Journal of Ergonomics" and about 18% articles are in "Applied Ergonomics". Applied Ergonomics is also a promising avenue for release of the research. After this (12%) of materials are published in human factors related journals (Human factors and Human factors & Ergonomics Association), human factors are the best avenue for publication. Around (6%) of the research is published in safety related journals out of which safety and health at work is the best avenue for publication.

3.3 *Characteristics of the Included Studies*

Table 1 shows the various characteristics considered in the analysis.

Table 1 Summary of articles reviewed

S. No.	Reference	Participants, N, gender	Factors other than handle height
1	[4]	n = 46; 35 (m)/11 (f) (np)	Foot placement, the hands' role in the deployment of the forces, body weight and height
2	[6]	n = 6; 3 (m)/3 (f) (np)	Tight-fitting clothing, foot placement, body postures and handle height
3	[11]	n: 1	Bodyweight, velocity, hand force
4	[2]	n = 8; 8 (m)/0 (f) (np)	Treadmill speed (2 km/h), resting time, climatic conditions, weight and shoulder height of the body

(continued)

Table 1 (continued)

S. No.	Reference	Participants, N, gender	Factors other than handle height
5	[8]	n = 19; 19 (m)/0 (f) (np)	Age, weight, height, shoulder height, arm length
6	[12]	n = 6; 4 (m)/2 (f) (np)	Bodyweight, height, velocity, hand force
7	[13]	n = 20; 10 (m)/10 (f) (np)	Knuckle/knee/hip height and bodyweight
8	[14]	n = 4; 2 (m)/2 (f) (p)	Force, hand height, shoulder height, knuckle Height and cart load
9	[15]	n = 40; 20 (m)/20 (f) (np)	Isokinetic velocity, anthropometric data, time limit for max effort
10	[16]	n = 8; 0 (m)/8 (f) (p)	Bodyweight, height and age
11	[9]	–	Distance, pace, weight of the cart, distance, velocity of the foot, bodyweight
12	[17]	n = 8; 5 (m)/3 (f) (np)	Ankle support shoes, height of the seat pan, bodyweight, height, age
13	[18]	n = 10; 5 (m)/5 (f) (np)	Wheel diameter, cart loads (73,181 kg) and anthropometric data
14	[19]	n = 7; 7 (m)/0 (f) (p)	Cart load, height, weight, joint loading, forces
15	[20]	n = 8; 8 (m)/0 (f) (np)	Direction of force, magnitude of force, body posture
16	[21]	n = 24; 11 (m)/13 (f) (np)	Handle diameter, grip, meal tray size, the height of eye over a cart, cart weight and strength
17	[22]	n = 148; 65 (m)/83 (f)	Knob diameter, cylindrical bar dia and length
18	[23]	n = 7; 7 (m)/0 (f) (p)	Cart weight (85, 135, 320 kg), material properties of cart (Rubber wheels, wheels dia)
19	[24]	n = 12; 12 (m)/0 (f) (p)	Handle bar diameter, Steel chain with the load, Stretching requirement
20	[25]	n = 60; 29 (m)/31 (f) (np)	Age, stature, weight, shoulder height, shoulder circumference, waist size, arm size, forearm size, thigh size, leg size, and skid-free wooden surface
21	[26]	n = 9; 9 (m)/0 (f) (np)	Cart weight, wheel diameter, cart height, rolling resistance and loads applied
22	[27]	n = 20; 10 (m)/10 (f) (np)	Type of device, the magnitude of load, the level of control needed, and the push speed

(continued)

Table 1 (continued)

S. No.	Reference	Participants, N, gender	Factors other than handle height
23	[28]	n = 920; 604 (m)/316 (f) (np)	Body weight, height and age
24	[29]	n = 11; 11 (m)/0 (f) (np)	Cart movement and shoulder height, anthropometric data
25	[30]	n = 19; 9 (m)/10 (f) (np)	Handlebar dia/length and handle cover
26	[31]	n = 30; 8 (m)/22 (f) (p)	Weight on wheelchair, anthropometric data (bodyweight, age and height)
27	[32]	n = 86; 46 (m)/40 (f) (np)	Height, weight, age, handle diameter
28	[10]	–	Slope, wheels, friction, cart weight, feet distance, pushing/pulling frequency, distance between foot
29	[33]	n = 24; 12 (m)/12 (f) (np)	Handle orientation: vertical and horizontal and bodyweight, height, age
30	[34]	n = 100; 50 (m)/50 (f) (np)	Handle diameter, anthropometric data
31	[35]	n = 13; 1 (m)/12 (f) (p)	The handle grip material and size
32	[36]	–	Weight/ load of the cart, location and nature of the handle
33	[37]	n = 31; 17 (m)/14 (f) (np)	Age, gender, height, weight of the body, floor mat, shoes
34	[38]	n = 200; 100 (m)/100 (f); (p)	Age, weight, stature, acromial height, foot positions and handle heights
35	[39]	n = 20; 8 (m)/11 (f) (np)	Two loads (32.5 and 42.5 kg), anthropometric data

n total no. of participants, *np* non-professional, *p* professional, *m* male, *f* female

4 Conclusions

Some conclusions can be drawn from this review: (1) Anthropometric data are certainly the most influential parameter. (2) Handle height at the knee, elbow, and shoulder was mostly used and can reduce at least some measures of musculoskeletal strain and thus used for the management of the risk of injury. (3) Whenever the professional participants were invited for the experiment, they can bear a higher load at different handle height. Thus, we can say that professionals have the task experience and technique that helps them to reduce the strain. (4) Some important factors other than anthropometric data are certainly reducing the musculoskeletal disorders, they are: handle design, exerted forces, cart mass, task experience, foot placement, type of pushing (one-handed or two-handed). Studies have also shown that lowering cart mass reduces stress on musculoskeletal disorders as well.

References

1. Workfit Blog page. https://www.work-fit.com/blog/10-ways-to-stop-workplace-injuries-related-to-pushing-pulling. Last accessed 2020//11//9
2. Bimand AS, Grady RM (1983) Industrial workplace layout design an application of engineering anthropometry. Ergonomics 26(5):433–447
3. Chaffin DB (1972) Biomechanical computerized simulation of human strength in sagittal-plane activities. AIIE Trans 4(1):19–28
4. McDaniel JW (1974) Effects of operator stance on pushing and pulling tasks. AIIE Trans 6(3):185–195
5. Snook SH (1978) The ergonomics society the society's lecture 1978. The design of manual handling tasks. Ergonomics 21(12):963–985
6. Chaffin DB, Andres RO, Garg A (1983) Volitional postures during maximal push/pull exertions in the sagittal plane. Hum Factors 25(5):541–550
7. Lee KS (1982) Biomechanical modelling of cart pushing and pulling. Doctoral dissertation, University of Michigan, Ann Arbor
8. Garg A, Beller D (1990) One-handed dynamic pulling strength with special reference to speed, handle height and angles of pulling. Int J Ind Ergon 6(3):231–240
9. Hoozemans MJM, Van Der Beek AJ, Frings-Dresen MHW et al (1998) Pushing and pulling in relation to musculoskeletal disorders: a review of risk factors. Ergonomics
10. Garg A, Waters T, Kapellusch J et al (2014) Psychophysical basis for maximum pushing and pulling forces: a review and recommendations. Int J Ind Ergon
11. Lee KS, Chaffin DB, Herrin GD (1984) Simulation of cart pushing and pulling. In: Proceedings of the Human Factors Society
12. Lee KS, Chaffin DB, Herrin GD et al (1991) Effect of handle height on lower-back loading in cart pushing and pulling. Appl Ergon
13. Kumar S (1995) Upper body push-pull strength of normal young adults in sagittal plane at three heights. Int J Ind Ergon 15(6):427–436
14. Resnick ML, Chaffin DB (1995) An ergonomic evaluation of handle height and load in maximal and submaximal cart pushing. Appl Ergon
15. Kumar S, Narayan Y, Bacchus C (1995) Symmetric and asymmetric two-handed pull-push strength of young adults. Hum Factors 37(4):854–865
16. Van Der Woude LHV, Van Konlngsbruggen CM, Kroes AL et al (1995) Effect of push handle height on net moments and forces on the musculoskeletal system during standardized wheelchair pushing tasks. Prosthet Orthot Int
17. Mackinnon SN (1998) Isometric pull forces in the sagittal plane. Appl Ergon 29(5):319–324
18. Al-Eisawi KW, Kerk CJ, Congleton JJ et al (1999) The effect of handle height and cart load on the initial hand forces in cart pushing and pulling. Ergonomics
19. Hoozemans MJM, Jansen JP, Van Dieën JH et al (2000) Back compressive and shear forces during cart pushing and pulling. In: Proceedings of the XIVth Triennial Congress of the International Ergonomics Association and 44th Annual Meeting of the Human Factors and Ergonomics Association, "Ergonomics for the New Millennium"
20. De Looze MP, Van Greuningen K, Rebel J et al (2000) Force direction and physical load in dynamic pushing and pulling. Ergonomics
21. Das B, Wimpee J, Das B (2002) Ergonomics evaluation and redesign of a hospital meal cart. Appl Ergon
22. Peebles L, Norris B (2003) Filling, "gaps" in strength data for design. Appl Ergon 34(1):73–88
23. Hoozemans MJM, Kuijer PPFM, Kingma I, Van Dieën JH et al (2004) Mechanical loading of the low back and shoulders during pushing and pulling activities. Ergonomics
24. Lee TH (2004) Static lifting strengths at different exertion heights. Int J Ind Ergon 34(4):263–269
25. Cheng TS, Lee TH (2004) Human pulling strengths in different conditions of exertion. Percept Mot Skills 98(2):542–550

26. Hoozemans MJM, Slaghuis W, Faber GS et al (2007) Cart pushing: the effects of magnitude and direction of the exerted push force, and of trunk inclination on low back loading. Int J Ind Ergon
27. Marras WS, Knapik GG, Ferguson S (2009) Loading along the lumbar spine as influence by speed, control, load magnitude, and handle height during pushing. Clin Biomech
28. Tiwari PS, Gite LP, Majumder J et al (2010) Push/pull strength of agricultural workers in central India. Int J Ind Ergon 40(1):1–7
29. Lee YJ, Hoozemans MJM, van Dieën JH (2011) Handle height and expectation of cart movement affect the control of trunk motion at movement onset in cart pushing. Ergonomics
30. Hoffman SG, Reed MP, Chaffin DB (2011) A study of the difference between nominal and actual hand forces in two-handed sagittal plane whole-body exertions. Ergonomics 54(1):47–59
31. Lee S-Y, Kim S-C, Lee M-H et al (2013) Comparison of shoulder and back muscle activation in caregivers according to various handle heights. J Phys Ther Sci 25(10):1231–1233
32. Lin JH, McGorry RW, Maynard W (2013) One-handed standing pull strength in different postures: Normative data. Appl Ergon 44(4):603–608
33. Chow AY, Dickerson CR (2016) Determinants and magnitudes of manual force strengths and joint moments during two-handed standing maximal horizontal pushing and pulling. Ergonomics 59(4):534–544
34. Or C, Lin JH, Wang H et al (2016) Normative data on the one-handed static pull strength of a Chinese population and a comparison with American data. Ergonomics 59(4):526–533
35. Wallius M-A, Rissanen SM, Bragge T et al (2016) Effects of mop handle height on shoulder muscle activity and perceived exertion during floor mopping using a figure eight method. Ind Health 54(1):58–67
36. Argubi-Wollesen A, Wollesen B, Leitner M et al (2017) Human body mechanics of pushing and pulling: analyzing the factors of task-related strain on the musculoskeletal system. In: Safety and health at work
37. Yu D, Xu X, Lin JH (2018) Impact of posture choice on one-handed pull strength variations at low, waist, and overhead pulling heights. Int J Ind Ergon 64:226–234
38. Jain R, Meena ML, Sain MK, Dangayach GS (2019) Pulling force prediction using neural networks. Int J Occup Saf Ergon 25(2):194–199
39. Li KW, Yi C, Liu M (2020) Maximum endurance time modeling for push and pull tasks considering gender and handle height. Hum Facto Ergon Manuf Serv Ind hfm.20865

Development of an Optimal PID Controller for the 4-DOF Manipulator Using Genetic Algorithm

Shyam Prasad Kodali, Ravi Kumar Mandava, and Boggarapu Nageswara Rao

Abstract Tuning of the PID controller is a time consuming and laborious task and several researchers are working on providing better solutions. In this research article, the authors implemented an optimal PID controller for the 4-DOF planar robotic manipulator which will aid in smooth control of joint motions. The dynamics of the manipulator are calculated following the Lagrange-Euler formulation. After the controller is designed, the controller gains have been tuned utilizing manual tuning procedure reported earlier. As the manual procedure does not guarantee optimal parameters, the authors used a population-based optimization algorithm, that is, a Genetic algorithm for achieving better results. The procedure of tuning controller gains using genetic algorithms is presented and further, the performance of the developed algorithm is verified and compared with that obtained using manual procedure using computer simulations.

Keywords 4-DOF manipulator · PID controller · Manipulator dynamics · Genetic algorithm

1 Introduction

Nowadays, robotic manipulators are introduced in most industrial applications to reduce human effort and exposure to hazardous environments. In general, the manipulator configuration contains three sub-assemblies, namely electrical, mechanical, and control components. In a mechanical context, the manipulator consists of several links connected either in serial or parallel form to an end effector with the help of various prismatic and/or revolute joints. To date, researchers have established several methods for defining the accuracy and performance of manipulators. Soomro [1] and Fu et al. [2] proposed a mathematical model for the 3-DOF robotic manipulator using

S. P. Kodali · B. N. Rao (✉)
Department of Mechanical Engineering, Koneru Lakshmaiah Education Foundation, Vaddeswaram 522502, India

R. K. Mandava
Department of Mechanical Engineering, MANIT Bhopal, Bhopal 462003, India

R. Agrawal et al. (eds.), *Recent Advances in Industrial Production*, Lecture Notes in Mechanical Engineering, https://doi.org/10.1007/978-981-16-5281-3_3

kinematics analysis, for determining the velocity, acceleration, and moment of inertia of different joints of the manipulator. In [3] the authors established a kinematic and dynamic model for a 3-DOF manipulator to control the noise factors. The authors in [4] discussed a generalized algorithm for exact dynamic modeling of RRR flexible spatial manipulator. The goodness of this algorithm being that it uses minimum number of equations for solving manipulator dynamics. Further, the proposed algorithm is also used to control the manipulator with minimum computational cost. Goyal et al. [5] analysed the workspace of the robotic manipulator using the singularities and the Jacobian matrix of the robotic manipulator. Of the many problems associated with tracking of the manipulator's end effector in different applications, systematically designing and governing of the various robotic manipulators joint motions is a challenging task. Tien-Dung et al. [6] implemented a PD controller with online gravity compensation for a 2-DOF manipulator. Later, Titov et al. [7] developed two control strategies for a 2-DOF manipulator, the first one utilizing force-torque control and the second one utilizing impedance control. The authors in [8] implemented a traditional PID controller for a flexible robotic manipulator with six links connected with six joints. So far, researchers around the world are continuing the research on the design of various types of tuning algorithms such as manual, Ziegler and Nichols (Z-N), and soft computing based algorithms. Mandava et al. [9] implemented a PID controller for a 4-DOF serial manipulator and used the manual tuning method to tune the gains. It is observed that the manual tuning method is a time-consuming process. In [10], Ziegler and Nichols established a tuning method for PID controller that was complex and time consuming in tuning the gains. Later, many researchers focussed on genetic algorithm (GA) [11], particle swarm optimization algorithm (PSO) [12], invasive weed optimization (IWO) algorithm [13], fuzzy logic [14], and neural network [15] based control algorithms. Yurni et al. [16] designed a fuzzy PID controller for the 4-DOF industrial robotic arm which was used to identify objects. To control the trajectory of the 3-RRR parallel manipulator, Sheng and Wei [17] designed a GA-based controller that was an improved version in terms of the PID controller's robustness and precision. Nasr and Badr [18] proposed a novel PID controller algorithm for a two-link manipulator. The authors used artificial bee colony (ABC) algorithm to tune the developed controller's gains. In [19] the authors implemented a PSO and ABC-based PID control algorithm for a two-link flexible robotic arm and compared it with the Z-N controller. The authors used PSO and ABC algorithms to tune the K_P, K_D, and K_I of the said developed controller, and studied the controller's performance and accuracy. The results show that the PSO-PID controller is superior to the ABC-PID controller. Recently, Loucif and Kechida [20] designed a sliding mode control (SMC) with a PID surface for a manipulator and the gains were tuned using antlion optimization (ALO) and gray wolf optimizer (GWO) algorithms. The goodness of the algorithm was checked in terms of integral time square error and absolute error. Alongside, Jishnu et al. [21] developed an observer-based PID controller for a four-link planar manipulator. The authors used the PSO algorithm to optimize the controller's pole values and minimize the error. In the present article, the authors implemented a GA-based PID controller for a 4-DOF planar robotic manipulator.

2 Mathematical Formulation

The kinematics and dynamics describing the 4-DOF planar robotic manipulator shown in Fig. 1 are discussed in this section, Using these the respective position and the torque needed at each of the manipulator joints can be estimated. The lengths of the robotic manipulator links are represented by L_1, L_2, L_3, L_4 and the links' masses are designated as m_1, m_2, m_3, m_4 respectively.

Inorder to estimate the robot end effectors' position along with its orientation, the angles made by various links (i.e. L_1 to L_4) of the manipulator are represented as $\theta_1, \theta_2, \theta_3$, and θ_4. Further, a systematic procedure following the D-H notations is used to model the manipulators' forward kinematics, which enables the description of the end effector with respect to the manipulators' base in terms of the spatial Cartesian coordinates. The D-H parameters obtained for the said manipulator are noted in Table 1, and the homogeneous transformation matrix obtained using the forward kinematics approach is given in Eq. (1).

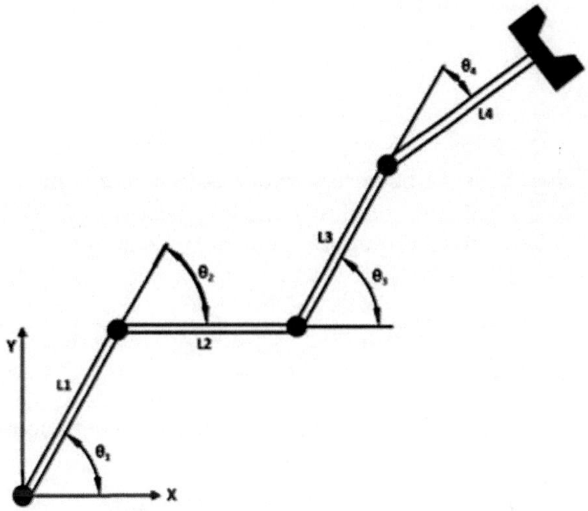

Fig. 1 4-DOF planar manipulator

Table 1 D-H notations

Link (i)	a_i	α_i	d_i	θ_i
1	L_1	0	0	θ_1
2	L_2	0	0	θ_2
3	L_3	0	0	θ_3
4	L_4	0	0	θ_4

$$
{}^{0}T_{4} = \begin{bmatrix} C_{1234} & -S_{1234} & 0 & L_1C_1 + L_2C_{12} + L_3C_{123} + L_4C_{1234} \\ S_{1234} & C_{1234} & 0 & L_1S_1 + L_2S_{12} + L_3S_{123} + L_4S_{1234} \\ 0 & 0 & 1 & 0 \\ 0 & 0 & 0 & 1 \end{bmatrix} \tag{1}
$$

where,

$S_1 = \sin(\theta_1)$, $C_1 = \cos(\theta_1)$, $S_{12} = \sin(\theta_1 + \theta_2)$, $C_{12} = \cos(\theta_1 + \theta_2)$, $S_{123} = \sin(\theta_1 + \theta_2 + \theta_3)$, $C_{123} = \cos(\theta_1 + \theta_2 + \theta_3)$, $S_{1234} = \sin(\theta_1 + \theta_2 + \theta_3 + \theta_4)$ and $C_{1234} = \cos(\theta_1 + \theta_2 + \theta_3 + \theta_4)$.

The Lagrange-Euler formulation is used to model the 4-DOF planar robotic manipulator dynamics. The developed dynamic equations play a significant role while executing the given task in real-time. It is essential to note that limitations arise due to the inertia of moving masses introduced as a result of the motion of various links the system is comprised of. The manipulator dynamics is modeled taking into consideration the gravitational, centrifugal (Coriolis), and inertial force contributions. Thus, required torque at different joints of the manipulator is obtained using the dynamic equation of motion given by Eq. (3).

$$
\tau_i = \sum_{j=1}^{n} M_{ij}(q)\ddot{q}_j + \sum_{j=1}^{n}\sum_{k=1}^{n} C_{ijk}\dot{q}_j\dot{q}_k + G_i \quad i, j = 1, 2, \ldots n \tag{3}
$$

where τ_i is the necessary torque at joint i, q is the displacement of joint, $\dot{q}_j$ and $\dot{q}_k$ are the velocities of joint j and k respectively, $\ddot{q}_j$ is the acceleration of joint j.

M_{ij} is the inertia force coefficient given by:

$$
M_{ij} = \sum_{p=\max(i,j)}^{n} Tr\left[d_{pj} I_p d_{pi}^T\right], \quad i, j = 1, 2, \ldots n
$$

C_{ijk} is the Coriolis or centrifugal force coefficient given by:

$$
C_{ijk} = \sum_{p=\max(i,j,k)}^{n} Tr\left[\frac{\partial(d_{pk})}{\partial q_p} I_p d_{pi}^T\right], \quad i, j = 1, 2, \ldots n
$$

G_i is the gravity force given by:

$$
G_i = -\sum_{p=i}^{n} m_p g d_{pi} {}^{p}_{e}\overline{r}_p, \quad i, j = 1, 2, \ldots n
$$

Designing a PID controller is relatively simpler compared to other controllers, with its performance primarily dependent on the values of three parameters referred to as the proportional gain (K_P), the derivative gain (K_D), and the integral gain (K_I) which, are chosen so as to reduce the magnitude of error signal (e). Therefore, the necessary torque at different joints of the manipulator in order to reach the desired final position starting from an initial position is determined by using Eq. (4).

$$T = K_P e + K_D \dot{e} + K_I \int e dt \tag{4}$$

Thus, PID controller designed for a 4-DOF planar manipulator can be expressed as Eq. (5).

$$\tau_i = K_{Pi}\left(\theta_{if} - \theta_i\right) - K_{Di}\dot{\theta}_i + K_{Ii}\int e(\theta_i)dt \tag{5}$$

where, i = 1, 2, …, n links of the manipulator.

3 Formulation of the Problem

The problem of tuning the three gains of the designed PID controller, viz., the proportional gain (K_P), the derivative gain (K_D), and the integral gain (K_I), is formulated as an optimization problem of minimizing the integral error as defined by Eq. (6).

$$\text{Minimize } Z = \sum_{i=1}^{4} e(\theta_i) \tag{6}$$

Subject to constraints

$$50 \leq K_{P1}, K_{P2}, K_{P3}, K_{P4} \leq 130$$

$$40 \leq K_{D1}, K_{D2}, K_{D3}, K_{D4} \leq 100$$

$$20 \leq K_{I1}, K_{I2}, K_{I3}, K_{I4} \leq 70$$

where $e(\theta_i)$ is the respective error at joint *i*. The remaining terms carry their traditional meaning. Further, the optimization problem is solved using conventional genetic algorithms, following the procedure outlined in the following section.

3.1 Tuning of PID Controller by Using Genetic Algorithm

For the past few decades, genetic algorithms have emerged as a powerful and very popular optimization techniques compared to other optimization algorithms. Initially, the algorithm starts with a population containing several chromosomes. Each chromosome represents a solution to the problem, and its performance is calculated based on the fitness function. A group of randomly selected chromosomes undergoes three major stages, namely selection, crossover and mutation. By using these three basic operations the new chromosomes or solutions are generated which hopefully are better solutions. The procedure adopted to tune the PID controller using genetic algorithm is outlined below:

- Initially, GA starts with an initial set of random guesses or solutions for the PID controller parameters K_P, K_D, and K_I. The entire set of guesses is referred to as the population with population size equal to the total number of guesses. Each set of guesses is represented as a binary string and referred to as a population member or chromosome.
- The fitness defined using Eq. (6) as the inverse of $(1 + Z)$, is evaluated for each population member or chromosome after converting the binary string to its equivalent real value. After determining the fitness function each set of decoded parameters K_P, K_D, and K_I is passed to the PID controller.
- Based on the evaluated fitness the population is subjected to selection operation, wherein chromosomes having lower fitness value are selected with higher chances for undergoing crossover and mutation operations subsequently. Tournament selection is used in the current work.
- Now the selected chromosomes called the parents are subjected to crossover, yielding new chromosomes or child solutions that may be better than one or both of the parents. A single point crossover operation is implemented here.
- Finally, mutation operator is applied to prevent the process ending up in a local minimum and further maintain the population's diversity.
- The process is repeted until convergence is achieved. The convergence criteria is either a specified number of GA generations are completed or the fitness of population has reached a specified minimum value.

4 Results and Discussion

The necessary torque at each of the manipulator joints and its control analysis is simulated in this work. The tuning of PID controller gains K_P, K_D, and K_I is achieved using the manual method i.e., by trial and error and a metaheuristic optimization method, GA. The tuned PID controller gains obtained applying the manual and GA procedures are compared as given in Table 2. Recognizing the fact that the performance of GAs is influenced by the algorithm parameters, systematic studies, as illustrated in Fig. 2, are conducted to estimate the optimal parameter values. It can be seen that, crossover rate of 0.5, mutation rate of 0.045, population size of 50, and maximum generations equal to 40, are suitable values of principal GA parameters. The identified GA parameters are then utilized in the estimation of the optimal gains of the developed PID controller.

Table 2 Comparison of manual and GA tuned gains of the PID controller

Joints	Manual tuned			GA tuned		
	K_P	K_I	K_D	K_P	K_I	K_D
1	120	50	80	115.05	52.58	75.23
2	100	45	74	105.26	47.25	68.32
3	80	40	65	82.35	45.98	65.34
4	60	25	62	58.32	32.56	60.48

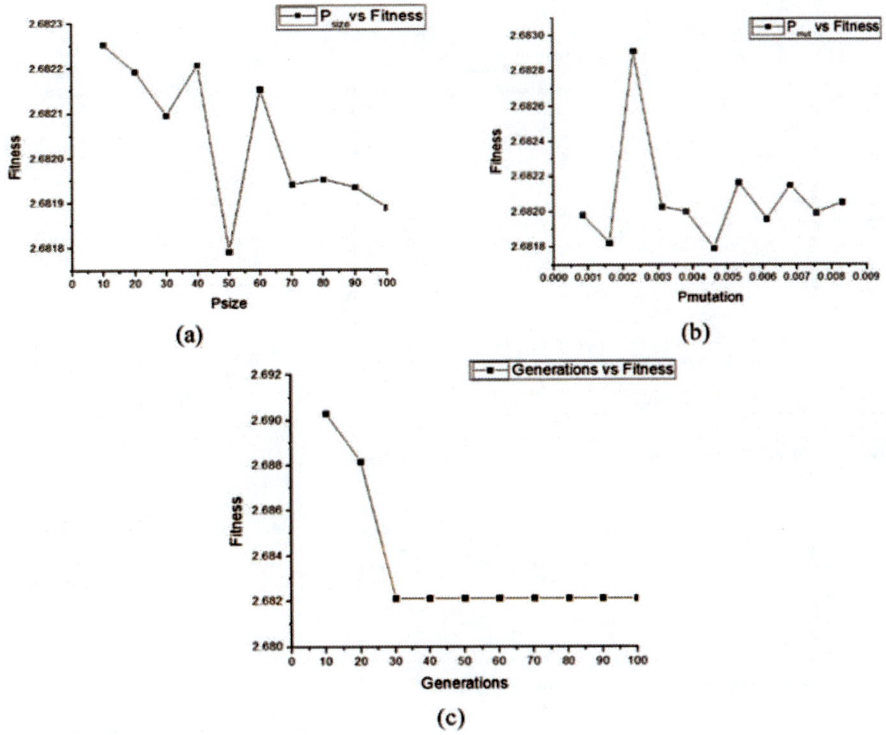

Fig. 2 Studies showing influence of GA parameters: **a** mutation rate, **b** population size, and **c** number of generations

Next, a comparative study of resulting convergence error with the manually tuned and GA tuned controller gains is performed, which is illustrated in Fig. 3. It is observed that initially, the deviation or error is high at all joints, and after a few seconds, the error slowly settles down. It can also be observed that the percentage of error is less when the gains are tuned with the help of a genetic algorithm when compared to the manual tuning method. Figure 4 shows the torque necessary at several joints of the robotic manipulator using the resulting optimal gains from manual tuning and GA tuning method. The results show that the torque essential at joint 1 is higher

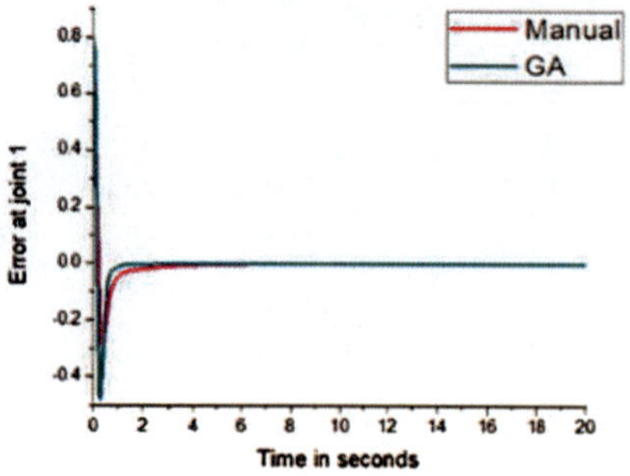

(a) joint 1

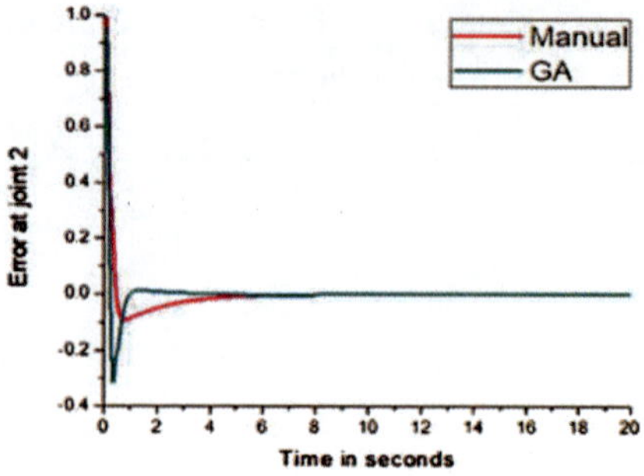

(b) joint 2

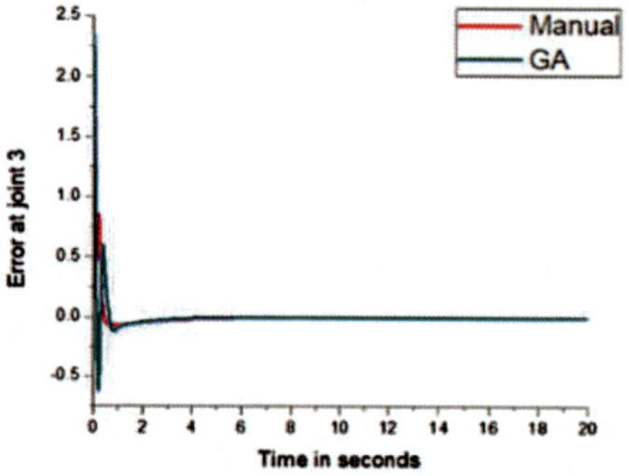

(c) joint 3

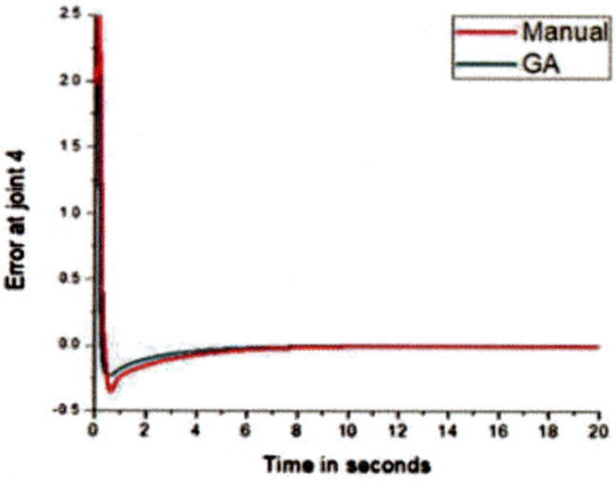

(d) joint 4

Fig. 3 Variation of error at various joints

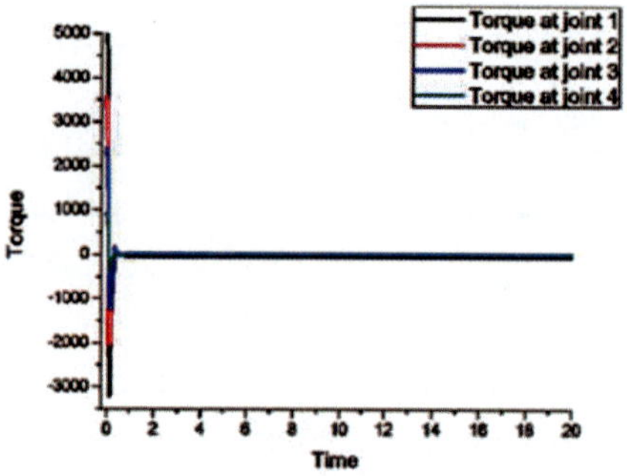

(a)

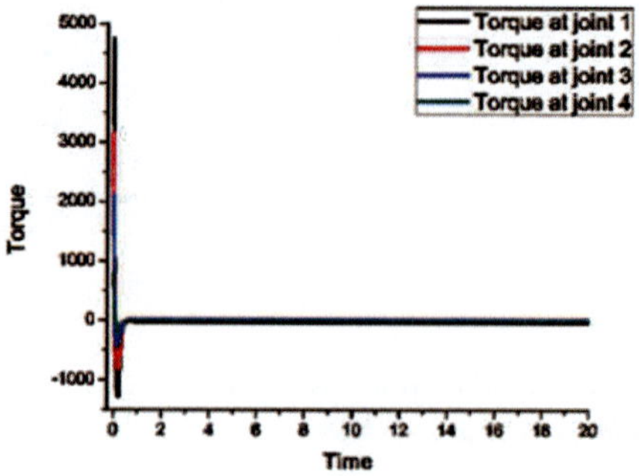

(b)

Fig. 4 Joint torque variations: **a** using manually tuned gains, and **b** GA tuned gains

compared to the other three joints. This is because the first link drives the other three links and joints of the manipulator. Based on the results of convergence of error and required torque at different joints of the manipulator, it can be concluded that a PID controller tuned using GA performs better than the manually tuned PID controller.

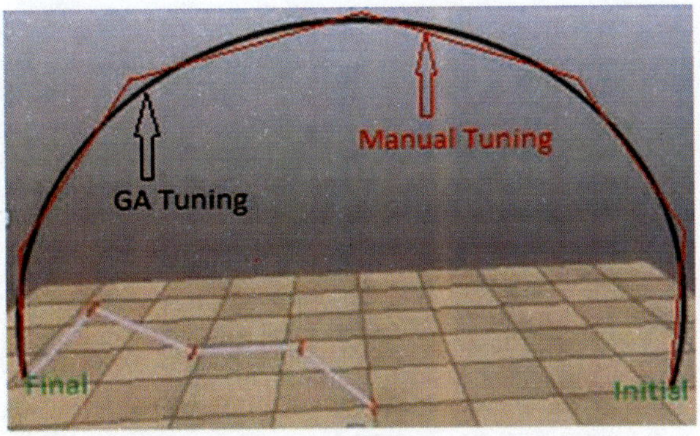

Fig. 5 Comparison of path tracking capability of the developed PID controllers

Finally, a simulation has been conducted on V-REP software to identify the manipulator's path tracking capability. Figure 5 illustrates the paths tracked by the manipulator's end-effector utilizing the gains obtained by the manual tuning procedure and GA-based tuning of the PID controller. Observe that the manipulator follows very closely the desired trajectory when using GA tuned gains. In contrast when using manually tuned gains, there is not only a clear deviation from the desired trajectory but also the motion is not smooth. All simulations in the reported study are performed on a 64-bit operating system PC with Intel i3 processor, and 4 GB RAM.

5 Conclusions

A PID controller has been designed to control the 4-DOF planar manipulator and tuning of the controller gains is accomplished with the help of both manual and with the application of genetic algorithm procedures. The optimal values of the controller gains are estimated following a systematic procedure. The simulation results show that, use of GA tuned PID controller results in better performance compared to that achieved when using manually tuned PID controller, in terms of the percentage of error and torque required to manipulate the robot within the boundary conditions.

References

1. Soomro ZA (2013) Kinematic modeling and simulation of 2-D Link and 3-R pendulum serial manipulator robotic arm. Int Res 2(4):169–175
2. Fu KS, González RC, Lee CSG (1987) Robotics-control sensing vision and intelligence. McGraw-Hill, New York
3. Al-Dois H, Jha A, Mishra RB (2011) Investigations into the parameters influencing the dynamic performance of 3-RRR planar & articulated robot manipulators. Tamkang J Sci Eng 14(4):313–322

4. Moolam RK, Braghin F, Vicentini F (2013) Dynamic modeling and simulation of spatial manipulator with flexible links and joints. In: Dimitrovová Z et al (eds) 11th International Conference on Vibration Problems 2013, Lisbon
5. Goyal K, Sethi D (2010) An analytical method to find workspace of a robotic manipulator. J Mech Eng 41(1):25–30
6. Dung LT, Kang H-J, Ro Y-S (2010) Robot manipulator modeling in Matlab-SimMechanics with PD control and online gravity compensation. In: International Forum on Strategic Technology 2010 Proceedings. IEEE, Ulsan, pp 446–449
7. Titov V, Shardyko I, Isaenko S (2013) Force-torque control implementation for 2 DoF manipulator. Procedia Eng 69:1232–1241
8. Sarkhel P, Banerjee N, Hui NB (2013) Analysis and control of a six link serial manipulator with flexible joints. In: iNaCoMM2013 Proceedings. AMM, Roorkee, pp 992–998
9. Mandava RK, Vundavalli PR (2015) Design of PID controllers for 4-DOF planar and spatial manipulators. In: International conference on robotics, automation, control and embedded systems (RACE). IEEE, Chennai, pp 1–6
10. Ziegler JG, Nichols NB (1942) Optimum setting for automatic controllers. Trans ASME 64:759–768
11. Abu-Dakka FJ, Assad IF, Alkhdour RM, Abderahim M (2017) Statistical evaluation of an evolutionary algorithm for minimum time trajectory planning problem for industrial robots. Int J Adv Manuf Technol 89:389–406
12. Mandava RK, Manas KS, Vundavilli PR (2015) Optimization of PID controller parameters for 3-DOF planar manipulator using GA and PSO. New developments in expert systems. In: Bennet A (ed) Nova science publishers, New York
13. Mandava RK, Vundavilli PR (2018) Tuning of PID controller parameters of a Biped Robot using IWO algorithm. In: Proceedings of the 4th international conference on mechatronics and robotics engineering, Valenciennes, pp 90–94
14. Dehghani A, Khodadadi H (2015) Fuzzy logic self-tuning PID control for a single-link flexible joint robot manipulator in the presence of uncertainty. In: 15th International conference on control, automation and systems (ICCAS). IEEE, Busan, pp 186–191
15. Sun F, Sun Z, Woo PY (2001) Neural network-based adaptive controller design of robotic manipulators with an observer. IEEE Trans Neural Networks 12(1):54–67
16. Oktarina Y., Fradina S, Tresna D, Pola R, Muhammad N (2019) Fuzzy-PID controller design of 4 DOF industrial arm robot manipulator. Comput Eng Appl J 8(2):123–136
17. Sheng L, Wei L (2018) Optimization design by genetic algorithm controller for trajectory control of a 3-RRR parallel robot. Algorithms 11(7):1–13
18. Nasr AE, Badr RI (2017) Novel PID tracking controller for 2DOF robotic manipulator system based on artificial bee colony algorithm. Electr Control Commun Eng 13(1):55–62
19. Annisa J, Darus IZM, Tokhi MO, Mohamaddan S (2018) Implementation of PID based controller tuned by evolutionary algorithm for double link flexible robotic manipulator. In: International conference on computational approach in smart systems design and applications (ICASSDA). IEEE, Kuching, pp 1–5
20. Loucif F, Kechida S (2020) Sliding mode control with pid surface for robot manipulator optimized by evolutionary algorithms. In: Farouk M, Hassanein M (eds) Recent advances in engineering mathematics and physics. Springer, Cham
21. Jishnu AK, Mandava RK, Vundavilli PR (2020) Design of optimal state observer-based controller for 4-DOF planar manipulator using PSO. In: Li L, Pratihar D, Chakrabarty S, Mishra P (eds) Advances in materials and manufacturing engineering 2020, LNME. Springer, Singapore, pp 151–162

Assessing the Carbon Foot Print of an Ayurveda Medical Institute: A Case of National Institute of Ayurveda, Jaipur, India

Gaurav Gaurav, Tejas Kumar, Chandni Khandelwal, Alok Bihari Singh, M. L. Meena, Sundeep Kumar, and G. S. Dangayach

Abstract The According to the "Paris Agreement", India has committed to greenhouse gas emission. By 2030, India's greenhouse gas emission intensity per unit gross domestic product (GDP) will be reduced by 33–35% compared to the 2005 baseline. In this context, similar goals are encouraged to be followed by all fields of business, including hospitals and academic institutions. In order to contribute to the country's efforts on this issue, this study presents the 2020 greenhouse gas emission account of the Ayurveda medical institute at the National Institute of Ayurveda (NIA) Jaipur, India, based on the academic and hospital activities established in the year of 2020. This study was conducted using IPCC methodology with scope 1, scope 2, and scope 3. Scope 1 includes in-house transportation, refrigerants, chemical, and consumption of liquefied petroleum gas. In Scope 2, electricity is the only source of emissions, while Scope 3 includes personnel commuting, water supply, cloths, surgical accessories, wastewater treatment, oils, chemicals, stationery items, papers, polyethylene, biomedical waste, and solid waste, etc. The results show that at the Ayurveda medical institute, commuting, water, and wastages (Scope 3) generates 71.73% of carbon emissions, which is the highest. Liquefied petroleum gas consumption, refrigerant usage, in-house transportation and chemicals usage emissions (Scope 1) are 16.57% and direct emissions (Scope 2) are 11.57%. Refrigerant usage is the most influential activity in Scope 1 emissions. Commuting is the most dominant activity in Scope 3 emissions, accounting for more than 38% of the overall carbon footprint of the NIA, Jaipur, India.

G. Gaurav · A. B. Singh · M. L. Meena · G. S. Dangayach (✉)
Department of Mechanical Engineering, Malaviya National Institute of Technology, Jaipur 302017, India

T. Kumar
Jayshree Periwal High School, Jaipur 302021, India

C. Khandelwal
Department of Management Studies, Malaviya National Institute of Technology, Jaipur 302017, India

S. Kumar
Department of Management Studies, Government Engineering College, Ajmer 305025, India

R. Agrawal et al. (eds.), *Recent Advances in Industrial Production*, Lecture Notes in Mechanical Engineering, https://doi.org/10.1007/978-981-16-5281-3_4

Keywords Carbon footprint · Greenhouse gas emissions · Ayurveda medical institute · Scope-based carbon footprint

1 Introduction

Due to global warming, climate change is a key issue attracting global attention. This is a challenge for all developing countries that reduce greenhouse gas emissions. Global warming has been not only an environmental problem for decades but also one of the world's biggest challenges. The key causes of GHG emissions are rapid population growth and the consumption of energy. Therefore, many summits on carbon emissions have been organized in the international community, such as COP and Intergovernmental Panel on Climate Change (IPCC), and various agreements and regulations to reduce carbon emissions have been proposed. Carbon dioxide is the main cause of global warming. To limit the increase in future global warming at 2 °C, it is essential that the atmospheric CO_2 equivalent concentration stable to approximately 450 ppm [1, 2]. To this end, the measures of the "Paris Agreement" must be implemented to reduce greenhouse gas emissions. To achieve this goal, according to the Paris Agreement, India pledged to create a cumulative carbon sink of 250–300 million tonnes of carbon dioxide equivalent by 2030. To make environmental performance develop in the direction of sustainable development, every organization, enterprise, and government agency must pay attention to greenhouse gas emissions. In the past ten years, the organization has made many efforts to reduce greenhouse gas emissions [3]. In this perspective, the involvement of autonomous establishments with high community trustworthiness, such as universities, hospitals, institutions can be seen as essential and strategic. Sustainability, especially environmental sustainability, has become a key issue for governments. To address this issue, Government working on various components like Life cycle assessment, carbon foot print, Minimum quantity lubrication, Water footprint, Energy footprint, Emission footprint, Land footprint, Biodiversity footprint, Social footprints, Economic footprints [4–9].

According to Cortese [10], universities have a responsibility to add value, awareness, knowledge, and skills to a sustainable future. However, although more and more universities are improving education for sustainable development [11], it is necessary to make sustainable development an integral part of the institutional framework of Lozano et al. [12]. In this sense, although some Indian universities/ institutions have been working hard [3, 13, 14]. However, no academic medical institute/ Ayurvedic institute in India has published a report on greenhouse gas emissions.

Facts have proved that health care itself can contribute to climate change [15]. In particular, hospitals are highly energy-intensive, consume a lot of resources, and generate a lot of waste [16]. The environmental footprint of health care activities is increasingly common; in particular through the use of life cycle analysis (LCA), a tool for quantifying environmental effects. In 2008, health care in the United States contributed 8% of the country's total carbon footprint (updated to 10% in 2016) [15,

17], while in 2012, England reported that only 4% of its carbon dioxide emissions were attributed to health care [16]. Only one study available at international level on carbon footprint of Australian health care [18].

In India, no national carbon footprint study on Ayurvedic healthcare institute has been conducted so far. Understanding the carbon footprint attributed to healthcare will indicate the severity of this concern, identifying potential hotspots that may allow for more targeted approaches to reducing CO_2e emissions in a world that is generating more and more carbon. In this article, according to the "GHG Protocol" standard (IPCC, 2006), a 100-year time frame was used to calculate the greenhouse gas emissions of the Ayurvedic Medical Institute of the National Institute of Ayurveda (NIA) Jaipur, India. The Carbon Footprint is calculated considering scopes 1 to 3, and several categories of emission sources (transportation, energy purchase, energy generation, and solid waste generation) were considered to calculate emissions of a particular category.

2 Materials and Methods

The National Institute of Ayurveda (NIA) Jaipur was established on February 7, 1976. It is the top research institute of the Indian government under the Ministry of AYUSH. It aims to promote the growth and development of Ayurveda and become a continuously developing high-level teaching and training, research, and patient care demonstration institutions and arouse scientific knowledge of the Ayurvedic medical system. Since its establishment in 1976, the institute has made tremendous developments in the fields of education, training, research, and patient care. NIA is known for its excellent teaching, training, and patient care activities in the field of Ayurveda, as well as its unparalleled academic level (at undergraduate, graduate, post-doctorate, diploma, and certificate levels). The main campus of the college is composed of multi-story buildings with 14 teaching departments, its affiliated laboratories, teacher clubs, offices, seminar halls, museums, lecture halls, equipped with DLP projectors, audio-visual aids. There are also 280 bedding hospitals approved by the national accreditation board for Hospitals and Healthcare (NABH), OPD, Panchakarma troops, central laboratories, luxury wards, bungalow wards, yoga troops, and other hospital complexes. There are 5 independent multi-story dormitories for boys and girls. The pharmacy is equipped with heavy-duty stoves and machinery for manufacturing various medicines, staff dormitories, guest houses, water tanks, and reservoirs. There is also a well-equipped auditorium with a capacity of 500 seats.

2.1 Assessment Methodologies of Carbon Footprint

This article highlights the relevance of carbon footprint inventories for medical institutes wishing to measure their full scope 1 to 3 emissions based on the 2006 IPCC

Guidelines [19]. The estimation method of the basic carbon footprint is the same in all analysis categories, as shown below:

Stage 1: Measured the activity data of each category (electricity (kWh) used, Liter's fuels, gas consumption, kilometers traveled, and Kilograms/tonnes).

Stage 2: Find the relevant greenhouse gas emission factors (kg CO_2e/activity unit).

Stage 3: Multiply the activity data by the relevant emission factors to estimate the emissions (kg CO_2 equivalent) of each category, and then add all emissions of each department to know the overall carbon footprint of the institute (GHG emission = activity data × emission factor).

As per the 2006 IPCC Guidelines, Greenhouse gas emissions are allocated into three group's i.e. Scope-1 to 3. Scope-1 encompasses all single direct sources of greenhouse gas emissions from sources owned or controlled by your institute, including electricity, heat, or steam generation; transportation of materials, goods, waste. Scope-2 relates to indirect emissions from the consumption of electricity, steam, cooling, and heating that are related to the production of imported energy sources, and scope 3 relates to all other types of indirect emissions that may come from the establishment's activities but come from sources that are operated by another organization.

2.2 *System Boundary and Scope of Assessment*

According to WRI, in 2015, the "Greenhouse Gas Protocol" corporate standards provided carbon footprint guidelines by classifying emission activities into three scopes (Scope 1, Scope 2, and Scope 3). The scope of operation of this study includes: Scope 1: direct emissions from diesel burning (generator sets), Steam Generation, Electricity Generation by Diesel set, Refrigerant Usage, Chemicals Usage, In house Transportation and Fertilizer; Scope: indirect energy emissions from purchased electricity from the state power administration; Scope 3: Other indirect emissions include: commuting of faculty, doctors, staff, patients and students at NIA campus the paper used for pamphlet, poster, intend book, answering papers, prescription slip, test papers, notices, note sheet, announcements, teaching notes, patient manuals, stock register, laboratory manuals, medicine manuals, course manuals, and many other paper materials used in printed and non-printed forms in the NIA campus; Various wastes generated in institute and hospital premises, including biomedical waste and solid waste like metal, glass, paper, and plastic; food waste emissions; refrigerants used in air conditioners and refrigerators; institutions Procurement: Cleaning essentials, chemicals, electrical and electronic components, computers, printers, glassware purchased for research labs and hospital; hospital procurement such as Pharmaceutical, Medical Equipment, Laundry Service, Cotton and cotton cloths, different form of Polyvinyl-chloride (PVC), Plastic Bags; others types of materials and product required for maintenance and daily essentials services for the NIA Campus. Table 1 provides information for various categories under three scopes.

Table 1 Emission scope and their categories

S. No.	Scope	Categories
1	Scope 1	Steam generation
		Electricity generation by diesel set
		Refrigerant usage
		Chemicals usage
		In house transportation
		Fertilizer
2	Scope 2	Purchased electricity
3	Scope 3	Staff commuting
		Bus travel
		Solid waste
		Paper usage
		Biomedical waste
		Computers
		Printer ink
		Water consumption
		LPG consumption
		Patient travels
		Visitor travels
		Student commuting
		Pharmaceutical
		Medical equipment
		Laundry service
		IT
		Food
		Cotton and cotton cloths
		Polyvinyl-chloride (PVC)
		Plastic bags
		Metals
		Glassware

2.3 *The Carbon Footprint Framework*

The Kyoto Protocol has identified a variety of gases with global warming potential, although only three gases are usually stated [carbon dioxide (CO_2), methane (CH_4), nitrous oxide (N_2O)]. Carbon dioxide is most commonly used as a reference gas, its emissions are expressed in units of carbon dioxide equivalent (CO_2e). This study estimates the gross CO_2e emissions per year in tonnes. Figure 1 shows the carbon footprint framework which is used to measure carbon footprint in this study.

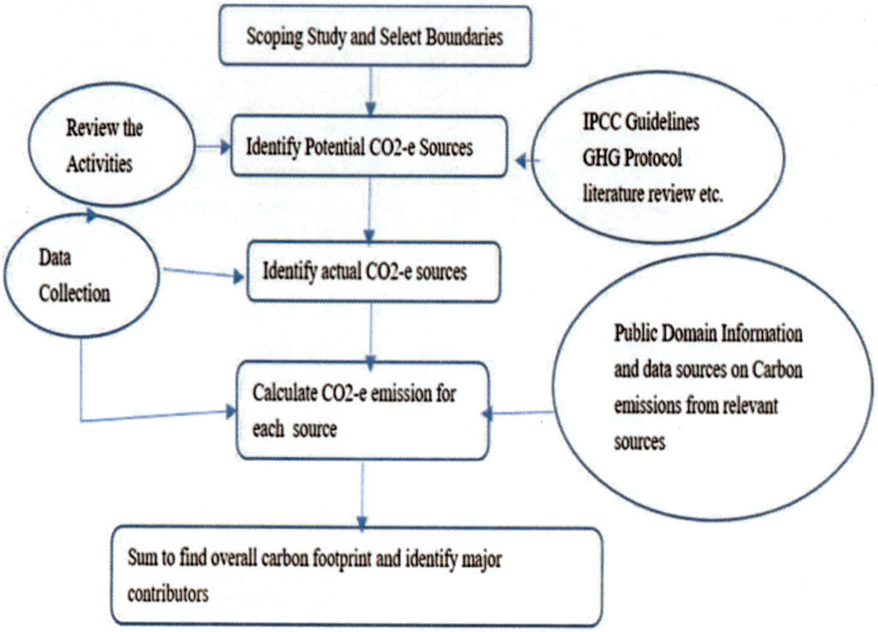

Fig. 1 Framework to calculate carbon footprint [3]

2.4 Emission Factors

The selection of emission factors is the most difficult task of this study. Since the emissions factor is the foundation of the evaluation, they may affect the results. Therefore, the emission factors play important role in any carbon footprint study. In order to select emission factors for this study, we review different sources such as IPCC emission factors, Eco invent datasets, some publications, and country-specific carbon footprint standards. Due to the low availability of data under Indian conditions, data with global validity is always preferred. If these data are not available, European data are taken into account. Table 2 provides emission factors applied for different activities under scope 1 to scope 3.

2.5 Data Collection

Activity data of the NIA were collected from different departments of the institute, hospital, and residence site and added together to calculate the carbon footprint. Primary data were collected from the central store of the institute and hospital. A "bottom-up" approach, has been implemented to bring together data relating to the areas of bio-waste, travel, solid waste, food, sanitation products, and information

Table 2 Emission factors used in this study

Scope	Activity	Emission source
Scope 1	LPG consumption	Defra 2016 [20]
	Refrigerant usage	Defra 2016 [20]
	In house transportation	Defra 2016 [20]
	Chemicals usage	Winnipeg [21]
Scope 2	Purchased electricity	CEA—All India electricity statistics—general review 2012 [22]
Scope 3	Commuting	Defra 2016 [20]
	Water	Defra 2016 [20]
	Waste	Yaman 2020 [23]

technology, and wastewater. Pharmaceutical and medical equipment data have been collected mainly on a 'top-down' approach.

3 Result and Discussion

This section introduces the research results acquired from scope-wise emissions and activity-wise carbon footprint analysis, identification of carbon footprint hot spots, and some strategies to reduce overall carbon footprint. The main goal of this study was to measure the carbon footprint of NIA Jaipur. Table 3 indicates the carbon footprint of the NIA campus based on the scope and different activities. The study shows that the total emissions recorded for the year 2020 are 1689.44 tons of CO_2 equivalent.

Table 3 Carbon footprint for NIA campus

Scope	Activities	Tonne of CO_2e	Emissions (%)
Scope 1	LPG consumption	35.51	2.10
	Refrigerant usage	236.25	13.98
	In house transportation	5.32	0.31
	Chemicals usage	2.86	0.17
Scope 2	Purchased electricity	197.75	11.71
Scope 3	Commuting	653.98	38.71
	Water	347.62	20.58
	Waste	210.15	12.44
Total		1689.44	100

GHG emissions are reported for each category in terms of total emissions in tons of CO_2 eq. and total emission in percentage. From this analysis, it was found that the largest contributors are commuting (38.71%), which has 653.98 tonnes of CO_2 equivalent carbon footprint. In house transportation and chemicals, usage accounts for very fewer percentages of emissions i.e. 0.31% and 0.17% respectively. Scope 3 is the highest carbon footprint contributor as compare to scope 1 and scope 2. In Scope 3, commuting is the highest carbon emission activity which accounts for 653.98 tonnes of CO_2e followed by water 347.62 tonnes of CO_2e, and a different type of waste 210.15 tonnes of CO_2e.

3.1 Scope Wise Emissions

This subsection introduce the research results acquired from scope-based carbon footprint analysis, identification of carbon footprint hot spots. Figure 2 shows the carbon footprint contribution based on the scope of the NIA campus. The results of the analysis showed that scope 3 has the highest carbon footprint emissions (71.73%) followed by scope 1 (16.57%) and scope 2 (11.57%). Scope 3 emissions of NIA campus (70%) are greater than that of scope 1 (16.57%) and scope 2 (11.57%). This is because the commuting of staff, students, and patients/visitors are high contributors to Scope 3 emissions of NIA Campus. Scope 1 is directly connected to the emission sources while scope 2 and scope 3 do not include the sources that are directly connected to the inputs.

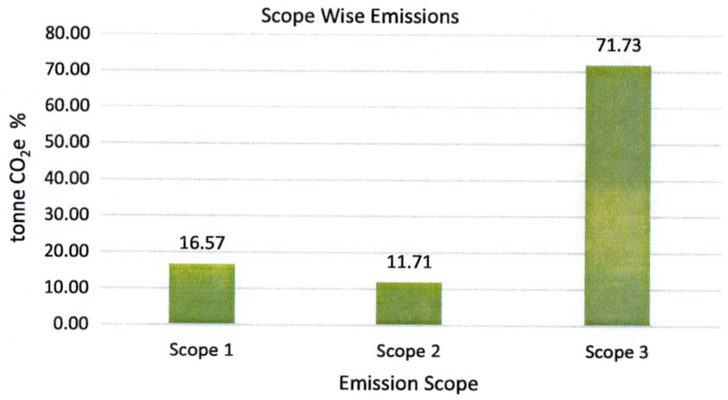

Fig. 2 Scope wise emissions

3.2 Activity Wise Emission

Although scopes are an essential indicator, it is important to concentrate on the key contributor's carbon footprint activities within the scopes. Figure 3 shows the carbon footprint percentage distribution of NIA campus activities. As shown in Fig. 3, the top five emission activities are refrigerant usage, purchased electricity, commuting, water and wastages account for 13.98%, 11.71%, 38.71%, 20.58%, and 12.44% of the total carbon footprint, respectively. Obviously, commuting (staff, students, and patient/visitors commuting) is the driving factor for the total carbon footprint of NIA Campus.

In summary, commuting contribution in carbon footprint is 38.71% of total emission followed by water (tap water and wastewater) 20.58% of total emissions and refrigerant usage 13.98% of the total emissions. Different types of waste account for 12.44% of the total emissions. Some uncertainty arises since this does not plant-specific emission data and there is also an uncertainty due to demand changes of different activities. Figure 4 shows the carbon footprint of commuting. Commuting is the first area to focus on because it is an important contributor to the total carbon footprint of the NIA campus. The extensive use of petrol and diesel for the vehicle for commuting is an important factor leading to a large amount of greenhouse gas emissions. Commuting accounts for 38.71% of the total emissions because staff, students, and patients/visitors are traveled by their own/rented vehicle to avail the medical facility from the NIA campus. However, as the use of renewable energy (electrical vehicle) and natural gas grows, the commuting mode is gradually shifting to low-carbon options. Switching to the electrical vehicle has great potential to reduce

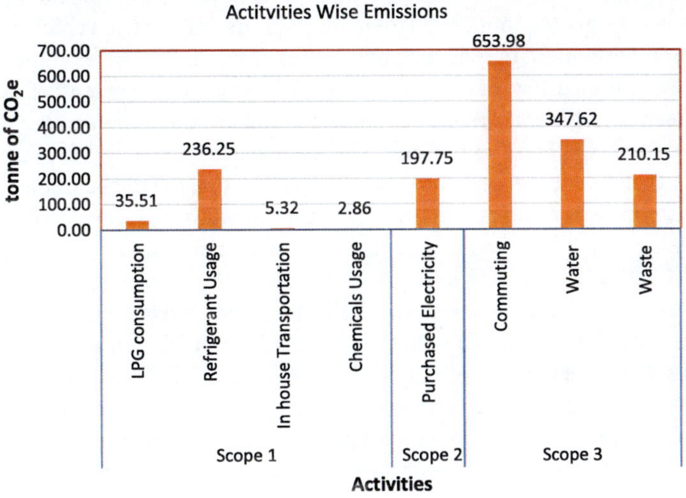

Fig. 3 Activities wise emissions

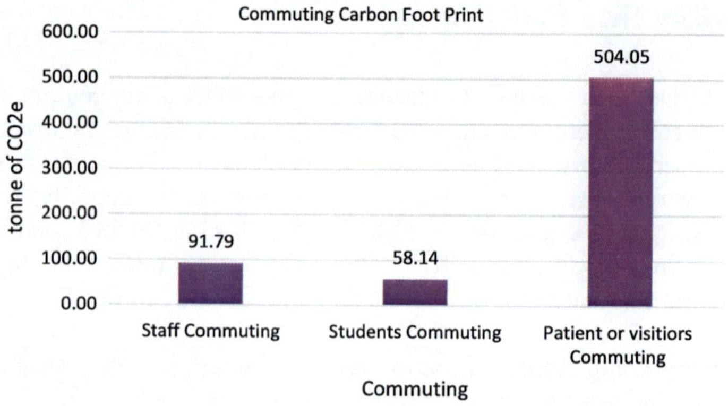

Fig. 4 Commuting emissions

dependence on fossil fuels and reduce the rate of greenhouse gas emissions in the transportation mode.

4 Conclusion

This article analyzes the carbon footprint of the NIA campus based on 2006 IPCC carbon footprint guidelines. The results of this study contain important insights that can guide decision-makers to focus on reducing the carbon footprint of the medical institute. The results show that Scope 3 emissions have the highest carbon footprint overall. It is strongly recommended to increase the use of electric vehicles and public transport for commuting to reduce Scope 3 emissions. And also recommended using recycled water for a different purpose on the NIA campus. The extensive use of fossil fuels in electricity generation is an important feature of the Indian power production mix that increases Scope 2 emissions. The results also show that Scope 1 emissions are greater than Scope 2 emissions. The result shows that the total carbon footprint of the NIA campus was estimated at 1689.44 tonnes of CO_2e. The results of the analysis showed that scope 3 has the highest carbon footprint emissions (71.73%) followed by scope 1 (16.57%) and scope 2 (11.57%). From this analysis, it was found that the largest contributors are commuting (38.71%), which has 653.98 tonnes of CO_2 equivalent carbon footprint. In house transportation and chemicals, usage accounts for very fewer percentages of emissions i.e. 0.31% and 0.17% respectively. In Scope 3, commuting is the highest carbon emission activity which accounts for 653.98 tonnes of CO_2e followed by water 347.62 tonnes of CO_2e, and different types of waste 210.15 tonnes of CO_2e. Establishing an inventory of greenhouse gas emissions is very important for developing strategies to improve NIA's carbon footprint. To reduce the carbon footprint, priority should be given to sources with higher emissions. In scope 2 emissions, the purchased electricity accounts for 11.71% of the total emissions. This emission can be reduced by using solar photovoltaic panels. NIA has a large roof area that can be used for solar photovoltaic panels. Therefore, by using renewable resources as energy, electricity, emissions can be greatly reduced. In Scope 3, commuting and wastages account for 71.73% of total emissions. This can be reduced by using the electrical vehicle and public transport for commuting and also recommended to use recycled water for a different purpose in the NIA campus.

Acknowledgements We acknowledge the financial support of Malaviya National Institute of Technology Jaipur, India. We also thank National Institute of Ayurveda (NIA) Jaipur, India for contributing their data.

References

1. Rogelj J, Den Elzen M, Höhne N, Fransen T, Fekete H, Winkler H, Meinshausen M (2016) Paris agreement climate proposals need a boost to keep warming well below 2 °C. Nature 534(7609):631–639
2. Güereca LP, Torres N, Noyola A (2013) Carbon footprint as a basis for a cleaner research institute in Mexico. J Clean Prod 47:396–403
3. Dangayach GS, Gaurav G (2020) Assessment of greenness through carbon footprint. MATTER Int J Sci Technol 6(1)
4. Dangayach GS, Gaurav G, Gupta S (2020) Development of footprint framework of performance measurement system for SMMOs. J Adv Manag Res 17(5):727–756. https://doi.org/10.1108/JAMR-05-2020-0070
5. Gaurav G, Sharma A, Dangayach GS, Meena ML (2020) Assessment of jojoba as a pure and nano-fluid base oil in minimum quantity lubrication (MQL) hard-turning of Ti–6Al–4V: a step towards sustainable machining. J Clean Prod 272:122553
6. Gaurav G, Sharma A, Dangayach GS, Meena ML (2020) Bibliometric analysis of machining of titanium alloy research. Mater Today Proc (in Press). https://doi.org/10.1016/j.matpr.2020.10.217
7. Gaurav G, Sharma A, Dangayach GS, Meena ML (2020) A review of minimum quantity lubrication (MQL) based on bibliometry. Curr Mater Sci (in Press). ISSN: 2666-1462
8. Mishra H, Gaurav G, Khandelwal C, Dangayach GS, Rao PN (2020) Environmental assessment of an Indian municipal wastewater treatment plant in Rajasthan. Int J Sustain Eng (in Press). https://doi.org/10.1080/19397038.2020.1862349
9. Cucek L, Klemes JJ, Kravanja Z (2012) A review of footprint analysis tools for monitoring impacts on sustainability. J Clean Prod 34:9–20
10. Cortese AD (2003) The critical role of higher education in creating a sustainable future. Plan High Educ 31(3):15–22
11. Ozawa-Meida L, Brockway P, Letten K, Davies J, Fleming P (2013) Measuring carbon performance in a UK University through a consumption-based carbon footprint: De Montfort University case study. J Clean Prod 56:185–198
12. Lozano R, Lukman R, Lozano FJ, Huisingh D, Lambrechts W (2013) Declarations for sustainability in higher education: becoming better leaders, through addressing the university system. J Clean Prod 48:10–19
13. Sangwan KS, Bhakar V, Arora V, Solanki P (2018) Measuring carbon footprint of an Indian university using life cycle assessment. Procedia CIRP 69:475–480
14. Bantanur S, Mukherjee M, Shankar R (2015) Emerging dimensions of sustainability in institutes of higher education in India. Int J Sustain Built Environ 4(2):323–329

15. Eckelman MJ, Sherman J (2016) Environmental impacts of the US health care system and effects on public health. PloS One 11(6):e0157014
16. SDU (Sustainable Development Unit, UK National Health Service) (2013) Carbon footprint update for the NHS in England 2015. NHS Sustainable Development Unit, Cambridge
17. Chung JW, Meltzer DO (2009) Estimate of the carbon footprint of the US health care sector. JAMA 302(18):1970–1972
18. Malik A, Lenzen M, McAlister S, McGain F (2018) The carbon footprint of Australian health care. Lancet Planet Health 2(1):e27–e35
19. IPCC (2006) Intergovernmental panels for climate change. Available from: https://www.ipcc.ch/report/ar4/
20. DEFRA (2016) (Department for Environment, Food and Rural Affairs). Available from: https://www.gov.uk/government/publications/greenhouse-gas-reporting-conversion-factors-2016
21. Emission factors in kg CO_2-equivalent per unit. https://www.winnipeg.ca/finance/findata/matmgt/documents/2012/682-2012/682-2012_Appendix_H-WSTP_South_End_Plant_Process_Selection_Report/Appendix%207.pdf. Last accessed 2020/11/21
22. Central Electricity Authority (CEA), India. https://cea.nic.in/reports/annual/annualreports/annual_report-2013.pdf. Last accessed 2020/11/21
23. Yaman C (2020) Application of sterilization process for inactivation of *Bacillus stearothermophilus* in biomedical waste and associated greenhouse gas emissions. Appl Sci 10(15):5056

The Linkages Between Spare-Parts Management and Maintenance Management in Army Supply Chain of Vehicles

Chander Sheikhar and Rajesh Matai

Abstract This paper endeavours to provide an insight into linkages between Spare-parts Management and Maintenance Management in Army Supply Chain of Vehicles operating in challenging, under-developed and tough areas. There is hardly any research work carried out in this niche field. But there is a great scope of improvement in army supply chains by application of the latest management techniques. The paper demystifies the salient aspect of Army supply chains for spareparts and maintenance chain. It focuses on data analysis aspects of the supply chain of spareparts for commercial pattern vehicles (CPVs) or trucks used by various Armies including VED, ABC and FNS analysis. It also carries out data analysis of various repairs of heavy and light CPVs executed by maintenance chain and brings out salient repair indices. Comparative analysis of army supply chain of spareparts and maintenance chain brings out certain voids and establishes linkages between these two chains. The synergy between supply chains of spareparts and maintenance is necessary to enhance efficiency and responsiveness of army supply chains.

Keywords Army supply chains · Spare-parts and maintenance management

1 Introduction

1.1 Unique Challenges of Army Supply Chains

Army is a self-sufficient organization and has all functions including maintenance and repairs of equipment and inventory management of all stores pertaining to Army requirement, integral to it. Army supply chains operate in tough and challenging areas having treacherous weather and terrain conditions. Sometimes road connectivity is virtually non-existent, and internet or mobile telephone connectivity is a rare

---

C. Sheikhar (✉) · R. Matai
Department of Management, Birla Institute of Technology and Science, Pilani, India
e-mail: p20170504@pilani.bits-pilani.ac.in

R. Matai
e-mail: matair@pilani.bits-pilani.ac.in

R. Agrawal et al. (eds.), *Recent Advances in Industrial Production*, Lecture Notes in Mechanical Engineering, https://doi.org/10.1007/978-981-16-5281-3_5

luxury. Another challenge of managing Army Supply Chains is the uniqueness of the inventory which has limited or No civil end-use. The army supply chain of spare-parts cannot function on classic 'Just in Time' model of production/manufacturing company as stock out costs may be catastrophic in case of operational emergency when suppliers may not be responsive enough. Therefore, 'Just in Case' becomes a necessary inventory control technique which increases the stock level and associated cost. Higher responsiveness results in higher associated cost and compromises efficiency [1].

1.2 Overview

This paper is an exploratory study to establish linkages between spare-parts management and maintenance management in the Army supply chain of vehicles. Section 2 brings out the salient aspect of Army supply chains for spareparts and maintenance chain. Section 3 covers detailed data analysis aspects of the supply chain of spareparts for heavy and light CPVs including VED, ABC and FNS analysis. Section 4 carries out data analysis of various repairs of heavy and light CPVs executed by maintenance chain and brings out salient repair indices. Section 5 dwells upon comparative analysis of army supply chain of spareparts and maintenance chain. Section 6 discusses the possible linkages between these two chains. Section 7 finally concludes the paper and comments on avenues for future research work.

2 Demystifying Army Supply Chains

The amount of research work carried out in army supply chains has been very limited. The authors have referred to the models for repair and maintenance chain published by Jones [2] and Sharma et al. [3] and also carried out protracted interaction with various Subject Matter Experts dealing with Army Supply Chains to demystify the structure and process flow of Army Supply Chain for Spare-parts and Maintenance. Figure 1 gives out the purpose of Army Supply Chain for Spare-parts and Maintenance, various facilities and flow of information and material within these supply chains.

Army maintenance chain is dependent on the support of spare-parts supply chain for executing repair and maintenance activities. Spare-parts support section (SSS) is an integral part of Forward Workshops (FWS) and stocks fast-moving spare-parts. For balance spare-parts, maintenance chain is supported by spare-parts facilities like Material Support Unit (MSU), Regional Warehouse (RWH) and Central Warehouse (CWH). Some of the repairs are simple in nature, take less time and can be undertaken near the operational locations of Army units at Forward Workshops (FWS). It avoids logistics involved in the back-loading of vehicles for repairs. Intermediate Workshops (IWS) caters for the overflow of repair and maintenance load and also carry out

Fig. 1 Army supply chains of spare-parts and maintenance. *Source* Jones [2] and Sharma et al. [3] and interaction of authors with Subject Matter Experts

specialized repairs. Some facilities like test equipment, skilled manpower are located at Regional Workshops (RWS) which cater for overflow of repairs of FWS and IWS dependent on them. RWS are designed to undertake component level repairs of printed circuited boards, a complete overhaul of major assemblies etc. Repaired printed circuited boards, major assemblies etc. are supplied by RWS to FWS for carrying out repairs by replacement. Central Workshops (CWS) have sophisticated machinery and various test equipment to execute high technology complex repairs and also carry out a complete overhaul to give a new lease of life to vehicles and equipment. CWS are the ultimate echelon of repairs and take care of overflow of all RWS and scheduled life extension and product improvement tasks of complete range of vehicles and equipment.

The locations of facilities of spare-parts and maintenance chains are movable. FWS are highly mobile, and frequency for movement depends on operational requirements. IWS and MSU are also mobile but the frequency of movement is comparatively less. RWS and RWH are more of less static and sends small detachments only. CWS and CWH are completely static. Inventory level of each warehouse varies based on the range and depth of equipment sustained. At an average, it is three-month stock held based on past consumption data, and review/procurement takes place based on monthly consumption figures. MSU and RWS cater for a mix of inventory, i.e. spare-parts for a variety of vehicles and equipment, clothing, general stores etc. Whereas, CWH store one type of inventory/items only, i.e. spare-parts for vehicles or armament or electronic equipment, etc.

The interaction between spare-parts maintenance chains is by an exchange of reports and returns in the form of spreadsheets and sharing of data through Local Area or Autonomous Networks. Real-time asset visibility between supply and maintenance claims is lacking. James and Cook bring out that the problem in the communication system has an impact on asset visibility and inventory management [4]. However, in a large number of Armies, the efforts are being made to ensure total asset visibility to all echelons of spare-parts and maintenance chains. True integration of supply and maintenance chains takes place at FWS wherein SSS is part of it, Interactions between IWS and MSU is also good being co-located and not much separated by geographical distance. The inter-se distance between RWS and RWH and CWS and CWH is more. It is based on geographical spread of country, availability of infrastructure and contractor support. Contractor Support (CS) is readily available for MSU, RWH and CWH. But it may or may not be available to SSS depending upon the remoteness of operational location. There are no issues of vertical communication within spare-parts and maintenance chains, but it is hierarchy based. However the horizontal communication between spare-parts and maintenance supply chains, in most of the armies' world over, has lot of scope for improvement.

3 Data Analysis of the Supply Chain of Spareparts

3.1 Analysis of Demand Pattern

Since SSS is the final customer of the supply chain of spare-parts, data of demands over a period of one year was captured from the open domain for commercial pattern vehicles (CPVs) or trucks used by various Armies. The vehicles were categorized into heavy CPVs (or heavy trucks) and light CPVs (or light trucks). Spare parts are required for ensuring the operational readiness of vehicles. The spare-parts can be of two categories, i.e. required for scheduled maintenances and for corrective maintenance or repairs. The forecasting for scheduled maintenance is deterministic, while for repairs, forecasting may be stochastic with more complexity in provisioning process [5]. Figure 2 depicts the demand pattern of spare-parts separately for heavy CPVs and light CPVs. Out of 8069 spare-parts of heavy CPVs demanded in one year, 3233 spare-parts (i.e. 40%) were supplied by MSU, 2943 spare-parts, (i.e. 37%) were procured ex contractors support (CS) and balance demands of 1893 spare-parts were under processing for issue through MSU/RWH/CWH. For light CPVs, 7244 spare-parts were demanded in one year. 2094 spare-parts (i.e. 29%) were supplied by MSU, 4630 i.e. 64% spare-parts were proceed through CS and balance demands for 520 spare-parts (i.e., 7%) were under processing. It indicates that spare-parts availability for heavy CPVs was better through MSU. However, for light CPVs, the availability of spareparts had to be met through CS.

Analysis of demands of heavy and light CPVs was also carried out various systems of the CPVs. The spare-parts were divided into six categories based on vehicle system,

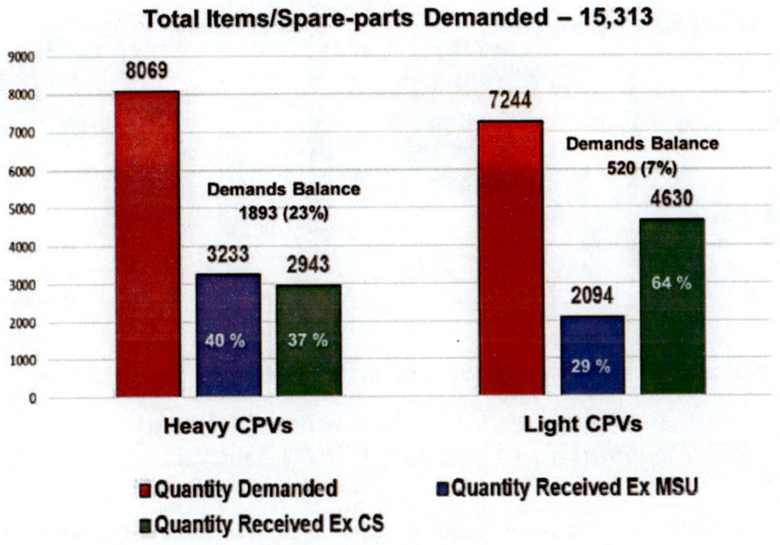

Fig. 2 Demand pattern of spare-parts of heavy CPVs and light CPVs

i.e. Transmission, Suspension, Engine (to include Air Intake, Exhaust, Fuel, Cooling and Lubrication system), Electric, Brake (including Pneumatic System) and Body and Miscellaneous (to include Steering System and various fitment items on the CPVs). Data of demand details of spare-parts of heavy and light CPVs are given at Fig. 3. Transmission System has a maximum percentage share of spare-parts demands followed by Electric System, both for heavy and light CPVs.

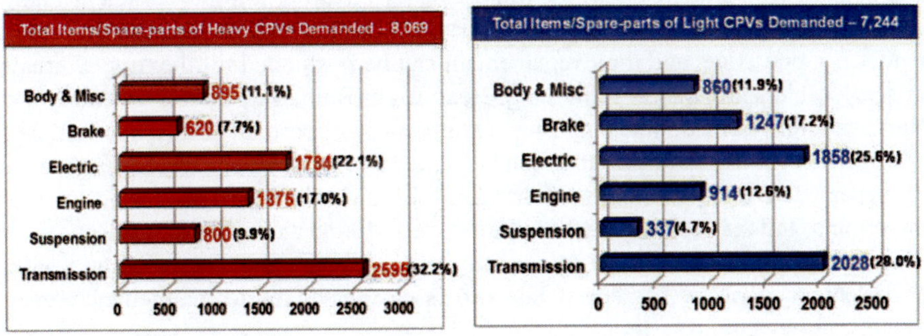

Fig. 3 The demand for spare-parts of heavy and light CPVs

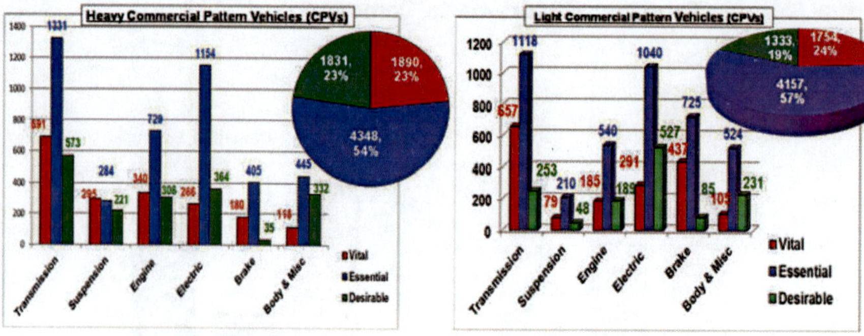

Fig. 4 VED analysis: spare-parts of heavy and light CPVs

3.2 *Vital, Essential and Desirable (VED) Analysis*

VED Analysis is a qualitative method [6, 7] and it categorizes the spare-parts based on their functional importance. The categorization is based on knowledge of experts, and sometimes, it is published in maintenance scales or catalogue of spare-parts of various CPVs. In order to remove any subjectivity, VED classification can also be carried out based on the Analytic Hierarchy Process (AHP) process [8]. Figure 4 depicts VED analysis carried out for the heavy and light CPVs. It can be seen that number of vital spare-parts of heavy CPVs (23%), and light CPVs (24%) are less than one-fourth of total spare-parts and most of the spare-parts fall into 'Essential' and 'Desirable' category.

Vital spare-parts are those spare-parts whose un-serviceability or non-availability completely renders the vehicle system non-operational e.g. Fuel Pump, Fan Belt etc. Vital spare-parts are the most important for ensuring a highly efficient maintenance chain. Essential spare-parts reduces the vehicle's performance, but don't render it inoperative, and their replacement can be postponed till the time alternate or serviceable spare part is found, e.g. Head Light Bulb, Self Starter assembly etc. Non-availability of essential spare-parts temporarily or partially affects the vehicle's performance, and their importance can be ranked as secondary from the maintenance chain perspective. Desirable spare-parts don't significantly impact the performance of the vehicle, and their replacement can be delayed without any urgency, e.g. Rear View Mirror, Wiper Blade etc. VED categorization of spare-parts not only has an impact on inventory levels of spareparts but also is closely linked to maintenance/repair aspects of the vehicle system.

3.3 *Always Basic Control (ABC) Analysis*

However, the inventory level of spareparts does not depend only on the criticality of the spare-parts from a functional point of view. Shrinking budget or financial

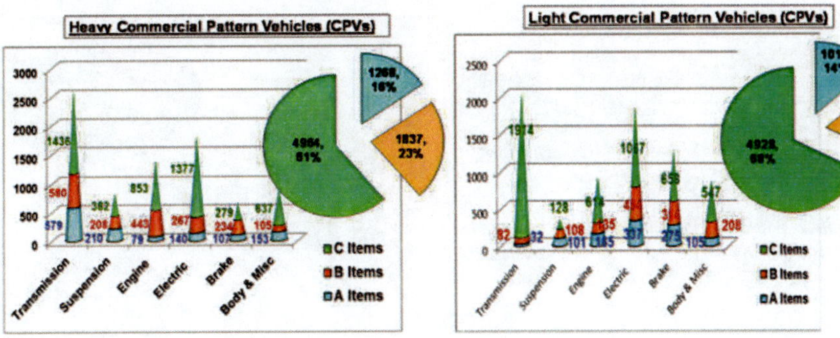

Fig. 5 ABC analysis: spare-parts of heavy and light CPVs

allocations has a cascading effect on the performance of army supply chains of spare-parts and maintenance. Therefore, ABC Analysis was carried out based on annual demand and average unit price or annual consumption cost of spare-parts [9]. In a traditional ABC Analysis, Category A Items are only 10–20% of total items but accounts for approx 50% value of items. Therefore, a greater amount of inventory control is required to be exercised for A Items. Category B Items constitute 20–30% of the total volume and account for approx 30% of the total value. B Items require a moderate degree of inventory control. Category C Items are approx 70–80% of total volume and constitute approx 20% total cost value. Therefore, routine or relaxed control can be exercised on C Items which are a major chunk of total inventory. ABC analysis was carried out for spare-parts of heavy and light CPVs and is depicted at Fig. 5. For heavy CPVs, Category A spare-parts were 16%, B as 23% and Category C of spare-parts were 61%. For light CPVs, A, B and C spare-parts were 14%, 18% and 68% respectively.

ABC Analysis helps to achieve optimum utilization of scarce budgetary allocations, by controlling inventory carrying cost of costly Category A Items. Diallo et al. suggest different inventory policies for Category A, B and C Items. S, S-1 System for Category A items, wherein inventory is maintained at S level, and order is placed when inventory level becomes S-1. The s, Q policy for Category B items, i.e. the order is placed for s quantity to ensure Q level of inventory. Category Class C items follow Economic Order Quantity, i.e. EOQ System, which ensures minimum ordering cost and inventory carrying costs [10].

3.4 *Fast, Normal and Slow (FNS) Analysis*

In order to gain a greater insight into the frequency of usage of spare-parts of heavy and light CPVs, FNS Analysis [6] was carried out (see Fig. 6). Fast spare-parts move in/out of stock on fast and frequent basis. For heavy CPVs, it constituted 21% of total spare-parts. For light CPVs, the figure of fast-moving spare-parts had risen to

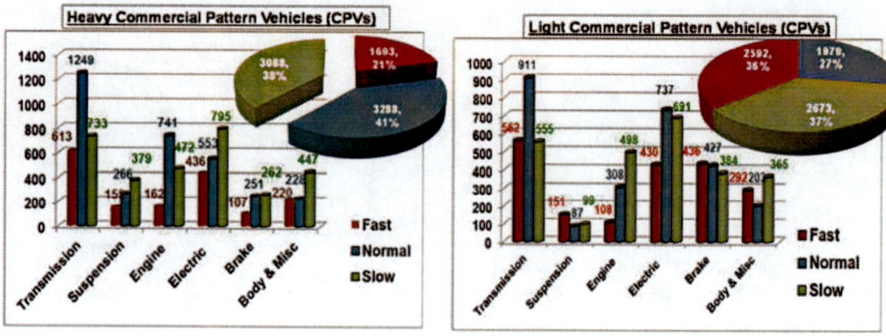

Fig. 6 FNS analysis of spare-parts of heavy and light CPVs

36%, indicating a faster turnover of inventory than in comparison with heavy CPVs. Approximately, slightly more than one-third of inventory of spareparts for heavy and light CPVs is slow-moving. Therefore, the reduced inventory level of slow-moving spare-parts should be kept at SSS. Sople had observed in a supply chain of repairs of defence ships that a large chunk of spare-parts inventory is non-moving and is carried for 5–6 years [11].

4 Data Analysis of Repair/Maintenance Chains

Most of the repair/maintenance activities are dependent on the replacement of spareparts. There are very few repairs in which spare-parts consumption is Nil. To analyze the linkages between spare-parts supply chain and maintenance chain, data of repairs undertaken from open domain for commercial pattern vehicles (CPVs) or trucks used by various Armies at FWS for the duration of one quarter, i.e. three months were analyzed. Figure 7 depicts system-wise repairs undertaken for heavy and light CPVs. Quantum of repairs for light CPVs (590 repairs) is more than heavy CPVs (396 repairs), as light CPVs operate in more tough and challenging areas than

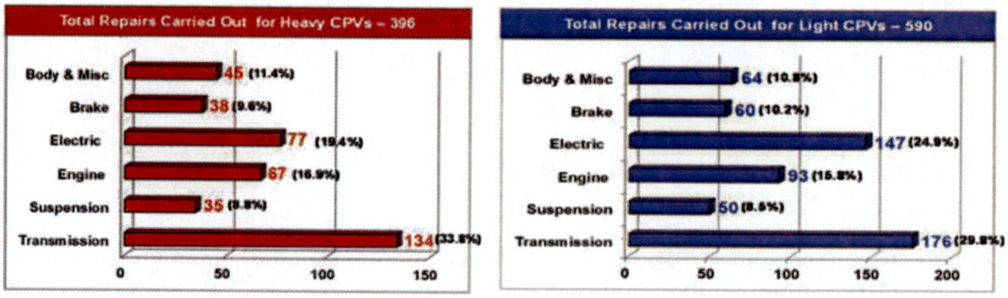

Fig. 7 System wise repairs data of heavy and light CPVs

Repair Indices

Heavy CPVs		Light CPVs	
MKBF	717 Kms	MKBF	604 Kms
MTBF	225 Days	MTBF	178 Days
MTTR	5.1 Days	MTTR	5.8 Days

Fig. 8 Repair indices of heavy CPVs and light CPVs

in comparison with heavy CPVs. As a result, a number of defects and consequent repairs are more for light CPVs playing on tough roads/tracks conditions.

Repair indices in terms of Mean Kilometers between Failures (MKBF), Mean Time between Failures (MTBF) and Mean Time to Repair (MTTR) were also captured from the open domain for commercial pattern vehicles (CPVs) or trucks used by various Armies over a span of one year. The deductions for heavy and light CPVs are given in Fig. 8. MKBF data is linked with MTBF data for heavy and light CPVs. MKBF and MTBF data is also indicative of the quality of repair and maintenance activities which is dependent on the quality of spareparts. Better is the quality of spareparts used in repairs; the MKBF and MTBF data is improved. MKBF and MTBF data of light CPVs are less than heavy CPVs. It further validates the fact that defect occurrence is more in light CPVs due to operation in tough and under-developed roads and tracks. MTTR is a function of the availability of spareparts. In case of non-availability of spare-parts in SSS, the vehicle has to wait for repairs due to lead time involved in provisioning or procurement of spare-parts through MSU or Contractor Support (CS). MTTR is also dependent on the complexity of repairs. MTTR data of light CPVs is more than that of heavy CPVs due to increase occurrence of defects. MKBF/MTBF parameters can be increased, and MTTR can be reduced if there is a synergy between the functioning of supply chains of spare-parts and maintenance.

5 Comparative Analysis of Army Supply Chains of Spare-Parts and Maintenance Chain

Figure 9 depicts spare-parts provisioning for various systems of heavy CPVs vis-à-vis quantum of repairs carried out in these systems. The pattern of spare-parts provisioning and system-wise repair is maximum for Transmission System accounting for 32.2% spare-parts demands and 33.8% repairs, followed by Electric System with spare-parts provisioning of 22.1 and 19.4% of repairs quantum. There is a good correlation between data of spare-parts provisioning and repairs carried out for four out of six systems of heavy CPVs. There is scope of improvement in spareparts provisioning of Brake and Electric System of heavy CPVs commensurate to requirement for repairs.

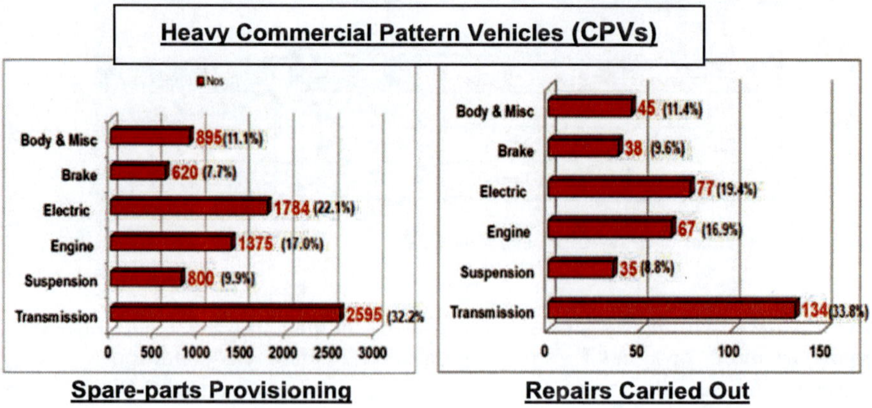

Fig. 9 Comparative analysis of heavy CPVs

For light CPVs, the pattern of spare-parts provisioning for three out of six systems is in consonance with repairs carried out in various systems (See Fig. 10). Spare-parts provisioning in Brake System (17.2%) is more than quantum of repairs (10.2%) due to criticality of Brakes in tough and challenging areas and spare-parts of Brake System fall in 'Vital' category. For Suspension System, spare-parts provisioning (4.7%) is less than a percent share of repairs (8.5%) and needs improvement. Since spare-parts of Suspension System primarily fall in 'Essential' and 'Desirable' categories, the replacement action is sometimes delayed. For Engine System, spare-parts provisioning is less, as a number of assemblies, i.e. Fuel Pump, Water Pump etc. are made available during replacement of complete Engine assembly during repairs of CPVs. These assemblies facilitate the number of other repairs by resorting to cannibalization of defective assembly with a serviceable and retrieved one.

When we view spare-parts management and maintenance management in Army Supply Chain of vehicles, there appears to synchronization as far as percentage of

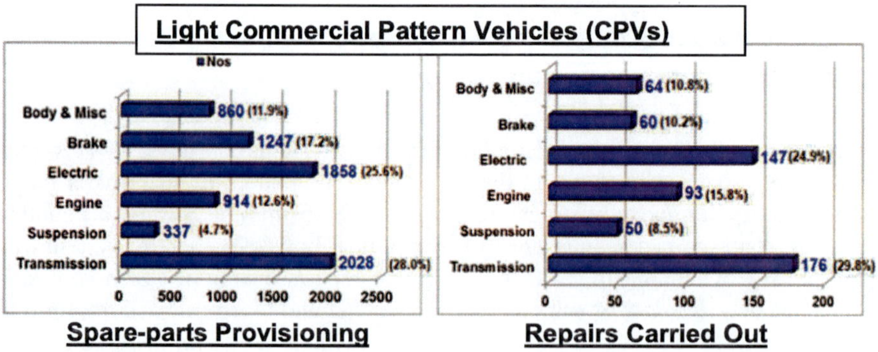

Fig. 10 Comparative analysis of light CPVs

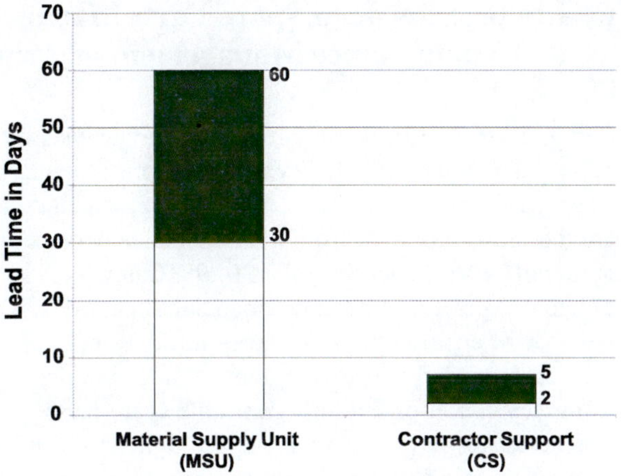

Fig. 11 Lead time analysis for provisioning of spare-parts through MSU and CS

spareparts and repairs in various systems of CPVs are concerned. But repair indices, as given in Fig. 8 indicate that there is certainly scope of improvements. Increased MTTR is a pointer towards a lack of coordination between supply chains of spare-parts and maintenance. If the supply chain of spare-parts is completely aligned to maintenance/repair requirements, the dependence on contractor support (which is more or less an emergent procurement) should not have been 37% for heavy CPVs and 64% for light CPVs (Fig. 2). The supply chain of spare-parts through MSU, RWH and CWH should have been in a position to meet most of the requirements. In order to gain further insight, lead time analysis was carried out from open domain for commercial pattern vehicles (CPVs) used by various Armies for provisioning of spare-parts in SSS through MSU and CS (see Fig. 11).

Lead time for provisioning of spareparts for heavy and light CPVs used by various armies varies from one to two months. Whereas, lead time through CS works out to be two to five days. Provisioning of spare-parts through MSU aggregates demand of large No of SSS and exploits economy of scales to achieve cost-saving. Being an integral element of Army Supply Chains, it ensures responsiveness in tough areas or mobile operations, where CS may not be in a position to supply spareparts. Due to reduced cost in provisioning of spare-parts by exploiting central procurement system and economy in scales, Army Supply Chain is efficient also. At the same time, CS is an important tool to take care of uncertainty in demand of spare-parts for executing repair/maintenance activities of CPVs. Therefore, CS enhances the responsiveness of Army Supply Chain of spare-parts, albeit increased cost of spare-parts compromises efficiency. The reduction in lead times, control of variability and collaboration with the suppliers are necessary for ensuring optimum safety stocks [12]. Ford Motor Company has controlled variability by reducing the number of platforms for its CPVs [13]. It has ensured economies of scale and resulted in huge cost saving.

6 Linkage Between Spare-Parts Management and Maintenance Management of Army Supply Chains

Figure 12 summarizes the complete data analysis of army supply chains of spare-parts and maintenance of CPVs and explains linkages between army supply chains of spare-parts and maintenance. Vital spares, which constitute 23.8% of total volume, are the most important from a maintenance perspective to ensure operational availability of CPVs. Strict control on 14.9% Category A spareparts of vehicles will ensure efficient utilization of scarce financial allocations. The focus on 24% fast-moving spares will ensure quick inventory turnover of spare-parts and faster repairs. Over provisioning of slow-moving items will lead to dead stock and blocking of financial resources in non-essential or desirable spare-parts.

The spare-parts procurement through MSU, RWH and CWH aggregates the demand and tends to focus primarily on long-term planning and exploiting economy of scales. However, Lead Time in provisioning of spare-parts is more, and it is less responsive to changing requirements of maintenance chain that emerge during near term execution. CS, which is also integral to maintenance chain, provides the much-needed flexibility to cater for the unforeseen requirement of spare-parts. There is a need to combine VED, ABC and FNS analysis along with lead time information to decide about the optimum level of each spare-part. Braglia et al. [14] have

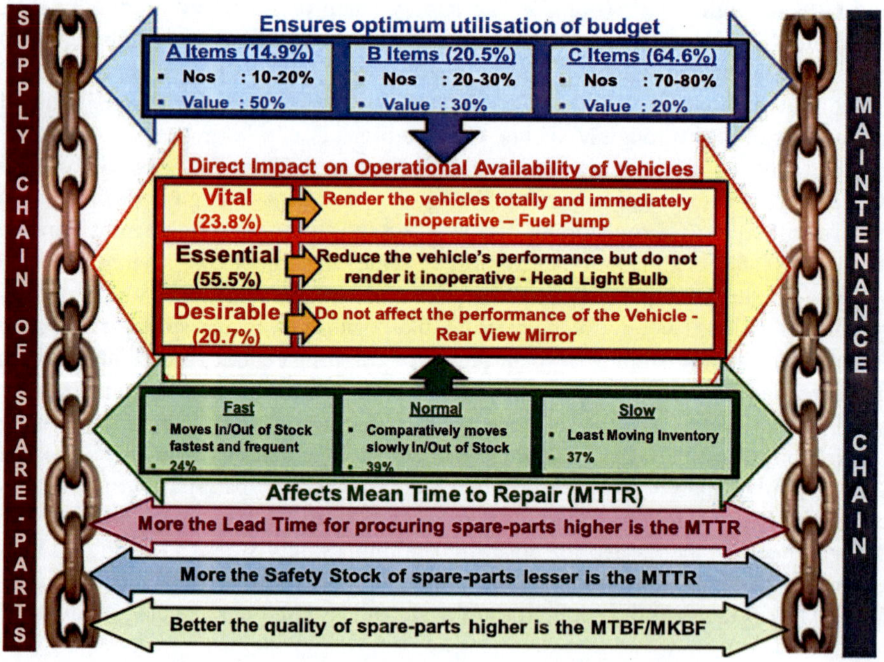

Fig. 12 Linkages between army supply chains of spare-parts and maintenance chains

brought out a classification criteria Multi-Attribute Spare Tree Analysis (MASTA) which have two successive steps. The first step identifies four criticality classes of the spare part to analyze provisioning methods and the next step is about the best control strategy of inventory management for each criticality class [14]. Teixeira et al. [15] suggest a multi-criteria classification for management of different spare parts. The multi-criteria classification tool will facilitate the organization to take decisions regarding inventory level of spareparts based on quantitative analysis and accurate information. Enhanced focus on the quality of spare-parts will ensure defect-free operation of vehicles for a longer duration and will enhance MKBF and MTBF indices of maintenance chain. The procurement of spareparts should not be based on the Lowest-Cost selection criterion and focus should be quality [16]. Spare-parts replenishment with current practices of three Months Maintenance Figures (MMF) cannot be universally applied for various types/categories of spare-parts. The responsiveness of army supply chains can be improved by reducing overall lead time for procurement of spare-parts and execution time for repairs. Spare-parts availability needs to be tuned to fluctuating requirement for repairs/maintenance, i.e. increasing availability when demand is less and vice-versa. It will ensure a reduction in safety stocks and investment in unnecessary inventory. Stocking of frequently required fast moving spare-parts should be increased in SSS to meet frequent repair's requirements. Less frequently used spare-parts can be centrally stored at MSUs and RWHs. A paper on the spare-parts supply chain of heavy trucks [17] considers four facility locations i.e. supplier, central warehouse, regional warehouses and service warehouses. It suggests more inventory in service warehouses to improve responsiveness. In order to reduce inventory, many firms stock more inventory in central warehouses and back it by an efficient transport system.

Spare-parts management and maintenance management of army supply chain are closely linked and require complete synergy and integration. In some armies, both the supply chains are managed by one department to ensure seamless integration. Whereas, in some armies, both functions are considered as a specialized one with spare-parts provisioning linked to procurement function clubbed with other types of inventories. However, the integration is missing between the department managing the supply chain of spareparts and the department carrying out maintenance and repair activities [7]. As per a seminar report of The Centre for Land Warfare Studies (CLAWS), New Delhi [18], the existing system of compartmentalized procurement, repair and maintenance needs to be replaced with integrated Supply and Maintenance Chains. Automation, Strong linkages, in planning, coordination and execution of spare-parts provisioning and repair/maintenance activities, are vital to enhance responsiveness and efficiency of army supply chains and ensure operational availability and mission reliability of vehicles. To economize the costs involved in traditional maintenance practices, many organizations are adopting a Condition Based Maintenance (CBM) approach, as it can forecast or predict failures [19]. Aviation Sector is extensively following the CBM approach, and it is finding its way in the vehicles also. With Intelligent Maintenance Systems (IMS) and CBM, the preventive and corrective maintenance requirements are easily forecasted and effectively integrated with the management of spare-part [20]. IMS are used in the industrial sector to overcome sporadic and erratic demand of spare-parts and avoid low equipment availability. In order to ensure cost reduction in supply chain spare-parts, Israel et al. has proposed a method to integrate information provided by IMS into the operational planning of a spare parts supply chain [21]. There is ample scope of application of best management practices of corporate or industry supply chains like enhancing asset visibility, functional integration, Performance Based Logistics, Velocity Based Logistics, Strategic Sourcing, Collaborative Planning Forecasting and Replenishment, Vendor Managed Inventory etc. Yaoa et al. [22] have developed an analytical model to explore how parameters of supply chain affect the cost savings by adopting collaborative initiatives of vendor-managed inventory. It shows that integration will result in inventory cost reductions based upon the ratio of Order Costs of the Supplier to the buyer and the ratio of the carrying charges of the supplier to the buyer.

7 Conclusion

Spare-parts and maintenance management in Army supply chains of vehicles are closely linked with each other. There is a need to ensure seamless communication and synergy between supply chains of spareparts and maintenance to enhance the efficiency and responsiveness. There is hardly any research work carried out in this niche field. But there is a great scope of improvement in army supply chains operating in tough and challenging areas to ensure cost-effectiveness and operational responsiveness. Some of the areas requiring future research are enumerated below:-

- Transformation of procurement processes to cut down lead time and enhance inventory availability at optimum cost.
- Management of supply chains of spare-parts and repair/maintenance activities by single agency/ department or different ones.
- Application of modern supply chain management practices of industry to enhance functioning of army supply chains.

References

1. Chopra S, Kalra DV (2019) Supply chain management: strategy, planning and operation, 7th edn. Pearson India Education Services Pvt Ltd
2. Jones JV (2006) Integrated logistics support handbook. Sole Logistics Press, McGraw-Hill Publication
3. Sharma P, Kulkarni MS (2016) Framework for a dynamic and responsive time separated—lean-agile spare parts replenishment system in army. Int J Prod Perform Manag 65(2):207–222
4. Cook MJML (2010) Army ammunition management information system challenges. Army Sustain 40–41
5. Sekaran Nair MK, Mohamad H, Abdul Jamil H, Mat Radzi Z (2018) Forecasting military vehicle spare parts requirement using neural networks followed by application of tacit knowledge. Int J Eng Technol 7(4.29):13–17

6. Cavalieri S, Garetti M, Macchi M, Pinto R (2008) A decision-making framework for managing maintenance spare parts. Prod Plann Control 379–96
7. Arya V, Sharma P, Singh A, De Silva PTM (2017) An exploratory study on supply chain analytics applied to spare-parts supply chain. Benchmarking Int J 24:1571–1580
8. Gajpal PP, Ganesh LS, Rajendran C (1994) Criticality analysis of spare parts using the analytic hierarchy process. Int J Prod Econ 35:293–297
9. Ramanathan R (2006) ABC inventory classification with multiple-criteria using weighted linear optimization. Comput Oper Res 33(3):695–700
10. Diallo C, Ait-Kadi D, Chelbi A (2009) Integrated spare-parts management, handbook of maintenance management and engineering. Springer, Berlin, pp 191–222
11. Sople V (2017) MRO supply chain striving for operational efficiency. Int J Bus Insights Transf 10(1):40–45
12. Haraburda Col. (Retd.) SS (2016) Transforming military support processes from logistics to supply chain management. Army Sustain 12–15
13. Ramsey M (2011) Ford SUV Marks new world car strategy. Wall Street J Online
14. Braglia M, Grassi A, Montanari R (2004) Multi-attribute classification method for spare parts inventory management. J Qual Maint Eng 10(1):55–65
15. Teixeira C, Figueiredo ILM (2017) Multi-criteria classification for spare parts management: a case study. In: 27th International conference on flexible automation and intelligent manufacturing, FAIM2017, 27–30 June 2017, Modena, Italy. 1560–1567
16. Au-Yong CP, Ali AS, Ahmad F (2016) Enhancing building maintenance cost performance with proper management of spare parts. J Qual Maint Eng 22(1):51–61
17. Sun J, Qu T, Duxian Nie R, Li P (2019) Research on 'location-inventory' problem of spare parts supply chain based on product service system. In: 11th CIRP conference on industrial product-service systems
18. CLAWS (2018) Revolution in military logistics: lean, sustainable, reliable supply and maintenance chain (S&MC). Seminar Report 17 Mar 2018. Centre for Land Warfare Studies, New Delhi
19. Andersson J, Jonsson P (2018) Big data in spare parts supply chains—the potential of using product-in-use data in aftermarket demand planning. Int J Phys Distrib Logist Manag 48(5):524–544
20. Saalmann P, Zuccolotto M, da Silva TR, Wagner C, Giacomolli A, Hellingrath B, Pereira CE (2019) Application potentials for an ontology-based integration of intelligent maintenance systems and spare parts supply chain planning. In: 48th CIRP conference on manufacturing systems
21. Israel EF, Albrecht A, Frazzon Bernd Hellingrath EM (2017) Operational supply chain planning method for integrating spare parts supply chains and intelligent maintenance systems. In: Proceedings of the 20th World Congress The International Federation of Automatic Control Toulouse, France, 9–14 July 2017, pp 12428–12433
22. Yaoa Y, Eversb PT, Dresnerb ME (2007) Supply chain integration in vendor-managed inventory. Decis Support Syst 43:663–674

An Integrated ISM-AHP Computing Framework for Evaluating Supply Chain Competitiveness

Ajay Verma and Nisha Singhal

Abstract In this paper a comprehensive method for evaluating supply chain competitiveness has been presented. The method is an integrated approach of Interpretive structural modeling and analytic hierarchy process. Using ISM the enablers are prioritized and then four supply chains are compared using AHP. The most important enablers are identified using ISM and then the supply chains are compared using the priorities given by the experts. This paper presents not only a way to establish interrelationships among the enablers of SCC but also presents an integrated methodology to evaluate various supply chains using this method.

Keywords Interpretive structural modelling (ISM) · Self-structured matrix · Reachability matrix · Pairwise comparison

1 Introduction

Supply chain competitiveness has not been explored adequately for the last one decade hence it requires a due attention [1]. SCC is a philosophy to incorporate the ways to improve competitiveness or in other words, we can say that to consider the competitiveness issue in the measurement and evaluation of SCC [2].

A model of competitiveness factors is shown in Fig. 1. There are forces which are driving companies and eventually the supply chains to adopt competitiveness strategies to achieve supply chain competitiveness as shown in figure. These forces are to be focused on by the supply chain mangers also to handle this throat cutting competition in the global era [3].

In this paper a comprehensive integrated methodology using ISM and AHP is provided to take care of this competition issue.

A. Verma (✉) · N. Singhal
Maulana Azad National Institute of Technology, Bhopal, M.P, India

R. Agrawal et al. (eds.), *Recent Advances in Industrial Production*, Lecture Notes in Mechanical Engineering, https://doi.org/10.1007/978-981-16-5281-3_6

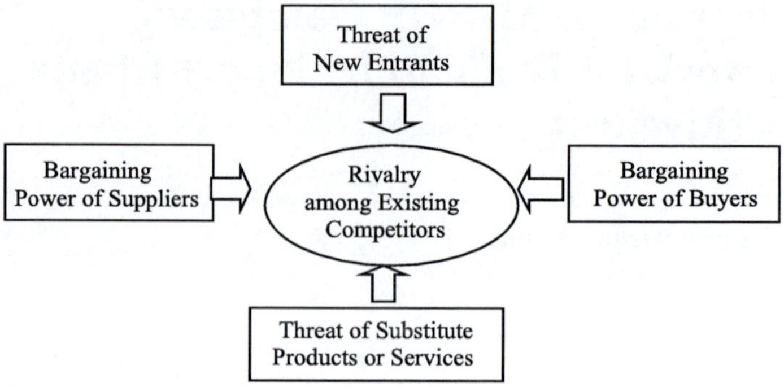

Fig. 1 Forces of competitiveness [4]

2 Literature Review

Porter stated that in this global scenario, there is a competition between supply chains and not between the individual industries [4]. In last 2 years, there is an inadequate exploration of supply chain performance measurement and only concepts and theories were developed [5, 6]. And SCC is a new buzzing word too [6]. Recently some of the strategies and variables are identified and derived for SCC. Like mass customization [7], Information and communication, internet use etc. [8–10] emphasizes on supply chain flows and twelve drivers for supply chain competitiveness were presented by him. On the other hand, [11] stressed on customer service to gain SCC. In this paper, a comprehensive evaluation method for SCC has been presented which is an integration of ISM and AHP methodologies. Using ISM, the most valuable variables are identified and arranged in a hierarchy while using AHP these variables are used to analyses and compare four supply chains using expert choice, 2000 software [12–14].

3 Application of ISM-AHP

The integrated ISM-AHP framework explores interrelationships among the variables and assists the decision making. The whole ISM-AHP methodology is explained here. ISM was proposed by Warfield in 1974. The various steps of the ISM-AHP framework is given below.

1. Identify the criteria.
2. Contexual relations are then find out.
3. SSIM development
4. Generation of Reachability matrix
5. Level Partitioning
6. Transitive links are removed

7. Node diagraph construction
8. ISM model construction
9. ISM Model is then reviewed.
10. Modify the model
11. Making a Goal for the AHP model
12. Weightages obtained from the experts.
13. Pairwise comparison
14. Synthesis
15. Sensitivity Analysis

The self SSIM is shown in Table 1.

Performing ISM methodology, the following rankings are obtained (Table 2).

From this hierarchy, enabler 10 makes base of the model hence it is important followed by enabler 13 and 1, 11 and 8 which are necessary for achieving SCC. Variables 4 and 2 are in the mid of the model and are crucial, while 9, 15 and 14 are also correlated and important. Enabler 12 has come up to be the most important enabler to achieve SCC. The managers and supply chain professionals should focus on this enabler to achieve SCC. Then the AHP methodology is used to compare four

Table 1 The self SSIM (Verma et al. 2011)

Sr. No.	Enablers	15	14	13	12	11	10	9	8	7	6	5	4	3	2
1	Cooperation and coordination	V	V	V	V	V	A	V	O	V	V	V	V	V	V
2	Agility	V	V	V	V	A	A	V	A	A	V	A	A	A	
3	Delivery performance	V	V	V	V	A	A	V	A	A	V	A	X		
4	Customer orientation and responsiveness	V	V	V	V	O	V	O	O	V	V	O			
5	Supply chain Synergy-collaboration	V	V	V	V	A	A	V	V	A	V				
6	Demand and product management	O	A	A	V	O	A	O	O	A					
7	Integration of key elements	V	V	V	V	V	A	V	O						
8	Inventory management	V	V	V	V	O	A	V							
9	Flexibility	V	V	A	V	A	A								
10	Info. Tech	V	A	V	V	V									
11	SA	V	V	V	V										
12	Mass Custo	A	A	A											
13	Cost	A	X												
14	Quality	A													
15	SCFC														

Table 2 Rankings of the enablers

Enabler No	Enabler	Level
1	Cooperation and coordination	XII
2	Agility	VI
3	Delivery performance	VII
4	Customer orientation and responsiveness	VII
5	Supply chain synergy-collaboration	IX
6	Demand management	II
7	Integration	XI
8	Inventory management	VIII
9	Flexibility	V
10	IT capabilities	XIII
11	Strategic alliances	X
12	SCC collaboration	I
13	Mass customization	III
14	Cost efficiency	III
15	Quality management	IV

alternative SCM. The main enablers are divided into sub variables through literature review and formed a model as shown in the Fig. 2

The model is analyzed using expert choice 2000 software and the final shows priorities in AHP methodology. SC 3 is the is performing best among all the four supply chains that we have considered. Evaluation.

Figure 3 shows the comparison among four SCs. As shown in the figure, Supply chain 3 has come out to be the most competitive one in this model. The modeling is

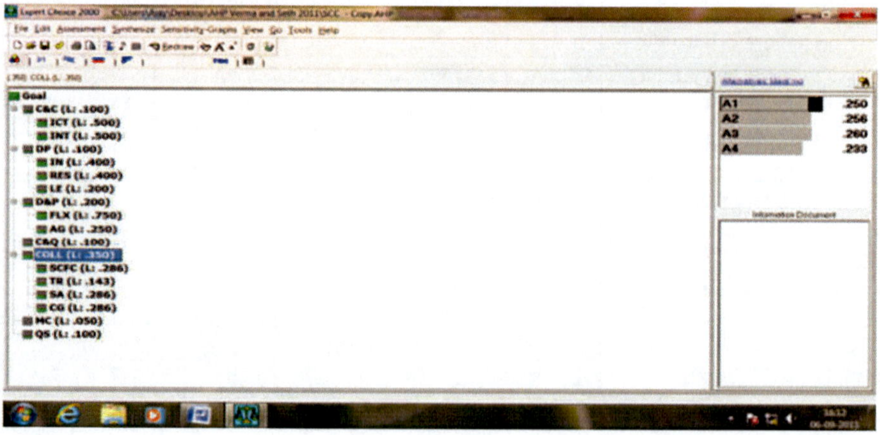

Fig. 2 Expert choice model

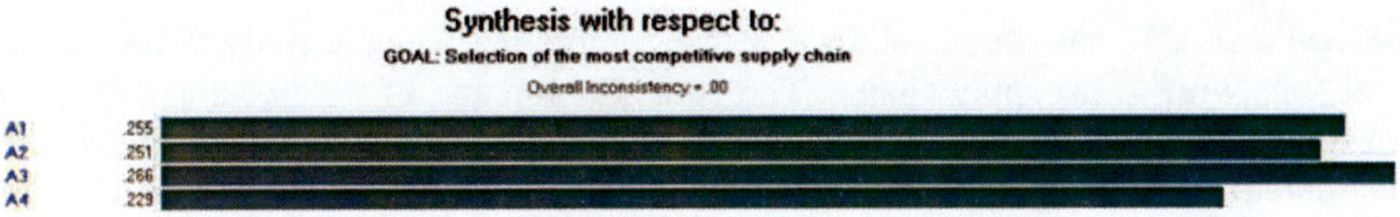

Fig. 3 Comparison and synthesis

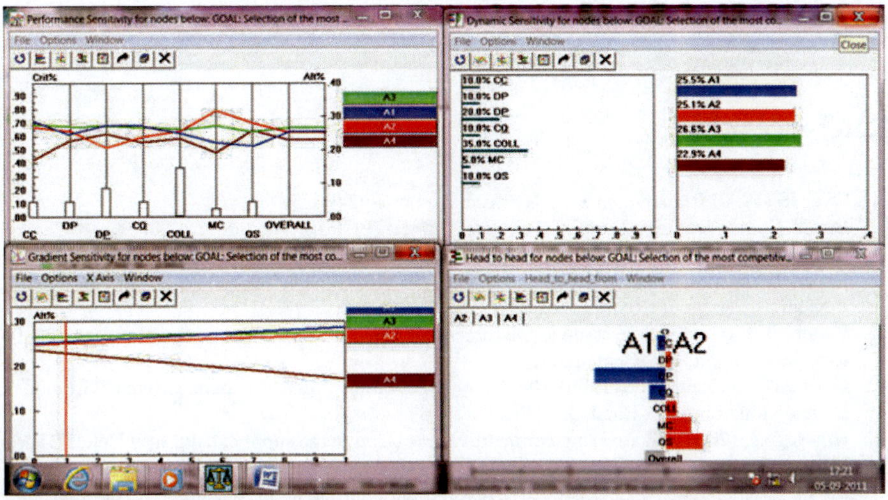

Fig. 4 Sensitivity analysis

based on the priorities attached to the variables by experts and then the model was analysed using expert choice 2000.

Figure 4 shows sensitivity analysis of the four company's comparison. In the figure it is clear that the results are changing when any of the parameters are changed. The sensitivity analysis is used to show the sensitivity or responsively of the method with respect to the changes heaping in the variables. As for as the results are concerned, it is shown in the model that collaboration is the most important variable to achieve SCC followed by information technology and supply chain flows with cost and quality as the prime ones. Mass customization and inventory management are also the important aspects to be taken care of.

4 Conclusions

Recently, supply chain competitiveness is a vital philosophy to be leader in the global markets. The paper focused on the importance of SCC and its analysis. SCC also impacts the profitability of any organization. In this paper, an integrated methodology of ISM and AHP has been explained with a comprehensive case study of four organizations with their supply chains. The collaboration and IT with cost and qualigy and flexibility are the most important enablers to achieve SCC as per this study. The integrated modelling may also be used in some other areas having number of parameters and their comparative values. Interpretive Structural Modelling is used to correlate the parameters with AHP has been used to compare the supply chains. This method provides SC professionals and decision makers to make decision regarding best supply chain.

References

1. Dyer JS (1990) Remarks on the AHP. Manage Sci 36:249–258
2. Expert Choice Decision Support Software Tutorial (ECPro), Version 9, 2000. Expert Choice, Inc., Pittsburgh
3. Expert Choice Software (2000) Produced by Expert Choice, Inc., 4922 Ellsworth Avenue, Pittsburgh, PA. www.expertchoice.com
4. Porter ME (2008) The five competitive forces that shape strategy. Harvard Bus Rev 86(1):78–93 (Special Issue on HBS Centennial)
5. Hauser D, Tadikamalla P (1996) The Analytic Hierarchy process in an uncertain environment: a simulation approach. Eur J Oper Res 91:27–37
6. Howgego C (2002) Maximizing competitiveness through the supply chain. Int J Retail Distrib Manag 30(12):603–605
7. Levi DS, Kamansky P, Levi ES (2012) Designing and managing the supply chain-concepts, strategies and case studies, 2nd edn. Tata McGraw-Hill publishing company Ltd.
8. Mentzer JT (2004) Fundamentals of supply chain management—twelve drivers of competitive advantages. University of Tennessee, Knoxville, Response books, New Delhi
9. Miller JG, Roth AV (1998) Manufacturing strategies: executive summary of the 1987 North American futures survey. Oper Manage Rev 6(1):8–22
10. Murths TP et al (1998) Country capabilities and the strategic state: how national political institutions affect MNC strategies. Strateg Manag J 15:113–129
11. Ray T, Triantaphyllou E (1999) Procedures for the evaluation of conflicts in rankings of alternatives. Comput Ind Eng 36:35–44
12. Saaty TL (1994) Fundamentals of decision making and priority theory with the analytic hierarchy process. RWS Publications, Pittsburgh
13. Saaty TL (1980) The analytic hierarchy process. Planning, priority setting, resource allocation. McGraw-Hill, New York
14. Saaty TL (2000) Fundamentals of decision making and priority theory with the analytic hierarchy process, vol 6. RWS Publications, Pittsburgh
15. Gunasekaran A, Patel C, Tirtiroglu E (2001) Performance measures and metrics in a supply chain environment. Int J Oper Prod Manag 21(1/2):71–87

Experiencing Life Cycle Assessment in Indian Additive Manufacturing Industries: Needs, Challenges and Solutions

Alok Yadav, Anbesh Jamwal, Rajeev Agrawal, and Sundeep Kumar

Abstract Indian manufacturing industries are now more focused on the adoption of new business models over traditional business models to achieve sustainability in business practices. Now the industries are assessing their environmental impacts to promote sustainability and life cycle assessment applications. This study focused on the semi-structured interview with the Indian additive manufacturing experts from the Southern and northern region of India. The study was conducted on email interviews, in-depth phone calls and a multiple-choice questionnaire survey circulated through emails. The interviewees represented the different stakeholders in the Indian additive manufacturing industries with varying knowledge of life cycle assessment including the additive manufacturing product manufacturers, consultants for AM industries, research institutes and academia. The experts suggested that there is a need for good understanding the different factors in additive manufacturing industries which will help them to lower their carbon emissions and conduct the LCA in the industries with proper LCA tools and databases available for additive manufacturing industries. This study will be helpful for the researchers and R&D in industries to provide insights that how Indian additive manufacturing industries experience the use of life cycle assessment. In the study, several awareness issues are addressed which will be beneficial for Indian additive manufacturing industries to expand and accelerate the life cycle assessment applications in additive manufacturing industries.

Keywords Life cycle assessment · Sustainability · Additive manufacturing · Industries · Survey · Challenges

A. Yadav · A. Jamwal · R. Agrawal (✉)
Department of Mechanical Engineering, Malaviya National Institute of Technology, J.L.N. Marg, Jaipur, Rajasthan 302017, India
e-mail: ragrawal.mech@mnit.ac.in

S. Kumar
Centre for Electronic Governance, Department of Technical Education, Govt. of Rajasthan, Jaipur 302004, India

R. Agrawal et al. (eds.), *Recent Advances in Industrial Production*, Lecture Notes in Mechanical Engineering, https://doi.org/10.1007/978-981-16-5281-3_7

1 Introduction

The manufacturing industries are one of the largest sectors which are responsible for CO_2 emission [1]. Each stage and individual activity inside the manufacturing industries i.e. taking raw materials to produced finish goods release carbon in the form of gasses like CO_2, CH_4, CFC, N_2O (called GHG) etc. [2]. These gasses are play role in global warming and climate change [3]. In this way to measure the print of emission of gasses at each step-in manufacturing industries measurement of carbon, footprinting is necessary for the manufacturing industry. In recent year's environmental pollution is the major problem that drastically affecting living organism on the earth and worsening the climatic conditions worldwide [4]. By the carbon footprinting in manufacturing industries, we analyse the CO_2 emission and capable to reduce the emission. Manufacturing industries are working towards cleaner production to minimize the pollution level [5].

The whole countries in this world nowadays are facing the problems of pollution, global warming ozone layer depletion, acidification [6]. These problems are due to deforestation, use of non-degradable materials, use of toxic chemicals in industry, emission of various toxic gases from industries, only humans are responsible for this and only humans can renew [7]. According to a survey, about 40% of death is due to water or air pollution. Many political actions and countries meeting is held to find a solution [8]. The concentration of CO_2 is 340 ppm in 1980 and now its level increased to 415 ppm till December 2020 [9]. Which is a vast increment also the hottest month since 1880 is July 2019, by these data we can imagine how the future will be? Among all these problems global warming, a major source of which is CO_2 emissions has been attracting attention from the international community [10].

All over the world manufacturing industries helps in economic growth for both developed and developing nations. In developing, nations industry sectors like SMEs boost the financial condition and provided a luxurious lifestyle to people [11, 12].

The manufacturing industries is the largest contributor to CO_2 emissions and energy consumption [13]. Manufacturing industries consist big part of primary energy use which produced CO_2 emission, global warming is one of the major problems [14]. The most influencing gas for global warming is CO_2 which contribute 80% out of four major gases (CO_2, CH_4, N_2O, and chlorofluorocarbon). In the whole world one-third of primary energy is used in manufacturing industries and produced two fifth GHG emission. [15].

1.1 Life Cycle Assessment (LCA)

LCA is a procedure to recognize the environment impact or environmental emissions of a product through every step of its life-cycle [16]. Life-cycle assessment (LCA) is a process in which the overall energy and material flow of a system is calculated also the impact of emission and wastes released during the whole life cycle is analyzed and

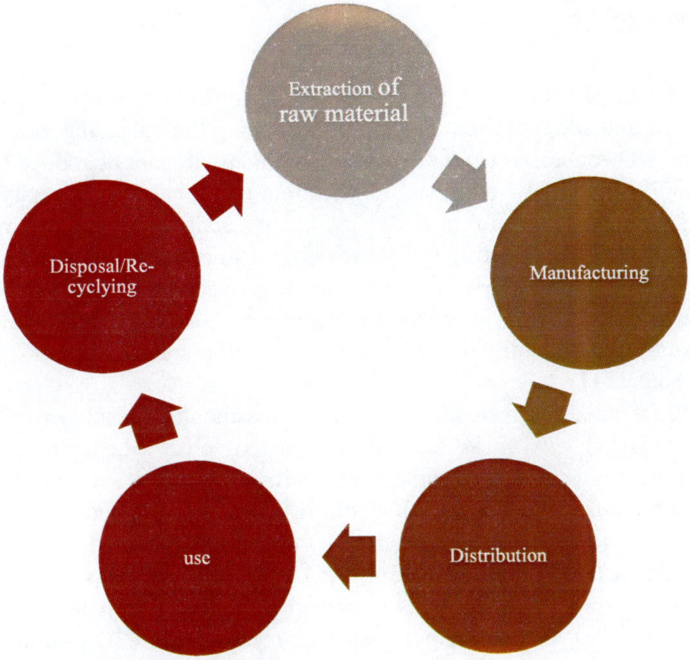

Fig. 1 Stage of LCA

strategies to reduce the emissions are provided [17]. LCA involves all the steps such as extraction phase which includes raw materials extraction from the natural resources, processing phase of raw materials, manufacturing phase, use-phase, maintenance phase, and disposal phase or recycling of a product [18]. Stages for LCA is presented in Fig. 1.

"In the past few years, initiatives have been taken to standardize the life-cycle assessment methodology which has been published by ISO (International standards of organizations).

- ISO 14040: launched in 1997 and discussed LCA principles, environmental management and framework of LCA.
- ISO 14041: Launched in 1998 discuss the goal definition and inventory analysis with environmental management.
- ISO 14042: launched in 2000 discuss the Life cycle impact assessment and environmental management (2000).
- ISO 14043: launched in 2000 discuss the cycle interpretation and environmental management."
- ISO 14040: launched in 2006 discussed the important phases for any LCA which are as follow: first is to define the goal of your research work second is to collect the data then calculation to find total environmental impact and finally last one to study the data and find some optimal conclusion[19].

1.2 Needs of LCA

Today's green manufacturing forced the manufacturing industries to change the traditional method and adopt optimal method of producing the finished goods with high quality [20]. CO_2 emission from manufacturing industries can be reduced either by process modification or by additive modification. But it should necessary to know the new process used is useful or not? Now a day's additive manufacturing is used in industries that replace the traditional processes. But is necessary to know additive manufacturing is beneficial or not from various perspectives, specifically when it is manufacturing all the environmental impact factor, solid or liquid waste, energy requirement these all factor need to be studied so that there is clear transparency of emission levels [21].

These all factor can be found by Life cycle assessment; its result provides a clear picture of emission level make the industry person select a better process which provides optimize the quality of finish goods with a lower level of emissions [22].

Additive manufacturing plays a big role in manufacturing industries currently. AM includes a wide range of application 3D printing is one of them it is used for rapid manufacturing for pattern making. AM process is a three-step process. The first is CAD file formation second is to convert in STL file to read the 3D printer and lastly it provides the final output. Emission level varies in 3D printing by using different material. It may be higher or maybe lower than the traditional method [23].

The Indian Commission concluded the Integrated Product Policy that LCA is the best currently available technique for measuring the future environmental impacts of goods [23]. LCA is a commonly used method for assessing the environmental effects of all aspects of the manufacturing industry's life cycle [24]. LCA can be used in manufacturing on a variety of levels, including product, part, and element level, whole manufacturing, and even entire communities [25]. LCA is used in manufacturing certification systems as well as environmental labelling (for example, Environmental Product Declarations) to measure, communicate, and control environmental emissions from the whole manufacturing process or individual manufacturing products [26]. Manufacturing knowledge can be a useful tool for solving some of the challenges of obtaining required manufacturing data during the early stages of design [27]. While the conclusions of an LCA study should be accurate and true for policymakers to make the best decisions possible, the lack of clarity and/or inconsistency of the data used in the analysis affect the quality of the findings [28]. Furthermore, conducting an LCA is a difficult and time-consuming process [29]. The approach is continuously being improved by LCA specialists and testing communities to address concerns such as uncertainty, transparency, comparability, and data quality and availability.

In recent years, there has been an increase in the use of LCA in assessing the environmental efficiency of manufacturing goods and manufacturing in all Indian states [30]. States have good testing environments that allow for long-term studies of the built environment's environmental impacts. In cooperation with research institutes and the Finnish manufacturing sector, India's Technical Research Centre has conducted a significant number of LCA-related studies [31]. Recent research has focused on the importance of embodied carbon, the environmental effects of near-zero-energy production, energy-efficient refurbishment at the manufacturing and district level, and the use of the LCA approach in construction processes [32].

There are good testing environments in India, as well as rising market demand for LCA in manufacturing [33]. Based on LCA considerations, a national Research Centre on Zero Emission Business is setting zero-emission goals. LCA studies on insulation materials manufacturing with advanced photovoltaics and LCAs on pilot manufacturing within the zero-emission manufacturing centre are examples of recent findings from India [34]. The new Zero Emission Communities in Smart Cities (ZEN) research centre seeks to support the transition to a low-carbon world by creating healthy neighborhoods with zero greenhouse gas emissions. Even though LCA paperwork is not included in any existing Indian manufacturing code or legislation, India has seen an increase in market demand for EPDs on manufacturing goods [35]. When EPDs are used to promote the use of goods with lifecycle detail, demand for Environmental Product Declarations (EPDs) skyrockets. EPDs started by the packaging sector are in high demand today, according to manufacturers. The Indian government department in charge of overseeing publicly owned industrial facilities, which is lobbying for lower greenhouse gas emissions in their programmers [36]. Furthermore, the Indian standardization agency is currently working on an Indian standard for manufacturing greenhouse gas emissions measurements. Several initiatives have been launched to improve overall manufacturing LCA awareness and understanding of the core life cycle impacts of traditional modern manufacturing and refurbishment projects [37].

LCA across Indian states can be adopted as forerunners in sustainable manufacturing in light of the above-mentioned effort on the manufacturing sector. Deeper cooperation and the establishment of common methods for the creative use of LCA in the development of sustainable manufacturing industry and refurbishment will strengthen state collaboration, their position as strong players in sustainable manufacturing, and the Indian manufacturing industry's market opportunities [38].

The Indian network "development of sustainable production and refurbishment strategies" aims to expand and enhance the application of life cycle assessment in the Indian manufacturing sector. States have banded together under the Indian network project to finance and encourage the manufacturing sector to expand applicative on of life cycle assessment [39]. The Indian networking project will hold four workshops to enable network participants to meet, share perspectives, and learn more about the most recent results on manufacturing industry LCA from national and international studies. Furthermore, the different players in the industry's familiarity with LCA are extremely useful knowledge for the Indian network. In the development of the manufacturing sector Life Cycle Assessment, the LCA network has researched the problems and needs of the manufacturing industry (LCA) [40].

This study aims to foster research-industry collaboration and recognize the challenges and needs of the Indian manufacturing industry in the production and application of manufacturing LCA, to identify solutions and facilitate groundbreaking LCA applications. LCA study group surveyed the Indian network, Collect data from 57 interviewees, to find additional information from various stakeholders on the status of LCA implementation and challenges in various Indian states.

2 Methodology

The research was carried out using data gathered from stakeholders in the Indian manufacturing industry. From April 2019 to July 2020, through a survey data was collected for this study. This survey includes an interview, phone call, email. The survey was sent to a wide range of stakeholders in the Indian manufacturing industry, with a focus on those who know about the application of LCA operation. This search was conducted using Survey Monkey, an online tool.

The interviewees were given the option of responding to the survey online mode or via email to the respective industry expert. Email surveys were gathered and entered into the survey applications like google form and survey monkey and other platforms. Some of the responses were received in the paper form. Any of the interviewees were asked to participate in a Google meet, zoom video call and longer telephone interview in which the same questions from the survey were addressed in greater detail. Some of videocalls were done on Facetime application. With the aid of the survey's debate and feedback, the findings of the survey were further studied. The data were also examined to see whether there was a connection between the participants' backgrounds and their responses.

Participants' backgrounds (including the state and form of company they served, as well as their familiarity with manufacturing LCA), awareness gaps and problems in manufacturing LCA, and the need for cooperation among Indian states were all covered in the survey.

A total of 57 people from Indian states took part in the survey, with 16 of them responding via email and the rest responding directly online. Twelve interviewees decided to engage in a more in-depth phone interview. With differing levels of awareness of LCA, the respondents, represented stakeholders in the Indian manufacturing sector, including manufacturing component producers, developers, architects, contractors, organizations, and research institutes.

The possibility of misinterpretation or varying understanding of the surveys obtained online and via emails is one of the study's major weaknesses. Interpretations differ from one state to the next, from one enterprise to the next, and from one city to the next. However, there is reason to assume that, in contrast to a questionnaire on a wider sustainability subject, differing understanding is a minimum problem in a questionnaire with a strong emphasis on the LCA-method. The longer phone interviews reduce the chance of misinterpretation, but they have another drawback in that they were conducted in various Indian language and only translate into English.

This study's questions cover a wide range of problems, requirements, and solutions in manufacturing LCA, and there's a chance that some of the more similar questions will get mixed up. This is not regarded as a significant issue, however, since the writers closely evaluate the answers, and the questionnaire is structured in such a way

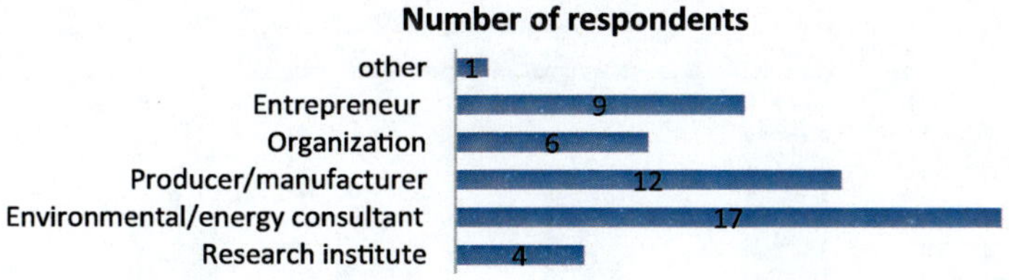

Fig. 2 Category versus respondents

that combining questions would not result in contradictory or incorrect conclusions. Another drawback of the research is the lack of literature references to related studies conducted in other states.

3 Survey and Result

This segment discusses the survey's findings and how they were analyzed. It should be remembered that the number of respondents in some of the survey results exceeds 49 since some of the respondents provided more than one response.

3.1 Background of Interviewees

The interviewees' backgrounds and familiarity with LCA are first gathered in the questionnaire. Environmental (energy) contractors, total, 17, producers (manufacturers) 12, entrepreneurs 9, associations 6, and research institutes are among the interviewees, 4. One replied chose "other" and mentioned in the reply section that she or he works for a municipality.

The repliers were asked for experiences with LCA in the study. According to the survey responses and the respondents' general background experience, respondents have experienced LCA in the context of research, consulting, sourcing, production programmed, environmental labelling of their goods, and/or waste management, as well as in educational settings. Respondents details is shown in the Fig. 2.

3.2 Manufacturing Industry LCA Knowledge Gaps

The survey was used to assess the most significant information gaps in manufacturing LCA. Figure 3 shows the main information about the gap in manufacturing LCA which is collected by interviewees. Here is a lack of awareness for recycling and reuse of energy given by 17 respondents, similarly. The lack of approximate default

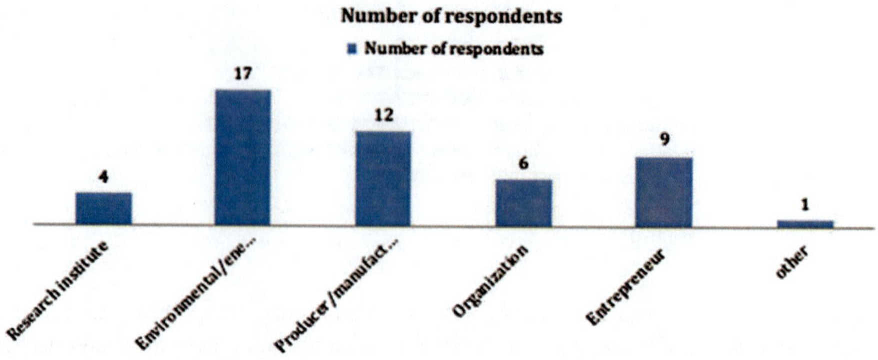

Fig. 3 Consultants versus respondent graph

by 17 participants. The respondents also identified the environmental effect of the manage and repair category and industrial background, the environmental impact of destruction and re-manufacture, pollution factors for energy supplies and power other than greenhouse emission, and negligible evidence for some item groups as significant obstacles to applying LCA in manufacturing.

The findings reveal that the profit of recycling and reuse of energy in the atmosphere is a significant awareness deficit in the manufacturing sector.

Respondents identified insufficient data for energy production transportation and modes of transportation, and technology used in manufacturing in the reply to the online survey and phone interviews. "It would be helpful to provide an outline of standard transport distances and modes of transportation, including transport distances for different waste fractions," one respondent said. Furthermore, a few respondents criticized the existing available databases, claiming that they are outdated and in need of updating. Finally, it was suggested that other impact groups should take precedence over global warming prospects. The consideration of hazardous substances/chemicals, for example, was defined as a field that requires further study.

The participants wanted to define the aspects of the product life cycle that lacked evidence and evaluation criteria. Life cycle phases with missing data or instructions were defined as the usage stage 30 respondents, end-of-life 26, advantages beyond device boundary 15, transportation to manufacturing site 9 and commodity stage 7 (see Fig. 4).

Fig. 4 Respondent's versus stages of LCA

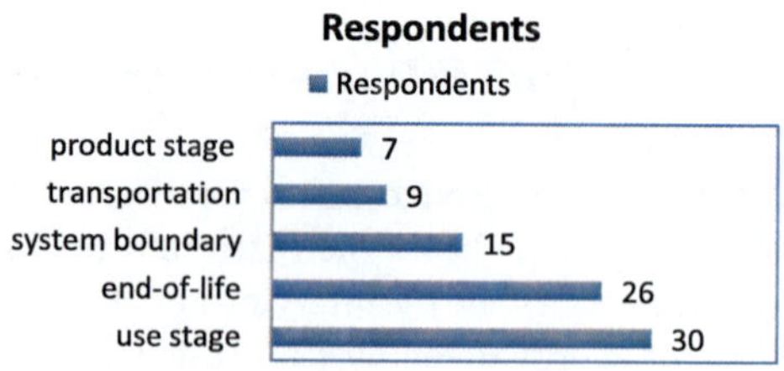

3.3 *Why Collaboration Needed Between the Indian States*

The last part of this survey asks for the importance of collaboration between Indian states and the problem catch by Indian corporation. The responses are as follows: very significant 29, important 14, moderately important 9, marginally important 2, and not important 2.

4 Conclusions

The results of present research work study allow us to understand how the Indian manufacturing industries are experiencing the knowledge and role of LCA in the additive manufacturing industries. Also, we investigate the relation of the industry knowledge related to the environmental certification schemes by state governments, local laws and laws of Govt. of India. The study focuses on major problems and their solutions that are present in the industry that needs to be addressed on the expansion of LCA. The study also emphasizes the collaboration of Indian states. Based on the analysis of responses and results obtained from our study we have recommended following conclusions which will help the managers and practitioners. Also, our findings will be helpful for the researchers for the future research work related to this area.

(a) During the application of LCA in the manufacturing industry, the main problem is how to collect the data.
(b) Performing Life cycle assessment in the Indian manufacturing industry is time-consuming as well as costly.
(c) The creation of an Indian database for data collection for LCA in the Indian manufacturing industry reduce the waste, time and cost for this purpose there are needs for collaboration between Indian states.
(d) How to manage the end/recycling phase in the manufacturing industry is found to be difficult.
(e) It was found that there is a need for effective study for the application of LCA during manufacturing phases.
(f) Manufacturing LCA is time-consuming and costly for the Indian manufacturing industry. One approach to this problem may be to use popular, easy-to-use resources and databases of digital data.
(g) Further research into the time-aspect of manufacturing LCA, as well as the related use of weighting factors or discount rates, is required.
(h) When it comes to implementing the findings of LCA trials, comparability and transparency are the most pressing issues. In India's manufacturing sectors, LCA findings are commonly used as a historical data source.
(i) A dialogue about steering instruments for sustainable production in Indian states should be initiated; common steering approaches, as well as harmonized environmental documents and measurement standards, will be beneficial.

References

1. Cabeza LF, Rincón L, Vilariño V, Pérez G, Castell A (2014) Life cycle assessment (LCA) and life cycle energy analysis (LCEA) of buildings and the building sector: a review. Renew Sustain Energy Rev 29:394–416
2. Khasreen M, Banfill PF, Menzies G (2009) Life-cycle assessment and the environmental impact of buildings: a review. Sustainability 1(3):674–701
3. Jamwal A, Agrawal R, Sharma M, Kumar V (2021) Review on multi-criteria decision analysis in sustainable manufacturing decision making. Int J Sustain Eng 1–24
4. Sharma A, Saxena A, Sethi M, Shree V (2011) Life cycle assessment of buildings: a review. Renew Sustain Energy Rev 15(1):871–875
5. Chau CK, Leung TM, Ng WY (2015) A review on life cycle assessment, life cycle energy assessment and life cycle carbon emissions assessment on buildings. Appl Energy 143:395–413
6. Ramesh T, Prakash R, Shukla KK (2010) Life cycle energy analysis of buildings: an overview. Energy Build 42(10):1592–1600
7. Rashid AFA, Yusoff S (2015) A review of life cycle assessment method for building industry. Renew Sustain Energy Rev 45:244–248
8. Karimpour M, Belusko M, Xing K, Bruno F (2014) Minimising the life cycle energy of buildings: Review and analysis. Build Environ 73:106–114
9. Jamwal A, Agrawal R, Sharma M, Kumar V, Kumar S (2021) Developing A sustainability framework for Industry 4.0. Procedia CIRP 98:430–435
10. Vilches A, Garcia-Martinez A, Sanchez-Montanes B (2017) Life cycle assessment (LCA) of building refurbishment: a literature review. Energy Build 135:286–301
11. Jamwal A, Agrawal R, Sharma M (2021) Life cycle engineering: past, present, and future. In: Sustainable manufacturing. Elsevier, pp 313–338
12. Jamwal A, Agrawal R, Sharma M, Kumar A, Kumar V, Garza-Reyes JAA (2021) Machine learning applications for sustainable manufacturing: a bibliometric-based review for future research. J Enterp Inf Manage
13. Islam H, Jollands M, Setunge S (2015) Life cycle assessment and life cycle cost implication of residential buildings—a review. Renew Sustain Energy Rev 42:129–140
14. Wong JKW, Zhou J (2015) Enhancing environmental sustainability overbuilding life cycles through green BIM: a review. Autom Constr 57:156–165
15. Haapio A, Viitaniemi P (2008) A critical review of building environmental assessment tools. Environ Impact Assess Rev 28(7):469–482
16. Hendrickson C, Horvath A, Joshi S, Lave L (1998) Peer reviewed: economic input–output models for environmental life-cycle assessment. Environ Sci Technol 32(7):184A-191A
17. Weidema BP, Thrane M, Christensen P, Schmidt J, Løkke S (2008) Carbon footprint: a catalyst for life cycle assessment? J Ind Ecol 12(1):3–6
18. Bribián IZ, Usón AA, Scarpellini S (2009) Life cycle assessment in buildings: state-of-the-art and simplified LCA methodology as a complement for building certification. Build Environ 44(12):2510–2520
19. Pfister S, Koehler A, Hellweg S (2009) Assessing the environmental impacts of freshwater consumption in LCA. Environ Sci Technol 43(11):4098–4104
20. Jamwal A, Agrawal R, Sharma M, Giallanza, A (2021) Industry 4.0 technologies for manufacturing sustainability: a systematic review and future research directions. Appl Sci 11(12):5725
21. Yadav A, Jamwal A, Agrawal R, Kumar A (2021) Environmental impacts assessment during sand casting of Aluminium LM04 product: a case of Indian manufacturing industry. Procedia CIRP 98:181–186
22. Kylili A, Fokaides PA (2016) Life cycle assessment (LCA) of phase change materials (PCMs) for building applications: a review. J Build Eng 6:133–143
23. Bribián IZ, Capilla AV, Usón AA (2011) Life cycle assessment of building materials: comparative analysis of energy and environmental impacts and evaluation of the eco-efficiency improvement potential. Build Environ 46(5):1133–1140

24. Nikolaou E (2019) Environmental assessment of construction and demolition waste with the use of life cycle analysis: a case study of multi-floor residential building
25. Tukker A (2000) Life cycle assessment as a tool in environmental impact assessment. Environ Impact Assess Rev 20(4):435–456
26. Burgess AA, Brennan DJ (2001) Application of life cycle assessment to chemical processes. Chem Eng Sci 56(8):2589–2604
27. Ibn-Mohammed T, Greenough R, Taylor S, Ozawa-Meida L, Acquaye A (2013) Operational vs. embodied emissions in buildings—a review of current trends. Energy Build 66:232–245
28. Finnveden G (2000) On the limitations of life cycle assessment and environmental systems analysis tools in general. Int J Life Cycle Assess 5(4):229–238
29. Srdić A, Šelih J (2011) Integrated quality and sustainability assessment in construction: a conceptual model. Technol Econ Dev Econ 17(4):611–626
30. Hellweg S, i Canals LM (2014) Emerging approaches, challenges and opportunities in life cycle assessment. Science 344(6188):1109–1113
31. Volk R, Stengel J, Schultmann F (2014) Building information modeling (BIM) for existing buildings—literature review and future needs. Autom Constr 38:109–127
32. Peng J, Lu L, Yang H (2013) Review on life cycle assessment of energy payback and greenhouse gas emission of solar photovoltaic systems. Renew Sustain Energy Rev 19:255–274
33. Dixit MK (2017) Life cycle embodied energy analysis of residential buildings: a review of literature to investigate embodied energy parameters. Renew Sustain Energy Rev 79:390–413
34. Wong JK, Li H, Wang SW (2005) Intelligent building research: a review. Autom Constr 14(1):143–159
35. Malmqvist T, Glaumann M, Scarpellini S, Zabalza I, Aranda A, Llera E, Díaz S (2011) Life cycle assessment in buildings: the ENSLIC simplified method and guidelines. Energy 36(4):1900–1907
36. Chastas P, Theodosiou T, Bikas D (2016) Embodied energy in residential buildings-towards the nearly zero energy building: a literature review. Build Environ 105:267–282
37. Dixit MK, Fernández-Solís JL, Lavy S, Culp CH (2012) Need for an embodied energy measurement protocol for buildings: a review paper. Renew Sustain Energy Rev 16(6):3730–3743
38. Dixit MK, Fernández-Solís JL, Lavy S, Culp CH (2010) Identification of parameters for embodied energy measurement: a literature review. Energy Build 42(8):1238–1247
39. Guinee JB, Heijungs R, Huppes G, Zamagni A, Masoni P, Buonamici R, Rydberg T (2010) Life cycle assessment: past, present, and future
40. Rossi B, Marique AF, Glaumann M, Reiter S (2012) Life-cycle assessment of residential buildings in three different European locations, basic tool. Build Environ 51:395–401

Analysis of Barriers in Sustainable Supply Chain Management for Indian Automobile Industries

Anbesh Jamwal, Akshay Patidar, Rajeev Agrawal, Monica Sharma, and Vijaya Kumar Manupati

Abstract The manufacturing industries of India are facing the global pressure to adopt the sustainable supply chain management practices. While there are some studies related to manufacturing industries across the world to examine the barriers but these studies are from developed nations. There are very fewer studies which are on manufacturing industries and from developing nations. To bridge this gap, it is very important to identify the critical barriers which influence the implementation of sustainable supply chain management in the manufacturing industries of India. Therefore, this study identifies the critical barriers in the implementation of sustainable supply chain management practices. There are total nine critical barriers found out after literature review and experts opinions from the industries and academia. Total ten manufacturing industries dealing in the automobile sectors were targeted to collect the data and responses collected from the both academia and industry. Further data is synthesized and AHP method is applied to rank the barriers and evaluate the effect of the barriers on the SSCM implementation. Cost of the sustainability and economical conditions in the manufacturing industries is found to be most influencing barrier among all the barriers. The findings of this study aims to support the Indian manufacturing industries in the structural way so that policy makers and managers from the manufacturing industries can identify the most influencing barrier and work in future to eliminate that for the implementation of SSCM in their business practices to achieve business excellence. Also, this study can be helpful to the stake-holders to achieve the sustainable development goal 2030.

A. Jamwal · R. Agrawal (✉)
Department of Mechanical Engineering, Malaviya National Institute of Technology, J.L.N. Marg, Jaipur, Rajasthan 302017, India
e-mail: ragrawal.mech@mnit.ac.in

A. Patidar · M. Sharma
Department of Management Studies, Malaviya National Institute of Technology, J.L.N. Marg, Jaipur, Rajasthan 302017, India

V. K. Manupati
Department of Mechanical Engineering, National Institute of Technology Warangal, Warangal, India

R. Agrawal et al. (eds.), *Recent Advances in Industrial Production*, Lecture Notes in Mechanical Engineering, https://doi.org/10.1007/978-981-16-5281-3_8

Keywords Sustainable development · Sustainable supply chain management · Sustainable manufacturing · Manufacturing industries · Analytical hierarchy process · Barriers · India

1 Introduction

The rapid change in the manufacturing and production systems of Indian industries has increased the supply chain practices in the industries which results in the environmental degradation. Sustainable supply chain practices are the integration of all three sustainability factors i.e. economical, social and environmental which can help to achieve the sustainability in the supply chain of the industries [1]. Global pressure on the Indian manufacturing industries has forced them to adopt the sustainable supply chain practices to achieve the sustainable development goal of 2030 [2]. Therefore, sustainable supply chains are becoming popular in the Indian manufacturing industries over the traditional supply chains to compete with the developed nations. In the developing nations like India and Bangladesh implementation of sustainable supply chain practices in manufacturing industries is still a great challenge due to lack of proper framework [3]. Also, policies related to the environmental conservation and sustainable development has forced the industries to integrate the pollution prevention strategies during the supply chain practices to minimize the effect of supply chains on the environment [4]. Sustainable manufacturing practices in the manufacturing industries allows the industries to include all the phases of product life cycle manufactured in the industries by incorporating the all three sustainability pillars i.e. social, economical and environmental. Implementation of sustainable supply chain practices in the industries enables the organizations to consider the adverse the supply chain practices on the sustainability development goals of industry [5].

In the last few years development and advancement in supply chain practices in manufacturing industries in developed nations have forced the developing nations to consider the important aspects such as economic, social and environmental during the implementation of supply chain practices which are known as sustainable supply chain practices [6]. Manufacturing firms are now considering issues in supply chain such as "green purchasing", "environmental certification", green human resource management", "sustainable supplier selection", "social sustainability", "Industry 4.0″, and "green marketing". In the last due to the customer pressure and Govt. policies many industries in the world have implemented the sustainable supply chain practices to achieve the sustainability in their business practices. Implementation of sustainable supply chain practices in the developing nations is still a challenge for the industries due to the lack of proper framework [7]. This study focuses on the identification of sustainable supply chain implementation barriers in the Indian manufacturing industries over the traditional supply chain practices. Since, Indian manufacturing industries are now working on implementation of sustainable supply chain practices in their business practices. This study tries to address the following research question on Implementation of SSCM practices in manufacturing industries:

- RQ1: What are the main barriers in implementation of SSCM for Indian manufacturing industries?
- RQ2: What is their impact on the SSCM practices?
- RQ3: How can identification and ranking of SSCM implementation barriers can help the managers of Indian manufacturing industries to achieve the excellence in their business practices?

To answer these research questions the following objectives are set which are discussed below:

Identification of SSCM barriers on the basis of literature survey.

Propose a conceptual framework for the Indian manufacturing industries.

Formulation of both practical and managerial utility of proposed framework for decision makers and managers to achieve the excellence in business.

Analytical hierarchy process methodology is adopted in the study to develop a framework for the manufacturing industries. From the literature survey 9 critical barriers were found out which is further classified into the three main criteria's i.e. social, economical and environmental. Section 2 of the paper represents the literature review and Sect. 3 represents the methodology adopted with the case illustration. Further, results and discussion basis on the methodology is discussed with the practical and managerial implications with the limitations of the study.

2 Literature Review

In any business organization the term "sustainability" is defines as the consideration of TBL concepts which deals with the economical factors, social concerns and environmental issues. The incorporation of TBL concept in the manufacturing industries can help the industries to achieve the sustainability in their business practices [8]. Over the years sustainable supply chain management is evolve as an emerging research area in the manufacturing industries over the traditional supply chain due to the adverse effect of supply chain practices on the environment which results in higher carbon emissions and green house effect [9]. Previously researchers have focused on the integration of economic issues with the environmental issues or green issues which results in the development of green supply chain management. Studies in past few decades have shown that how green marketing, green materials sourcing, green purchasing, green human resource management and green manufacturing helps to protect the environment by the incorporation of green supply chain practices in the industries [10]. But now industries are focusing on the customer satisfaction and completing the volatile demands of customer which can be fulfill by the SSCM by considering the all three pillars of sustainability. SSCM is the emerging research area in past few years in the supply chain [11]. Figure 1 shows the research publication trends in the area of SSCM over the years which shows that in last few years researchers are now focusing on the SSCM.

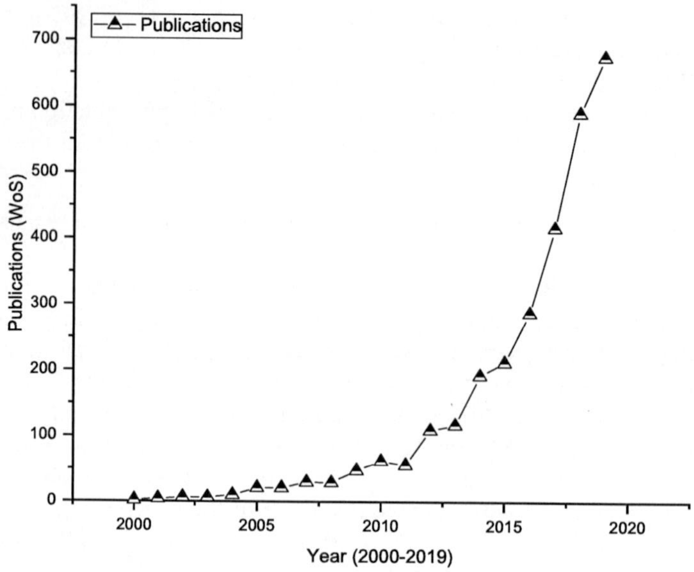

Fig. 1 Publications on sustainable supply chain management on WoS (2000–2019)

Publication trend shows that researchers are now focusing on sustainable supply chain management. Industries are now adopting the both environmental management systems and cleaner production strategies to improve sustainability in their supply chain practices [12].

To find the barriers among the sustainable supply chain practices paper from databases i.e. Scopus, WoS and Google scholar were taken into consideration. The barriers found out from the papers were further refined and finalized by the brainstorming sessions with the Indian manufacturing industries experts, academia experts working in the sustainable supply chain area. In the next phase framework is developed with the Analytical hierarchy process. Table 1 shows the all nine critical barriers in the implementation of SSCM for Indian manufacturing industries along with their respective criteria's and description.

2.1 *Problem Description*

Due to the customer pressure, volatile demands and Govt. regulations Indian manufacturing are now started the implementation of sustainable supply chain practices in their organizations. India is now considered as largest manufacturing hub for automobile components and expected to be grown more by 2025. In early stages of automobile sector development industries have faced external barriers such as lack of proper organization infrastructure and higher investment costs. Now India is

Table 1 Barriers in SSCM implementation for Indian industries

Barrier in SSCM implementation for Indian manufacturing industries	Description of barrier	Criteria	Authors reference
(B1)—Lack of environmental policies	Industries are not aware about the environment related policies in SSCM	C1 ENV	[13–15]
(B2)—Lack of awareness in local customers among green products	Customer are now aware about the new technologies and green products manufacturing by the industries	C1 ENV	[16–19]
(B3)—Lack of reverse logistics practices	Industries are not focusing on the reverse logistics practices	C1 ENV	[20, 21]
(B4)—Absence of society pressure on industries	Lack of society awareness for SSCM which results in absence of pressure from society to implement SSCM practices	C2 SOC	[22, 23]
(B5)—Lack of Govt. support in adoption of SSCM	Govt. is not supporting in the adoption of SSCM practices, there is political pressure on the industries	C2 SOC	[24–26]
(B6)—Lack of demands and pressure for low prices	Customer demands are not regular or customers are pressuring industries to lower the prices of their products	C2 SOC	[27, 28]
(B7)—Capacity constraints	Capacity constraints for the industries in the SSCM	C3 ECO	[29, 30]
(B8)—Cost for sustainability and economical conditions	Cost to improve sustainability and economical conditions are too high	C3 ECO	[31–33]
(B9)—Lack of funds for SSCM	Industries don't have much funds for the SSCM implementation	C3 ECO	[34, 35]

known as automobile market base for south East Asian region. Firstly, industries are focused on the green purchasing, green marketing and green manufacturing but as the customer demands are volatile and market scenario has been changed to mass customization from mass production industries in India are now considering the social aspects in supply chain with the environmental and economical aspects. The major industries in India manufacture engine valves and steering components and distribute them in nearby states or to export to other countries. The major clients are Tata motors, Ford India, Mahindra and Hyundai motors. It is found that Indian manufacturing industries are still facing an issue with the proper framework to overcome from the SSCM implementation challenges with the consideration of all three sustainability aspects. In this study 9 critical barriers are found out related to all three sustainability aspects and classified in the three criteria. This paper discusses the barriers collected from the literature survey and proposes a framework based on

AHP. In the present study data is collected from 10 automobile industries from the North India.

3 Solution Methodology

In the present study the objective of the research study is achieved by the Analytical hierarchy process methodology. Analytical hierarchy process methodology was developed by Saaty to integrate the industrial and academia experts' opinion collected through the questionnaire survey and interviews and then evaluating their scores into a hierarchy by decomposing the problem from the higher order hierarchy to lower hierarchy [36].

AHP is widely used MCDM in the decision making problems because of its usability and effortlessly reasonable system over other MCDM techniques [37].

AHP has certain limitations when the measurement scale for the study is unstable, subjective nature and nonexistence of impreciseness which needs the fuzzy environment to solve such type of problems. Fuzzy-AHP uses the vagueness and uncertainty of expert's judgment in the terms of linguistic variables. In such type of cases uncertainty can be reduced by using Fuzzy-AHP method [38]. In this study we have considered Saaty fundamental scale for the pairwise comparisons. The crisps input have been calculated on the basis of the responses collected from the industry experts and academia inputs.

Consistency ratio can be calculated as:

$$\mathrm{CI} = \mathrm{CI}/\mathrm{RI} \tag{1}$$

where RI is random index.

Crisps inputs for each criteria and barrier is calculated with AHP approach.

3.1 Computing Normalized Weights of Each Criteria

In AHP approach if more than two levels are exists then the PV (priority vectors) can be combined with the priority matrices. This gives the one final priority vector. Normalized weight is calculated by the dividing the sum of particular column of criteria with each criteria. Then the final weight for is calculated by taking the average row-wise.

3.2 Computation of Global Weights

Table 2 represents the global weights calculated for the present study by Analytical hierarchy process. All the nine barriers found out for the SSCM implementation for the Indian manufacturing industries is distributed into three criteria's on the basis of the experts inputs from the industry and academia. The global weight for each barrier is normalized and represented in the Table 2. Global effect of each barrier in the SSCM implementation is calculated by multiplication of normalized weight of the barrier with its particular criteria. Ranking of the each barrier with its normalized weight is shown in the Table 3.

Table 2 Global weights of barriers

Barrier	Global weights	Normalized weights	%
B1	0.4237	0.0471	4.7076
B2	0.9348	0.1039	10.3863
B3	0.2805	0.0312	3.1165
B4	1.6476	0.1831	18.3070
B5	0.4876	0.0542	5.4173
B6	0.2855	0.0317	3.1724
B7	0.8455	0.0939	9.3943
B8	2.8177	0.3131	31.3080
B9	1.2772	0.1419	14.1906
SUM	9.0000	1.0000	100.0000

Table 3 Ranking of barriers

Barrier	Name	%	Rank
B1	**(B1)**—Lack of environmental policies	3.1586	7
B2	**(B2)**—Lack of awareness in local customers among green products	6.9686	5
B3	**(B3)**—Lack of reverse logistics practices	2.0910	9
B4	**(B4)**—Absence of society pressure on industries	15.6460	3
B5	**(B5)**—Lack of Govt. support in adoption of SSCM	4.6299	6
B6	**(B6)**—Lack of demands and pressure for low prices	2.7113	8
B7	**(B7)**—Capacity constraints	11.0889	4
B8	**(B8)**—Cost for sustainability and economical conditions	36.9555	1
B9	**(B9)**—Lack of funds for SSCM	16.7503	2

4 Results and Discussions

In the present study nine critical barriers in the implementation of sustainable supply chain management practices were found out from the literature survey on the basis of papers collected from various databases such as Taylor and Franics, Inderscience, Springerlink, Scopus, Wiley online library, Emerald, WoS and Google scholar. Further, the barriers were refined with the brainstorming session with industry experts and the academia working in the area of sustainable supply chain management. Analytical hierarchy process is adopted for the framework development for the manufacturing industries. Ten major automobile industries were targeted and 174 responses were collected from the industries and academia. Ranking of barriers and their effect in implementation is calculated with the help of AHP method. It is found that Cost for sustainability and economical conditions is the most influencing barriers for the Indian manufacturing industries. Indian industries are still following the traditional supply chain concepts. Adoption of a new system over a existing one is never be an easy task for any organization. In the developing nations, industries are facing an issue with the funding. To achieve the sustainability in the supply chains higher investment cost is required and there is lack of financial support from both investors and management for the sustainability and economical conditions improvements. At 2nd lack of funds for SSCM implementation. In most of the manufacturing industries management are still focusing on the lower investments and higher returns. To achieve sustainability in the business higher investment cost required which is a critical barrier for the Indian manufacturing industries. Indian manufacturing industries are facing an issue related to the investment for sustainability development which affects the SSCM implementation in the Indian manufacturing industries. At 3rd absence of society pressure on the industries for the SSCM adoption. This is due to the lack of customer awareness. Customers in developing nations are not much aware about the sustainability goals and benefits of the SSCM to the society. This also affects the SSCM implementation in the manufacturing industries.

Lack of reverse logistic practices having least affect on the SSCM implementation in manufacturing sector.

5 Conclusion

SSCM implementation practices in the developing nations are becoming popular research trends in the industrial sectors. Manufacturing industries are now implementing the sustainable supply chain management practices to achieve business excellence. Implementation of the SSCM over the traditional supply chain is never easy task as it is always influenced by some critical barriers. Before, implementation of new framework it is necessary to evaluate the effect of each barrier on the SSCM implementation on the particular sector of industry. Therefore this study focuses on the identification of critical barriers in the implementation of the SSCM.

Besides the identification of implementation challenges in sustainable supply chain management for Indian manufacturing industries this paper also focuses on the calculate the effect of particular barrier on the implementation. There were total nine critical barriers found out after refining the brainstorming with the industry and academia experts. Total of 10 industries were targeted for the study from North region of India and responses were collected from the industries and academia. Based on the data and AHP methodology following conclusions can be drawn:

1. Cost for the sustainability and economical conditions in the manufacturing industries is the most critical barrier in SSCM implementation the Indian manufacturing industries which can be resolve by financial support for the economical and sustainability conditions improvement from the Govt. funding agencies.
2. This study is the unique as there is very fewer studies have been carried out in the developing nations on the sustainable supply chain for manufacturing industries by considering the barriers on the basis of all three sustainability aspects.
3 The proposed research framework is a unique application and helps the decision-policy makers to use as benchmark in the context of the India for manufacturing industries

It is found that AHP approach is the very effective method for ranking the implementation barriers for the Indian manufacturing industries. However, this method depends upon the larger number of responses from the industry and academia experts.

5.1 Practical Implications

The possible SSCM barriers have been identified by the literature review from the various databases and experts opinions. There may be some human biasness in responses. The manufacturing industries need to consider the highest weight barrier for the successful implementation of SSCM practices.

5.2 Managerial Implications

The proposed model considers the sustainable supply chain management implementation challenges. Managers and policy-makers from the Indian manufacturing industries have freedom to include the customized the sustainability barriers based on the particular industry sector.

5.3 *Future Scopes of the Study*

Further this study can be extended by developing the Hybrid Fuzzy model for the Sustainable supply chain management implementation barriers to find most influencing barriers. MCDM techniques such as VIKOR, ISM, TISM and DEMETAL can be used in the future studies in the fuzzy environment to remove the uncertainty.

References

1. Koberg E, Longoni A (2019) A systematic review of sustainable supply chain management in global supply chains. J Clean Prod 207:1084–1098
2. Yusuf Y, Menhat MS, Abubakar T, Ogbuke NJ (2019) Agile capabilities as necessary conditions for maximising sustainable supply chain performance: an empirical investigation. Int. J. Prod. Econ. 107501
3. Xiao C, Wilhelm M, van der Vaart T, van Donk DP (2019) Inside the buying firm: exploring responses to paradoxical tensions in sustainable supply chain management. J Supply Chain Manag 55(1):3–20
4. Manavalan E, Jayakrishna K (2019) A review of Internet of Things (IoT) embedded sustainable supply chain for industry 4.0 requirements. Comput Ind Eng 127:925–953
5. Jamwal A, Agrawal R, Sharma M, Kumar A (2022) Sustainable material selection for Indian manufacturing industries: a hybrid multi-criteria decision-making approach. In Proceedings of the International Conference on Industrial and Manufacturing Systems (CIMS-2020), pp 31–43. Springer, Cham
6. Yu M, Cruz JM (2019) The sustainable supply chain network competition with environmental tax policies. Int J Prod Econ 217:218–231
7. Reinerth D, Busse C, Wagner SM (2019) Using country sustainability risk to inform sustainable supply chain management: a design science study. J Bus Logist 40(3):241–264
8. Nayak R, Akbari M, Far SM (2019) Recent sustainable trends in Vietnam's fashion supply chain. J Clean Prod 225:291–303
9. Modica PD, Altinay L, Farmaki A, Gursoy D, Zenga M (2020) Consumer perceptions towards sustainable supply chain practices in the hospitality industry. Curr Issue Tour 23(3):358–375
10. Yun G, Yalcin MG, Hales DN, Kwon HY (2019) Interactions in sustainable supply chain management: a framework review. Int J Logist Manag
11. Allaoui H, Guo Y, Sarkis J (2019) Decision support for collaboration planning in sustainable supply chains. J Clean Prod 229:761–774
12. Rajeev A, Pati RK, Padhi SS (2019) Sustainable supply chain management in the chemical industry: evolution, opportunities, and challenges. Resour Conserv Recycl 149:275–291
13. Jamwal A, Agrawal R, Sharma M, Kumar V, Kumar S (2021) Developing a sustainability framework for industry 4.0. Procedia CIRP 98:430–435
14. Hsu CC, Tan KC, Zailani SHM (2016) Strategic orientations, sustainable supply chain initiatives, and reverse logistics. Int J Oper Prod Manag
15. Patidar A, Sharma M, Agrawal R, Sangwan KS, Jamwal A, Gonçalves M (2021, June) Sustainable supply chain research and key enabling technologies: a systematic literature review and future research implications. In International Conference Innovation in Engineering, pp 305–319. Springer, Cham
16. Su CM, Horng DJ, Tseng ML, Chiu AS, Wu KJ, Chen HP (2016) Improving sustainable supply chain management using a novel hierarchical grey-DEMATEL approach. J Clean Prod 134:469–481
17. Khan M, Hussain M, Saber HM (2016) Information sharing in a sustainable supply chain. Int J Prod Econ 181:208–214

18. Hong J, Zhang Y, Ding M (2018) Sustainable supply chain management practices, supply chain dynamic capabilities, and enterprise performance. J Clean Prod 172:3508–3519
19. Jamwal A, Agrawal R, Sharma M, Giallanza A (2021) Industry 4.0 technologies for manufacturing sustainability: a systematic review and future research directions. Appl Sci 11(12):5725
20. Ansari ZN, Kant R (2017) A state-of-art literature review reflecting 15 years of focus on sustainable supply chain management. J Clean Prod 142:2524–2543
21. Reefke H, Sundaram D (2017) Key themes and research opportunities in sustainable supply chain management–identification and evaluation. Omega 66:195–211
22. Agrawal V, Mohanty RP, Agrawal AM (2020) Identification and analysis of enablers of SCM by using MCDM approach. Bench: An Int J
23. Fallahpour A, Olugu EU, Musa SN, Wong KY, Noori S (2017) A decision support model for sustainable supplier selection in sustainable supply chain management. Comput Ind Eng 105:391–410
24. Matthews L, Power D, Touboulic A, Marques L (2016) Building bridges: toward alternative theory of sustainable supply chain management. J Supply Chain Manag 52(1):82–94
25. Bechtsis D, Tsolakis N, Vlachos D, Iakovou E (2017) Sustainable supply chain management in the digitalisation era: the impact of Automated Guided Vehicles. J Clean Prod 142:3970–3984
26. Bastas A, Liyanage K (2018) Sustainable supply chain quality management: a systematic review. J Clean Prod 181:726–744
27. Matić B, Jovanović S, Das DK, Zavadskas EK, Stević Ž, Sremac S, Marinković M (2019) A new hybrid MCDM model: sustainable supplier selection in a construction company. Sym 11(3):353
28. Busse C, Meinlschmidt J, Foerstl K (2017) Managing information processing needs in global supply chains: A prerequisite to sustainable supply chain management. J Supply Chain Manag 53(1):87–113
29. Bendul JC, Rosca E, Pivovarova D (2017) Sustainable supply chain models for base of the pyramid. J Clean Prod 162:S107–S120
30. de Camargo Fiorini P, Jabbour CJC (2017) Information systems and sustainable supply chain management towards a more sustainable society: Where we are and where we are going. Int J Inf Manage 37(4):241–249
31. Mariadoss BJ, Chi T, Tansuhaj P, Pomirleanu N (2016) Influences of firm orientations on sustainable supply chain management. J Bus Res 69(9):3406–3414
32. Raut RD, Narkhede B, Gardas BB (2017) To identify the critical success factors of sustainable supply chain management practices in the context of oil and gas industries: ISM approach. Renew Sustain Energy Rev 68:33–47
33. Montabon F, Pagell M, Wu Z (2016) Making sustainability sustainable. J Supply Chain Manag 52(2):11–27
34. Jamwal A, Agrawal R, Sharma M, Kumar V (2021) Review on multi-criteria decision analysis in sustainable manufacturing decision making. Int J Sustain Eng 1–24
35. Fontes CHDO, Freires FGM (2018) Sustainable and renewable energy supply chain: a system dynamics overview. Renew Sustain Energy Rev 82:247–259
36. Al-Harbi KMAS (2001) Application of the AHP in project management. Int J Project Manage 19(1):19–27
37. Chang DY (1996) Applications of the extent analysis method on fuzzy AHP. Eur J Oper Res 95(3):649–655
38. Kahraman C, Cebeci U, Ulukan Z (2003) Multi-criteria supplier selection using fuzzy AHP. Logistics Infor Manage

Optimization of Injection Timing for a C.I. Engine Fuelled with Gomutra Emulsified Diesel

Amit Jhalani, Dilip Sharma, Digambar Singh, and Pushpendra Kumar Sharma

Abstract High temperatures conditions are responsible for smoke and NO_x emissions in compression ignition engines. Use of water emulsified diesel fuel facilitates low-temperature combustion of diesel which is found to be a prominent option for emission reduction. Aligning to this technology, gomutra emulsified diesel (GMD) fuel has been tested in this research for improved results. The experimental study has been performed on a stationary, constant rpm, direct-injection C.I. engine. 15% gomutra emulsified diesel with 4% emulsifier was utilized for optimization of the injection timing with GMD emulsions. The effect of injection timing at various IT 19°, 21°, 23° and 25° btdc was then analyzed. It was observed that retardation by 2° btdc, i.e. at IT 21 engine performance got improved with lower diesel knock. It was observed that the BTE was improved up to 24.2% at IT 21° btdc as compared to 23.9% at IT 23° btdc. In addition, NOx was also found to be reduced considerably. Overall, it is concluded that the performance of engine with emulsion fuel could be improved by retarding the fuel injection timing.

Keywords Water · Emulsion · Gomutra · Cow-urine · Emissions · Injection Time

1 Introduction

BS-VI emission norms have been implemented in India in the year 2020 [1]. It has created various challenges to surpass BS-V and meet the requirements in such a short duration. These fast-changing standards and increased environmental concern is stimulating fuel researchers to explore for cleaner and efficient methods of combustion. Out of the several emission reduction techniques, water emulsification in diesel has been found as an important alternate for improved performance and lower exhaust

A. Jhalani (✉) · D. Sharma · D. Singh · P. K. Sharma
Department of Mechanical Engineering, Malaviya National Institute of Technology Jaipur, Jaipur, Rajasthan, India
e-mail: 2015rme9521@mnit.ac.in; amit.me@dknmu.org

A. Jhalani
Department of Mechanical Engineering, Dr. K.N. Modi University, Newai, Rajasthan, India

R. Agrawal et al. (eds.), *Recent Advances in Industrial Production*, Lecture Notes in Mechanical Engineering, https://doi.org/10.1007/978-981-16-5281-3_9

emissions [2, 3]. However, the improvement shown in the papers are quite inconsistent and exhibits a wide difference in the results. Optimum blend ratio, HLB value and surfactant proportion, BSFC, brake thermal efficiency, and engine emissions varied with the authors. Some researchers have reported decreasing BSFC trend s with an increase of water proportion while some others showed increased BSFC with use of water-emulsified diesel in increased proportions [4, 5]. Exhaust emissions like CO and hydrocarbon are found to be increased in many papers with water-emulsified diesel due to the low-temperature combustion while some other researchers reported decreased CO and HC emissions with water-diesel emulsion due to micro-explosion phenomenon [6, 7]. Fan et al. [8] found optimum emulsion with 8% surfactant blend whereas Attia and Kulchitskiy [7] used only 0.5% of emulsifier. Huo et al. [9] showed better stability with a HLB value of 5 while El-Din et al. [10] reported a stable emulsion formulation with HLB value of 10.

The process adopted for emulsion formation, droplet size, the test set up and the complicated combustion behaviors due to micro-explosion in water-diesel emulsion fuels are the possible factors which have been reported by the researchers for the inconsistency in the results [11]. GMD emulsion fuel is a similar technology as water-diesel emulsion [12]. The engine operating parameters also play a major role in the engine performance and exhaust emission characteristics [13]. Compression ratio, injection pressure (IP) and injection timing are the parameters which must be analyzed before feeding any new fuel in diesel engine. Hence, in this work, the injection timing has been optimized for the GMD emulsion to assess its effects on the performance of the engine. The performance of the engine is assessed on the basis of the brake thermal efficiency, smoke and NOx emissions.

2 Test Setup

The current experimental work is performed on a Kirloskar make C.I. engine (TV1 model). It is a direct injection, single cylinder, stationary, compression ignition engine. The test engine is augmented to a dynamometer (eddy current type) for load application. Two rotameters with control valves were connected to regulate water flow for cooling of engine and dynamometer. For the temperature measurement at different locations, a number of thermocouples are installed. During the whole experimental work, the compression ratio and injection pressure were kept at same setting as shown in Table1. The engine exhaust emissions were tested by an AVL 5-gas analyzer machine. Some important engine specifications are shown in Table 1.

Table 1 Test engine specifications

Particular	Detail
Engine category	1 cylinder, 4 stroke, compression ignition engine
Model	Kirloskar TV1 engine facilitated by Apex Innovations, India
Cylinder capacity	661 cc
Engine speed	Constant speed 1500 RPM
Rated power	3.5 kW
Cooling method	Water cooling
Compression ratio	20:1
Injection pressure	210 bar
Linked software	'Enginesoft' developed by Apex Innovation, India

3 Gomutra Emulsified Diesel Fuel

Gomutra consists of 95% water along with 2.5% urea, a very small proportions of Na, K, Ca, Mg and some other micro-constituents [14]. The density of diesel and water differs sufficiently. The interfacial tension within the oil and water is very high which does not allow mixing of these two liquids for making of a stable solution [15]. In this technique, for the emulsion to be used as a fuel, formation of a stable and homogeneous emulsion is a vital aspect. Hence, surfactants were used to form the homogeneous and stable emulsions under high energy processing methods.

3.1 Chemistry of Emulsifiers

There are a number of non-ionic surfactants available which could be utilized for making gomutra-diesel emulsion viz. sorbitol sesquioleate, sorbitan monolaurate, Triton X-100, detergents, solgen 40, TDS-30, and more along with other adequate proportional combinations of the relevant surfactants [2]. Most of the researchers have observed better performance with a mixture of surfactants rather than a single emulsifier having same value of HLB [16]. Henceforth, the common non-ionic emulsifiers Span 80 and Tween 80 have been used in this work for the emulsion making (Table 2).

3.2 Emulsion Processing

The study was performed by adding different quantities of gomutra in diesel (5%, 10%, 15% and 20%). 40 ml of surfactant (made up of 34 ml of Span 80 and 6 ml of Tween 80) was used for preparation of 1 L of emulsion with constant HLB value (5.9).

Table 2 Emulsifiers used to make the emulsions [17, 18]

Emulsifier name	Span 80	Tween 80
Appearance	Pinkish yellow	Amber-liquid
HLB value	4.3	15
Chemical behavior	Non-ionic	Non-ionic
Molecular form	$C_{32}H_{60}O_{10}$	$C_{24}H_{44}O_6$
Molecular wt	604.82	428.61
Made by	Merck Life Science (Pvt) Ltd	Loba Chemie

Table 3 HLB values of different blends of tween and span

Span 80 (%)	Tween 80 (%)	HLB value
100	0	4.3
83	17	6
65	35	8
46	54	10
28	72	12
9	91	14
0	100	15

All the emulsion samples were made keeping the percentage of surfactant (4%) and the HLB value (5.9) constant. Table 3 shows the HLB Values of various blends of Span and Tween. Emulsion was formed by adding the specific quantity of surfactant in the diesel and then processing in a mechanical homogenizer at 15,000 rpm. Thereafter, the necessary volume of cow-urine was mixed in it gradually. All the emulsions were obtained as a milky-white opaque in appearance (Fig. 1). Most of the authors have observed similar emulsions in their work when experiments were carried out with water-diesel emulsions [19, 20].

The processed emulsions were observed to be almost homogeneous and stable for more than 15 days. It was observed that creaming phenomenon occurred after that at the bottom which could be observed clearly after that duration. The complete phase separation was seen after a duration of approximately 40 days.

3.3 *Emulsified Fuel Properties*

Density and viscosity of the emulsified fuel was observed to be increased as compared to the diesel. Gomutra addition in diesel lowered down the overall heating value of emulsion as gomutra does not have its own heating value. The properties of the 15% GMD emulsion are compared with the standard diesel fuel in Table 4.

Fig. 1 Picture of emulsion (GM15)

Table 4 Physicochemical properties of GM15 and diesel

Property	Diesel	Emulsion
Cow-urine proportion	0%	15%
Density (kg/m^3)	830	865
Kinematic viscosity (cSt)	2.98	4.9
Lower calorific value (kJ/kg)	42,500—Standard 42,900 – Experimentally Observed	35,100—Experimentally Observed 34,425—estimated (81% of diesel)
Summarized molecular formula	$C_{15.243}$ $H_{27.548}N_{0.009}$	$C_{15.243}$ $H_{27.548}N_{0.009}$ + 15% GM

4 Performance Analysis

With the introduction of cow-urine in diesel, various combustion factors get affected more or less which have their impact on the efficiency and emissions. To carry out the analysis and assess the influence of injection timing (IT) variation on the performance of engine fuelled with cow-urine emulsion, IT was varied. The engine showed its optimum performance 21.9% BTE with neat diesel fuel at IT 23° btdc. Hence the various blends of 5%, 10%, 15% and 20% of emulsion were first tested at IT 23° btdc. A significant improvement in the BTE is observed with emulsion fuels. Maximum 23.9% BTE was obtained with GM15. Due to longer ignition delay with 15% and higher proportions knocking was observed. Elongated ignition delay allows fuel to get accumulated inside the combustion chamber before the combustion starts. This high amount of accumulated fuel burns instantaneously to produce a higher RoPR and high HRR [21, 22]. The viscosity value of 20% GMD emulsion is also very high which lowers down the probability of better fuel atomization and homogeneous charge preparation. Hence lower efficiency was observed with GM20.

$$BTE = \frac{BP*3600}{mf*CV} \tag{1}$$

Here, m_f = fuel consumption rate in kg/hr; CV = lower heating value (kJ/kg). However, the bsfc of emulsion was found to be more as compared to the diesel but a little attention will reveal that the gomutra content does not have its own heating value, hence its addition in the diesel drops down the overall CV. It can easily be observed from Eq. (1) that the influence of a lower calorific value supersedes the effect of high bsfc and as a result efficiency improves.

The optimized 15% GMD emulsion was then used to optimize the injection timing. The composition of emulsion and all the other emulsion parameters were kept constant and only IT was varied. The experimental results were plotted for the brake thermal efficiency at different load conditions and for different settings of injection timing 19°, 21°, 23°, 25° btdc (Fig. 2). It was found that efficiency was slightly improved up to 24.2% at IT 21 as compared to 23.9% at IT 23. Along with the improvement in efficiency, the diesel knock reduced considerably due to retardation in injection timing at IT 21. High GM emulsification creates low combustion temperature conditions which tends to make longer chemical delay and require more time in vaporization [10]. The high viscosity of emulsion prolongs physical delay. Retarding of the injection timing facilitated the fuel to enter in the relatively higher temperature and pressure combustion zone [23]. It resulted in the early vaporization

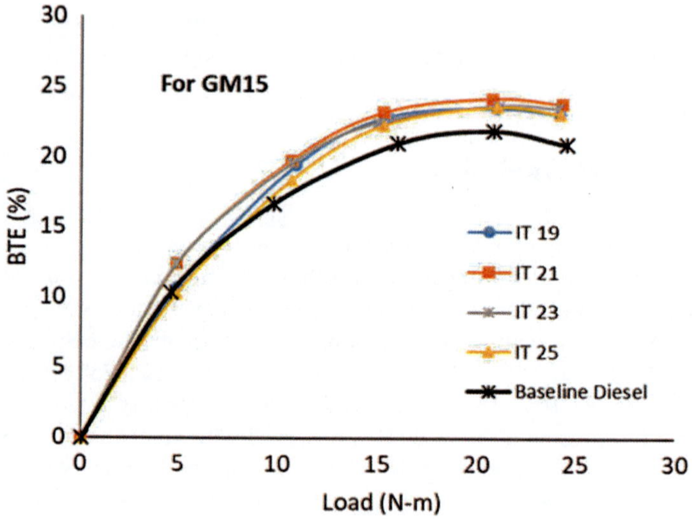

Fig. 2 Effect of injection timing over BTE

and mixing of the fuel. The tendency of fuel accumulation got reduced in this case and thus providing a smoother engine run with improved efficiency.

5 Emission Analysis

The main pollutants of diesel engines are black smoke, particulate matter (PM) and nitrogen oxides (commonly called NOx) which are emitted in troublesome quantity. These emissions are very harmful for human health but as well as it degrades the complete ecology [24]. Physicochemical properties of fuels do also influence the engine exhaust emissions significantly.

5.1 NOx Emissions

The important factors which influence the extent of NO_x emissions formed includes oxygen concentration, combustion temperature, and the stay period (duration for which the burnt gases remain inside the combustion chamber). The latent heat of evaporation and sensible heat of cow-urine helps to cut down the temperature of combustion zone and hence lower NO_x formation occurred [25]. Around 21.3% maximum reduction in the NO_x was observed with 15% GMD emulsion at IT 23. High ignition delay makes a favorable condition for high NO_x emission but the water content of fuel reduces the combustion temperature and balances the effect of temperature rise due to high pressure [21]. Retarding of the injection timing facilitated the fuel to enter in the relatively higher temperature and pressure combustion zone. It resulted in early vaporization and charge preparation. The tendency of fuel accumulation got reduced in this case and thus providing a smoother engine run. It also facilitates low NOx emissions as shown in Fig. 3.

5.2 Smoke

Smoke emissions were measured in terms of opacity. It reflects the amount of smoke particles confined in the exhaust emission i.e. the opacity is proportional to the particulate matter. Adequate air–fuel mixing situations are encouraging for low smoke emission. Therefore, on increasing the GM proportion, smoke opacity got reduced due to the micro-explosion phenomenon. Smoke emissions were found reduced continuously with the adding of gomutra but retardation in the injection timing led to increase the smoke. Retardation allows lesser time for charge preparation and results in fuel rich zones. When the IT was retarded from IT25°btdc to 19°btdc, the smoke emissions got increased 54.9 to 60.3% at full load (Fig. 4).

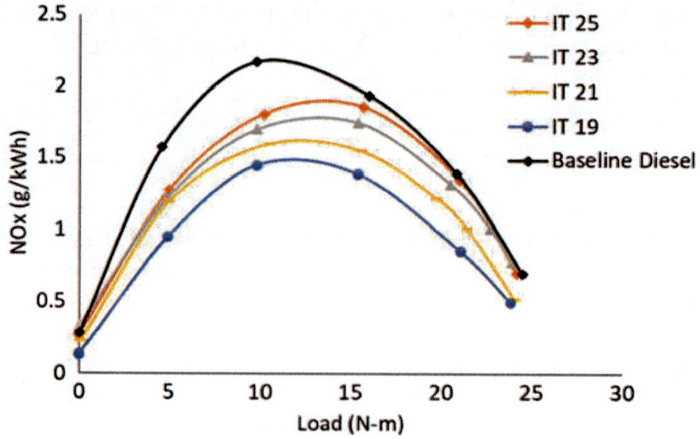

Fig. 3 Effect of Injection Timing on NO_x emissions

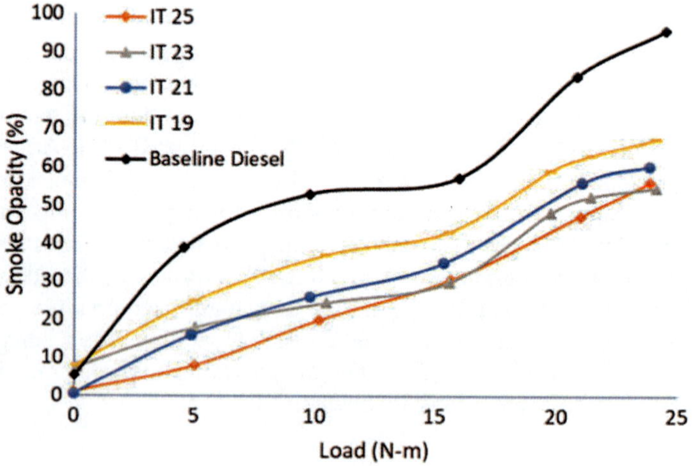

Fig. 4 Influence of injection timing on smoke emissions

6 Conclusion

In this experimental work, the influence of injection timing over the engine performance and exhaust emissions have been analyzed critically under low-temperature combustion conditions with GMD emulsion fuel. It was found that retardation in the injection by 2° btdc helps to enhance the performance of engine. The maximum efficiency obtained with optimized GM15 at IT 21° btdc was 24.2% as compared to maximum 23.7% at IT 23° btdc. Low-tech non-road diesel engines are frequently used in India for agricultural application. Emulsified gomutra-diesel fuels could prove in improved performance of these engines along with reduced exhaust emissions. Retardation in the injection timing could bring positive improvement in the engine performance and emissions characteristics.

References

1. Sharma PK, Sharma D, Soni SL, Jhalani A, Singh D, Sharma S (2020) Characterization of the hydroxy fueled compression ignition engine under dual fuel mode: experimental and numerical simulation. Int J Hydrogen Energy 45:8067–8081. https://doi.org/10.1016/j.ijhydene.2020.01.061
2. Jhalani A, Sharma D, Soni SL, Sharma PK, Sharma S (2019) A comprehensive review on water-emulsified diesel fuel: chemistry, engine performance and exhaust emissions. Environ Sci Pollut Res. https://doi.org/10.1007/s11356-018-3958-y
3. Wang Z, Wu S, Huang Y, Huang S, Shi S, Cheng X (2018) Experimental investigation on spray, evaporation and combustion characteristics of ethanol-diesel , water-emulsion filed diesel and neat diesel fuels. Fuel 231:438–48. https://doi.org/10.1016/j.fuel.2018.05.129
4. Soudagar MEM, Nik-Ghazali N-N, Abul Kalam M, Badruddin IA, Banapurmath NR, Akram N (2018) The effect of nano-additives in diesel-biodiesel fuel blends: a comprehensive review on stability, engine performance and emission characteristics. Energy Convers Manag 178:146–177. https://doi.org/10.1016/j.enconman.2018.10.019
5. Khatri D, Goyal R (2020) Performance, emission and combustion characteristics of water diesel emulsified fuel for diesel engine: a review. Mater Today Proc. https://doi.org/10.1016/j.matpr.2020.04.560
6. Subramanian KA (2011) A comparison of water-diesel emulsion and timed injection of water into the intake manifold of a diesel engine for simultaneous control of NO and smoke emissions. Energy Convers Manag 52:849–857. https://doi.org/10.1016/j.enconman.2010.08.010
7. Attia AMA, Kulchitskiy AR (2014) Influence of the structure of water-in-fuel emulsion on diesel engine performance. Fuel 116:703–708. https://doi.org/10.1016/j.fuel.2013.08.057
8. Fan X, Hu W, Yang J, Xu X, Gao J (2008) A new emulsifier behavior of the preparation for micro-emulsified diesel oil. Pet Sci Technol 26:2125–2136. https://doi.org/10.1080/10916460701429100
9. Huo M, Lin S, Liu H, Lee CFF (2014) Study on the spray and combustion characteristics of water-emulsified diesel. Fuel 123:218–229. https://doi.org/10.1016/j.fuel.2013.12.035
10. Noor El-Din MR, El-Hamouly SH, Mohamed HM, Mishrif MR, Ragab AM (2014) Investigating factors affecting water-in-diesel fuel nanoemulsions. J Surfactants Deterg 17:819–831. https://doi.org/10.1007/s11743-013-1533-6
11. Jhalani A, Sharma D, Soni SL, Sharma PK (2019) Effects of process parameters on performance and emissions of a water emulsified diesel fueled compression ignition engine. Energy Sources, Part A Recover Util Environ Eff. https://doi.org/10.1080/15567036.2019.1669739
12. Jhalani A, Sharma D, Soni S, Sharma PK, Singh D (2020) Feasibility assessment of a newly prepared cow-urine emulsified diesel fuel for CI engine application. Fuel 288:119713. https://doi.org/10.1016/j.fuel.2020.119713
13. Kumar Sharma P, Sharma D, Lal Soni S, Jhalani A (2019) Characterization of the nonroad modified diesel engine using a novel entropy-VIKOR approach: experimental investigation and numerical simulation. J Energy Resour Technol 141. https://doi.org/10.1115/1.4042717
14. Jerry Kaneko J (2008) JWH and MLB. clinical biochemistry of domestic animals. Elsevier. https://doi.org/10.1016/B978-0-12-370491-7.X0001-3
15. Carpenter J, Saharan VK (2017) Ultrasonic assisted formation and stability of mustard oil in water nanoemulsion: effect of process parameters and their optimization. Ultrason Sonochem 35:422–430. https://doi.org/10.1016/j.ultsonch.2016.10.021

16. Peng L, Liu C, Kwan C, Huang K (2010) Optimization of water-in-oil nanoemulsions by mixed surfactants. Colloids Surf A Physicochem Eng Asp 370:136–142. https://doi.org/10.1016/j.colsurfa.2010.08.060
17. National Center for Biotechnology Information (2014) Sorbitan monooleate. PubChem Compd Database, 1–22
18. National Center for Biotechnology Information (2017) Tween® 80. PubChem Compd Database 1–30
19. Mehta RN, More U, Malek N, Chakraborty M, Parikh PA (2015) Study of stability and thermodynamic properties of water-in-diesel nanoemulsion fuels with nano-Al additive. Appl Nanosci 891–900. https://doi.org/10.1007/s13204-014-0385-3
20. Forgiarini A, Esquena J, González C, Solans C (2001) Formation of nano-emulsions by low-energy emulsification methods at constant temperature. Langmuir 17:2076–2083. https://doi.org/10.1021/la001362n
21. Jhalani A, Soni SL, Sharma D, Sharma PK (2018) Comparative performance analysis of an SI engine with treated and raw biogas. Int J Renew Energy Technol 9:39. https://doi.org/10.1504/IJRET.2018.090103
22. Debnath BK, Saha UK, Sahoo N (2015) A comprehensive review on the application of emulsions as an alternative fuel for diesel engines. Renew Sustain Energy Rev 42:196–211. https://doi.org/10.1016/j.rser.2014.10.023
23. Mondal PK, Mandal BK (2019) A comprehensive review on the feasibility of using water emulsified diesel as a CI engine fuel. Fuel. https://doi.org/10.1016/j.fuel.2018.10.076
24. Chow JC (2001) TAC. Diesel engines: environmental impact and control. J Air Waste Manage Assoc 51:1258–70. https://doi.org/10.1080/10473289.2001.10464354
25. Park JW, Huh KY, Lee JH (2001) Reduction of NOx, smoke and brake specific fuel consumption with optimal injection timing and emulsion ratio of water-emulsified diesel. Proc Inst Mech Eng Part D-Journal Automob Eng 215:83–93. https://doi.org/10.1243/0954407011525476

Extraction of 3D Solid Model of Decaying Tooth from 2D DICOM Images

Vaishnavi V. Gejji, Ravi Yerigeri, and C. M. Choudhari

Abstract Different dental procedures can be optimized by carrying out force analysis on the affected tooth. For this, a 3D solid model of the tooth is required. Moreover, it should be in a format which can be imported into any design and analysis software such as SolidWorks, Ansys, etc. However, as the shape of the tooth is arbitrary, developing a 3D model from scratch would not yield the required level of accuracy. Hence, here, DICOM images of the decaying tooth have been used to produce a 3D model. The open-source software platform Slicer 4.10.2 is used for developing a 3D tooth model of any person just based on their 2D DICOM images.

Keywords 3D solid model of tooth · DICOM images · Slicer 4.10.2 · STL format

1 Introduction

Due to their primary function of chewing food, teeth are frequently prone to decay. Various dental procedures have been developed over time to stop this decay at the early stages. One such common procedure is fitting an artificial cap on the tooth using a binder. However, low retention strength of these binders is a significant issue. Quality of dental implants highly depends on the materials used [1, 2]. Experiments can be performed to check the best possible composition. However, experimental procedures require many tooth samples which, in turn, require substantial surface preparation and thermocycling. Hence, using analytical methods such as Finite Element Analysis before experimental procedures can save both time and efforts [3–6]. However, for this, an accurate 3D model of the decayed tooth is required. Out of the various 3D modelling file formats available, we select the STL format [7].

V. V. Gejji (✉)
Terna Engineering College, Nerul, Navi Mumbai, India

R. Yerigeri · C. M. Choudhari
Department of Mechanical Engineering, Terna Engineering College, Nerul, Navi Mumbai, India
e-mail: raviyerigeri@ternaengg.ac.in

C. M. Choudhari
e-mail: cmchoudhari@ternaengg.ac.in

R. Agrawal et al. (eds.), *Recent Advances in Industrial Production*, Lecture Notes in Mechanical Engineering, https://doi.org/10.1007/978-981-16-5281-3_10

2 Literature Review

In this paper, Slicer 4.10.2 [8, 9] has been used. However, there are many software packages that accept DICOM images as input and help build a 3D model from it. Cheng et al. [10] and Kamio et al. [11] have provided a comparison between such commonly available software. Additionally, D Argüello et al. [12] have specified the semi-automatic process of Region growing available in 3D Slicer software, which helps in accurate organ construction from 2D slices of DICOM images. Fedorov et al. [13] highlight the advantages of 3D Slicer, which make it the most user-friendly software for medical imaging. Pichon et al. [14] describe the segmentation process as one of the most critical processes in 3D volume extraction. Egger et al. [15] have suggested two more methods of segmentation available in 3D Slicer. Yip et al. [16] have shown the effectiveness of the semi-automatic segmentation algorithm over manual segmentation. Also, Chalupa et al. [17] have presented an open-source Supervised Segmentation toolbox extension for 3D Slicer.

3 Extraction of 3D Model in Slicer

3.1 Opening DICOM Images in Slicer

After opening the Slicer software, click on the DCM icon on the top left corner. Then, click 'import' in the new window. Then select the folder consisting of all the DICOM images and click 'Load' on the bottom left corner of the same window [18]. The images now open into the Slicer window (Fig. 1).

The GUI of Slicer is divided into four parts. These divide a body into differ- ent sections, which helps in identifying the location of specific body parts or organs, movement of any part, or in understanding the structure of those organs [19] (Fig. 2)

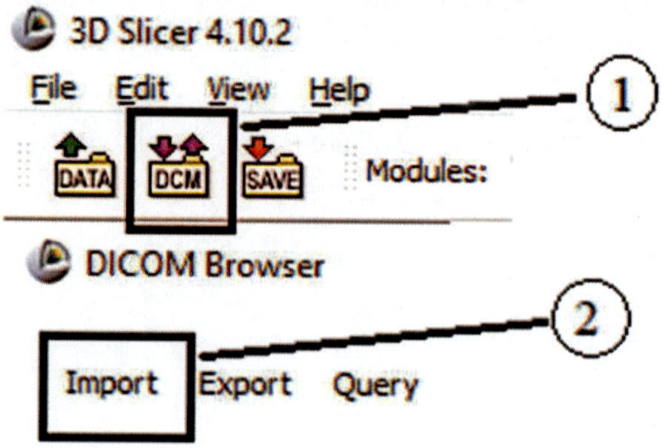

Fig. 1 Opening DICOM images in Slicer

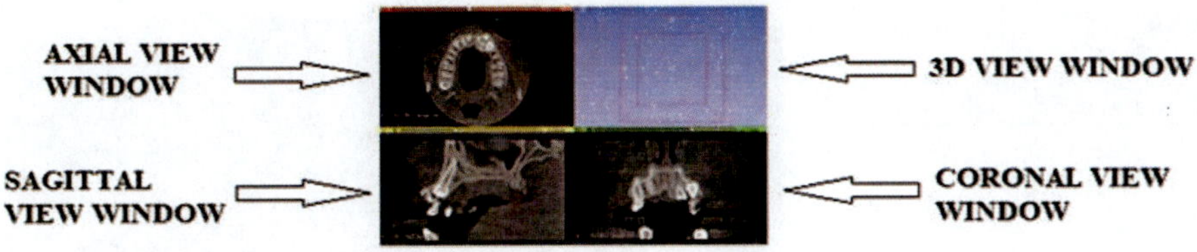

Fig. 2 DICOM images of Case Study 2 opened in Slicer

3.2 Volume Rendering

The volume rendering command of this software can be found under the 'Modules' search bar. This helps generate a 3D volume in the 3D view window [20, 21]. This volume is just a visualization of the 3D model. It forms a cloud volume that is not solid. Hence, saving this will not generate the required solid model. This command is used to assign preset and crop the volume (Fig. 3).

Selecting a Preset. Since our volume represents a set of teeth, we select the CT-Bone preset. Bone structures in the volume appear white and the surrounding gum region, called gingiva, appears pink. Furthermore, adjust the intensity of the Preset effect using the adjuster provided. Once, the required editing is done, we select the 'Display ROI' command. Now, the volume is ready for cropping (Fig. 4).

Crop Volume. Select 'Display ROI' in Volume Rendering. Now, edit the ROI such that it covers the region of the required tooth. Once the required volume is obtained, select 'Apply' and the cropped volume is created. After this, go back to

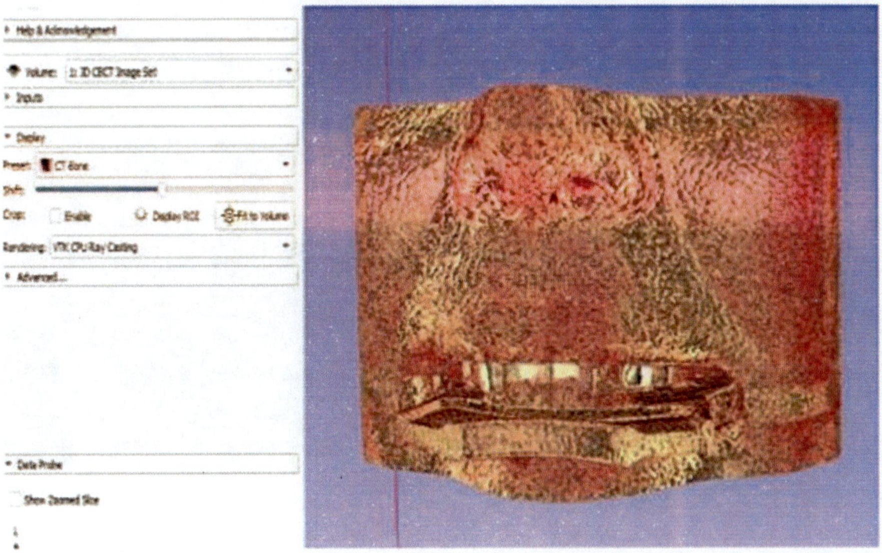

Fig. 3 Volume rendered from DICOM images with 'CT-Bone' Preset

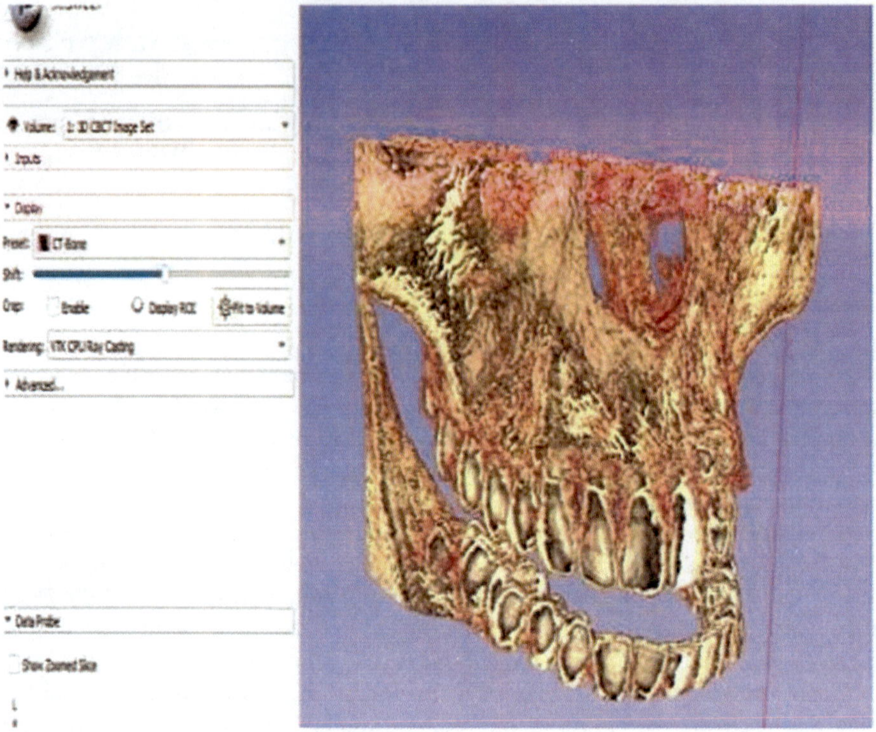

Fig. 4 3D volume after adjusting preset value

the Volume Rendering tab. Here, de-select the 'Display ROI' and 'Display Volume' feature. This is to provide visual simplicity for further operations (Fig. 5).

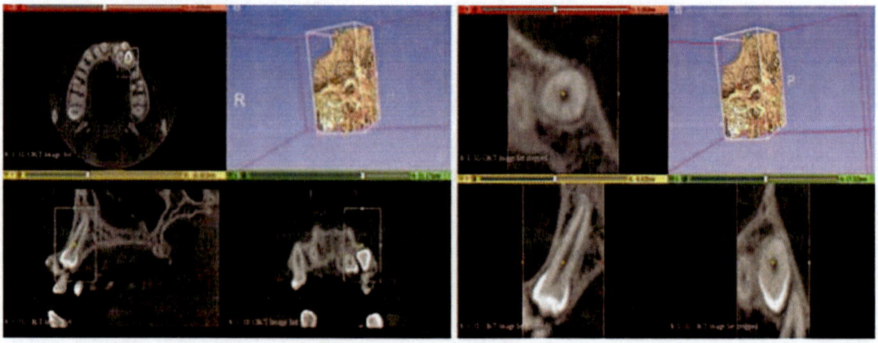

Fig. 5 Adjusting ROI to get the final cropped views

Fig. 6 Multiple segments in the segment editor

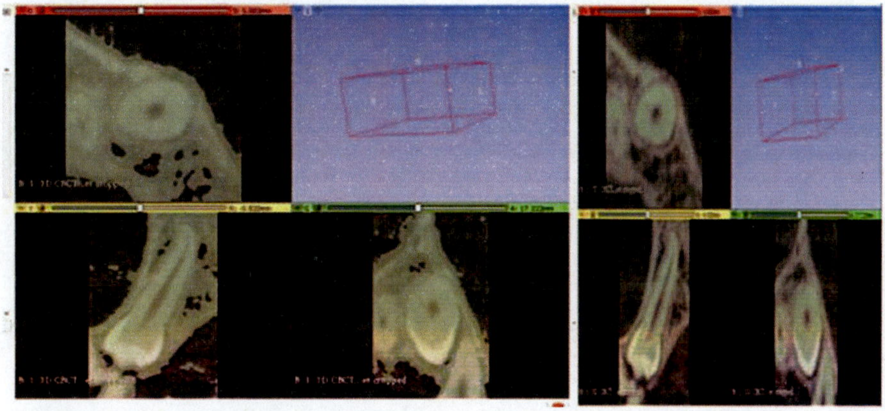

Fig. 7 Before and after adjusting the threshold limits

3.3 *Segmentation*

The tooth structure is to be separated from the surrounding Gingiva. For this, we do the following.

Creating multiple segments. Here, we create two separate segments [22, 23]. These segments are assigned different colours. Then, rename these segments as Tooth and Gingiva respectively. Now select the Tooth segment. The main factor which helps separate these two is the threshold value (Fig. 6).

Selecting Threshold Value. Figure 7 shows the default threshold value. However, here, gingiva, jaw area and tooth are highlighted. Hence, the lower limit of threshold needs to be increased. Hence, the range of threshold value is to be manually adjusted such that the maximum part of the calcified bone is highlighted with a minimum amount of surrounding gingiva. Once this is ensured, apply the effect and choose the 'Show 3D' command to see the 3D volume. Based on this, either move forward or adjust the range of threshold again till a satisfactory result is obtained.

3.4 *Using the Multiple Segments*

The two segments that have been initially formed are to be utilized here. First, select the 'Tooth' segment. Then select the threshold operation. Once that is selected, adjust

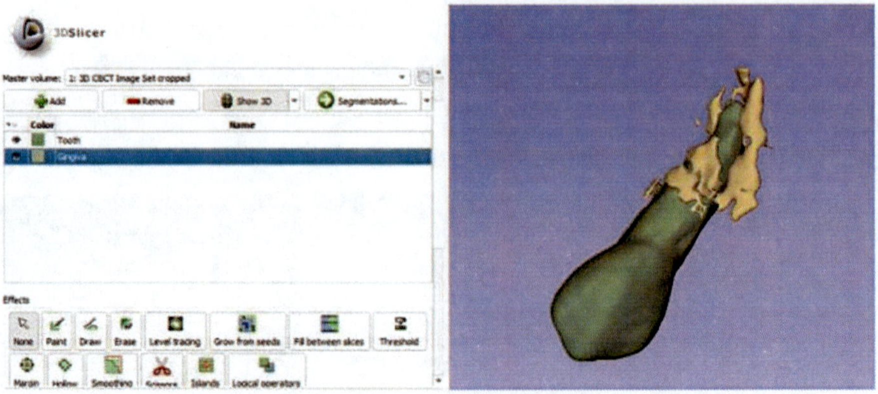

Fig. 8 Masking of the volume

the threshold value by adjusting the lower and upper limit as described above. Then, use the option called 'Use for masking'.

3.5 *Masking*

This process is used to overcome the limitations of the threshold process. Some portion of the roots remains unselected after thresholding. Hence, the masking process is used to paint over the areas of the tooth that are required, but have not been thresholded. It has to be done carefully without highlighting the neighbouring regions. 'Tooth' segment is assigned green colour and hence, the thresholded regions and the painted region appear in the same colour. Now, select the second segment 'Gingiva' and repeat the thresholding and masking process, as mentioned above. But now, the region of our interest is the region in the exterior of the tooth. This region is then marked in yellow. Now, 'apply' this and click the 'Show in 3D' command. In the 3D view window, a two-coloured volume can be seen.

As seen from Fig. 8, the masking feature helps in visually separating the gingiva from the tooth. After this, the two segments defined before, have two separate respective volumes associated with them. The 'Tooth' segment has the green coloured volume associated with it whereas the 'Gingiva' segment has the yellow one.

3.6 *Obtaining the Required Segment Volume*

Out of the above, we require only the 'tooth' volume. For this, simply turn off the visibility of the 'Gingiva' segment by clicking on the eye symbol next to it. Now, only a green coloured volume can be seen (Fig. 9).

Fig. 9 Turning off/on visibility of a segment volume

Fig. 10 Tooth model after segmentation

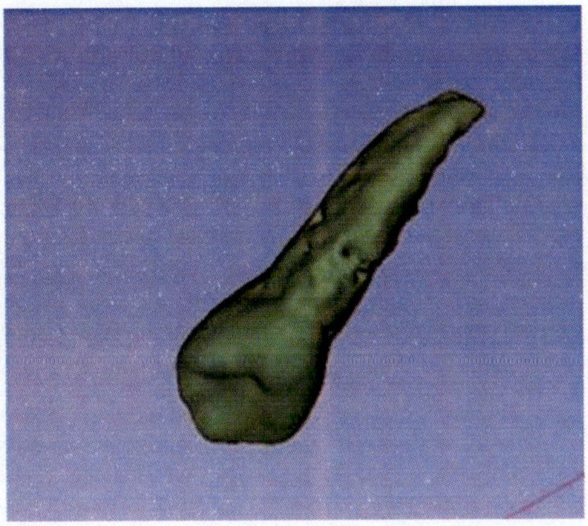

As seen in the Fig. 10 given below, a single tooth structure is obtained. But, this 3D structure has some visible surface irregularities. Hence, in order to minimize these, the smoothening feature as mentioned in 3.8 is used.

3.7 Fine Editing of Volume

The volume obtained can have some irregularities and some extruded unwanted features. To eliminate these, following can be used:

Islands. This feature is used to remove any small independent volume pieces surrounding the tooth. For dental applications, use 'Keep the largest Island'. This keeps the tooth and removes gingiva from the surrounding (Fig. 11).

3.8 Smoothening

We use the 'Median' mode of smoothening for all the case studies.

Selecting Kernel Size. The kernel size here refers to the width x height of the filter mask. It basically determines the intensity of the smoothening effect. In this case, for both the case studies, we define a kernel size of 3.00 mm (Fig. 12).

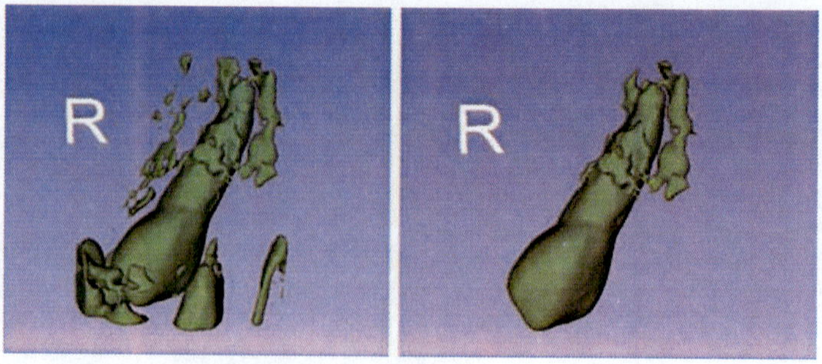

Fig. 11 Before and after applying Islands feature

Fig. 12 Final 3D tooth model after smoothening

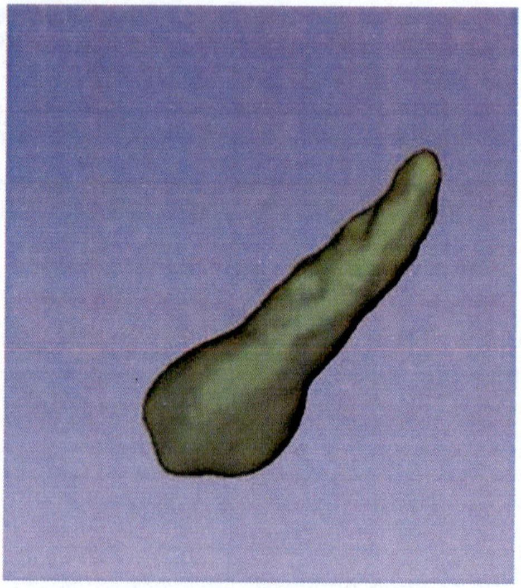

3.9 *Saving File*

Select the 'Segmentation' button next to the 'Show in 3D' option. Then, select 'export to files' option. This gives an option of saving the current file into a format which is not the default format of Slicer software. Select the STL format and define the directory for saving the file. Then, select 'export', and the file gets saved in the STL format in the predefined folder.

4 Conclusion

The paper shows how to extract 3D model of tooth from 2D DICOM images. This model can now be opened and edited in any of the design or analysis software successfully and 3D printed if required. This process can now be used for different patients and dental analysis can be easily and accurately done.

References

1. Sachin B (2013) Comparison of retention of provisional crowns cemented with temporary cements containing stannous fluoride and sodium fluoride-an in vitro study. J Indian Pros- thodont Soc. 13(4):541–545. https://doi.org/10.1007/s13191-012-0162-5
2. Osman RB, Swain MV (2015) A critical review of dental implant materials with an emphasis on titanium versus zirconia. Materials (Basel) 8(3):932–958. Published 2015 Mar 5. https://doi.org/10.3390/ma8030932
3. Oladapo BI, Zahedi SA, Vahidnia F, Ikumapayi OM, Farooq MU (2018) Three-dimensional finite element analysis of a porcelain crowned tooth. Beni-Suef Univ J Basic Appl Sci
4. Mahajan S, Patil (2019) Application of finite element analysis to optimizing dental implant
5. Eijden V (1991) Three dimensional analysis of human bite force magnitude and moment. Arch Oral Biol 36:535–539
6. Graf H (1975) Occlusal forces during function. In: Rowe NH (ed) Occlusion: research on form and function. University of Michigan
7. Lee G, Nam J, Han H, Kwon S (2019) A study on 3D file format for web-based scientific visualization. Int. J. Adv. Culture Technol. 7(1):243–247
8. Slicer [online]. 2020 [cit. 2020–01–15] https://www.slicer.org/
9. Slicer 4.10.2 [online]2020 [cit. 2020–01–15] Fedorov A, Beichel R, Kalpathy-Cramer J, Finet J, Fillion-Robin J-C, Pujol S, Bauer C, Jennings D, Fennessy FM, Sonka M, Buatti J, Aylward SR, Miller JV, Pieper S, Kikinis R (2012) 3D slicer as an image computing platform for the quantitative imaging network. Magn Reson Imaging 30(9):1323–41. PMID: 22770690. PMCID: PMC3466397. https://doi.org/10.17703/IJACT.2019.7.1.243
10. Cheng G, San Jose Estepar R, Folch E, Onieva J, Gangadharan S, Majid A (2016) Three-dimensional printing and 3D slicer: powerful tools in understanding and treating structural lung disease. Chest 149(5)
11. Kamio T, Suzuki M, Asaumi R et al (2020) DICOM segmentation and STL creation for 3D printing: a process and software package comparison for osseous anatomy. 3D Print Med 6:17. https://doi.org/10.1186/s41205-020-00069-2
12. Argüello D, Sánchez Acevedo HG, González-Estrada OA (2019) Comparison of segmentation tools for structural analysis of bone tissues by finite elements. J Phys Conf Series 1386:012113
13. Fedorov A, Beichel R, Kalpathy-Cramer J et al (2012) 3D Slicer as an image computing platform for the quantitative imaging network. Magn Reson Imaging 30(9):1323–1341. https://doi.org/10.1016/j.mri.2012.05.001
14. Pichon E, Tannenbaum A, Kikinis R (2004) A statistically based flow for image segmentation. Med Image Anal 8(3):267–74. [PMC free article] [PubMed] [Google Scholar]
15. Egger J, Kapur T, Fedorov A, Pieper S, Miller J, Veeraraghavan H, Freisleben B, Golby A, Nimsky C, Kikinis R (2013) GBM volumetry using the 3D slicer medical image computing platform. Sci Rep 3:1364
16. Yip S, Parmar C, Blezek D, Estepar R, Pieper S, Kim J, Aerts H, Ooijen P (2013) Application of the 3D slicer chest imaging platform segmentation algorithm for large lung nodule delineation. PLoS ONE 12(6)

17. Chalupa D, Mikulka J (2018) A novel tool for supervised segmentation using 3D slicer. Symmetry 10:627
18. Spin-Neto R, Marcantonio E Jr, Gotfredsen E, Wenzel A (2011) Exploring CBCT-based DICOM files. A systematic review on the properties of images used to evaluate maxillofacial bone grafts. J Digit Imaging 24:959–966
19. Wikipedia contributors (20203). Anatomical plane. In: Wikipedia, The Free Encyclopedia. Retrieved 11:03, November 5, 2020, from https://en.wikipedia.org/w/index.php?title=Anatomical_plane&oldid=976570213
20. Pieper Steve, Finet Julien, Yarmarkovich Alex, Aucoin Nicole Volumes.
21. https://www.slicer.org/wiki/Documentation/4.10/Modules/Volumes. [CrossRef]
22. Finet J, Yarmarkovich A, Liu Y, Freudling A, Kikinis R. Volume Rendering. https://www.slicer.org/wiki/Documentation/4.10/Modules/VolumeRendering. [CrossRef]
23. Csaba P, Andras L, Steve P, Wendy P, Ron K, Jim M. Segment Editor. https://slicer.readthedocs.io/en/latest/user_guide/module_segmenteditor.html. [CrossRef]

Analyzing the Drivers for Lean and Green Manufacturing Using ISM Approach

Sarita Prasad, A. Neelakanteswara Rao, and Krishnanand Lanka

Abstract Customer awareness towards the sustainability, environmental performance and government policies have enforced the manufacturing firms to implement best techniques and practices to fulfil customer expectations. In this study, the drivers for lean and green manufacturing have analyzed which will help the organizations to reduce the waste emissions, improve the environmental performance and increase the profit. A total of sixteen drivers are identified from the literature and consulted with experts' team for further study. These drivers are further analyzed with interpretive structural modelling (ISM) approach to see the relationship among drivers. The study shows that cost reduction, increase in market share, government support and adequate technology development are the key drivers for the lean and green manufacturing. In the end, paper concludes the study with some implications. These implications will guide the management and policymakers to focus on the main drivers for lean and green implementation in the manufacturing industry.

Keywords Lean manufacturing · Green manufacturing · Drivers · Interpretive structure modelling · MICMAC analysis

1 Introduction

Competition in global and local market has encouraged the manufacturing organizations to use new advanced green techniques and achieve economic profits [1]. Due to increase in awareness of environmental impact among customers, many firms are trying to achieve sustainability in supply chain [2]. Global markets are focusing for standard norms, rules and regulations to reduce environmental effect. These rules and regulations encouraging managers to adopt new advanced techniques in their firms which will reduce the environmental effect [3]. Lean and green manufacturing is an integrated approach which help the firms to achieve economic profits with the application of environmental friendly practices and techniques [4, 5]. The lean

S. Prasad (✉) · A. N. Rao · K. Lanka
Department of Mechanical Engineering, National Institute of Technology Warangal, Warangal 506004, India

R. Agrawal et al. (eds.), *Recent Advances in Industrial Production*, Lecture Notes in Mechanical Engineering, https://doi.org/10.1007/978-981-16-5281-3_11

and green integrated approach will guides the management in firms to implement new techniques and achieve sustainability in supply chain. The connection between lean and green manufacturing has been discussed by various researchers in past few years. The integrated approach of lean and green manufacturing will help the organizations to achieve social, economic and environmental sustainability [4, 6, 7]. Micheli [8] analyzed the effect of drivers such as firm size, ISO 14,001 certification, past performance etc. on green drivers in Chinese manufacturing companies. Luo [9] explored the synergies between lean and green manufacturing and factors which can affect the implementation process in Malaysian manufacturing firms. Various factors can motivates and encourage the firms to adopt lean and green manufacturing in their firms. These factors are also known as 'drivers'. The main aim of this study is identification and modelling of drivers for lean and green manufacturing. This will guide the management of various firms to implement lean and green strategy in Indian industries. In the first phase, with the help of expert's suggestions and literature review, a total of sixteen drivers have identified for further study. These drivers have analyzed with the help of interpretive structure modeling (ISM) approach and categorized with MICMAC (Matrice d' Impacts croises multiplication applique' an classment) approach.

In summary, the modelling of lean and green manufacturing drivers for Indian industries will help the organizations to achieve economic, social and environmental sustainability. The paper has arranged into six sections which include introduction. Section 2 covers the literature review for the drivers of lean and green manufacturing. Section 3 shows the methodology adapted in this study. Section 4 covers the results and discussion of the study. Section 5 and 6 presents the managerial implications and conclusion of the study with future scope and limitations.

2 Literature Review

Many researchers have shown the integrated approach of lean and green manufacturing implemented in various firms to achieve economic and environmental efficiency [6, 10–12]. Song [13] analyzed the effect of factors which influence the adoption of lean and sustainable manufacturing system in automotive industries. Compos [14] developed a model for integrated lean and green approach to implement hybrid lean sustainable system in firms. The drivers selected from literature survey and expert's opinions have shown in Table 1.

Table 1 Drivers for lean and green manufacturing

Drivers	References
Cost reduction	[6, 15, 16]
Competitiveness	[8, 16–18]
To increase market share	[16, 19, 20]
Government Support	[2, 8, 10, 14–17, 19]
Public pressure	[6, 8, 17, 18]
Green brand/company image	[6, 8, 10, 15–18]
Environmental awareness of customers	[2, 8, 10, 15–17, 19, 21]
ISO 14,001 certification	[15, 18, 21]
Potential use of energy resources	[2, 10, 15, 16, 18]
Reusing and recycling materials and packaging	[16, 18, 21]
Reverse logistics	[16, 18]
Green design/green purchasing/green innovation	[10, 18, 21]
Top management commitment	[2, 6, 10, 14, 16, 17, 19]
Stakeholders Pressure	[2, 10, 15, 16, 18]
Environmental collaboration with suppliers	[15, 16, 19, 21]
Adequate technology development	[2, 8, 17, 19]

3 Methodology

3.1 Interpretive Structural Modelling

ISM approach is generally used to develop a comprehensive model which is used to define the relationship among a set of factors. ISM method is primarily based on the data collected from the experts' opinions and suggestions. A total of eight member team was selected for this study. The experts participated in the study belongs to manufacturing industries and academic institutes. The steps followed in ISM approach are explained below:

- Identification of the set of driving factors for lean and green manufacturing.
- Structural Self Interaction matrix (SSIM) development with the help of experts' opinions and suggestions.
- Development of Initial reachability matrix (IRM).
- Checking for transitivity in IRM and conversion to final reachability matrix.
- Level partitioning for factors.
- MICMAC analysis for the driving factors.

Table 2 SSIM matrix

S .No.	16	15	14	13	12	11	10	9	8	7	6	5	4	3	2	1
1	V	V	V	V	V	V	V	V	V	V	V	V	V	X	V	
2	A	X	A	V	V	V	V	O	V	O	X	A	A	A		
3	V	V	V	V	V	V	V	V	V	V	V	V	X			
4	V	V	V	V	V	O	V	V	V	V	V	V				
5	A	V	X	V	V	V	V	V	V	O	O					
6	A	X	A	V	V	V	V	V	V	O						
7	A	O	O	X	O	O	A	A	O							
8	A	A	A	V	X	X	V	V								
9	A	A	A	V	A	A	A									
10	A	A	A	V	X	X										
11	A	A	A	V	X											
12	A	A	A	V												
13	A	A	A													
14	A	V														
15	A															
16																

3.2 Development of SSIM

SSIM is developed with the help of 'Lead to' type relationship among drivers. The relationship among drivers are shown by four symbols to show the dominance relationship among drivers. The symbols used in the study are 'V', 'A', 'X' and 'O'. Symbol 'V' shows that driver 'a' leads to achieve driver 'b'. Symbol 'A' shows that driver 'b' leads to achieve driver 'a'. Symbol 'X' shows that driver 'a' and 'b' are helping to each other. Symbol 'O' shows that driver 'a' and driver 'b' are not related to each other. SSIM is shown in Table 2.

3.3 Initial Reachability Matrix

SSIM matrix is transformed into initial reachability matrix after the conversion of symbols into binary values. IRM matrix can be seen in Table 3. The rule of conversion of all symbols 'V', 'A', 'X' and 'O' are explained below:

In SSIM matrix if value (x, y) is V, then value of (x, y) coordinate will 1 and (y, x) coordinate will 0. Similarly, if in SSIM matrix value (x, y) is A, then then value of (x, y) coordinate will 0 and (y, x) coordinate will 1. If in SSIM matrix value (x, y) is X, then value of coordinate (x, y) and coordinate (y, x) will be 1. In SSIM matrix value (x, y) is O, then value of coordinate (x, y) and coordinate (y, x) will be 0.

Table 3 Initial Reachability Matrix

S. No.	1	2	3	4	5	6	7	8	9	10	11	12	13	14	15	16
1	1	1	1	1	1	1	1	1	1	1	1	1	1	1	1	1
2	0	1	0	0	0	1	0	1	0	1	1	1	1	0	1	0
3	1	1	1	1	1	1	1	1	1	1	1	1	1	1	1	1
4	1	1	1	1	1	1	1	1	1	1	0	1	1	1	1	1
5	0	1	0	0	1	1	0	1	1	1	1	1	1	1	1	0
6	0	1	0	0	0	1	0	1	1	1	1	1	1	0	1	0
7	0	0	0	0	0	0	1	0	1	0	0	0	1	0	0	0
8	0	0	0	0	0	0	0	1	1	1	1	1	1	0	0	0
9	0	0	0	0	0	0	1	0	1	0	0	0	1	0	0	0
10	0	0	0	0	0	0	1	1	1	1	1	1	1	0	0	0
11	0	0	0	0	0	0	0	1	1	1	1	1	1	0	0	0
12	0	0	0	0	0	0	0	1	1	1	1	1	1	0	0	0
13	0	0	0	0	0	0	1	0	0	0	0	0	1	0	0	0
14	0	1	0	0	1	1	0	1	1	1	1	1	1	1	1	0
15	0	1	0	0	0	1	0	1	1	1	1	1	1	0	1	0
16	0	1	0	0	1	1	1	1	1	1	1	1	1	1	1	1

3.4 Final Reachability Matrix and Level Partitions

In initial reachability matrix, rule of transitivity is applied to remove the transitivity in matrix. The rule of transitivity can be explained as if a variable ‘X’ is having relation with variable ‘Y’ and variable ‘Y’ is having relation with variable ‘Z’, then Variable ‘X’ will definitely have a relation with variable ‘Z’. Final reachability matrix is developed after the application of transitivity rules in the IRM. The FRM can be seen in the Table 4. These drivers are further levelled through level partitioning. The level partitioning is developed with the help of reachability and antecedent set of each driver.

3.5 Formation of ISM Model

Based on FRM, a diagraph is developed which includes transitivity links. These transitivity links have removed and drivers have replaced with the corresponding drivers numbers. The final model is known as ISM model for drivers. The model can be seen in Fig. 1.

Table 4 Final Reachability Matrix

S. No.	1	2	3	4	5	6	7	8	9	10	11	12	13	14	15	16	Driving Power
1	1	1	1	1	1	1	1	1	1	1	1	1	1	1	1	1	16
2	0	1	0	0	0	1	0	1	0	1	1	1	1	0	1	0	8
3	1	1	1	1	1	1	1	1	1	1	1	1	1	1	1	1	16
4	1*	1	1	1	1	1	1	1	1	1	0	1	1	1	1	1	15
5	0	1	0	0	1	1	0	1	1	1	1	1	1	1	1	0	11
6	0	1	0	0	0	1	0	1	1	1	1	1	1	0	1	0	9
7	0	0	0	0	0	0	1	0	1*	0	0	0	1	0	0	0	3
8	0	0	0	0	0	0	0	1	1	1	1	1	1	0	0	0	6
9	0	0	0	0	0	0	1	0	1	0	0	0	1	0	0	0	3
10	0	0	0	0	0	0	1	1*	1	1	1	1	1	0	0	0	7
11	0	0	0	0	0	0	0	1	1	1	1	1	1	0	0	0	6
12	0	0	0	0	0	0	0	1	1	1	1	1	1	0	0	0	6
13	0	0	0	0	0	0	1	0	0	0	0	0	1	0	0	0	2
14	0	1	0	0	1	1	0	1	1	1	1	1	1	1	1	0	11
15	0	1	0	0	0	1	0	1	1	1	1	1	1	0	1	0	9
16	0	1	0	0	1	1	1	1	1	1	1	1	1	1	1	1	13
Dependence Power	3	9	3	3	6	9	8	13	14	13	12	13	16	6	9	4	141

3.6 *MICMAC Analysis*

MICMAC analysis is used to analyze the relationship among dependency and driving power of drivers. The values of driver power and dependency power were calculated in final reachability matrix. The MICMAC analysis is shown in Fig. 2. In MICMAC analysis, drivers are divided into 4 parts or quadrants. These quadrants are autonomous, dependent, independent and linkage drivers.

Drivers with low value of driving and low value of dependence power comes under autonomous driver's category. Drivers with high value of dependence power and low value of driving power comes under dependent driver's category. Drivers with high value of dependence power and high value of driving power comes under linkage driver's category. Drivers with high value of driving power and low value of dependence power comes under independent driver's category. The division of these drivers into four categories are important for firms, as it will help to define the role of each driver. The MICMAC analysis will define the effect of each driver on the other category drivers. Drivers belong to independent category will support the linkage drivers. Similar way, independent drivers and linkage drivers will support the dependent drivers.

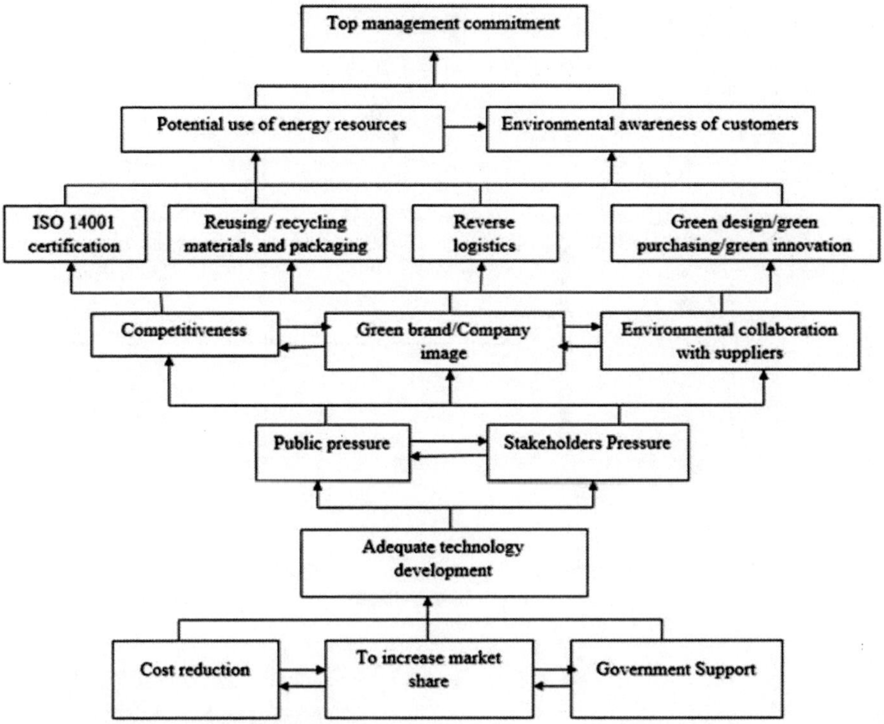

Fig. 1 ISM model

4 Results and Discussion

The outcomes of this study are explained below:

- 'Cost reduction', 'to increase market share' and 'Government support' found as primary drivers for lean and green manufacturing implementation. For any manufacturing firm, cost of the product is an important point of concern for manufacturer and customer as well. Price strategy plays an important parameter in global and local market for any organization. Lean and green practices help the firm to focus on the overall cost of manufacturing which includes administrative cost, capital cost, sales and cost associated with environmental related techniques. Researchers [15, 22] have identified the cost reduction as the most important driver for lean and green manufacturing. Manufacturing organizations generate a large amount of waste in terms of waste water, gas emission and solid waste. Government can encourage the organization through various incentive schemes and assistance to adapt new advanced green techniques. Government can also implement strict rules, laws and regulations which strongly encourage firms to adapt new green and lean techniques in the manufacturing system.

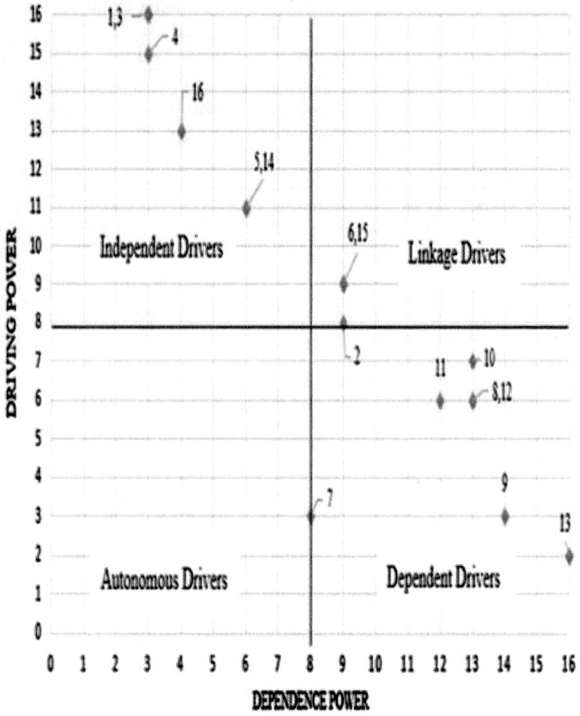

Fig. 2 MICMAC analysis

- 'Adequate technology development', 'Public pressure', 'Stakeholder pressure' found as important drivers for lean and green manufacturing. Manufacturing organizations need to be upgrade the system with advanced and highly energy efficient technology. The adaption of these advanced technologies encourages the firms to achieve economic, environmental and operational efficiency. Study [8] suggested that adequate technology development can help the various firms to reduce the waste emissions and utilize the resources efficiently. Public pressure by various organizations such as NGOs, local authorities, social activities and media can play an important role towards adaption of green practices by the firms. Public pressure and opinions will help to maintain a reputation by organizations [22]. MICMAC analysis suggest that no drivers were found as autonomous category. This mean that, the study confirms that all drivers selected for study are playing an important role for lean and green manufacturing. All the drivers analyzed in the study are essential factors towards the implementation of lean and green in the manufacturing process.

5 Managerial Implications

The management from organization can take this model as a road map for understanding and assessing the firm's performance and activities for greenness and leanness. This modeling approach provide flexibility to decision makers towards implementation of lean and green manufacturing. Study will guide the management to analyze the most influential drivers for lean and green manufacturing in the early stage of strategy adaption. The study can help the researchers to compare the results of the study with other modelling techniques and expand the results. The study can also help the practitioners to analyze the implementation benefits of lean and green manufacturing.

6 Conclusions

Environmental performance of a manufacturing firm plays a significant role in the decision making process [16]. Environmental awareness among customer and supplier will help to reduce the consumption of natural resources and help the manufacturing process activities to be sustainable. The adaption and implementation process of lean and green activities in firm are challenging task for management as it needs a high level of responsibility, skill and financial support. This can be achieved with the help and support from government agencies by providing funds, incentive schemes, conducting assistance programs for skill development etc. The study develop a hierarchical level ISM based model for lean and green manufacturing drivers. This study provides a brief view of drivers for lean and green manufacturing. Sixteen drivers have identified from literature review and expert's opinions. These drivers are cost reduction, competitor green strategies, to increase market share, government support, public pressure, green brand/company image, environmental awareness of customers, ISO 14,001 certification, the potential use of energy resources, reusing and recycling materials and packaging, reverse logistics, green design/green purchasing/green innovation, top management commitment, stakeholders pressure, environmental collaboration with suppliers and adequate technology development. The contextual relationship among the drives are analyzed by the ISM modelling approach to determine the most important driver for lean and green manufacturing. ISM model is used in this study, as it help to know and analyze the dependency among drivers with the help of expert's opinions and suggestions. According to finding of the study, cost reduction, increase market share and government support are most important drivers for the lean and green implementation process. Cost reduction is always a primary important driver for organization. The adaption of lean and green techniques require adequate funding and guidance from high skilled management.

The modelling approach of drivers help the industrial practitioners to implement lean and green strategies in their firms. This study will provide a valuable input to researcher to expand the results with more structured techniques and real data. This study is having some limitations, the drivers are primarily identified from literature and expert opinions in the study. Survey approach can be used to identify more realistic drivers which are not reported in literature.

References

1. Yadav G, Desai TN (2017) A fuzzy AHP approach to prioritize the barriers of integrated Lean Six Sigma. Int J Qual Reliab Manag
2. Orji J, Liu S (2020) A dynamic perspective on the key drivers of innovation-led lean approaches to achieve sustainability in manufacturing supply chain. Int J Prod Econ 219:480–496
3. Garza-Reyes JA, Jacques GW, Lim MK, Kumar V, Rocha-Lona L (2014) Lean and green–synergies, differences, limitations, and the need for six sigma. In: IFIP international conference on advances in production management systems, 71–81
4. Thanki S, Govindan K, Thakkar J (2016) An investigation on lean-green implementation practices in Indian SMEs using analytical hierarchy process (AHP) approach. J Clean Prod 135:284–298
5. Diaz-Elsayed N, Jondral A, Greinacher S, Dornfeld D, Lanza G (2013) Assessment of lean and green strategies by simulation of manufacturing systems in discrete production environments. CIRP Ann 62(1):475–478
6. Gandhi NS, Thanki SJ, Thakkar JJ (2018) Ranking of drivers for integrated lean-green manufacturing for Indian manufacturing SMEs. J Clean Prod 171:675–689
7. Verrier B, Rose B, Caillaud E, Remita H (2014) Combining organizational performance with sustainable development issues: the Lean and Green project benchmarking repository. J Clean Prod 85:83–93
8. Micheli GJ, Cagno E, Mustillo G, Trianni A (2020) Green supply chain management drivers, practices and performance: a comprehensive study on the moderators. J Clean Prod 121024
9. Luo Y, Jie X, Li X, Yao L (2018) Ranking Chinese SMEs green manufacturing drivers using a novel hybrid multi-criterion decision-making model. Sustainability 10(8):2661
10. Thanki SJ, Thakkar J (2014) Status of lean manufacturing practices in Indian industries and government initiatives: a pilot study. J Manuf Technol Manag 25(5):655–675
11. Garza-Reyes JA (2015) Lean and Green—a systematic review of the state of the art literature. J Clean Prod 102:18–29
12. Kurdve M, Zackrisson M, Wiktorsson M, Harlin U (2014) Lean and Green integration into production system models Experiences from Swedish industry. J Clean Prod 85:180–190
13. Vinodh S, Ramesh K, Arun CS (2016) Application of interpretive structural modelling for analyzing the factors influencing integrated lean sustainable system. Clean Technol Environ Policy 18(2):413–428
14. Campos LM, Vazquez-Brust DA (2016) Lean and green synergies in supply chain management. Supply Chain Manag
15. Li Q, Found P (2016) Lean and green supply chain for the product-services system (PSS): the literature review and a conceptual framework. Procedia CIRP 47:162–167
16. AlManei M, Salonitis K, Xu Y (2017) Lean implementation frameworks: the challenges for SMEs. Procedia Cirp 63:750–755
17. Chiet CW, Ching NT, Huat SL, Fathi M, Tzuu TJ (2019) The integration of lean and green manufacturing for Malaysian manufacturers: a literature review to explore the synergies between lean and green model. IOP Conf Series: Earth Environ Sci 268(1):012066
18. Zhou Y, Xu L, Muhammad Shaikh G (2019) Evaluating and prioritizing the green supply chain management practices in Pakistan: Based on Delphi and fuzzy AHP approach. Symmetry, 11:1346

19. Jaiswal P, Kumar A (2018) Assessment of drivers to implement integrated lean green manufacturing system in Indian SMEs through IF-TOPSIS approach. Int J Manag Decis Making 17(2):224–243
20. Govindan K, Azevedo SG, Carvalho H, Cruz-Machado V (2015) Lean, green and resilient practices influence on supply chain performance: interpretive structural modeling approach. Int J Environ Sci Te 12(1):15–34
21. Gaikwad L, Sunnapwar V (2020) An integrated lean, green and six sigma strategies. Total Qual Manag
22. Mollenkopf D, Stolze H, Tate WL, Ueltschy M (2010) Green, lean, and global supply chains. Int J Phys Distrib Logistics Manag

Quantifiable Contribution of Sustainable Manufacturing Enablers in Indian SMEs

Deepak Sharma, Pravin Kumar, and Rajesh Kr Singh

Abstract The cost constraints are higher comparatively for Small and Medium Enterprises (SMEs) in India. In view of recurring competitive developments in product demand, design, values, life cycle and increasing environment & government guidelines, SMEs are required to plan the manufacturing on sustainable basis as to remain in business for long run. In this study we have considered twelve important enablers for SMEs. We have shortlisted the enablers after lot of literature review and then they are characterize into four dimensions considering expert's opinion. Author expects that these identified valuable enablers will help the management of SMEs in India to plan their business on sustainable basis as to remain competitive viable and growing.

Keywords Sustainable manufacturing (SM) · Small and medium enterprises (SMEs) · Enablers · Analytical hierarchy process (AHP)

1 Introduction

In present scenario the growth of nation and manufacturing organization lies in the path of sustainable development as it bounds all the vital three aspects economic, environmental and social [1]. India's manufacturing sector, which is a fundamental part, contributes 14–18% of India's GDP which can clearly push the growth of country economy [2]. The growth engine is with small and medium enterprises (SMEs) which employ 107 millions approx 40% of Indian work force. The employment generation and gross output makes SMEs prime pillar for economic activities and social matter. On the other hand they are consuming huge amount of money, different materials,

D. Sharma (✉)
GLA University, Mathura, India
e-mail: deepak.sharma@gla.ac.in

D. Sharma · P. Kumar
Delhi Technological University, Delhi, India

R. K. Singh
Management Development Institute (MDI), Gurgaon, India

R. Agrawal et al. (eds.), *Recent Advances in Industrial Production*, Lecture Notes in Mechanical Engineering, https://doi.org/10.1007/978-981-16-5281-3_12

energy etc. This consumption is giving rise to alarming situation for stock levels of natural resources which at other part also results in generation of lots of wastes, gases, poisoning of biosphere. In view of it is necessary for manufacturing organizations to pursue those activities which should cause minimal effect to environment while holding economic and social benefits [3].

In current situation manufacturing companies' prime responsibility is to succeed for sustainable development without affecting productivity and profitability [4–6]. Sustainable development was introduced in a widespread way by the Brundtland Commission, which defined it as development that "meets the needs of the present without compromising the ability of future generations to meet their own needs" [7]. In view of emerging frequently variable demand patterns, compulsive government and environmental guidelines and financial constraints in the highly competitive SMEs it is required to planned for sustainable manufacturing (SM) in India. Thus to excel in SM practices twelve vital enablers have identified and priotized.

The rest of the paper has been arranged as: Sect. 2 presents the literature review of SM practices in India. Section 3 is of problem novelty. Section 4 presents the methodology of AHP. Section 5 presents data analysis and results. Section 6 is of discussion and concluding remarks.

2 Literature Review

In this increasing competitive scenario SM is turning out to be the most researched and emerging domain among researchers and practitioners. Frequent changing demand pattern, government regulatory bodies' pressure for adopting changing environmental guidelines has a deep impact on the performance of manufacturing SMEs. Thus to keep them live and competitive in long term, system have to be transform according to sustainability concept. Gunasekaran and Spalanzani [8] have made a significant contribution to sustainable business growth in manufacturing and services by proposing various research directions, namely sustainable supply chain, sustainable product and process development product recovery operations, etc. Dhull and Narwal [9] in his investigation considered corporate social & environmental responsibilities as one of the prime enablers for the adoption of protective environmental practices. Malviya and Kant [10] depict that environmental education and training is one of the key enablers of sustainable practice however Prasad et al. [11] find them significant for employees' growth and development which motivates for maximum involvement. Bhanot et al. [12] highlighted that government regulations and promotions helps manufacturing industries in adopting SM practices. Malviya and Kant [10] depicted employee empowerment as an essential enabler while Singh and Debnath [13] prompted that employee empowerment has negligible link with system sustainability. According to Bhanot et al. [12], E-economy and green organization image are important enablers as it helps in integration of regional market with global markets and to keep up sustainable inputs to the industries.

In this study twelve enablers which will support and accelerate sustainable development in SMEs are depicted and shown in Table 1 with its illustration and references.

Table 1 Description of enablers for sustainable manufacturing

S. No	Enablers	Description	References
E_1	Changing market pressure	Customer satisfaction, competitors trade and commercial practices	Kulatunga et al. [14], Amrina and Yusof [15], Mittal et al. [16], Lin [17], Garg et al. [18]
E_2	Regulations and promotions by Government	Accountability and private–public participation, judicial regulations and law enforcement	Amrina and Yusof [15], Mittal et al. [16], NMCC [19], Garg et al. [18], Lin [17], Florida and Davison [20]
E_3	Investment in Innovation & Technology	Use of advance Technological practices for enhancement of performance, speed and cost control	Mittal et al. [16], NMCC [19]
E_4	Remanufacturing	Manufacturing products to original dimensions from available used repaired or new spares as to save cost and to avoid loss of resources	Nasr and Thurston [21]
E_5	Lowering the cost of manufacturing	Efficient process management with minimum waste outputs	NMCC [19]
E_6	Stakeholder engagement	Commitment of entrepreneurs, employees, financers, vendors and customers	Koho et al. [22]
E_7	Education and training system	Upgradation of technological and manufacturing standards in reference to required training and awareness on environmental practices	NMCC [19]
E_8	Increased customers environmental awareness	Customers are regularly made aware by government and NGOs on critical environmental issues to make them conscious on future humanity needs	Amrina and Yusof [15]
E_9	Attracting foreign direct investment (FDI)	Inorder to sustain global manufacturing at economical terms and appreciate the growth on long term basis FDIs are desired	NMCC [19]

(continued)

Table 1 (continued)

S. No	Enablers	Description	References
E_{10}	Increased International environmental standards	Ever increasing deterioration of environment the existence of all living beings is threatened and thus standards are updated regularly to monitor it	Amrina and Yusof [15]
E_{11}	Increasing demands of corporate social responsibility	Contribution of corporate is necessary to the society as to make their products and services acceptable on long term basis	Amrina and Yusof [15]
E_{12}	Incentives on Development in E-Economy	Deployment of E-Technology in Manufacturing Sector	Mittal et al. [16], NMCC [19]

Based on literature and experts' opinion, these enablers are classified into four dimensions. These are then summarized in "Fig. 1".

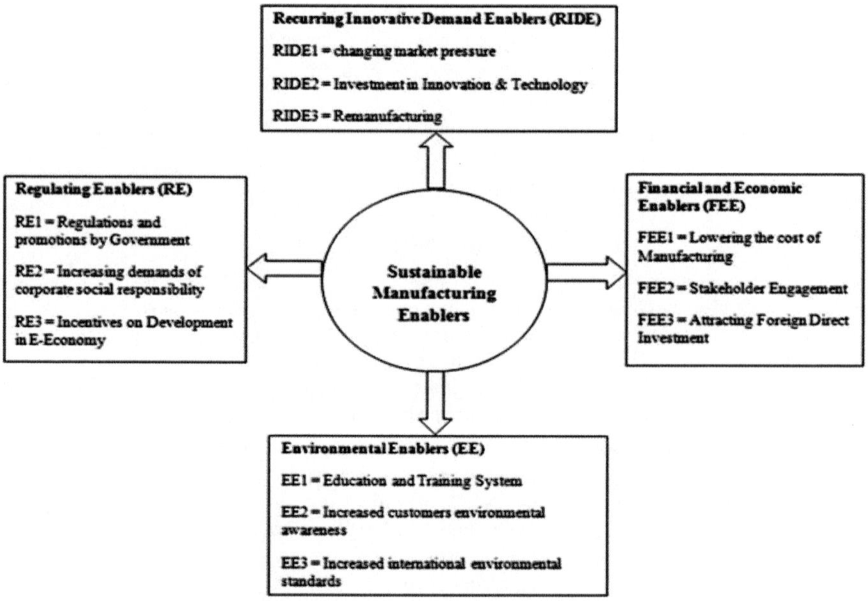

Fig. 1 Framework for assessing enablers of sustainable manufacturing

3 Problem Novelty

Although there have been lots of researches going on towards sustainability for manufacturing industries yet very few researchers have been actually done from SMEs point of view.

The main objectives of this research are to:

- Identification of vital enablers for SMEs towards sustainability.
- Evaluation of these enablers by using AHP technique.

4 Methodology

In this study total twelve enablers which act as a motivation for manufacturing Industries towards sustainability are considered. For further computation these enablers are grouped into four main dimensions for easy calculation. Similar sorting of factors by using AHP technique has been done by other researchers in his study (Singh et al. 2019). The methodology framework for ranking of enablers by AHP is shown in 'Fig. 2'.

- **Analytical Hierarchy Process (AHP)**

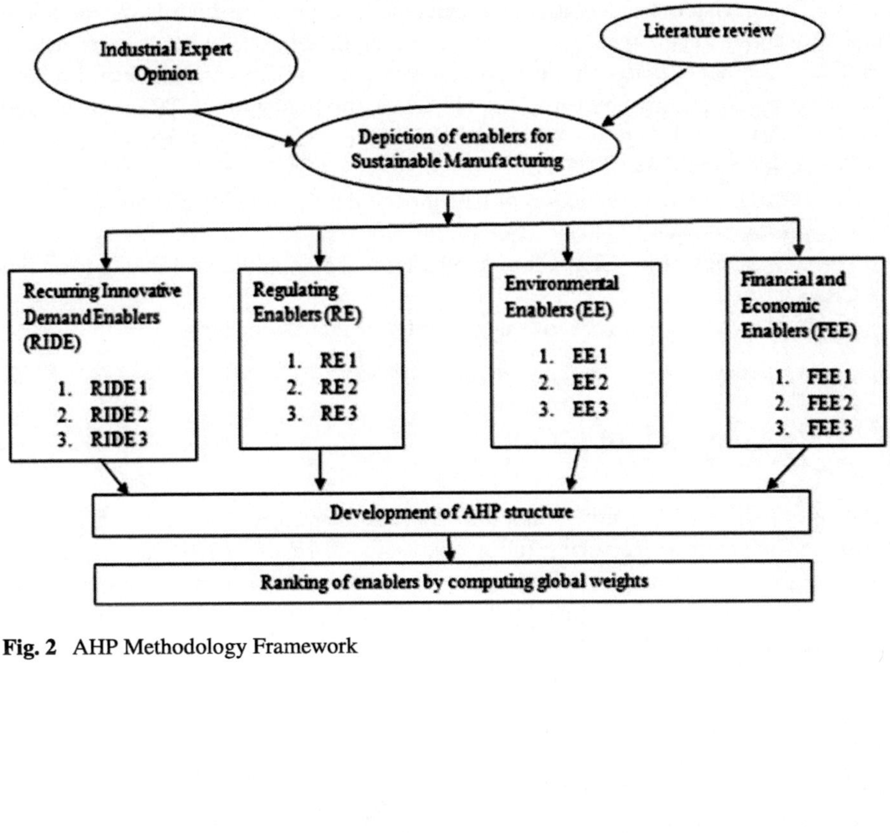

Fig. 2 AHP Methodology Framework

The evaluation of enablers has been done to find their priorities by using Analytic Hierarchy Process (AHP). AHP is a multi-attribute decision support tool. It was developed by Saaty (1977). The objective is defined at first and according to that the hierarchical structure of criteria and alternative is formed. Sub criteria can also be included if necessary. The criteria are analyzed in pair wise mode and according to that, the rank is awarded to all the alternatives. The AHP is a highly regarded decision making technique which uses a predefined fundamental scale of predefined numbers. The quantitative and qualitative attributes are utilized to capture the fundamental scale [23–26]. The individual likings are converted into ratio scale weights which later are combined to linear additive weights for each and every alternative. The final result of weight can be used to give the rank to alternatives. By ranking the alternatives, the AHP process finally assists the decision maker to take a final decision of any problem. The implementation of AHP is shown in the following sequence of steps [27–29]:

Step 1: Creating the hierarchical structure, first step for the AHP application is to construct the hierarchical structure as per the decision elements. The pair wise comparison between criteria and decision alternatives is formed by the experts using a nine point scale as per the standard AHP methodology. The matrices are developed and all pair wise comparisons are drawn.

Step 2: Building the pair wise comparison matrix, these matrices are constructed to compare the elements according to their relative importance compared to each other.

Step 3: Consistency calculation, the calculation is done for finding the values of relative weights. The Eigen values are also calculated to check the consistency of priority of elements. After that the consistency index (CI) is calculated for all the matrices (matrix order n) by using Eq. (1). The consistency ratio (CR) is calculated using Eq. (2) according to the CI and RI (Random Consistency index). The CI and CR are defined as follows [28]:

The formula for the calculation of CI for each matrix of order n is given as:

$CI = (\lambda max - n)/(n - 1)$ (1).

The consistency ratio (CR) is computed after that by applying following formula:

$CR = CI/RI$ (2).

It should be noted that RI varies as the order of matrix changes.

5 Data Analysis and Results

First of all, the enablers are grouped in four main groups. The matrix is formed for these four groups [Recurring Innovative Demand Enablers (RIDE), Regulating Enablers (RE), Environmental Enablers (EE) and Financial and Economic Enablers (FEE)]. The weights are assigned by the experts. The pair wise comparison matrices have been shown in tabular forms. These Tables (Table 2, 3, 4, 5, 6 and 7) show the weight assigned by the experts and also show the maximum Eigen value and

Table 2 Pair wise comparison matrix for main groups enablers of SMS

Enablers	Recurring innovative demand enablers (RIDE)	Regulating enablers (RE)	Environmental enablers (EE)	Financial and economic enablers (FEE)	Relative weights	Rank
Recurring Innovative Demand Enablers (RIDE)	1	7	5	3	0.5579	1
Regulating Enablers (RE)	0.14286	1	0.33333	0.20000	0.0569	4
Environmental Enablers (EE)	0.20000	3	1	0.33333	0.1219	3
Financial and Economic Enablers (FEE)	0.33333	5	3	1	0.2633	2
Maximum Eigen Value = 4.11846, CI = 0.0394866						

Table 3 Pair Wise comparison matrix for dimension-1, RIDE enablers of SMS

Enablers	Investment in Innovation and Technology (RIDE 1)	Changing market pressure (RIDE 2)	Remanufacturing (RIDE 3)	Relative Weights	Rank
Investment in Innovation and Technology (RIDE 1)	1	0.50000	4	0.3238	2
Changing market pressure (RIDE 2)	2	1	6	0.5869	1
Remanufacturing (RIDE 3)	0.25000	0.16667	1	0.0893	3
Maximum Eigen Value = 3.00946, CI = 0.00473					

the value of C.I. for each matrix. The AHP methodology is used to finally solve the problem.

6 Discussions and Concluding Remarks

The enablers are identified and firmed by going through literature review and Industrial experts opinions. These enablers are vital from SMEs sustainability point of view. Twelve enablers are then finalized and grouped into four main dimensions

Table 4 Pair wise comparison matrix for dimension-2, RE enablers of SMS

Enablers	Regulations and promotions by Government (RE 1)	Increasing demands of corporate social responsibility (RE 2)	Incentives on Development in E-Economy (RE 3)	Relative Weights	Rank
Regulations and promotions by Government (RE 1)	1.00000	7.00000	3.00000	0.6434	1
Increasing demands of corporate social responsibility (RE 2)	0.14286	1.00000	0.20000	0.0738	3
Incentives on Development in E-Economy (RE 3)	0.33333	5.00000	1.00000	0.2828	2
Maximum Eigen Value = 3.06551, CI = 0.032755					

Table 5 Pair wise comparison matrix for dimension-3, EE enablers of SMS

Enablers	Trained and educated manpower (EE 1)	Increased customers environmental awareness (EE 2)	Increased International environmental standards (EE 3)	Relative Weights	Rank
Trained and educated manpower (EE 1)	1	7	3	0.6555	1
Increased customers environmental awareness (EE 2)	0.14286	1	0.25000	0.0796	3
Increased International environmental standards (EE 3)	0.33333	4	1	0.2648	2
Maximum Eigen Value = 3.03186, CI = 0.02746					

according to the kind of impact they exert on the objective. The dimensions are Recurring Innovative Demand Enablers, Regulating Enablers, Environmental Enablers and Financial and Economic Enablers.

Table 6 Pair wise comparison matrix for dimension-4, FEE enablers of SMS

Enablers	Lowering the cost of Manufacturing (FEE 1)	Stakeholder Engagement (FEE 2)	Attracting Foreign Direct Investment (FEE 3)	Relative Weights	Rank
Lowering the cost of Manufacturing (FEE 1)	1	2	7	0.6267	1
Stakeholder Engagement (FEE 2)	0.50000	1	2	0.2634	2
Attracting Foreign Direct Investment (FEE 3)	0.14286	0.50000	1	0.1099	3
Maximum Eigen Value = 3.03508, CI = 0.03024					

AHP technique has been used to find out the priority of enablers in order to make SMES sustainable. The pair wise comparisons adopted are based on the experts opinions. The findings of this study can be summarized as follows:

- The pair wise comparison of main groups shows that the highest priority dimension is Recurring Innovative Demand Enablers. The second important dimension is Financial and Economic Enablers, third important dimension is Environmental Enablers and fourth important is Regulating enablers. This assessment can be utilized for sustainable development in manufacturing industries.
- Changing market pressure has been found as the most important enabler from the highest priority dimension (Recurring Innovative Demand Enablers). Investment in Innovation & Technology follows as the second and Remanufacturing as the least important enabler in this dimension.
- The second ranked dimension, financial and economic has following enablers as per their ranks in descending order—lowering the cost of manufacturing, stakeholder engagement and attracting foreign direct investment.
- Third ranked dimension, i.e. environmental enablers has its sub factors in following order priorities—Trained and educated manpower followed by increased International environmental standards and then increased customer's environmental awareness.
- The fourth ranked dimension, regulating enablers has following enablers as per their ranks in descending order—regulations and promotions by Government, incentives on development in E-Economy and Increasing demands of corporate social responsibility.
- All the enablers are also arranged as per their global rank. The top five most prioritized enablers, according to their global rank, in the descending order are—changing market pressure, investment in innovation & technology, lowering

Table 7 Overall ranking of enablers for SMS

S. No	Main Dimensions	Preference weights	Ranks	Sub Dimensions for SMS	Relative preference weights	Relative Ranking	Global preference weights	Global Ranks
1	Recurring Innovative Demand Enablers (RIDE)	0.5579	1	Investment in Innovation & Technology (RIDE 1)	0.3238	2	0.1806	2
				Changing market pressure (RIDE 2)	0.5869	1	0.3275	1
				Remanufacturing (RIDE 3)	0.0893	3	0.0498	6
2	Regulating Enablers (RE)	0.0569	4	Regulations and promotions by Government (RE 1)	0.6434	1	0.0366	7
				Increasing demands of corporate social responsibility (RE 2)	0.0738	3	0.0042	12
				Incentives on Development in E-Economy (RE 3)	0.2828	2	0.0161	10
3	Environmental Enablers (EE)	0.1219	3	Trained and educated manpower (EE 1)	0.6555	1	0.0799	4

(continued)

Table 7 (continued)

S. No	Main Dimensions	Preference weights	Ranks	Sub Dimensions for SMS	Relative preference weights	Relative Ranking	Global preference weights	Global Ranks
				Increased customers environmental awareness (EE 2)	0.0796	3	0.0097	11
				Increased International environmental standards (EE 3)	0.2648	2	0.0323	8
4	Financial and Economic Enablers (FEE)	0.2633	2	Lowering the cost of Manufacturing (FEE 1)	0.6267	1	0.1650	3
				Stakeholder Engagement (FEE 2)	0.2634	2	0.0693	5
				Attracting Foreign Direct Investment (FEE 3)	0.1099	3	0.0289	9

the cost of manufacturing, trained and educated manpower and stakeholder engagement.

This paper may be useful for improving the present scenario of manufacturing industries and to make them competitive from sustainable point of view. This hierarchical framework results with its importance may be used by policy makers for developing an action plans for overcoming challenges for sustainable development in manufacturing industries. There are some limitations of this research. The opinions of experts are subjective and may be biased. The pair wise categorization also depends on the wisdom of experts. Analytical network process (ANP), Fuzzy DEMATEL processes and hybrid of MCDM techniques can be considered as a future scope of this study.

References

1. Vinodh S, Ramesh K, Arun C (2015) Application of interpretive structural modelling for analysing the factors influencing integrated lean sustainable system. Clean Technol Environ Policy 18(2):413–428
2. Kapoor R (2015) Creating jobs in India's organized manufacturing sector. Indian J Labour Econ 58(3):349–375
3. Joung CB, Carrell J, Sarkar P, Feng SC (2013) Categorization of indicators for sustainable manufacturing. Ecol Ind 24:148–157
4. Govindan K, Diabat A, Shankar KM (2015) Analyzing the drivers of green manufacturing with fuzzy approach. J Clean Prod 96:182–193
5. Mangla SK, Kumar P, Barua MK (2014) Flexible decision approach for analysing performance of sustainable supply chains under risks/uncertainty. Glob J Flex Syst Manag 15(2):113–130
6. Rosen MA, Kishawy HA (2012) Sustainable manufacturing and design: concepts. Pract Needs Sustain 4(2):154–174
7. Pezzoli K (1997) Sustainable development: a transdisciplinary overview of the literature. J Environ Planning Manage 40(5):549–574
8. Gunasekaran A, Spalanzani A (2012) Sustainability of manufacturing and services: investigations for research and applications. Int J Prod Econ 140(1):35–47
9. Dhull S, Narwal M (2018) Prioritizing the drivers of green supply chain management in Indian manufacturing industries using fuzzy TOPSIS Method: government, industry, environment, and public perspectives. Process Integr Optim Sustain 2:47–60
10. Malviya RK, Kant R (2017) Modeling the enablers of green supply chain management: an integrated ISM—fuzzy MICMAC approach. Benchmarking Int J 24:536–568
11. Prasad DS, Pradhan RP, Gaurav K, Chatterjee PP, Kaur I, Dash S, Nayak S (2018) Analysing the critical success factors for implementation of sustainable supply chain management: an Indian case study. Decision 45:3–25
12. Bhanot N, Rao PV, Deshmukh SG (2017) An integrated approach for analysing the enablers and barriers of sustainable manufacturing. J Clean Prod 142:4412–4439
13. Singh R, Debnath RM (2012) Modeling sustainable development: India's strategy for the future. World J Sci Technol Sustain Dev 9:120–135
14. Kulatunga A, Jayatilaka PR, Jayawickrama M (2013) Drivers and barriers to implement sustainable manufacturing concepts in Sri Lankan manufacturing sector. In: Proceedings of 11th global conference on sustainable manufacturing. Berlin, Germany, pp 171–176
15. Amrina E, Yusof SM (2012) Drivers and barriers to sustainable manufacturing initiatives in Malaysian automotive companies. In: Kachitvichyanukul V, Luong H, Pitakaso R (eds)

Proceedings of the Asia pacific industrial engineering & management systems conference, pp 629–634
16. Mittal V, Egede P, Herrmann C, Sangwan K (2013) Comparison of drivers and barriers to green manufacturing: a case of India and Germany. In: Nee AYC, Song B, Ong SK (eds) Re-engineering manufacturing for sustainability, Springer Singapore. pp 723–728
17. Lin RJ (2013) Using fuzzy (DEMATEL) to evaluate the green supply chain management practices. J Clean Prod 40:32–39
18. Garg D, Luthra S, Haleem A (2014) An evaluation of drivers in implementing sustainable manufacturing in India: Using dematel approach. Int J Soc Educ Econ Manag Eng 8:3517–3522
19. NMCC (2006) The national strategy for manufacturing. Technical Report. http://nmcc.nic.in/pdf/strategy%20paper%200306.pdf
20. Florida R, Davison D (2001) Why do firms adopt advanced environmental practices (and do they make a difference)? Resources for the Future. Wasington, DC
21. Nasr N, Thurston M (2006) Remanufacturing: a key enabler to sustainable product systems. In: Proceedings of 13th CIRP international conference on life cycle engineering, pp 15–18
22. Koho M, Torvinen S, Romiguer AT (2011) Objectives, enablers and challenges of sustainable development and sustainable manufacturing: views and opinions of Spanish companies. IEEE Int Sympos Assembly Manuf (ISAM) 2011:1–6. https://doi.org/10.1109/ISAM.2011.5942343
23. Saaty TL (1980) The analytic hierarchy process. McGraw-Hill Book Co, New York
24. Saaty TL (1986) Axiomatic foundation of the analytic hierarchy process. Manage Sci 32:841–855. https://doi.org/10.1287/mnsc.32.7.841
25. Saaty TL (1994) Fundamentals of decision making. RWS Publications, Pittsburgh, PA
26. Saaty TL (1994) How to make a decision: the analytic hierarchy process. Interfaces 24(6):19–43
27. Luthra S, Garg D, Haleem A (2013) Identifying and ranking of strategies to implement green supply chain management in Indian manufacturing industry using analytical hierarchy process. J Ind Eng Manage 6(4):930
28. Saaty TL (2000) Fundamentals of decision making and priority theory, 2nd edn. RWS Publications, Pittsburgh, PA
29. Saaty TL (2008) Decision making with the analytic hierarchy process. Int J Serv Sci 1(1):83–98

Circular Economy and Sustainable Manufacturing: A Bibliometric Based Review

Kiran Gundu, Anbesh Jamwal, Alok Yadav, Rajeev Agrawal, Jinesh Kumar Jain, and Sundeep Kumar

Abstract By the arrival of the fourth industrial revolution overall transformation using intelligent engineering and digital transformation has changed the production system. Now all devices are equipped with automation and machine learning which has become a priority most of industries. There is a need of address the opportunity of sustainable manufacturing. This will help to achieve the sustainability 2030 goals. This study maps to intend to identify that how sustainable manufacturing is research is contributing in the research of circular economy. In this study mapping different performance matrices extracted and top journals, top productive authors and highly cited papers top key words and are evaluated. This paper summarizes the sustainable manufacturing in circular economy during the year 2015–2020 and provides the future research scope on the basis study mapping.

Keywords Circular economy · Sustainable manufacturing · Bibliometric analysis · R studio · Review

1 Introduction

Now a day's both circular economy and sustainable manufacturing (SM) reshaping the industrial waves in the world. At present circular economy and sustainability manufacturing consider major trends in the industry now researches are integrating circular economy with the sustainable concepts to revolutionize the industry wave [1, 2]. Increase in digitalization in the global industry has transformed the industrial production system in last few years. This transformation changed the production from traditional to decentralized production system with intelligence [3]. Circular economy unlocks the sustainability in industries and moved towards the social

K. Gundu · A. Jamwal · A. Yadav · R. Agrawal (✉) · J. K. Jain
Department of Mechanical Engineering, Malaviya National Institute of Technology, J.L.N. Marg, Jaipur, Rajasthan 302017, India
e-mail: ragrawal.mech@mnit.ac.in

S. Kumar
Department of Management Studies, Engineering College Ajmer, Ajmer 305025, India

R. Agrawal et al. (eds.), *Recent Advances in Industrial Production*, Lecture Notes in Mechanical Engineering, https://doi.org/10.1007/978-981-16-5281-3_13

sustainability [4]. Circular economy is very successful tool as it delivers the message on reducing wasteful resources by implementing the effective design and integration of products and processes for improved resource-efficiency with circular material information flow involving recycling, recovery, reuse, and remanufacturing of products. Thus, circular economy becomes inevitable for continued economic prosperity and ecological balance to maintain the economic growth [5, 6]. Guidelines for circular economy driven by different government programs. In the developing countries concept is more popular than the developed nations. [7] Now manufacturing industries are focus on high revenue and lesser impact on environment by their manufacturing process [8, 9]. In the past few years authors had discussed opportunity of sustainable manufacturing and circular economy [10, 5]. But these researches have some limitations in literature. In this study we have tried to provide the bibliometric based review for circular economy. This study uses the Scopus data base for the analysis. In the present study following research questions are raised which are:

RQ1: What is the annual scientific production in this area?
RQ2: Who are top journals, most cited articles and author in this area?
RQ3: What are the mostly used keywords in this area?

The paper divided in to four sections of study, in Sect. 1 discuss about the introduction and various research of study. In Sect. 2 discussion on SM and circular economy relation is carried on them. In Sect. 3 we discuss Bibliometric analysis of system of SM and circular economy. In Final section the conclusion and implications of this study is mentioned.

2 Circular Economy and Sustainable Manufacturing

The aim of this research was to conduct a systematic and bibliometric review of scientific papers relating to the sustainable manufacturing and circular economy of the manufacturing sector from a sustainability perspective that were indexed in the Scopus databases

A systematic bibliometric literature review was conducted using a quantitative method in order to categories and interpreting the existing findings on sustainability manufacturing and circular economy in the current research. The paper summarizes in what manner and direction the past research has been carried out on this subject and provide insight knowledge into current and future research directions. Besides, this method of study aims to provide a scientific review of the most important and reputed publications on this topic.

SM used for decrease or removal of negative impact of manufacturing process by adoption of eco-friendly which includes the minimization of waste [11, 12]. Over the years of research there are many ways to define SM. SM is manufacturing in which minimizes the negative impact of manufacturing on environment and increase the efficiency [6]. The need of fast adaption of new technologies and never-ending customers demand also responsible for the industrial growth. Now the industries are more competitive and these are adopting new technologies in existing production system to satisfying the customers need [13, 14]. Circular economy is a promising choice which is the integration of manufacturing and business process as well as customers and suppliers [15]. The broader aspect of industrial inter net can be also seen in SM and circular economy which covers different sectors in industry like mining, power generation sector, health and manufacturing sectors. Autonomous robots, cyber security, horizontal and vertical systems integrations, simulation, industrial lot, big data analytics and augmented reality [16, 9]. These are different pillars of SM and circular economy. Now digitalization and sustainability concepts in the industries are increasing the competitiveness. Both the circular economy and SM approaches present practices converge [17, 18]. Examples remanufacturing, recycling, reverse logistics for the circular economy, resource efficiency, lean and green philosophies, elimination of toxic parts in the product and production process and sustainable designs which minimizes the health risk for the workers[19, 9]. SM and circular economy which results in improve efficiency, waste reduction, optimization of process, reduce energy consumption and employment generation for disabled workers [5, 7].

3 Methodology

Bibliometric analysis helps to evaluate the published books, articles, conference papers or chapters statistically [20]. It is an effective way to measure the influence of a scientific document in the scientific community. The 1st step is to build a bibliographic portfolio then, bibliometric studies are performed on theoretical, authors, countries and institute trends with the most publications and periodicals themes.

Over the past many years researches have used many databases such as Google scholar, Scopus and WOS for bibliometric analysis [20–22]. Now a day's authors mostly using Scopus and WOS database. Both web of science (WOS) and Scopus database are important for Bibliometric investigation in science and engineering field, both databases have consistent and accurate standardized records, also both databases have broad scope of SM and circular economy related publications and top journals related to same. That's the reason these databases are considered for analysis. It is easy to segregate the publications by author name and title of paper plus advance search options are available in both Scopus and web of science. For this research paper only the Scopus database is used to conduct the bibliometric analysis on SM and circular economy.

In Scopus databases, the search was conducted by incorporating the words Circular Economy; Sustainable Manufacturing as follows: "sustainability manufacturing" or "sustainability" and "circular economy", here "AND" and "OR" are called Boolean operators which are used for combining the keywords [20]. When the "and" operation is carried out in a database search, it includes all words that can appear in the results. While the use of "OR" operation indicate that at least one of the keywords

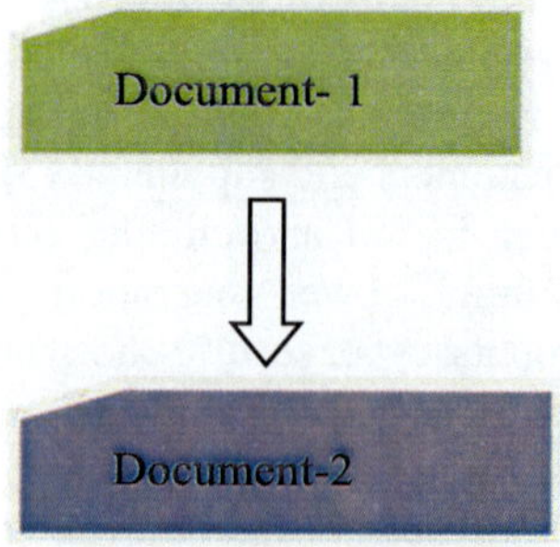

Fig. 1 Citation for document

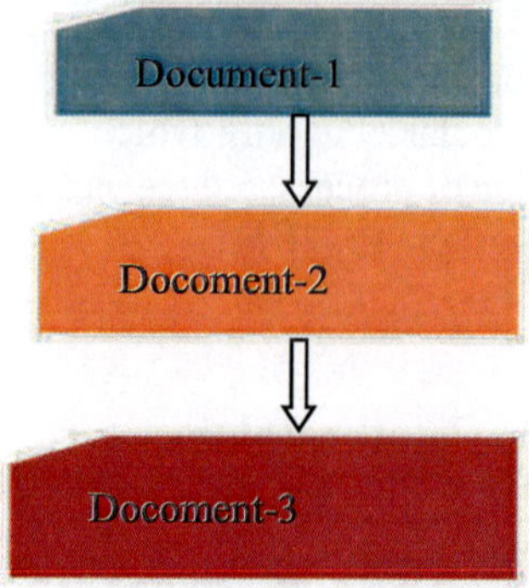

Fig. 2 Co-citation for document

appear in the search. Which is used for "articles, keywords, titles" in search, 612 documents were found.

Citation: Citation is defined as 'using the content of other articles' and write it in your words to getting help in researching your work [22].

Co-citation: The extent at which two documents of earlier literature are cited together by later literature is known as co-citation. It shows the degree of similarity between the documents; the greater the co-citation intensity of two documents, the more often they are semantically related [21].

Figure 1 given below explain about citation means as we write document 2 in which documents1is used as a reference then it is called as citation.

Figure 2 given below explain if we write document 3 for which we use document 1 and document 2 as a reference then document 2 and document 3 is called co-citation for document 1.

3.1 Document Wise Analysis

Table 1 shows the document wise analysis of SM and circular economy.

Total 34 papers are found after analysis. the conference papers 17.

Table.1 Document wise analysis

S. No.	Description	Results
1	No of documents	34
2	Total sources	17
3	Author keywords	71
4	Time period	2000–2021

3.2 *Annual Scientific Production in Circular Economy and Sustainable Manufacturing*

The data for the annual scientific production for articles related to circular economy and sustainable manufacturing is collected from Scopus database. Scopus is one of the largest database hubs for the scientific articles. The articles are Scopus database are more peer reviewed as compared to other scientific databases. The annual scientific production of articles can be seen in the Fig. 3. The results show that the research on circular economy was limited till 2015. However, the concept of sustainable manufacturing was coming into consideration in 1999 on Scopus database when due the pressure from government regulations and public pressure related to environmental protection programmes. After, 2015 researchers started linking both the area of sustainable manufacturing and circular economy. Hence, good growth in the no. of publications can be seen after 2015. In the last two years many articles in this area have been published in the reputed journals like *"Journal of cleaner production"* and *"International journal of production research"* which are gaining the interest from both academia and industries to adopt circular economy practices with sustainable manufacturing. It can be say that this research area is at its boom and expected to rise in more number of publications in upcoming years.

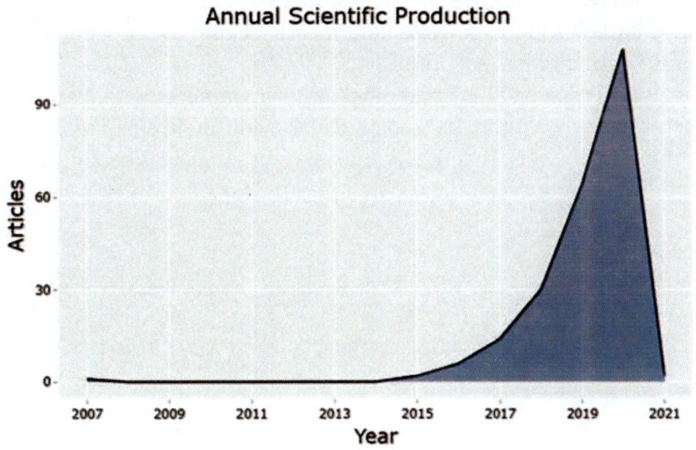

Fig. 3 Annual scientific production

Table.2 Top cited articles

S. No.	Article	Citations
1	Technological elements of circular economy and principles of 6R based closed loop material flow in sustainable manufacturing	101
2	Drivers to sustainable manufacturing practices and circular economy: a perspective of leather industries in Bangladesh	83
3	Cloud manufacturing as a sustainable process manufacturing route	68
4	Mapping industrial symbiosis development in Europe technologies of networks, characteristics, performance and contributions to circular economy	52
5	Manufacturing strategies for efficiency in energies and resources use: the role of metal shaping processes	36

3.3 Most Cited Articles

Most cited article are shown in the Table 2 in which we can see that articles published by "Procedia CIRP" has 101 citations and discuss about the 6R based closed loop supply chain. Now the authors are more focused towards the research in these two areas. Also, the authors are now linking the both circular economy and sustainable manufacturing to achieve the sustainability in Industry 4.0 and 2030 SDG agenda.

3.4 Most Productive Authors

Table 3 shows the productive authors with their total number of publications in the particular area. Jawahir,I.S. is most productive author with 4 documents. Jawahir IS has most of publication in this area and having a good contribution in sustainable manufacturing also. Now authors are linking the concept of Sustainable manufacturing with circular economy.

In the analysis Procedia CIRP had most of the documents with 4 articles. In some of bibliometric study's authors have excluded the conference papers to maintain the quality of. But it can be seen that in this analysis conference proceedings shouldn't be excluded while mapping the studies. We can see the top cited and top sources both are

Table.3 Top productive authors

S. No.	Author name	TP
1	Jawahir, I. S.	4
2	Bradly, R.	3
3	Ingarao, G.	3
4	Angioletti, C. M.	2
5	Bag, S.	2

from Procedia CIRP. Reputed publishers having sustainability, international journal of production research, cleaner production and journal of manufacturing systems still having less documents on SM and circular economy area which is expected to be increase in future.

3.5 *Most Relevant and Productive Sources*

It is essential to find out the most productive and relevant sources for the particular research area which will helps the authors to find out which sources are publishing more research articles in the particular research area. As, we know that the circular economy and sustainable manufacturing adoption is an emerging trend in both developing and developed nations. The research in these two research areas is at its boom. So it is required to find out the journals and other sources which are more relevant and publish articles related to sustainable manufacturing and circular economy. We have identified the most productive journals by the analysis done in R studio on macOS Big Sur. We have found that *"Journal of Cleaner Production"* having most of publications related to this area which is further followed by *"Sustainability"* and *"Resource conservation and recycling"*. All these journals are related to sustainability and publishing the articles related to sustainable manufacturing and other sustainability issues from a long time. Now with the development in Industry 4.0 these journals have started publishing articles related to topics like circular economy and Industry 4.0 technologies. We can also seen that few of sources are open access and these having a good citation structure because these articles are easily accessible to both academia and practitioners (Fig. 4).

3.6 *Top Countries and Institutes Working on SM and Circular Economy*

Top 8 countries working on SM and circular economy is shown below in Table 4. It is found that United states had 10 documents which followed by Italy, United Kingdom, India, South Africa, China, France, Netherlands. Only top 3 countries are developed remaining are developing countries working on SM and circular economy.

3.7 *Word Cloud in Circular Economy and Sustainable Manufacturing*

In this research work we have generated the word cloud of keywords on R studio software. Generally, Scopus bibliographic database having two types of keywords

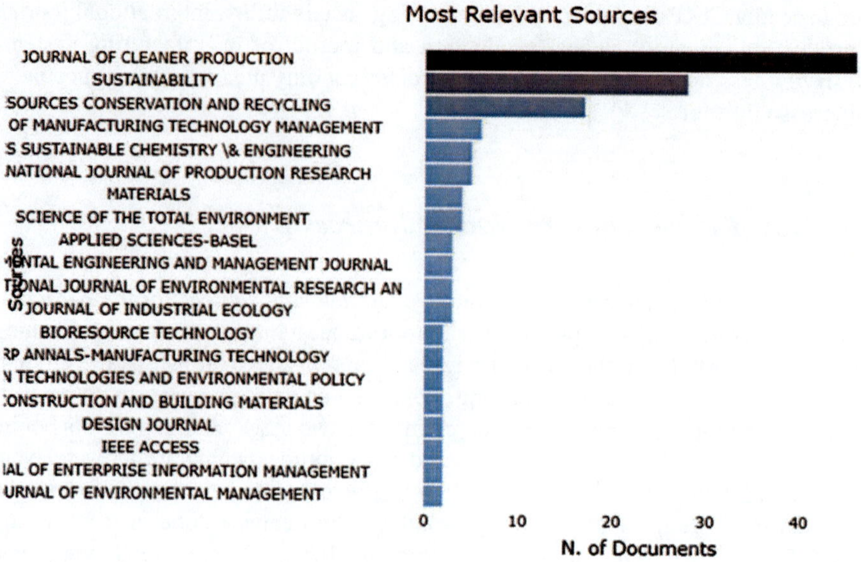

Fig. 4 Most productive journals and other sources

Table.4 Top productive countries

Country name	TP	Category
United States	10	Developed
Italy	9	Developed
United Kingdom	7	Developed
India	4	Developing
South Africa	4	Developing
China	3	Developing
France	3	Developed
Netherlands	3	Developed

(1) Keyword Plus (2) Author keywords. In this study we have used author keywords to map our study more specifically. The word cloud of author keywords is shown in the Fig. 5.

Here from the analysis we can see that circular economy is mostly used keyword by authors in their articles. Also, the *"Industry", "Industry", "production" and "sustainable manufacturing"* are widely used keywords in this area.

Fig. 5 Word cloud in circular economy and sustainable manufacturing

3.8 *Top Keywords*

Top keywords are used in the documents of sustainable manufacturing and circular economy shown in Table 5. It is found that mostly used key word is "sustainable manufacturing" with 40 times occurrence followed by "circular economy" with 31 occurrence, sustainable development with 18 occurrences, "manufacture" with 17 times occurrence and "economics" with occurrence of 10 times occurrence. However, other popular keywords in SM and circular economy such as "industrial economics "and "environmental impact" still less occurrence in the documents.

As discussed above network analysis will help the researchers to work in the sustainable manufacturing and circular economy. Now researchers are more focused

Table.5 Top key words with total occurrence

S. No	Key word name	Total occurrence
1	Sustainable manufacturing	40
2	Circular economy	31
3	Sustainable development	22
4	Sustainability	18
5	Manufacture	17
6	Economics	10
7	Life cycle	10
8	Recycling	10
9	Industrial economics	9
10	Environmental impact	7

towards the sustainability in the circular economy. Based on the study mapping following research questions for future research directions have been proposed.

RQ1: How can a SM and circular economy will help the Industry 4.0 to achieve 2030 SDG agenda?
RQ2: What are different common enablers and barriers in the adoption of sustainable manufacturing and circular economy? RQ3: What are the opportunities for these two in developing nations? These are some future research questions which need to address in the future studies these questions will help the researchers to work in this area.

4 Conclusion and Future Implications

This study helps to contribute the various research scopes in SM and circular economy is an emerging research area for most of the authors. It is found that research articles in this area are increasing rapidly.in 2019 no. of articles more than the no. of articles published 2015. This shows that researchers are now more focused towards the SM and circular economy. It is suggested that in future studies conference papers are should not be excluded from the bibliometric analysis. In top 5 authors are various research areas had strong research background, and some are started research career in this area and doing the productive research work. The research is going on both developed and developing countries. It is expected that these research gaps will be answered in next few years.

References

1. Jamwal A, Agrawal R, Sharma M, Kumar V (2021) Review on multi-criteria decision analysis in sustainable manufacturing decision making. Int J Sustain Eng 1–24
2. Machado CG, Winroth MP, da Silva EHD (2020) Sustainable manufacturing in Industry 4.0: an emerging research agenda. Int J Prod Res 58:1462–1484. https://doi.org/10.1080/00207543.2019.1652777
3. Jamwal A, Agrawal R, Sharma M, Giallanza A (2021) Industry 4.0 technologies for manufacturing sustainability: a systematic review and future research directions. Appl Sci 1(12):5725
4. de Sousa Jabbour ABL, Jabbour CJC, Foropon C, Filho MG (2018) When titans meet—can industry 4.0 revolutionise the environmentally-sustainable manufacturing wave? The role of critical success factors. Technol Forecast Soc Change 132:18–25. https://doi.org/10.1016/j.techfore.2018.01.017
5. Pieroni MP, McAloone TC, Pigosso DC (2019) Business model innovation for circular economy and sustainability: a review of approaches. J Cleaner Prod 215:198–216
6. Zhao H, Guo S, Zhao H (2018) Comprehensive benefit evaluation of eco-industrial parks by employing the best-worst method based on circular economy and sustainability. Environ Develop Sustain 20(3):1229–1253

7. D'Amato D, Droste N, Allen B, Kettunen M, Lähtinen K, Korhonen J, Toppinen A (2017) Green, circular, bio economy: A comparative analysis of sustainability avenues. J Cleaner Prod 168:716–734
8. Ghobakhloo M (2020) Industry 4.0, digitization, and opportunities for sustainability. J Clean Prod 252. https://doi.org/10.1016/j.jclepro.2019.119869
9. Haupt M, Hellweg S (2019) Measuring the environmental sustainability of a circular economy. Environ Sustain Indicat 1:100005
10. Jamwal A, Agrawal R, Sharma M, Kumar V, Kumar S (2021) Developing a sustainability framework for Industry 4.0. Procedia CIRP 98:430–435
11. Stock T, Seliger G (2016) Opportunities of Sustainable Manufacturing in Industry 4.0. In: Seliger G. Kohl H. MJ (eds) Procedia CIRP. Elsevier B.V., pp 536–541
12. Ejsmont K, Gladysz B, Kluczek A (2020) Impact of industry 4.0 on sustainability-bibliometric literature review. Sustain 12. https://doi.org/10.3390/su12145650
13. Korhonen J, Honkasalo A, Seppälä J (2018) Circular economy: the concept and its limitations. Ecolog Econ 143:37–46
14. Kirchherr J, Reike D, Hekkert M (2017) Conceptualizing the circular economy: An analysis of 114 definitions. Res Conserv Recycling 127:221–232
15. Prieto-Sandoval V, Jaca C, Ormazabal M (2018) Towards a consensus on the circular economy. J Cleaner Prod 179:605–615
16. Hong M, Chen EYX (2017) Chemically recyclable polymers: a circular economy approach to sustainability. Green Chem 19(16):3692–3706
17. Yadav G, Kumar A, Luthra S et al (2020) A framework to achieve sustainability in manufacturing organisations of developing economies using industry 4.0 technologies' enablers. Comput Ind 122. https://doi.org/10.1016/j.compind.2020.103280
18. Geissdoerfer M, Savaget P, Bocken NM, Hultink EJ (2017) The circular economy–a new sustainability paradigm? J Cleaner Prod 143:757–768
19. Melnyk SA, Smith RT (1996) Green manufacturing. Comput Automat Syst Soc Manuf Eng
20. Kamble SS, Gunasekaran A, Gawankar SA (2018) Sustainable Industry 4.0 framework: a systematic literature review identifying the current trends and future perspectives. Process Saf Environ Prot 117:408–425. https://doi.org/10.1016/j.psep.2018.05.009
21. Bhatt Y, Ghuman K, Dhir A (2020) Sustainable manufacturing. Bibliometrics and content analysis. J Clean Prod 260. https://doi.org/10.1016/j.jclepro.2020.120988
22. Jamwal A, Agrawal R, Sharma M, Kumar A, Kumar V, Garza-Reyes JAA (2021) Machine learning applications for sustainable manufacturing: a bibliometric-based review for future research. J Enterp Inf Manage

Identification of Challenges & Practices of Sustainability in Indian Apparel and Textile Industries

Amit Vishwakarma, M. L. Meena, G. S. Dangayach, and Sumit Gupta

Abstract Purpose of this study to identify the various challenges that Indian apparel industry is facing and these challenges are taken from beginning i.e., from growing of raw material. Moreover, this work include classification of these challenges into various dimensions of sustainability i.e., environmental, social, economic. It had been found out that some of have impact on all the three dimensions. To resolve these challenges some sustainability practices is suggested. These practices majorly focus upon environment protection. However, they are not sufficient in future much more sustainability practices significant initiatives required for better environment and sustainability.

1 Introduction

In its development, India's apparel sector plays a vital role. It provides not only the financial income, it also gives the people enormous jobs. It provides around 2.3% of India's GDP, while textile exports represent 12% of its foreign exchange profits. In this industry, almost 45 million people work directly or indirectly. This sector is suitable both for qualified and unqualified employees and provides inclusive growth. These data all illustrate the importance in the country's growth of the garment and textile sector. This sector, however, is very polluting for the environment. Greenhouse gas emission, hazardous waste generation, discharge into the environment of toxic wastewater containing dyes, etc. 20% of water contamination was determined to be caused by the textile sector [1]. The affluent contains various pollutants and dye colors [2]. When this water reached to the nearby water sources it negatively affect the aquatic life as well human life who consumed this water. Pollutants released by the global textile industry pollutes land and makes them useless as well as barren in

A. Vishwakarma (✉) · M. L. Meena · G. S. Dangayach
Department of Mechanical Engineering, Malaviya National Institute of Technology, Jaipur, India

S. Gupta
Department of Mechanical Engineering, Amity School of Engineering and Technology, Amity University Uttar Pradesh, Noida, India

R. Agrawal et al. (eds.), *Recent Advances in Industrial Production*, Lecture Notes in Mechanical Engineering, https://doi.org/10.1007/978-981-16-5281-3_14

the long run. This arises the need for green manufacturing of apparel is the need for present and the implementation of various sustainability practices. This will benefit our environment. Sustainability played a vital role in the development of industry like CSR (Corporate Social Sustainability) provide economic benefits to the employees [3]. Purpose of this study to identify various challenges from raw material till the end of life of the product in Indian apparel industry and suggest the sustainable solution for the challenges.

2 Literature Review

The work on sustainability started in 1980 but initially with different name i,e environment protection. It is very necessary to use natural resources in disciplined manner and prevent pollution. Apparel Industry release various pollutants in the surrounding. Microfibers from apparel and home textiles pollutes the ecosystem and harm the human health [4]. For social perspective, majorly unskilled workers are involved in this sector. They don't have any prior knowledge or professional training for this kind of work. Moreover, no proper wage system is followed in this sector. As in this sector a lot of female workers are involved and it had been seen that they got lower wages as compared to male worker. Practice section of this article include those practices which support the sustainability. These practices are related to prevention of environment pollution like reduction in the use of color dyes, reductions in the emission of greenhouse gases, recycled clothing, use of organic fiber. Consumer awareness practices like use of eco-label, reduction in fast fashion, elimination of throwaway culture of consumer. Table below contains the literature related challenges and various practices to of textile and apparel industries.

Table 1 shows the various connotations of sustainability.

2.1 Research methodology

This study identified some of the challenges in apparel and textile industry. Effect of these challenges in different aspect of sustainability i.e social, economic and environmental shown below in Fig 1. So the challenges identified are poor quality of raw materials, negative impact on the environment and society, low productivity, skill shortage, poor infrastructure, lack of sufficient governmental policies, fast fashion. These are seven in number. It can be seen from the table that every challenges had it effect on at least two dimensions of sustainability.

Table 1 Contribution of various authors in this field

Author	Year	Connotations
[5]	2016	Textile consumption and its effect on environment investigated by using the LCA (life cycle assessment) method. The industry sector approach is used for the assessment of different interventions. It contributes to environmental and social sustainability in the textile sector
[6]	2016	It utilizes a social constructionist approach to address the gap between sustainable fashion consumption and understanding of sustainable fashion
[7]	2017	This integrates and implements LP (lean practice) to Environmental sustainability in the apparel sector. Moreover, it also identified various barriers that arise due to this and how to tackle them effectively
[8]	2018	This paper identifies 14 barriers related to apparel and textile sustainability in India. Further, it identifies critical barriers among them with the use of the DEMATEL method. Lack of effective government policies and poor infrastructure came out to be significant barrier
[9]	2019	It identifies various challenges for incorporating sustainability in the design process and categorize it as internal and external challenges. Internal challenges include lack of consensus and knowledge regarding sustainable design, lack of design-led approaches implementing sustainability in fashion, and perceived trade-offs with other design criteria, such as aesthetic styles, costs, and fashion trends. External challenges include the complexity of sustainability issues, perceived insufficient consumer demand, attitudes and behaviour gaps in consumer purchasing decisions
[10]	2020	Some eco-friendly approaches followed for sustainable garment manufacturing from raw materials selection to the final stage of garment manufacturing
[11]	2019	This study using qualitative and quantitative analysis to identify sustainable incremental change in organizations. It found out that small life of apparel and textile products challenges the sustainability
[12]	2015	This study discusses about challenges faced by fashion industry. It particularly focusses on environmental and social challenges. Innovative ideas of sustainability can counter the challenges of fashion industry

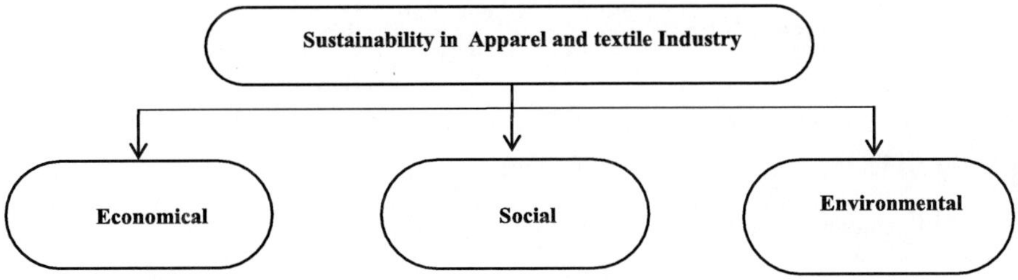

Fig. 1 Type of sustainability

2.2 *Classification of challenges*

In the below table some of the challenges are identified and their effect on which type of sustainability is shown (Table 2).

1. **Poor quality of raw materials** There is need to improve the quality of cotton. Poor quality of raw materials leads to low productivity. New and modern technology should be included to improve the quality of the raw material. Good quality of seed or hybrid seed are required to improve the quality of raw material ginning and pressing technique help to reduce the contamination and improve the quality of cotton [13].
2. **Negative impact on the environment and society** Textile and apparel Industry pollute the environment. It pollutes water soil, air. Due to the lack of environmental awareness and small enterprises discharge the polluted water directly into the water source. Polluted water of industry contains the hazardous chemical these are harmful for aquatic life as well as human health [14]. Globally, approx. 10% of air pollution is from this sector. So this sector is highly responsible for global pollution.
3. **Low productivity** There are so many reasons behind the low productivity. Limited use of modern technology, Power cuts, use of unskilled labour and the inadequate infrastructure are some of the possible reasons [15].
4. **Skill shortage** Labour working in this industry does not go any kind of training. Very few firm invest in training of manpower. Result of skill shortage leads to low productivity and it degrade the quality of the product. To overcome this problem workshop and training should be provided for the entire workforce [16]. Moreover, child labour practices is quite common in this sector because of the low wages. It is ethically and legally wrong practice but also responsible for the unskilled work.

Table 2 Categorization of sustainability into its dimension

S.N.	Challenges	Social sustainability	Environmental sustainability	Economical sustainability
1	Poor quality of raw materials	Yes	Yes	Yes
2	Negative impact on the environment and society	Yes	Yes	No
3	Low productivity	Yes	No	Yes
4	Skill shortage	Yes	Yes	Yes
5	Poor infrastructure	Yes	Yes	Yes
6	Lack of sufficient governmental policies	Yes	Yes	Yes
7	Fast Fashion	No	Yes	Yes

5. **Poor Infrastructure** Indian apparel and textile industry lack automation very much. Most of the industries using conventional way for manufacturing of apparel. Moreover, power cut is also frequent problem. To remain competitive in global market these problem need to be addressed. Further research and development and finance are also required to resolve this problem [17]. This poor infrastructure is responsible for the accident like building collapse, fire on industries. This cause loss of money and materials but sometimes it leads to loss of lives.1.
6. **Lack of sufficient governmental policies** For proper controlling of this sector good government policy required. Which can rectify its lacking like significant amount of unskilled labour working in this sector. Moreover, the government policy should address the following issues tough environment protect, waste reduction, recycle, handling of waste etc [18].
7. **Fast Fashion** This is originated because of lack of awareness of the consumer towards environment [19]. Consumer find cloths in low price so he purchases too many cloths because as per recent fashion and irrespective of the need. This cause consumer thrown away the older cloth without completing their product life. This created the problem of waste generation and disposal of clothing [20].

2.3 Practices to be adopted for sustainability

1. **Application of Life cycle assessment method (LCA)** Generally consumer compare product value in terms of money but LCA method provide the information about consumption of resources for manufacturing of product [21]. It becomes very essential for apparel and textile industry because it this industry directly affect the environment and natural resources like amount of water usage, carbon emission, eutrophication process. This should to design and develop the product which will not harm the environment after completing it life.
2. **Eco-labelling of the Apparel** Life cycle assessment of product should be accompanied by eco labelling of garment. Purpose of ecolabel to provide information to consumers about the apparels [21, 22]. This information consists of eco-friendly manufacturing process, safe disposal etc. This information help consumer to prefer eco-friendly and sustainable apparel. Ecolabelling has an important role in the development of sustainable fashion products globally.
3. **Environment friendly practices** This sector requires environment friendly practices at vast level. However, carbon dioxide gas emission and water pollution cannot be eliminated but can be minimized. Some of the environment friendly practices [23–25] are use of energy efficient process, use of renewable energy etc. and some more practices are shown below.

   - Use of low impact dies.
   - Reduction in the use of toxic chemical.
   - Optimized use of land and water.
   - Reduction in the use of emission of gases.

- Filter out of hazardous metal like lead, cadmium from used water.
- Prevent mixing of this used water in nearby water resources.

4. **Eco-friendly textile fibers** Cotton crop production consume high amount pesticide. About 55% of whole pesticide usage in India goes to cotton crop production [25, 26]. So there is need to use organic fibers. Cultivation of organic fiber do not use pesticide, fertilizers, synthetic agro –chemicals. So production of organic fibers do not harm the soil in any manner hence it is environmentally friendly although it is costlier than normal cotton.
5. **Recycled Clothing** Solid waste in the environment is goes on increasing and clothing industry plays major role in it. It has been found that approximately two third of manufacturing of clothing goes into the landfills after completion of product life. It is the fastest growing household waste [27]. Further young consumers have lack of awareness towards environment and they pay little attention towards it. They early thrown away the apparel. Recycling of cloths [28–30] present in this sector but it is not up to the mark. Much more work is required for the reduce, reuse and recycle technique.

3 Conclusion

This Study identified some of the challenges that Indian textile and apparel industry is facing. Further effect of these challenges on which dimension of sustainability is also mentioned i.e., economic, social and environmental sustainability. The challenges included in this research are poor quality of raw materials, negative impact on the environment and society, low productivity, skill shortage, poor infrastructure, Lack of sufficient governmental policies and fast fashion. Reduction or elimination of these challenges are essential for the growth of this sector. This will also increase the overall sustainability in this sector. This study also suggested some practices for the sustainability. These practices support and improve the sustainability. The included practices are application of life cycle assessment method, eco-labelling of the apparel, environment friendly practices, eco-friendly textile fibers, recycled clothing. All these practices have various benefits like environmental protection, bring consumer awareness etc.

4 Future Work

This work can be extended by including more barriers from different domain. These barriers can be from manufacturing process of apparel, supply chain or transportation, customer purchase for sustainable apparel etc. Criticality of challenges and its effect is also area of research as every challenge has the different effect on this sector. Moreover, there is possibility for the linkage among different challenges like dependence of challenges over one–another. So this is also the problem of researcher to find out relationship among various challenges. There is also need to bring awareness among consumers for buying sustainable apparel. More sustainability practises should be included in future to make this sector environmental economic and social efficient.

References

1. Nayak R, Panwar T, Nguyen LVT (2020) Sustainability in fashion and textiles: A survey from developing country. In: Sustainable technologies for fashion and textiles. Woodhead Publishing, pp 3–30
2. Kakull A, Risberg V (2012) A lost revolution? Empowered but trapped in poverty: women in the garment industry in Bangladesh want more. Swedwatch
3. Feng P, Ngai CSB (2020) Doing more on the corporate sustainability front: a longitudinal analysis of CSR reporting of global fashion companies. Sustainability 12(6):2477
4. Henry B, Laitala K, Klepp IG (2019) Microfibres from apparel and home textiles: prospects for including microplastics in environmental sustainability assessment. Sci Total Environ 652:483–494
5. Roos S, Zamani B, Sandin G, Peters GM, Svanström M (2016) A life cycle assessment (LCA)-based approach to guiding an industry sector towards sustainability: the case of the Swedish apparel sector. J Cleaner Prod 133:691–700
6. Henninger CE, Alevizou PJ, Oates CJ (2016) What is sustainable fashion? J Fash Market Manage Int J
7. Raj D, Ma YJ, Gam HJ, Banning J (2017) Implementation of lean production and environmental sustainability in the Indian apparel manufacturing industry: a way to reach the triple bottom line. Int J Fash Design Technol Educ 10(3):254–264
8. Gardas BB, Raut RD, Narkhede B (2018) Modelling the challenges to sustainability in the textile and apparel (T&A) sector: a Delphi-DEMATEL approach. Sustain Prod Consump 15:96–108
9. Hur E, Cassidy T (2019) Perceptions and attitudes towards sustainable fashion design: challenges and opportunities for implementing sustainability in fashion. Int J Fash Design Technol Educ 12(2):208–217
10. Patora-Wysocka Z, Sułkowski Ł (2019) Sustainable incremental organizational change—a case of the textile and apparel industry. Sustainability 11(4):1102
11. Pedersen ERG, Andersen KR (2015) Sustainability innovators and anchor draggers: a global expert study on sustainable fashion. J Fash Market Manage
12. Gupta RK (2006) Indian textile industry: prospects and challenges. http://www.indianmba.com/Faculty_Column/FC236/fc236.html. Accessed on 4 Mar 2017
13. You S, Cheng S, Yan H (2009) The impact of textile industry on China's environment. Int J Fash Design Technol Educ 2(1):33–43
14. Raichurkar P, Ramachandran M (2015) Recent trends and developments in textile industry in India. Int J Textile Eng Process 1(4):47–50
15. Taneja R (2012) Indian textile exports: past and present. Int J Multidis Academic Res 2(2):1–19
16. Aboagyewaa-Ntiri J, Mintah K, Aboagyewaa J (2016) Challenges and opportunities for the textile industry in Ghana: a study of the Adinkra textile sub-sector. Int Bus Res 9(2):127–136
17. Umarji V (2015) Dumping, lack of ftas, incentives hurting textile industry, Business standard http://www.businessstandard.com/article/economypolicy/dumping-lackof-ftas-incentives-hurting-textile-industry-115090900877_1.html
18. Albloushy H, Hiller Connell KY (2019) Purchasing environmentally sustainable apparel: the attitudes and intentions of female Kuwaiti consumers. Int J Cons Stud 43(4):390–401
19. Bhardwaj V, Fairhurst A (2010) Fast fashion: response to changes in the fashion industry. Int Rev Retail Distribut Cons Res 20(1):165–173

20. Gupta S, Dangayach GS, Singh AK, Meena ML, Rao PN (2018) Adoption of sustainable supply operation quality practices and their impact on stakeholder's performance and sustainable performance for sustainable competitiveness in Indian manufacturing companies. Int J Intell Enterp 5(1–2):108–124
21. Gupta S, Dangayach GS, Singh AK, Meena ML, Rao PN (2018b) Implementation of sustainable manufacturing practices in Indian manufacturing companies. Benchmark Int J
22. Jaiswal P, Kumar A, Gupta S (2018) Prioritization of green manufacturing drivers in Indian SMEs through IF-TOPSIS approach. U.P.B. Sci Bull Series D 80(2):277–292
23. Gupta S (2016) Some Issues in sustainable manufacturing: a select study of Indian manufacturing companies. Doctoral dissertation, MNIT Jaipur
24. Gupta S, Dangayach GS, Singh AK, Rao PN (2015) Analytic hierarchy process (AHP) model for evaluating sustainable manufacturing practices in Indian electrical panel industries. Procedia-Soc Behav Sci 189:208–216
25. Gupta S, Dangayach GS, Singh AK (2015) Key determinants of sustainable product design and manufacturing. Procedia CIRP 26(2):99–102
26. Jamwal A, Agrawal R, Gupta S, Dangayach GS, Sharma M, Sohag MAZ (2020) Modelling of sustainable manufacturing barriers in pharmaceutical industries of Himachal Pradesh: an ISM-fuzzy approach. In: Proceedings of international conference in mechanical and energy technology. Springer, Singapore, pp 157–167
27. Moore SB, Wentz M (2009) Eco-labeling for textiles and apparel. In: Sustainable textiles. Woodhead Publishing, pp 214–230
28. Higginson H, Saio N, Swinnerton A, Williams D (2010) Promoting sustainable Indian textiles: final report to the Department for Environment, Food and Rural Affairs (Defra). London, UK
29. Asif AKMAH (2017) An overview of sustainability on apparel manufacturing industry in Bangladesh. Sci J Energy Eng 5(1):1–12
30. Vadicherla T, Saravanan D (2014) Textiles and apparel development using recycled and reclaimed fibers. In: Roadmap to sustainable textiles and clothing. Springer, Singapore, pp 139–160

Challenges of Adoption of Blockchain Technology in Supply Chain: An Overview

Dnyaneshwar Jivanrao Ghode, Rakesh Jain, and Gunjan Soni

Abstract Blockchain Technology (BT) has grown noteworthy attention in recent years. Adoption of these technologies reforms supply chains (SC), produce an improved product quality within a lead time, and provide the customized products to the customer. BT has a characteristic of decentralized transactions and data management that has significant impact on traditional supply chain management. Integration of BT with SC is still in its early stages. Industries and academic research need to know the challenges in the adoption of BT in SC. It requires huge investments to investigate various challenges in adoption of BT in SC. Therefore, the aim of this study is to provide an overview about the challenges in adoption of BT in SC. A literature review has been conducted to find challenges in adoption of BT in SC. This paper recognizes some main challenges. Those challenges are to build trust among the SC members, prepare a policies that governs the implementation of BT in SC, maintain the balance between the transparency and confidentiality of the transaction data, make a hackproof system, enhance the communication within SC system, select a proper product in the early stage of adoption to recover the investment as early as possible, prove provenance of the product and train the people involved in implementing BT in SC. This work discussed potential implications based on challenges and provide platform for adoption of BT in SC.

Keywords Supply chain · Blockchain technology · Challenges

1 Introduction

Since past few years, there is propaganda of the cryptocurrencies like bitcoin, ether etc. BT is known in financial sector due to the evolution of bitcoin. However, BT has many applications in other sectors than financial [1]. BT consist of shared, and immutable ledger. BT enable to verify the transactions and immediately make available to all the stakeholders in the system. Real time updating of the business processes, monetary benefits, and reduction in frauds can be done with BT. The main cause of

D. J. Ghode (✉) · R. Jain · G. Soni
Malviya National Institute of Technology Jaipur, Jaipur 302017, India

R. Agrawal et al. (eds.), *Recent Advances in Industrial Production*, Lecture Notes in Mechanical Engineering, https://doi.org/10.1007/978-981-16-5281-3_15

the transformation of the business process using BT is SC [2]. The appearance of BT will have huge effect on present business system particularly SC of the various organizations [3]. Like internet, application of BT in SC has capability to reduce the cost of the transactions, increase in yield, authorize e-procurement, assimilate various businesses and give consent to the customized products and services. It now requires rigorous academic research which may support businesses to generate more yields [4].

Many companies are in the process to accept the adoption of blockchain in SC to enhance their performances. Adoption of BT in SC will disrupt the existing system to give birth to new blockchain-based SC system. As the various studies shows, BT offers many benefits to the businesses including SC. Very few studies have discussed the challenges in the adoption of the BT in SC. From the literature study, it came to know that there is inadequate knowledge of BT, it is an evolving nature of technology, and hence many establishments resist the transition from existing system to BT based SC system [4]. It is need to study blockchain and recognize the challenges in implementation of BT in SC; that will provide the information regarding the possible hurdles and help the organizations to prepare a strategy to overcome the identified challenges. From the literature study, this work identified some challenges under the administrative, technical, functional, and social heads. For the implementation of BT in SC, it is necessary to overcome the identified challenges. Furthermore, implications of the challenges have been discussed that will help researchers and practitioners in the implementation of BT in SC.

Rest of the paper covers: Sect. 2 describes the implementation of BT in SC includes the blockchain system and blockchain in SC management, Sect. 3 investigates the challenges in the adoption process of BT in SC. Section 4 discusses the implication of identified challenges on implementation of BT in SC. Finally, Sect. 5 provides concluding remarks with the future scope of the present research.

2 Implementation of BT in SC

2.1 Blockchain

BT is an entirely distributed system that stores cryptographed, reliable, and tamper proof information of transactions between the members of blockchain network. Like distributed network, BT has the function of supporting consensus, apprising and verifying the transaction data within the network. Blockchain networks are with transactions that are transparent, time-stamped and verified using consensus [5]. Figure 1 shows a typical diagram of blockchain. Blocks in blockchain includes a history of transaction data. For each transaction a new block with time-stamp is formed. New block is appended in the blockchain after the verification of transactions. Two blocks in the blockchain are connected with each other using hash. In Fig. 1, it can be seen that previous hash of the Block2 is the current hash of the Block1.

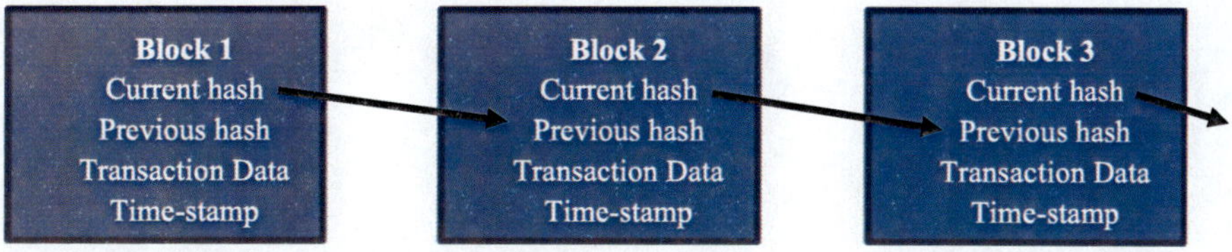

Fig. 1 Logical diagram of Blockchain

In this way Block1 and Block2 are connected with each other. Once the block is created and added in the blockchain, it cannot be reversed. Due to this feature, the term "Blockchain" is used. No one can change the data in the blocks of blockchain. Governance rule are used to verify the transactions in the blocks of the blockchain network [6].

2.2 *Blockchain in SC Management*

In recent time, due to mismatch between demand data and supply of products, performance of SC is sternly affected that increases the Bullwhip effect. Bullwhip effect can be minimized using the exact information of SC transactions [7]. Hence, many researchers focus on problem of uncertain SC information. Uncertainty of information can be minimized with the adoption of recent technology like BT. This improves the coordination among the various SC stakeholders to reduce the asymmetry in information sharing [8]. All the SC stakeholders have a crucial role in the adoption of BT in SC. Adoption of BT enhances the SC processes due to sharing of information transactions among the SC organizations [9]. In a global competitive era, SC practitioners have emphasis to improve the interorganizational trust and hence coordination between the SC stakeholders. With BT, SC can be transformed and make itself fit for the global competitive market.

BT has ability to provide immutable transparent, traceable and shared ledger system in the current business scenario. Due to which, may organizations want to adopt BT. Because of the extraordinary characteristics, bitcoin and many other cryptocurrencies has been widely used in many financial transactions. Hence, BT is useful in eradicating scam and provide fair transactions [10]. BT enable the organizations to perform transaction of products, data, and money without the intervention of third party that reduces the lead time and cost of the transaction and improve the trust among the stakeholders of SC [11]. Bitcoin is a cryptocurrency that does not require banking system to transfer from one party to another. It saved the time of transaction of payment drastically. Likewise, in SC system product can be transferred from one party to another. The record of transaction can be maintained using BT that is transparent to all the member of the SC. Hence, the product is trackable in the SC network using BT like Wallmart track the supply of food using BT [12].

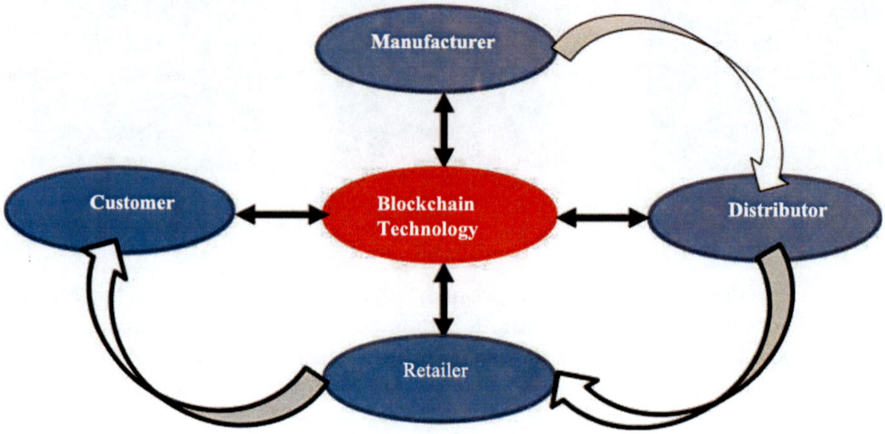

Fig. 2 Supply chain with Blockchain Technology

Also, pharmaceutical companies have keen watch on the temperature and humidity of the transportation facility of drugs [13]. BT provides the accurate data for forecasting the demand of the customers so that the supply can be maintained. It helps to reduce the inventory of the goods and minimizes the bullwhip effect. Transparency of the SC system has been improved with the accurate ledger of transactions using BT [14]. Figure 2 shows the SC with blockchain consisting of four nodes namely manufacturer, distributor, retailer, and customer. Blockchain makes available the transaction information to all the members in the SC network through distributed ledger. Exact data of transaction is visible to each node. This makes the BT based SC more trustful for the customers and assure them to provide the good quality product in time.

3 Challenges in Adoption of BT in SC

It is seen that reputed companies have the conventional SC system to supply goods and offer services. Due to the espousal of BT in the existing SC system may disrupt the existing SC activities. It is very costly affair [8]. Due to the advantageous features BT, many organizations want to implement it in their SC. Mostly, the process of adoption of BT in SC is in preliminary stage [2]. Many companies are fronting several barriers in the adoption process of blockchain in SC. Hence, investigators and practitioner have to be attentive on the hurdles in the implementation of blockchain in SC. Extant literature analyzed subsequent challenges and acknowledged them in SC adoption. All the identified challenges have been classified as administrative, technical, functioning and community type challenges as shown in Table 1.

Table 1 Challenges Recognized in the Adoption of BT in SC

Types	Challenges
Administrative	To develop trust between the various stakeholders in SC [6, 15]
	To maintain a governance policy of organization while achieving cooperation between the various partners of SC [15, 16]
Technical	To maintain transparency of data as per the requirement of business partners [6, 17]
	To avoid hacking, tampering, and fraud of records [17, 18]
Functioning	To enhance interoperability among the stakeholders [6, 19]
	To select a proper product type to recover the investment of blockchain in SC [19, 20]
Community	To provide a source of product to end-user [6, 21]
	To handle the behavior of supply chain stakeholders and train them to adopt in changing situation [16, 21]

3.1 Administrative Challenges

SC network consists of many stakeholders. Sharing of data among the various stakeholders is possible only when they have trust on each other [6]. It is necessary to make policy by the administration for sharing the data of transaction within SC network so that the trust among the stakeholders will be improved [15]. Hence, administrative challenges are the foremost challenge in the BT adoption in SC to make a policy to enhance the interorganizational trust between the members of SC.

3.2 Technical Challenges

BT has a characteristic of transparency. But, some business partners in the SC network does not want to disclose their secret information [17]. Maintaining privacy of the secret data is a vital challenge to provide a transparency in transaction while maintaining the confidentiality of its sensitive information. Subsequently, the transaction of SC with BT should be such that it should not be tampered by any of the SC member [18]. The technical team has a challenge to develop a blockchain in SC such that it should be immutable.

3.3 Functional Challenges

Various organizations in SC have different rules and regulations that prevent then to share their information with each other [19]. This resists the interoperability. Also, there is an impact of product type which affects the communication between the SC partners. It is intrinsic task before SC organizations to improve the communication between them and select proper product [20] for allowing of BT in their SC so that the financial investment will be recovered as soon as possible.

3.4 Community Challenges

Whenever there is adoption of new technology in the existing system, people of the society resist the changes. Hence, it is a critical task to train the people and provide environment such that the latest technology like BT can be adopted in the existing SC system [21]. Training to the people will improve the understanding about the advantage of BT in SC. BT helps to provide source of the product to the customer. Hence, changes in the behavior of the people working in SC helps to adopt the BT.

From the literature review, it is shown that there is ambiguity in the acceptance of blockchain in SC due to several challenges. It restricts the many companies from incorporating new technology in their SC. There are many hurdles that restrain the adoption of BT in SC such as unfaithfulness between the SC members, governing policies of individual organization, providing the transparency while keeping privacy of secrete information, development of a tamperproof system, lack of communication between the stakeholders, social barrier, and lock of information about the BT.

BT is able to revolutionize the SC by providing the visibility in transaction throughout the SC network, and maximizing the efficiency of SC by offering the accurate data of transactions. To achieve the benefits of BT in SC and for efficiently implementation of BT in SC, it is necessary to overcome the challenges in the initial stage. Then only BT will enable transparency in transactions and traceability of product in the SC.

4 Implications

Most of the businesses are at the stages of knowing the ability of the BT and its effect on their SC. Research of integration of blockchain with SC is in early-stage [21]. Consequently, this study identifies and distinguishes the various challenges of implementation of blockchain in SC that will help future practitioners.

The foremost challenge is to make adoption of BT in SC trustworthy about information, product and money flow. Use of BT in SC enhances network transparency and cut down monitoring cost of SC activities [22]. This encourages all the parties in SC to adopt blockchain. However, due to technical complexity of BT, SC parties are not having trust about BT application. Hence, it is a challenge for all the SC stakeholders to improve understanding and become confident about BT adoption in SC. Administratively; in blockchain all the activities are monitored by all the parties simultaneously to establish trust among themselves [6]. At the same time policies of particular organization are being followed [15] by developing consensus for particular transaction. So, extended traceability and visibility add great value to SC ensuring products genuineness and legality.

Next challenge is about maintaining the required level of data transparency. Every organization has its secreted information which cannot be disclosed to other organization. It is necessary to provide the transparency in the transaction data while maintaining the privacy of the secrete data using BT [23]. For this private and public blockchain can be used to connect all the concern parties in SC [24]; that allows maintaining the privacy of sensitive information and openness in some information like in food SC, where transparency brings marketing and branding. Data immutability plays significant role in preparing mindset of the organizations for use of BT in SC. BT keep the record of all the transactions with time. All these transactions are immutable and it cannot be altered after noted [25]. This feature of avoiding the fraud in the SC transactions allows blockchain adoption in SC.

To implement blockchain in SC, all the SC stakeholders should be in the panel. All the stakeholders of SC have a different rules and regulations [19]. In such situation, use of BT is a tremendous intricate task. BT plays crucial role in analysis and interpretation of the information shared by all the stakeholders of SC in shared ledger [26]. Distributed shared ledger of BT, enables SC to enhance interoperability; that explores SC business model to control the data and services visible to different SC partners. In functional type of challenges of BT in SC, along with interoperability, product type plays vital role. The function of BT in SC differs with product type. Blockchain in SC network has ability to bypass third party authentication [2]. While implementing BT in SC, it is necessary to select proper product SC so that investment of blockchain should be recovered in short period of time and earn profit. In blockchain, both interoperability and selection of proper product type can be able to provide monetary benefits to companies having complex SC.

Community type challenges cover the barriers of the social resistance and behavior of the people involved in the SC. The challenge is to provide the origin of the product to the society and train the people to adopt this novel technology. Owing to the societal changes from existing SC to the BT based SC, sustainability issues may appear. BT able to renovate the relations of customers and industries. At the same time, BT improves the trust among the stakeholders of SC that increases chances of suitability of BT [21]. As the data of the transactions using blockchain is verified and appended in the blockchain in time sequence, it cannot be tampered [6]. Accordingly, BT forms an unfailing network of SC to maintain the social and environmental sustainability [27]. To change from traditional working environment to blockchain environment, attitude of the people plays significant role. Acceptability of the new technology like BT is dependent on the knowledge and skill of the people involved in the SC. Training to the unskilled people will help to change their behavior and ease the adoptability of the BT in SC. Behavior intent is the people's feeling about the utility of BT [16]. There is need to provide training to the people involved in the adoption of the BT in SC that makes them expert and helps to make their mindset to accept the BT revolution [21]. Changes in the social and cultural environment along with behavior of the people to adopt the blockchain in SC will bring positive changes with time and experience.

It is implied from the study that challenges in the implementation of blockchain in SC can be overcome by strategic exertions of industries. The tactic should be to emphasis on to identify and analyze the influencing factors and its challenges that would be considered in developing SC with BT of particular organization.

5 Conclusion and Future Direction

Development of SC with BT is an evolving area of research [2]. Many practitioners and researchers are engaged on development of plan for the acceptance of blockchain in SC. Several studies of BT identified and discussed several challenges in the adoption of blockchain in SC. Identified foremost challenges are to build the interorganizational trust among the stakeholder of the SC, prepare and adopt the regulatory policies for the adoption of blockchain in SC, endure transparency of transaction data within SC system, offer immutability of the system to protect the data of the transaction in the SC network, improve the interoperability among the members of the SC, choose a proper product for implementation of BT in SC so that investments can be recovered as early as possible, boost the trust of the society on the brand of product by providing the provenance of the product, and train the workforce in the SC system and make them capable to tackle the blockchain application in the SC.

This study delivers a holistic approach for forthcoming study in the area of adoption of BT in SC and limited to the identification of the challenges. In future, the identified factors and its challenges can be prioritized using multi criteria decision making methodology. This will be useful to channelize the implementation of BT in SC. For researchers, this work provides valuable identifications of challenges for the development SC with BT.

References

1. Angelis J, Ribeiro da Silva E (2018) Blockchain adoption: a value driver perspective. Bus Horizons 62(3):307–314
2. Gupta M (2017) Blockchain for dummies. IBM Limited Edition. Wiley, NJ, USA
3. Mondragon AEC, Mondragon CEC, Coronado ES (2018) Exploring the applicability of blockchain technology to enhance manufacturing supply chains in the composite materials industry. In: Proceedings of 4th IEEE international conference on applied system innovation 2018, pp 1300–1303
4. Treiblmaier H (2018) The impact of the blockchain on the supply chain: a theory-based research framework and a call for action. Supply Chain Manage Int J 23(6):545–559
5. Risius M, Spohrer K (2017) A blockchain research framework: what we (don't) know, where we go from here, and how we will get there. Bus Inf Syst Eng 59(6):385–409
6. Wang Y, Hungh HJ, Paul B-D (2019) Understanding blockchain technology for future supply chains a systematic literature review and research agenda. Supply Chain Manage Int J 24(1):62–84
7. Zhao Y, Choi TM, Cheng TCE, Wang S (2018) Supply option contracts with spot market and demand information updating. European J Operat Res 266(3):1062–1071

8. Shen B, Choi TM, Minner S (2018) A review on supply chain contracting with information considerations: information updating and information asymmetry. Int J Prod Res 7543(May):1–39
9. Simchi-Levi D, Simchi-Levi E, Kaminsky P (1999) Designing and managing the supply chain: concepts, strategies, and cases. McGraw-Hill, New York
10. Iansiti M, Lakhani KR (2017) The Truth About Blockchain. Harv Bus Rev 95(1):118–127
11. Yeoh P (2017) Regulatory issues in blockchain technology. J Financ Regulat Compl 25(3):196–208
12. O'Byrne R (2019) Blockchain technology is set to transform the supply chain how blockchain can transform the supply chain. https://www.Logisticsbureau.Com/How-Blockchain-Can-Transform-the-Supply-Chain/. Last Accessed 01 July 2019
13. Lu Q, Xu X (2017) Adaptable Blockchain-Based systems: a case study for product traceability. IEEE Softw 34(6):21–27
14. Ivanov D, Dolgui A, Sokolov B (2019) The impact of digital technology and industry 4.0 on the ripple effect and supply chain risk analytics. Int J Product Res 57(3):829–846
15. Singh A, Teng JTC (2016) Enhancing supply chain outcomes through information technology and trust. Comput Hum Behav 54(2016):290–300
16. Kamble S, Gunasekaran A, Arha H (2019) Understanding the Blockchain technology adoption in supply chains-Indian context. Int J Prod Res 57(7):2009–2033
17. Lo SK, Xu X, Chiam YK, Lu Q (2018) Evaluating Suitability of Applying Blockchain. In: Proceedings of the IEEE international conference on engineering of complex computer systems, ICECCS, 2017, 158–161
18. Apte S, Petrovsky N (2016) Will blockchain technology revolutionize excipient supply chain management? J Excipi Food Chem 7(3):76–78
19. Hul R (2017) Blockchain: distributed event-based processing in a data-centric world. In: Proceedings of the 11th ACM international conference on distributed and event-based systems—DEBS '17, 2–4
20. Tian F (2016) An agri-food supply chain traceability system for China based on RFID & blockchain technology. In: 13th international conference on service systems and service management, ICSSSM 2016
21. Queiroz MM, FossoWamba S (2019) Blockchain adoption challenges in supply chain: an empirical investigation of the main drivers in India and the USA. Int J Inf Manage 46:70–82
22. Wu H, Li Z, King B, Miled ZB, Wassick J, Tazelaar J (2017) A distributed ledger for supply chain physical distribution visibility. Information 8(4):1–18
23. Boucher P, Nascimento S, Kritikos M (2019) How blockchain technology could change our lives—European parliamentary research service. Homepage, at: http://www.europarl.europa.eu/RegData/etudes/IDAN/2017/581948/EPRS_IDA%20(2017)581948_EN.pdf. Last Accessed 19 July 2019
24. Casino F, Dasaklis TK, Patsakis C (2019) A systematic literature review of blockchain-based applications: current status, classification and open issues. Telematics Inform 36(2019):55–81
25. Lee JH, Pilkington M (2017) How the blockchain revolution will reshape the consumer electronics industry [Future directions]. IEEE Consumer Electron Magz 6(3):19–23
26. Dobrovnik M, Herold D, Fürst E, Kummer S (2018) Blockchain for and in logistics: what to adopt and where to start. Logistics 2(18):1–14
27. Kshetri N (2018) 1 Blockchain' s roles in meeting key supply chain management objectives. Int J Inf Manage 39:80–89

A Bibliometric Analysis of Sustainable Supply Chain Management: Research Implications and Future Perspectives

Akshay Patidar, Monica Sharma, Rajeev Agrawal, Kuldip Singh Sangwan, and Anbesh Jamwal

Abstract Sustainable supply chain (SSC) is an emerging research area that focuses on the triple bottom line pertaining to all stakeholders and related activities. Important Business decisions relevant processes, and activities are focused on features in SSC articles. Using the bibliometric techniques, the author attempted to analyze the research area's impact, its associated eminent authors, along with their affiliated institutions and countries. Through conducting network analysis in VOSviewer software and Gephi software researchers focus on co-authorship, author specified keywords clustering, and countries-based bibliographic analysis. The research identifies the most influential research work/authors in the defined duration. Using network analysis, authors can identify knowledge groups their affiliations, and future research opportunities. In contrast to the existing literature, the author here used keywords occurrence as a criterion for clustering. The identified clusters define the research themes and keyword occurrence helps in identifying future research implications.

Keywords Sustainable supply chain (SSC) · Bibliometrics · Web of science (WoS) · VOSviewer · Gephi · Network analysis

1 Introduction

During the period of late 1980s and early 1990s concept of Supply chain management, as a strategic competence is receiving increased attention. [1, 2] have described SCM as, "the systemic, strategic coordination of the traditional business functions and the tactics across these business functions within a particular company and across businesses within the supply chain, to improve the long-term performance of the individual companies and the supply chain as a whole."

A. Patidar (✉) · M. Sharma
Department of Management Studies, Malaviya National Institute of Technology, Jaipur, India

M. Sharma · R. Agrawal · A. Jamwal
Department of Mechanical Engineering, Malaviya National Institute of Technology, Jaipur, India

K. S. Sangwan
Department of Mechanical Engineering, Birla Institute of Technology and Science, Pilani, India

R. Agrawal et al. (eds.), *Recent Advances in Industrial Production*, Lecture Notes in Mechanical Engineering, https://doi.org/10.1007/978-981-16-5281-3_16

According to Brundtland Report Sustainable development can be defined as "meeting the needs of the present generation without compromising the ability of future generations to meet their needs" [3]. This led to an approach where the focus should be on environmental economical and social-oriented development. Some of the leading Researchers, Academicians also called it a TBL approach [4]. True sustainability occurs with the nexus of economical, environmental, and social practices. Seuring and Müller [5] defined SSCM as "The management of material, information, and capital flow as well as cooperation among companies along the supply chain while taking goals from all three dimensions of sustainable development, i.e., economic, environmental and social into account which are derived from customer and stakeholders' requirements. "The concept of SSC revolves around the integration of the TBL practices like supplier selection, purchasing, manufacturing, distribution along with retailing [6]. Need of the hour is focusing and promoting sustainable habits more and more. To achieve these objective researches are required and for better researches getting insight into the previous research is required. A sustainable supply chain as a research domain is gaining attention these days. Considering the last 10 years number of research articles has grown significantly and growth is still in positive digits. To sustain this growth and to maintain the pace, researchers are required to know the insights of the research progress. Bibliometric analysis is proved to be a good tool to get insights into research progression [7]. The bibliometric analysis not only provides insights, but also lets researchers know the direction, and intensity of research along with the influence of one research on the other [7].

This article is an author's effort to know the development of the sustainable supply chain till 2020 beginning from 2016 and an effort to know the major contributor in this area along with the institution and country contributing maximum. The research work is divided into four sections. Section 1 deals with the introductory part followed by Sect. 2 dealing with methods and data collection. Later on, Sect. 3 describes the implications from the bibliometric analysis i.e. research trend, citations analysis, keyword analysis, clusters of keywords, co-authorship based on their affiliations, and occurrences of author-specified keywords. Section 4 describes the conclusion and future implications and Sect. 5 enlists the references.

2 Methods and Data

Bibliometric researches are required for getting deeper insights and knowing the influence of literature over the other. Generally, bibliographic analysis is done where a large volume of data is there [7]. Authors like [8, 9] have used this technique for getting insight and impact of the journals for a defined period. [10] explained scientific work as intellectual convergence based on their common sources and co-referencing. [11] explained cited document having concept and intellect similarities, by identifying similarity among references used as citations in another document.

Co-authorship and concurrence are also frequently discussed concepts in the bibliographic analysis. Co-authorship shows the connectivity among the collaborating authors [12], whereas co-occurrence shows the structure of the literature[13].

Going through the concepts and applications of bibliometric analysis, through this study researcher here wants to present a bibliometric overview of Sustainable Supply Chain (SSC) researches employing descriptive and network analysis. Descriptive analysis in this study considers all researches, all of cited researches, total citation, citation per research, citation per cited research, and citation per year. Bibliometric analyses include analysis of citations, occurrences, and co-authorships. Researchers here used VOSviewer and Gephi software to conduct mapping analysis in this study. To develop the figure and to show the nodes and the connecting links between nodes, VOSviewer uses standards such as the number and total occurrence in the links [14]. Similarly, Gephi uses some measures like modularity rank and page rank to visualize the bibliographic network [15].

For bibliographic analysis, researchers here accessed the literature database from Web of Science (WoS) one of the largest databases in social science having multidisciplinary peer-reviewed literature. WoS is a frequently and widely accessed database for such kind of purpose. Using "Sustainable Supply Chain" as keyword and time frame of 2016 to 2020, WoS revealed a total of 548 published articles including journal articles, editorials, notes, conference papers. The time frame of 2016–2020 is chosen as very few researches matching the insight of this research were found and many pieces of research evidence were conducted using bibliometric analysis for getting insight into the sustainable supply chain till 2016. After applying exclusive criteria as a language other than English, only journal articles, researchers got only 493 articles. For bibliometric analysis, all 492 articles were shortlisted and termed as researches.

3 Results

3.1 Citation Structure and Research Trends

A total of 24 articles were published in the initial year of analysis i.e. 2016. Gradually with the advent of the years' articles increased i.e. 102 in 2017,189 in 2018, 171 in 2019, and 6 in 2020. 2020 has the least number of researches because of its dawning phase. The sustainable supply chain saw maximum researches in 2018. Table 1 summarizes SSC research work along with citation patterns, and Fig. 1 shows the research trend in the research area and citations for cited research in the domain of SSC.

Table 1 shows that among the researches some research (189) and terms of referenced researches (169) 2018 witness maximum productivity in the domain of SSC. In total 3930 citations were seen of the articles for SSC and maximum citations were recorded in 2017 with 1790 citations. In terms of influencing citations [16, 17], 2017

Table 1 Citation structure of researches

						Researches with citations > =				
Year	TR	NCR	TC	C/R	C/CR	50	40	20	10	1
2016	24	24	541	22.54	22.54	2	4	11	19	24
2017	102	100	1790	17.55	17.90	3	14	28	54	100
2018	189	169	1599	8.46	9.46	0	2	19	62	169
2019	171	0	0	0.00	0.00	0	0	0	0	0
2020	6	0	0	0.00	0.00	0	0	0	0	0

Notes TR total researches; *NCR* referenced researches; *TC* All citations; *C/R* average citations for research; *C/CR* average citations for cited research

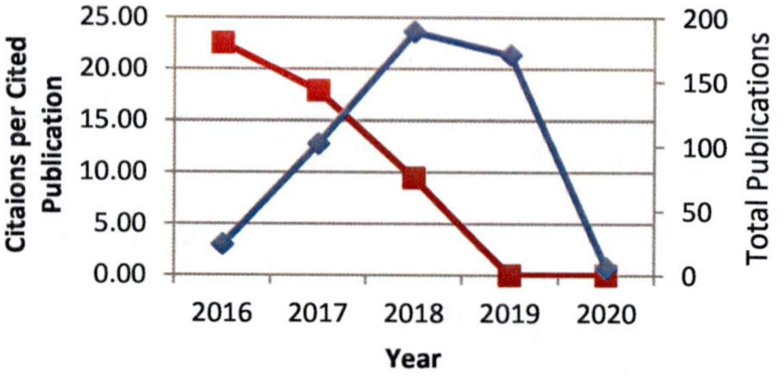

Fig. 1 Citation and Research trend. *Notes* Here blue line = *TR* Total Research; Red line = *C/CR* average citation per cited research

was the leading year. However focusing on citations per cited research in place of citations, 2016 was the most significant year with 22.54 (46.4%) citations followed by 2017 with 17.55(36.11%) citations. Genovese et al. [18] published article titled "Sustainable supply chain management and the transition towards a circular economy: Evidence and some applications" was the most influential article in this tenure with 162 total citations.

Figure 1 depicts the escalation pattern in the domain of SSC and citations per cited research spread during the defined time frame. Though a declining pattern in the citations for cited research is seen with afterward years, a drop can be obvious and not surprising as the afterward articles are newer. Figure 2 shows the collective efforts of the authors, several authors and relative research is plotted and elucidates that majority of research is done by the collaboration of three authors followed by 2 and 4 authors only one research article is found which is the result of the 11 authors

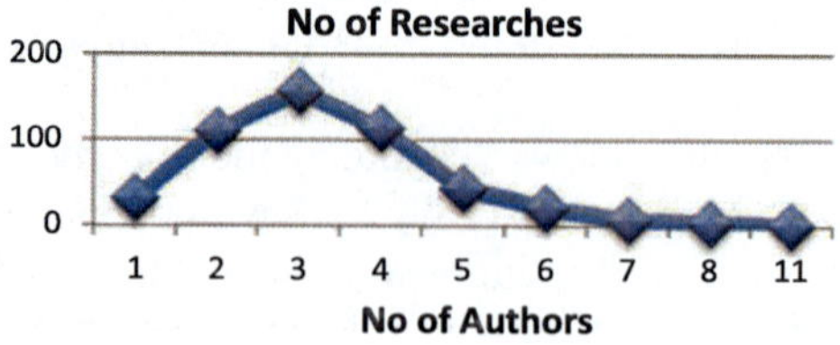

Fig. 2 Distribution of Research based on contributing authors

collaborative efforts. Thus in general, SSC has seen a robust and positive escalation pattern during the said tenure and SSC is witnessing a growth in 2020 dawn, which shows positivity in the area.

3.2 Most Cited Researches

Tsay [17] discussed the influential power of research by saying through citations reflect influence. Table 2 in Appendix enlists the influential researches in the domain of SSC between 2016 and 2020. All the top-cited researches are the flag holders of academic distinction in the discipline of SSC. For example, [18] discussed the transition of SSC towards a circular economy by providing evidence and applications. Also, this article leads the list of top researches in the domain with 162 total citations and with a leading average of 54 citations per year. Similarly, the work of [19] discussed the SSC framework and further research motivations, which proved to be a torchbearer to the future researcher and appears second in the list with a total citation of 97 and an average of 32.33 citations per year. Moving further in the list [20], appears on the third number with a total of 86 citations and focused on the improvement of SSCM using a Multi-criteria decision-making approach with an average of 21.50 citations per year. Further studies like [21] focused on the performance of SSCM and discussed a trade-off between environmental and cost performance with a sum of 57 citations and an averaging 14.25 citations per year. Discussing average cites per year Table represents [22–25] leads the position with 16.67, 16.33, 11.75, and 23.00 respectively.

The current study elaborates that, at least 40 citations were received by the top articles. The researches in Table 2 addresses various key issues about performance evaluation, supplier selection, carbon and water footprints, pricing economic and environmental performance, social performance in manufacturing and service originations thereby focusing on sustainable service supply chains and various industries like construction, agriculture, and food and with various methodologies like MCDM, game theory, optimization, integer linked programming. Articles also focused on various emerging technologies like blockchain technology, Big Data, IoT, Artificial Intelligence, etc. Surprisingly 33% of the recent paper or the researches during the 2020 dawn focused on blockchain technology which researchers frequently called the most disruptive technology of supply chain management. In this article's subsequent discussion, researchers recognized the top SSC area authors and their associated affiliations.

3.3 Leading Researchers and Their Affiliations

Table 3 focuses on the Authors on account of their researches, cited researches, their citations, citations for research, and citations for cited research. The table also

Table 2 Maximum cited articles

Most cited researches between 2016 and 2020					
TC	Title	Summary	Authors	Year	CPY
162	Sustainable supply chain management and the transition towards a circular economy: Evidence and some applications	Reuse of product to enhance the life cycle of the product by creating a circular economy to enhance the sustainability of the Supply Chain is discussed along with the evidences and applications from Industries	Genovese, A; Acquaye, AA; Figueroa, A; Koh, SCL	2017	54.00
97	Sustainable supply chain management: framework and further research directions	The author conducted a literature review to identify the future research directions and also suggested a detailed framework for creating further research opportunities	Dubey, R; Gunasekaran, A; Papadopoulos, T; Childe, SJ; Shibin, KT; Wamba, SF	2017	32.33
86	Improving sustainable supply chain management using a novel hierarchical grey-DEMATEL approach	The author using MCDM and grey theory integration proposed a hierarchical structure of aspects and criteria for supplier identification	Su, CM; Horng, DJ; Tseng, ML; Chiu, ASF; Wu, KJ; Chen, HP	2016	21.50
57	Sustainable supply chain management in emerging economies: Trade-offs between environmental and cost performance	Under the purview of Resource Dependence Theory, the Author conducted an empirical analysis and examined the impact on environmental and economical performance	Esfahbodi, A; Zhang, YF; Watson, G	2016	14.25

(continued)

Table 2 (continued)

Most cited researches between 2016 and 2020					
TC	Title	Summary	Authors	Year	CPY
50	Comparing linear and circular supply chains: A case study from the construction industry	Considering Industrial ecology and production economics, the author compared linear and circular supply chains and identified environmental gains using a case study in the construction Industry	Nasir, MHA; Genovese, A; Acquaye, AA; Koh, SCL; Yamoah, F	2017	16.67
49	Sustainable supply chain management with pricing, greening and governmental tariffs determining strategies: A game-theoretic approach	Using a game-theoretic approach a mathematical model is developed using Government as leader and two green and non-green supply chains as followers to identify pricing policies, greening strategies, and Govt tariffs	Madani, SR; Rasti-Barzoki, M	2017	16.33
47	A two-echelon sustainable supply chain coordination under cap-and-trade regulation	Considering Carbon emissions, the impact of unit emission trading price on the optimum decision variables is identified	Xu, JT; Chen, YY; Bai, QG	2016	11.75
46	Role of multiple stakeholders and the critical success factor theory for the sustainable supplier selection process	Considering the Critical Success Factor theory, a decision support system for sustainable supplier selection problems in the textile industry is resolved	Kannan, D	2018	23.00

(continued)

Table 2 (continued)

Most cited researches between 2016 and 2020

TC	Title	Summary	Authors	Year	CPY
46	Measuring the environmental sustainability performance of global supply chains: A multi-regional input–output analysis for carbon, sulphur oxide and water footprints	Measuring of environmental performance of global value chains using the MRIO modeling framework is proposed and a 15 years time-series study is highlighted	Acquaye, A; Feng, KS; Oppon, E; Salhi, S; Ibn-Mohammed, T; Genovese, A; Hubacek, K	2017	15.33
44	Institutional pressures, sustainable supply chain management, and circular economy capability: Empirical evidence from Chinese eco-industrial park firms	Using Institutional theory, authors created a conceptual model and conducted an empirical study to test the mechanism of circular economy, institutional pressure, sustainable supply chain design, and management	Zeng, HX; Chen, XH; Xiao, X; Zhou, ZF	2017	14.67
44	A decision support model for sustainable supplier selection in sustainable supply chain management	The study is conducted to know the sustainable supplier selection in an Iranian textile Industry and developed a hybrid model that can deal with inconsistency, uncertainty, and calculation complexity	Fallahpour, A; Olugu, EU; Musa, SN; Wong, KY; Noori, S	2017	14.67

(continued)

Table 2 (continued)

Most cited researches between 2016 and 2020					
TC	Title	Summary	Authors	Year	CPY
44	Key themes and research opportunities in sustainable supply chain management—identification and evaluation	The authors conducted a literature review followed by a Delphi method to gains insight into the themes in SSC research. Identified themes are integral to the management and performance of SSC	Reefke, H; Sundaram, D	2017	14.67
44	Manufacturer and retailer coordination for environmental and economic competitiveness: A power perspective	The study focuses on power relationships and coordination in SSCM. The author used a carbon emission-sensitive demand and using a game-theoretic approach a two-part tariff contract is designed	Chen, X; Wang, XJ; Chan, HK	2017	14.67
42	Fuzzy multi-objective sustainable and green closed-loop supply chain network design	The author addressed the design problem of a Closed-loop supply chain under the environmental considerations using a genetic algorithm and determined the optimal product flow among them	Soleimani, H; Govindan, K; Saghafi, H; Jafari, H	2017	14.00
42	Evaluation of the external forces affecting the sustainability of oil and gas supply chain using Best Worst Method	The author conducted a research survey in SSCM and O&G fields from American and European universities to obtain individual measures of importance and to identify collective importance Best Worst Method is used	Ahmad, WNKW; Rezaei, J; Sadaghiani, S; Tavasszy, LA	2017	14.00

(continued)

Table 2 (continued)

Most cited researches between 2016 and 2020					
TC	Title	Summary	Authors	Year	CPY
41	Lost in Transition? Drivers and Barriers in the Eco-innovation Road to the Circular Economy	The study analyzes evidence for different factors specially eco-innovation helping and hampering the development of circular economy and can contribute to the design of policy guidelines and organizational strategies	de Jesus, A; Mendonca, S	2018	20.50
41	Implementing sustainability in multi-tier supply chains: Strategies and contingencies in managing sub-suppliers	The author identified three main factors supply chain complexity, sustainability management capability, and type of sustainability to identify how firms extend their sustainability strategies to their sub-suppliers	Wilhelm, M; Blome, C; Wieck, E; Xiao, CY	2016	10.25
40	Revenue and promotional cost-sharing contract versus two-part tariff contract in coordinating sustainable supply chain systems with deteriorating items	The author proposes a revenue and promotional cost-sharing contract and a two-part tariff contract to coordinate this system and results show the both contract can lead to perfect coordination	Bai, QG; Chen, MY; Xu, L	2017	13.33
40	A strategic approach to social sustainability—Part 2: a principle-based definition	Overall two papers aim to define social sustainability and guide decision making and monitoring and propose better integration of social sustainability in many other fields	Missimer, M; Robert, KH; Broman, G	2017	13.33

(continued)

Table 2 (continued)

Most cited researches between 2016 and 2020					
TC	Title	Summary	Authors	Year	CPY
40	Sustainable supply chain network design: A case of the wine industry in Australia	The author proposes a generic model for sustainable wine supply chain network design encompassing economic, environmental, and social objectives. Non-dominated solutions are obtained and some balance scenarios are proposed	Varsei, M; Polyakovskiy, S	2017	13.33

Notes Here *TC* Total Citation; *CPY* Citation per year

Table 3 Institutions and countries associated with leading authors

								Research Citations > =				
Researcher	Affiliation	Country	TR	NCR	TC	C/R	C/CR	40	30	20	10	1
Saen, RF	Sohar University	Oman	15	13	107	7.13	8.23	0	0	0	5	13
Sarkis, J	Worcester Polytechnic Institute	USA	12	11	90	7.50	8.18	0	1	2	3	11
Jabbour, ABLD	University of Lincoln	UK	9	9	63	7.00	7.00	0	0	0	2	9
Jabbour, CJC	University of Lincoln	UK	9	9	43	4.78	4.78	0	0	0	1	9
Rasti-Barzoki, M	Isfahan University of Technology	Iran	9	7	136	15.11	19.43	1	2	0	6	7
Tseng, ML	Asia University	Taiwan	9	6	150	16.67	25.00	1	0	3	4	6
Govindan, K	University of Southern Denmark	Denmark	7	7	138	19.71	19.71	1	3	0	4	7
Seuring, S	University of Kassel	Germany	7	6	27	3.86	4.50	0	0	0	1	6
Caniato, F	Polytechnic University of Milan	Italy	6	4	46	7.67	11.50	0	0	0	2	4
Gunasekaran, A	California State University	USA	6	6	156	26.00	26.00	1	2	0	3	6
Izadikhah, M	Islamic Azad University	Iran	6	5	45	7.50	9.00	0	0	0	2	5
Jia, F	University of York Management School	UK	6	5	60	10.00	12.00	0	0	1	3	5
Lim, MK	Seoul National University	South Korea	6	5	58	9.67	11.60	0	0	1	3	5
Wu, KJ	Dalian University of Technology	China	6	3	97	16.17	32.33	1	0	0	2	3

(continued)

Table 3 (continued)

Researcher	Affiliation	Country	TR	NCR	TC	C/R	C/CR	Research Citations > =				
								40	30	20	10	1
Bai, CG	University of Electronic Science and Technology of China	China	5	4	11	2.20	2.75	0	0	0	0	4
Hong, ZF	Lanzhou University	China	5	5	57	11.40	11.40	0	0	1	2	5
Mangla, SK	University of Plymouth	UK	5	5	54	10.80	10.80	0	0	1	2	5
Sarkar, B	Yonsei University	South Korea	5	4	83	16.60	20.75	0	0	3	4	4

Notes TR total researches; *NCR* cited researches; *TC* all citations; *C/R* average citations for researches; *C/CR* average citations for cited researches differentiated researches under five citation categories viz. 40,30,20,10, and 1 cite, respectively.

In terms of articles contributed, Saen, RF affiliated with Sohar University, Oman are on the first rank, contributing 15 articles in the domain of SSC between 2016 to the beginning of 2020. The author has 13 cited articles and a total citation of 107 with 7.13 as citations per research and 8.23 as citations per cited research. The author also has 5 articles that are cited more than 10 times.

Discussing total citations, Gunasekaran, affiliated with California State University, leads the position with a total of 156 citations. The author published 6 articles during this tenure and also tops the list in citations per paper and citations per cited paper list with a score of 26. Tseng, ML associated with Asia University holds the second position with 150 total citations followed by Govindan, K. affiliated with the University of Southern Denmark with total citations of 138.

3.4 *Keywords Analysis*

According to Jones and Jackson, "Keywords are a list of words or phrases that are provided by the author and signify the meaning or main ideas presented in the paper" [26]. The author specified keywords that appeared were cleaned manually and frequency distribution was done initially to identify total keywords appearing uniquely and the number of their occurrence. Further analysis is summarized in the subsequent sections of the article.

Overview of Clusters

Table 4 shows six clusters of keywords formulated based on the clustering of the papers with the specified range of citations. Articles that were cited 5 or more than 5 times were shortlisted for the analysis. A total of six clusters were formed and named clusters 1, 2, 3, 4, 5, and 6 respectively. Cluster 1 contains keywords used in the articles that were cited between 5 and 9 times. Cluster 1 contains a total of 379 keywords in 89 articles. The average citation per keyword in cluster r 1 is 1.60

Table 4 Overview of clusters

Clusters	RC	TR	TK	TC	C/K	C/R
Cluster 1	5–9	89	379	607	1.60	6.82
Cluster 2	10–19	81	342	1161	3.39	14.33
Cluster 3	20–29	30	126	720	5.71	24.00
Cluster 4	30–39	11	57	360	6.32	32.73
Cluster 5	40–49	15	70	650	9.29	43.33
Cluster 6	50 and above	5	22	452	20.55	90.40

Note Here, *RC* Range of Citation; *TR* All researches; *TK* All keywords; *TC* All Citation; *C/K* citation for keyword; *C/R* citation for research

Fig. 3 Word cloud of six different clusters

per keyword and 6.83 per article. Similarly, cluster 2 contains 81 papers with 342 keywords and total citations equal to 1161. The average citations per keyword are 3.39 and per article is 14.33. Cluster 6 is having the least number of articles i.e. 5 with total keywords 453 and with an average citation per keyword equals 20.55 and 90.40 as average citation per article.

Clustering of Author Specific Keywords Based on Citations
Further for more clearance of clusters, six-word clouds were formed using their weights as a parameter to get a clear insight into the distribution of keywords along with their impacts. Figure 3 shows the six-word clouds formed by six clusters of keywords. Keywords mostly used like SSCM and sustainability forms the epicenter of word cloud and other keywords like performance, environment, development, supplier, selection shows the relation with epicenter and the font characteristics show the impact of the keyword.

3.5 Mapping with VOSviewer and Gephi

To analyze bibliographic data, VOSviewer and Gephi software for creating figures were used. [14] explains the importance of co-citation, multiple sources are cited

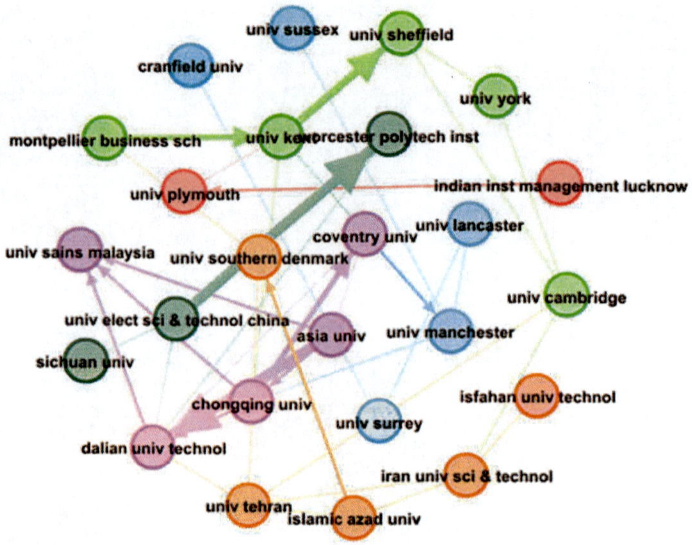

Fig. 4 Showing the co-authorship of author-affiliated institutions

in another source. Likewise, co-authorship shows the scholarly likeness amongst researchers with separate affiliations.

Co-authorship Structure

Figure 4 shows the network for co-authorship relation among SSC researchers and their affiliations publishing a minimum of five co-authored articles between 2016 and 2020. Institutions researching SSC can be divided into six clusters as pretty evident in the network diagram. Also, the network elucidates that institution like Indian Institute of Management Lucknow is collaborating with the University of Plymouth. Similarly, Dalian University, Chongqing University, Asia University, Coventry University, and the University of Sains Malaysia are collaborating for their researches. Arrow strength in the network represents the strength of co-authorship. Like the University of electronics science and technology is strongly collaborating with Worcester Polytechnic Institute. Similarly, Montpelier Business School is collaborating with the University of Sheffield indirectly. This co-authorship network can be more elaborated with author-affiliated countries as shown in Fig. 5.

Co-authorship of Author Affiliated Countries

Figure 5 displays the network for co-authorship based on author affiliations. From the figure, it is evident that People R China, England, USA, and France are forming a quadrilateral of research. Among the four of these quadrilaterals, England is collaborating with People R China, and the USA. Also, People R China is collaborating with the USA. One thing which is evident from this network is India is also coming closer to the research quadrilateral. England authors are co-authoring with Indian

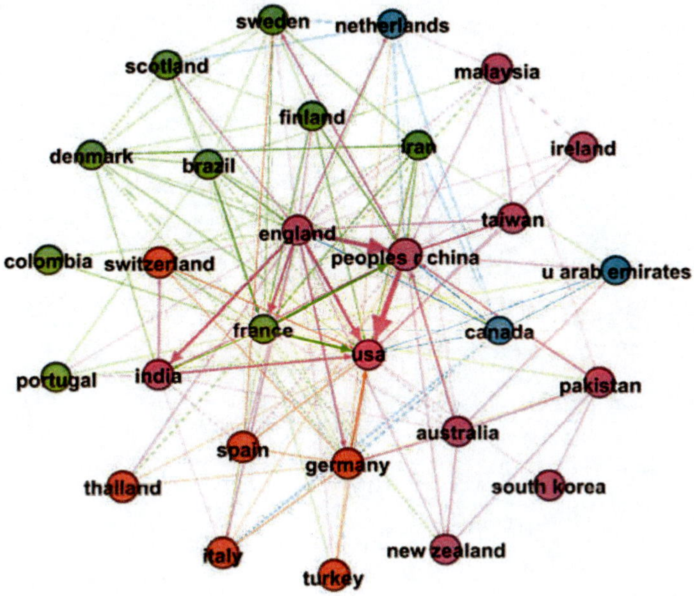

Fig. 5 Co-authorship network

authors. According to the network, countries can also be divided into six clusters as evident from the color of the nodes (countries) of the network.

Co-occurrence of Author Specified Keywords

Figure 6 displays the co-occurrence of author-specified keywords. It presents the most discussed themes during the tenure of analysis. Distance between the nodes represents the gaps between the researches. Sustainable supply chain and sustainability are close and the author frequently uses these keywords. However, keywords like food supply chain, uncertainty, multi-objective optimization are the themes where work is very less seen in the proximity of sustainable supply chain. From the network gaps for the research can also be predicted and themes can be used for future research. Like sustainable supply chain and food supply chain is a less researched area. Similarly talking about SC performance, it needs to be focused more on different industries. Also, very few kinds of research are seen utilizing integrated MCDM techniques in a fuzzy environment.

Environmental performance is the discussed theme but social performance lacks the research. Emerging technologies are also being used for creating a sustainable supply chain. But more focus is required for the said theme. Similarly, very few researches are seen for themes like optimizations, Machine learning, Blockchain, IoT devices. Therefore, author specified keyword networks are fruitful to analyze the research progression in a specified area.

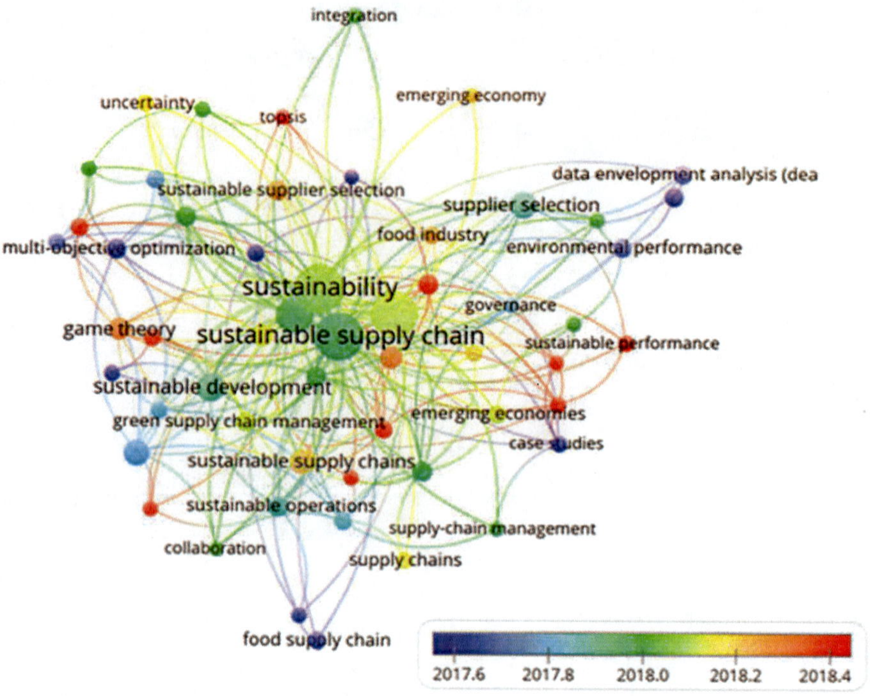

Fig. 6 Keyword Co-occurrence Network

4 Conclusions and Future Implications

The sustainable supply chain is the preferred research area for most of the authors. Research articles have grown at an increased pace with 24 articles in 2016 and 6 articles at the beginning of 2020. That clearly explains that 2020 will prove to be a promising year. Academic excellence will be seen by the completion of 2020. Research gaps among the diversified themes of Sustainable supply chain are expected to be filled in 2020 leading to a great collaborative outcome. The benefits of collaborative efforts can be easily seen in citation analysis and co-authorship analysis done in this research. Collaborative efforts are proved to be a good concept for ages. The academic influence of the researchers can be seen by the citations which are also increasing over time. In terms of productivity 2017 was proved to be the most productive year but it is anticipated that by the end of 2020 productivity will also rise.

An increased number of citations shows that Gunasekaran. A Govindan. K, Wu. KJ, Sarar. B, Tseng. ML has proven to be an eminent researcher in the domain. Prominent countries are People R China, England, USA and University of Plymouth, Worcester polytechnic institute, University of electronics science and technological China are appeared to be prominent research institutes. Networks for most eminent authors, institutions affiliated with them, and countries graphically show scholarly associations. Similarly, the co-occurrence of author-specified keywords for SSC shows the themes immediacy. Many keywords are very close to the sustainable supply chain and showed their association strength by their frequent appearance in the articles and many themes are there which lacks the association strength and therefore represents a research gap that needs to be filled by the future researches and is a boon to the researchers and scholars.

In conclusion, it can be said that researches for food supply chain, food industries, agriculture supply chains, Digitizing supply chains, blockchain technology, Data analytics, closed-loop supply chains, circular supply chains, social performance are the areas where if focused can prove to be fortified articles in the area. Creating a Circular economy to enhance the life cycle of the product along with enhancing the sustainable performance of SC proves to be a hotspot for future research. Targeting the proposed researches, an author can contribute to not only academia but can also contribute to the Industries where firms can achieve not only sustainability but can also work in the direction of cleaner production thereby acting as a bridge in between existing researches and cleaner production to achieve sustainable development. Discussing the Network design problem in the SSC domain is also proving to be an apple of the eye to researchers as most of the researchers have focused on the forward supply chain while conducting researches. Closed-loop supply chain is less evident in network design problem and reverse supply chain is the rarest of the rare research thereby reverse supply chain considering strategic, tactic and operational decisions proves to be a prominent future research direction. Focusing on uncertainty and resiliency, very few researches were evident as research domain in the above said time frame. This leads to the enhanced requirement of creating resilient supply chain catering uncertainties. The author hereby means of this bibliometric analysis proposed some future direction for research and propose their next agenda of research as to create a resilient sustainable supply chain network design catering uncertainty for solving network designing problem using MCDM and multi-objective programming approach. This study provides useful insights from the area of SSC. Since this study is confined to the data obtained from Web of Science (WoS) and for said time frame only, including other data sources and for different time frames may vary the results and can be a merit for another study.

Appendix

See Table 2.

References

1. Fahimnia B, Sarkis J, Davarzani H (2015). Green supply chain management: a review and bibliometric analysis. Int J Prod Econ 162:101–114. Author F, Author S (2016) Title of a proceedings paper. In: Editor F, Editor S (eds) Conference 2016, LNCS, vol. 9999. Springer, Heidelberg, pp. 1–13
2. Mentzer JT, DeWitt W, Keebler JS, Min S, Nix NW, Smith CD, Zacharia ZG (2002) Defining supply chain management. J Bus Logist 22(2):1–25
3. Schubert A, Lang I (2005) The literature aftermath of the Brundtland report our common future. A Scientometric study based on citations in science and social science journals. Environ Develop Sustain 7(1):1–8. LNCS Homepage, http://www.springer.com/lncs. Last Accessed 21 Nov 2016
4. Slaper TF, Hall TJ (2011) The triple bottom line: what is it and how does it work. Indiana Bus Rev 86(1):4–8
5. Seuring S, Müller M (2008) From a literature review to a conceptual framework for sustainable supply chain management. J Clean Prod 16(15):1699–1710
6. Delai I, Takahashi S (2013) Corporate sustainability in emerging markets: insights from the practices reported by the Brazilian retailers. J Clean Prod 47:211–221
7. Broadus R (1987) Toward a definition of bibliometrics. Scientometrics 12(5–6):373–379
8. Martínez-López FJ, Merigó JM, Valenzuela-Fernández L, Nicolás C (2018) Fifty years of the European journal of marketing: a bibliometric analysis. European J Market
9. Valenzuela LM, Merigó JM, Johnston WJ, Nicolas C, Jaramillo JF (2017) Thirty years of the journal of business & industrial marketing: a bibliometric analysis. J Bus Indust Market
10. Kessler MM (1963) Bibliographic coupling between scientific papers. Am Doc 14(1):10–25
11. Small H (1973) Co-citation in the scientific literature: a new measure of the relationship between two documents. J American Soc Inf Sci 24(4):265–269
12. Bales ME, Dine DC, Merrill JA, Johnson SB, Bakken S, Weng C (2014) Associating co-authorship patterns with publications in high-impact journals. J Biomed Inform 52:311–318
13. Lozano S, Calzada-Infante L, Adenso-Díaz B, García S (2019) Complex network analysis of keywords co-occurrence in the recent efficiency analysis literature. Scientometrics 120(2):609–629
14. Donthu N, Kumar S, Pattnaik D (2020) Forty-five years of journal of business research: a bibliometric analysis. J Bus Res 109:1–14
15. Cherven K (2015) Mastering Gephi network visualization. Packt Publishing Ltd
16. Ding Y, Cronin B (2011) Popular and/or prestigious? Measures of scholarly esteem. Inf Process Manage 47(1):80–96
17. Tsay MY (2009) Citation analysis of Ted Nelson's works and his influence on hypertext concept. Scientometrics 79(3):451–472
18. Genovese A, Acquaye AA, Figueroa A, Koh SL (2017) Sustainable supply chain management and the transition towards a circular economy: evidence and some applications. Omega 66:344–357
19. Dubey R, Gunasekaran A, Papadopoulos T, Childe SJ, Shibin KT, Wamba SF (2017) Sustainable supply chain management: framework and further research directions. J Clean Prod 142:1119–1130
20. Su CM, Horng DJ, Tseng ML, Chiu AS, Wu KJ, Chen HP (2016) Improving sustainable supply chain management using a novel hierarchical grey-DEMATEL approach. J Clean Prod 134:469–481
21. Esfahbodi A, Zhang Y, Watson G (2016) Sustainable supply chain management in emerging economies: trade-offs between environmental and cost performance. Int J Prod Econ 181:350–366
22. Nasir MHA, Genovese A, Acquaye AA, Koh SCL, Yamoah F (2017) Comparing linear and circular supply chains: a case study from the construction industry. Int J Prod Econ 183:443–457

23. Madani SR, Rasti-Barzoki M (2017) Sustainable supply chain management with pricing, greening and governmental tariffs determining strategies: a game-theoretic approach. Comput Ind Eng 105:287–298
24. Xu J, Chen Y, Bai Q (2016) A two-echelon sustainable supply chain coordination under cap-and-trade regulation. J Clean Prod 135:42–56
25. Kannan D (2018) Role of multiple stakeholders and the critical success factor theory for the sustainable supplier selection process. Int J Prod Econ 195:391–418
26. Jones KS, Jackson DM (1970) The use of automatically-obtained keyword classifications for information retrieval. Inf Storage Ret 5(4):175–201

Effect of Handle Orientation on Two-Handed Push Strength in Unorganized Sector Workers

Rahul Jain, K. B. Rana, Vikky Kumar, and M. L. Meena

Abstract Workers do labor–intensive exertion mostly by two hands. The existing two–handed push strength data discovered the substantial effect of multiple factors. Yet the impact of handle orientations on push strength data among Indian workers are not explored. For conducting laboratory experiments, 40 workers exerted two-handed push strengths. The effect of handle orientation has produced an increase in push/pull strengths at a lower angle. The interaction effects of modeled factors were suggested that unorganized sector workers produced higher strength at a 90° angle. The workers' training related to the orientation of handle during work is necessary for lessening the muscular fatigue of workers.

Keywords Handle orientation · Push strength · Handle height · Unorganized sector workers

1 Introduction

Push and pull Activities are commonly seen in both industrial and everyday life. For example, push and pull the trolley wheel of raw materials, push and pull the door in daily life, etc. Hence, pulling and pushing human strength would be essential for designing equipment or operating environment. Pushing and pulling activity may be described as the force of a hand pressure applied on a object horizontally or vertically [1]. An excellent way to differentiate among pushing and pulling is evident by the mean of application. Also, according to Lee et al. [2] for easier pulling and pushing, the basic guideline for designing task-based pulling and pushing work is necessary as per human capacity [3].

R. Jain (✉) · K. B. Rana
Department of Mechanical Engineering, University Departments, Rajasthan Technical University, Kota, Rajasthan, India
e-mail: rjain.npiu.me@rtu.ac.in

V. Kumar · M. L. Meena
Department of Mechanical Engineering, Malaviya National Institute of Technology Jaipur, Jaipur, Rajasthan, India

R. Agrawal et al. (eds.), *Recent Advances in Industrial Production*, Lecture Notes in Mechanical Engineering, https://doi.org/10.1007/978-981-16-5281-3_17

The high incidence and job-related accident costs due to overexertion at some stage in pushing and pulling activities in the industries are apparent [2, 4]. The primary cognizance of studies into the risk related to pushing and pulling tasks has been geared toward assessing the forces exerted on the palms. Various studies identified that manual arm strength functionality depends on the anthropometrics, postures, and shoe/floor friction [1–7]. Daams [8] established that posture performs an essential function in defining the greatest pressure exerted. Most pushing obligations will be dynamic, demanding employees to transport masses to a particular distance. Lee et al. [2] discussed that the estimation of dynamic pushing force is more complicated than static situations. In stationary pushing and pulling activities, the employee's body is in contact with the stationary item; however, this may not be the similar in the dynamic obligations. The stability concerns in workforces such as choosing tinier steps, or take up awkward postures during operations, increases the threat of health issues. The most beneficial height has been taken to the worker's shoulder height so that the forearm hand makes 90° angles to the frame, and the worker can practice maximum pushing pressure [2]. The real strength of the worker is calculated in this posture. Thus far, only a few studies have examined the influence of the handle angle on the pushing strength.

The principle related to the substance/configurations used in the different layout/operations and the frictional characteristics [9]. This frictional quality of surface can range with force exerted with the hand, the smoothness and the porosity of the floor, and contamination. The friction quality of human palmar skin depends on the surface place's scale in contact with the pores and skin. Depending on the work, perfect manage shapes of handle vary for the orthogonal push force, which most carefully matched pushing on a deal with a neutral forearm posture. While designing handles for manual materials dealing with, the handles have to distribute the force over the more significant possible floor area of the hand and arms. A few kinds of handle between the palms and the item are frequently required to push an item. The interface among hand and handles has been well documented, in particular on the subject of equipment. Numerous tool design concepts might be applied to the hand–handle coupling. Kumar et al. [9] noticed most pull electricity range via 25–30% in assessment to maximum push power in the course of exceeded maximal pull–push isometric and isokinetic modes at three heights (50, 100, and 150 cm).

The literature review presented suggests that all the research performed earlier were based on handle orientation positions. There has not been much investigation dealing with push and pull force on different angular positions of the handle in the unorganized sector workers. Current research aims to find the effect of handle orientation on push strength in unorganized sector workers.

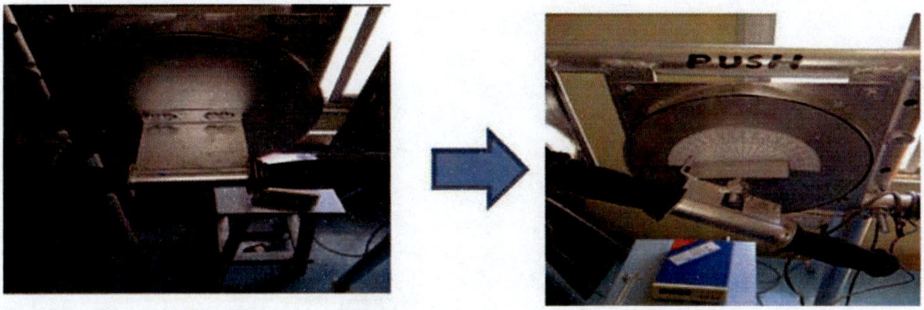

Fig. 1 Position of the handle **a** fixed **b** move and fixed

2 Methods

The current laboratory investigation was carried out on 40 subjects (23 Male, 17 Female) at the push strength measurement equipment developed in the institute's ergonomics lab. The whole subjects were divided into three categories: control group (unorganized sector workers), exposed 1 (student subjects), and exposed 2 (university staff). The subjects were randomly selected healthy persons (no health casualties like work-related health problems) between 18 and 40 years. All subjects signed consent before the experimental investigation, which was done as per the Helsinki declaration.

Subjects performed isometric exertion of push force on the different angular position of handle: 0°, 15°, 30°, 45°, 60°, 75°, 90°, 105°, 120°, 135°, 150°, 165° and last with 180°. During every interval, we have given one minute of rest (Fig. 1).

The subjects were requested to execute a two-handed push exertions on a handlebar in a standing pose (Fig. 2). They were educated to achieve the maximum push force in the horizontal plane calmly, without twitches as per the previous research protocol [6]. The subjects were asked to achieve the maximal strength in the first 2 s and uphold the maximal strength up to 3 s [9]. The all readings were noted and the average value of those evaluations was taken as a final value. Similarly, the same type of procedure will apply to the next conjugative angular position of the handle.

The push/pull strength data were statistically analyzed to determine the effect of various factors on the force application mode using the repeated measures analysis of variance test among the three categories.

3 Results and Discussions

Table 1 presents the mean and standard deviation (SD) for anthropometric dimensions of subjects. The mean age, elbow height, and weight of subjects were 36.2 ± 11.59 years, 1089.36 ± 37.61 mm, and 71.00 ± 13.94 kg. The unorganized sector

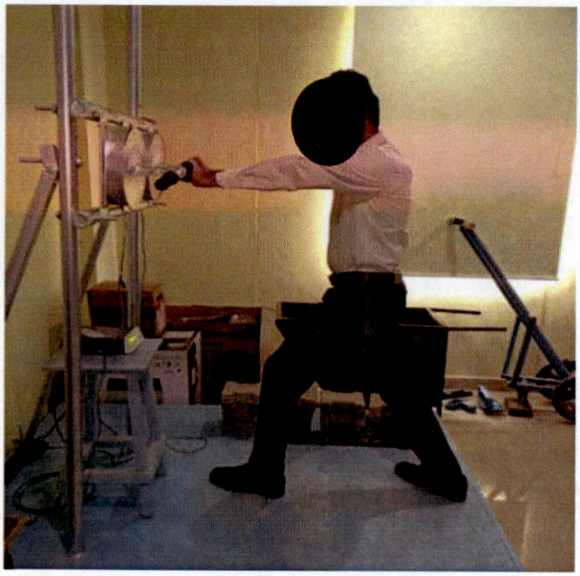

Fig. 2 Force exerted by the subject during pushing exertion

Table 1 Demographic data of the exposed and control group of subjects

Factor	Mean (SD)				*p*-value
	Overall	Exposed 1	Exposed 2	Control	
Age (in years)	36.92 (11.59)	37.00 (8.15)	48.46 (8.11)	25.31 (1.88)	*
Weight of subject (in kg)	71.00 (13.94)	64.77 (17.67)	74.00 (12.02)	74.15 (9.76)	0.14
Elbow height (in mm)	1089.36 (37.61)	1091.15 (33.16)	1076.54 (32.17)	1100.38 (45.07)	0.27

*$p < 0.001$, *SD* Standard deviation

workers had higher body weight to the industrial workers (65.8 kg) of Jaipur, as reported by Meena et al. [10].

Table 2 offers the mean and SD values of the push strengths of all subjects. It is also evident from Table 2 that the higher values of push strength can be achieved at 90° (increasing from 0°).

This finding is in line with the push strength findings of Kumar et al. [9], which is also found statistically significant as per the outcomes of repeated measures ANOVA findings presented in Table 3.

Table 2 Push strength data in the exposed and control group of the subject at different angles

Angles	Mean (SD)			
	Overall	Exposed 1	Exposed 2	Control
0	115.90 (23.45)	113.46 (14.47)	110.92 (28.53)	123.31 (25.10)
15	120.51 (25.09)	126.46 (25.25)	108.31 (26.05)	126.77 (20.87)
30	118.23 (24.13)	122.69 (17.12)	103.46 (24.79)	128.54 (23.73)
45	119.33 (24.16)	122.15 (20.23)	104.62 (24.61)	131.23 (20.90)
60	122.31 (21.67)	124.62 (18.42)	109.54 (22.74)	132.77 (18.17)
75	124.64 (26.67)	126.62 (22.02)	107.77 (24.42)	139.54 (24.84)
90	134.90 (30.42)	136.92 (29.35)	119.31 (32.01)	148.46 (24.04)
105	127.03 (29.08)	124.69 (20.97)	114.15 (33.24)	142.23 (26.39)
120	121.82 (29.14)	119.69 (21.55)	105.77 (28.15)	140.00 (28.09)
135	120.95 (30.41)	117.54 (24.02)	108.77 (31.18)	136.54 (30.71)
150	118.05 (30.41)	115.46 (20.05)	101.77 (23.95)	136.92 (35.67)
165	119.23 (28.46)	121.69 (20.02)	103.15 (24.81)	132.85 (32.58)
180	118.08 (22.33)	116.46 (13.63)	111.31 (26.59)	126.46 (23.59)

Table 3 ANOVA statistics for comparison within and between the different groups of subjects

Angles	Statistics	Sum of squares	df	Mean square	F-value	*p*-value
0	Between	1112.667	2	556.333	1.012	0.373
	Within	19,782.923	36	549.526		
	Total	20,895.590	38			
15	Between	2905.436	2	1452.718	2.488	0.097
	Within	21,020.308	36	583.897		
	Total	23,925.744	38			
30	Between	4475.692	2	2237.846	**4.563**	**0.017**
	Within	17,655.231	36	490.423		
	Total	22,130.923	38			
45	Between	4759.590	2	2379.795	**4.918**	**0.013**
	Within	17,421.077	36	483.919		
	Total	22,180.667	38			
60	Between	3611.692	2	1805.846	**4.566**	**0.017**
	Within	14,238.615	36	395.517		
	Total	17,850.308	38			
75	Between	6636.359	2	3318.179	**5.861**	**0.006**
	Within	20,382.615	36	566.184		
	Total	27,018.974	38			

(continued)

Table 3 (continued)

Angles	Statistics	Sum of squares	df	Mean square	F-value	*p*-value
90	Between	5604.667	2	2802.333	**3.41**	**0.044**
	Within	29,568.923	36	821.359		
	Total	35,173.590	38			
105	Between	5230.205	2	2615.103	**3.500**	**0.041**
	Within	26,900.769	36	747.244		
	Total	32,130.974	38			
120	Between	7704.667	2	3852.333	**5.648**	**0.007**
	Within	24,553.077	36	682.030		
	Total	32,257.744	38			
135	Between	5239.128	2	2619.564	**3.154**	**0.055**
	Within	29,898.769	36	830.521		
	Total	35,137.897	38			
150	Between	8163.436	2	4081.718	**5.447**	**0.009**
	Within	26,974.462	36	749.291		
	Total	35,137.897	38			
165	Between	5848.769	2	2924.385	**4.223**	**0.023**
	Within	24,930.154	36	692.504		
	Total	30,778.923	38			
180	Between	1543.538	2	771.769	1.597	0.216
	Within	17,395.231	36	483.201		
	Total	18,938.769	38			

Bold values are significant

4 Conclusions

The push force's magnitude increased from 0 to 90° and then decrease from 90 to 180°. At 90° angular position of handle the value of push force is maximum as compared to other angular position of the handle, the reason behind is at 90° angular position both hands of the subject is in a neutral position, and there is no any stress generated on forearm and wrist of the subject. The push force's minimum value is observed at 0° and 180° because the handle position is the same (horizontal position) in both the degree. This study can be used as a reference for further study to determine the pull force of different angular positions by using the dominant hand only.

Acknowledgements The authors would like to acknowledge the National Project Implementing Unit and the Ministry of Human Resources Development for funding support in this study.

Funding This research was supported by the Collaborative Research Scheme implemented by the National Project Implementing Unit (NPIU) funded by the Ministry of Human Resources Development, Government of India (No. 1-5727963012).

References

1. Baril-Gingras G, Lortie M (1995) The handling of objects other than boxes: univariate analysis of handling techniques in a large transport company. Ergonomics 38(5):905–925
2. Lee KS, Chaffin DB, Herrin GD et al (1991) Effect of handle height on lower-back loading in cart pushing and pulling. Appl Ergon 22(2):117–123
3. Hoozemans MJ, Van Der Beek AJ, Fringsdresen MH et al (1998) Pushing and pulling in relation to musculoskeletal disorders: a review of risk factors. Ergonomics 41(6):757–781
4. Warwick D, Novak G, Schultz A, Berkson M (1980) Maximum voluntary strengths of male adults in some lifting, pushing and pulling activities. Ergonomics 23(1):49–54
5. Jain R, Meena ML, Sain MK et al (2019) Impact of posture and upper-limb muscle activity on grip strength. Int J Occup Saf Ergon 25(4):614–620
6. Jain R, Meena ML, Sain MK et al (2019) Pulling force prediction using neural networks. Int J Occup Saf Ergon 25(2):194–199
7. Seo NJ, Armstrong TJ, Young JG (2010) Effects of handle orientation, gloves, handle friction and elbow posture on maximum horizontal pull and push forces. Ergonomics 53(1):92–101
8. Daams BJ (1993) Static force exertion in postures with different degrees of freedom. Ergonomics 36(4):397–406
9. Kumar S, Narayan Y, Bacchus C (1995) Symmetric and asymmetric two-handed pull–push strength of young adults. Hum Factors 37(4):854–865
10. Meena ML, Dangayach GS, Bhardwaj A (2013) Measuring anthropometric data for designing hand tools in handicraft industries. Int J Process Manage Benchmarking 3(3):334–351

Weight Optimization of Gears in the Transmission of an All-Terrain Vehicle

Shantanu Tiwari, Shikhar Verma, Shashwat Kulshreshtha, Naman Varshney, and Mayank Kushawaha

Abstract Power transmission in automobile plays a major role in the propulsion of an automobile by transferring the power as per the need for the terrain. In an automobile, the power generated by the engine is low in torque but an automobile requires a high amount of torque in comparison to the supplied torque. How can this be achieved? This is achieved with the help of different types of torque multipliers, one of the most common methods is the use of gears. What purpose do the gears serve? Gears can be used to multiply the torque or speed depending upon the gear ratio. With the help of gears, the power is transferred to the drive shaft or directly to the half shaft and then to the wheels. This increased amount of torque moves the vehicle in the initial stage of motion from a static position. During the motion of the gears, they encounter different kinds of forces like bending and dynamic loading for which they must have high strength and endurance limit. In most of the cases, the conventional design of gears is made neglecting the possible optimization options in terms of weight of the gears, since it is believed that higher the amount of material a component carries greater is the strength of that particular component, but at the same time it increases the weight drastically, which increases the inertia required to move the components and this directly reduces the efficiency of the whole system. The system proposed below, works on this issue and consequently leads to increased efficiency.

Keywords Gear design · Weight optimization · Dynamic loading · ATV

S. Tiwari · S. Verma · S. Kulshreshtha · N. Varshney
ABES Engineering College, Ghaziabad, India

M. Kushawaha (✉)
Mechanical Department, ABES Engineering College, Ghaziabad, India
e-mail: mayank.kushwaha@abes.ac.in

R. Agrawal et al. (eds.), *Recent Advances in Industrial Production*, Lecture Notes in Mechanical Engineering, https://doi.org/10.1007/978-981-16-5281-3_18

1 Introduction

1.1 Basic Introduction and Problem

Gears are an important part of the transmission system as they play a major role after the engine or motor in manipulating the torque transmitted. If any gear becomes dysfunctional, the whole vehicle comes to a halt, so it is important that they must possess the good dynamic load-bearing capacity as well as high bending strength [1]. During the motion of the gear under loading, there are various portions in the design of gear where the stress development is deficient, so those portions of the material should be removed in order to reduce the weight of the gear without affecting the overall strength and function of the gears.

1.2 Aim and Objective

What contribution does the given study provide to the modern automobile industry? With this research, we aim to provide weight optimization of gears in designs to improve the efficiency by reducing the mass of moment of inertia which will be achieved by removing the material from gear without affecting the actual functioning of gear.

2 Design and Analysis

2.1 Design

Few considerations are made before designing a gear which is assuming the mass of the vehicle, which helps to calculate the gear ratio required to propel the vehicle on the terrain, and then the module is decided or calculated whichever is feasible. This helps to decide the basic size and the number of teeth on the gear. The number of teeth is calculated on the basis of the Lewis equation [2] which is a relation between module, number of teeth, the load-bearing capacity of opted material. Before finalizing the CAD model of the gear, the teeth strength of gear is calculated using Lewis equation.

The CAD model is prepared on SolidWorks® and FEA is carried on Ansys® (Table 1).

Table 1 Specifications of gear [3]

Module (m)	2.5 mm
Number of teeth (z)	49
Width of gear (b)	25 mm
Applied force (F_t)	2716.44 N

Table 2 General material used for gears

Material/Properties	AISI 1045 (EN8)	AISI 4340 (EN 24)	20MnCr5 (EN 10084-2008)
Ultimate tensile strength (MPa)	1110	625	1158
Ultimate yield strength (MPa)	710	530	1034
Density (kg/m^3)	7850	7850	8000
Hardness (HRC)	35	32	60

2.2 *Material Selection [4–6]*

Gears are made of a wide variety of materials but a comparative study is made on the most commonly used materials for manufacturing of gears (Table 2).

The study and analysis are made on AISI 4340 (EN24).

2.3 *Analysis*

The Finite Element Analysis is conducted, in which static structural analysis is carried to compare the Factor of safety in order to support the possible material removal from the gear, to insure the working of the gear. The performed simulation incudes the extreme condition in which wheel is structed and maximum possible force that gear faces is **2716.44 N**. Also the mass of the both gears (i.e., before material removal and after) and mass moment of inertia calculated from SolidWorks® (Fig. 1; Tables 3 and 4).

Fig. 1 Gear 1 (without any weight optimization technique)

Table 3 Meshing parameters for both gears

Properties	Gear 1	Gear 2
Mesh size	1.0 mm	1.0 mm
Nodes	1,472,874	305,016
Elements	349,775	175,840

3 Results

It can be clearly seen that the factor of safety in both gear design is almost same but the mass of gear 2 is very less in comparison to gear 1, which shows that the strength of gear 2 is almost same as of gear 1 and stress flow has been improved. Also the comparison to mass moment of inertia has been given below which has been derived from mass properties feature of SolidWorks (Table 5 and 6).

4 Conclusion

From the results of the simulation it is evident that by following the given method of weight reduction, we can reduce the mass moment of inertia, this further reduces the losses in the powertrain as some portion of the power from engine is wasted in order to rotate the gears. If there are large number of gears and other rotational components, then the inertial losses are become significant and efficiency of whole system is affected. This method can not only be applied here but can also contribute in the heavy machinery industry too, increasing the efficiency and leading to increased overall profits.

The suggested method of weight reduction is as follows:

1. Provide the I-section in gears as shown in Fig. 2.
2. Provide the slots with web curved in the direction of motion as shown in Fig. 2 (Figs. 3, 4 and 5).

Table 4 Simulation conditions

	Gear 1		Gear 2	
	Image	Details	Image	Details
Fixture	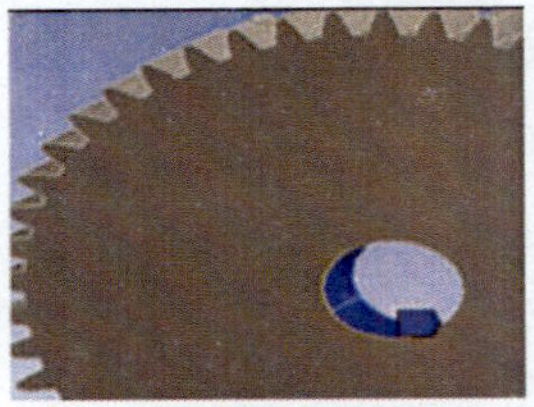	Single face fixed geometry		Single face fixed geometry
Load		A load of 2716.44 N perpendicular to face is applied		A load of 2716.44 N perpendicular to face is applied

Table 5 Comparison in mass of gears

	Gear 1	Gear 2
Mass (in grams)	2208.649	803.634
Factor of safety	2.3	2.1

Table 6 Mass moment of inertia of both gears

	Mass moment of inertia (grams * square millimeters)		
Gear 1	Ixx = 4,247,048.406	Ixy = 0.000	Ixz = 0.000
	Iyx = 0.000	Iyy = 2,583,659.390	Iyz = 0.000
	Izx = 0.000	Izy = 0.000	Izz = 2,583,659.390
Gear 2	Ixx = 1,947,366.08	Ixy = 62.82	Ixz = 1412.37
	Iyx = 62.82	Iyy = 1,129,423.45	Iyz = 984.25
	Izx = 1412.37	Izy = 984.25	Izz = 1,128,202.82

Fig. 2 Gear 2 (with weight optimization technique)

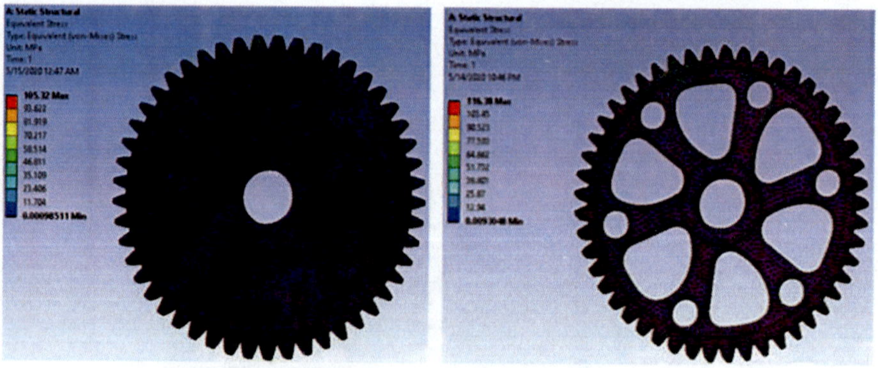

Fig. 3 Stress distribution in Gear 1 and Gear 2 respectively

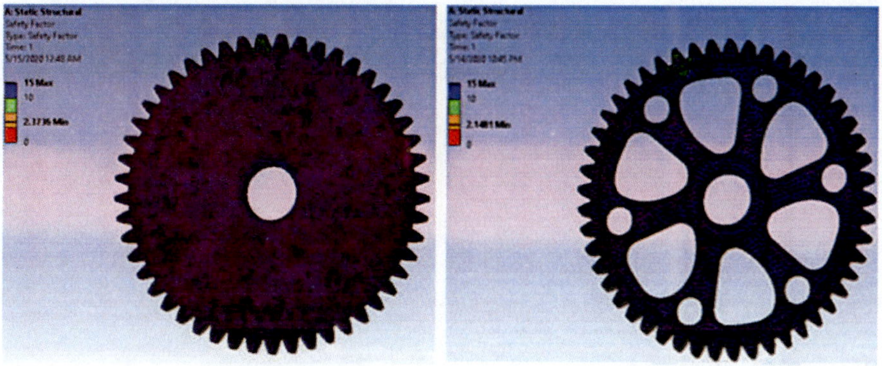

Fig. 4 Factor of safety of Gear 1 and Gear 2 respectively

Fig. 5 Shows the point and axis of mass moment of inertia in Gear 1 and Gear 2

References

1. Bhandari VB (2010) Design of machine elements, 3rd edn. The McGraw-Hill Companies, pp 330–348
2. Mahadevan K, Balaveera Reddy K (2013) Design data handbook, 4th edn
3. Chaudhari C, Pasalkar A, Gavade A, Jagtap A (2018) Design of two-stage single speed gearbox for transmission system of SAE BAJA vehicle. Int Res J Eng Technol (IRJET) 05(10):1682–1683. e-ISSN: 2395-0056; p-ISSN: 2395-0072
4. George B, Jose A, George AJ, Augstine A, Thomas AR (2016) Innovative design and development of transmission system of an off-road vehicle. Int J Sci Eng Res 7
5. Sunil VK, Kumar S, Singh A, Srivastava A, Sharma A, Dobriyal P (2017) Design and development of a transmission system for an all Terrain vehicle 04(05)
6. makeitfrom.com

Ergonomic Interventions in Maintaining Postural Stability in Pregnant Women at Their Workplaces

Nikhil Yadav, M. L. Meena, G. S. Dangayach, and Yashvin Gupta

Abstract This is an era of "Women Empowerment" spreading overall in various sectors. The factors like drastic change in economic conditions and other social perspective led to the employment of pregnant women rendering full time service. There has been a requirement aroused for the ergonomic design of the workplaces reducing the risk factors associated with the health of pregnant women and ensuring her greatest productivity. This paper reviews various ergonomic interventions that could be employed to maintain the postural stability of the pregnant women. A lot of researches have been done by people from medical backgrounds and physiotherapists in drafting out appropriate bodily postures leading to sustainability of the pregnant women in different work cultures. Center of mass of the body changes with uneven distribution of mass and shifting of center of gravity causing postural imbalance in the body. The work has mainly focused on prolong standing and sitting postures required at various tasks. Due to standing for long hours, the legs are prone to varicose veins that get exaggerated in pregnant women due to increase in weight of the body. In the same context sitting for long hours causes musculoskeletal disorder in hip and pelvic girdles. Some of the ergonomic interventions like designing footrest, antifatigue mats, sit-stand stools etc. designed for preoperative settings could be redesigned in relevance to workplace designing for pregnant women. The usability index of maternity support belt used to maintain proper posture is yet to be investigated. Since postural stability equilibrium decreases with the advent of pregnancy during its third trimester, a lot of tests along with the biomechanical solutions have to be manifested for the same. Due to decline of the equilibrium the musculoskeletal disorders are reported to be significantly increased which calls for the ergonomic interventions in workplaces for pregnant women.

Keywords Pregnant · Musculoskeletal · Ergonomics

N. Yadav (✉) · Y. Gupta
Department of Mechanical Engineering, Government Women Engineering College, Ajmer, India

M. L. Meena · G. S. Dangayach
Department of Mechanical Engineering, Malaviya National Institute of Technology, Jaipur, India

R. Agrawal et al. (eds.), *Recent Advances in Industrial Production*, Lecture Notes in Mechanical Engineering, https://doi.org/10.1007/978-981-16-5281-3_19

1 Introduction

"Women Empowerment, Make in India and *Atmanirbhar Bharat*", empowers the nation to contribute to the economy of the country and make oneself competent enough to face the challenges in the manufacturing, designing, production and other sectors as well. The participation of women in all the sectors is appreciably high. This follows the incorporations of ergonomic interventions at the workplaces to cater the needs of the women especially in pregnant conditions. Though the government have certain guidelines and norms for the working conditions at workplaces for pregnant women but yet these are practised at a small extent resulting in firing of the workforce or leave without pay. Doctors advise the restrictions on certain movements during pregnancy except a few, those become the integral parts of body balance. Standing and sitting are not the exceptions. Most of the pregnant women have grievances against the lower back pain arising due to false postures and improper workplace configuration. The postural load in the lower and upper extremity due to wrong practising of the posture is retained throughout the life and thus affecting the quality of living. The literature review has been done on the articles on postures at different stages like sitting, standing, walking, lifting a load, doing a repetitive task etc. Studies have also suggested that the postural changes have been varying individually with remarkable trend variations in lumber lordosis. The nature of jobs also affects the postural behaviour of pregnant women. The main body parts that used to get affected is trunk and arm posture. The constraints imposed to these parts affects the postures and behavioural pattern. The posture of the trunk segments is affected by the restriction imposed on anterior tilt of the pelvis by the increasing size of lower trunk. Researches are yet trying to find the effect of change in trunk dimensions on the upper part of the body especially while sitting. It would not be right to think that all the effects on posture would be reverted after postpartum. The musculoskeletal disorders caused due to change in the behavioural pattern would remain intact and could cause serious problems in the long run of life. The novelty of the work lies in digging up the factors and facts that actually affect the life of pregnant women postpartum. Women workers demand protection during the period of pregnancy as it leads to inability to cope up with many work factors requiring intense manual labor and awkward working conditions. Further the solution of the problems under study lies in ergonomic interventions that could be recommended at the workplaces for the smooth flow of the job without hampering the physical and mental health of the pregnant women.

2 Methodology

The preferred reporting items for systematic reviews and Meta-analyses (PRISMA) guidelines were used [1]. Keywords searched on databases were biomechanics, postural stability and anthropometry of pregnant women. One hundred and thirty-eight articles were selected based on the title of the articles and abstracts. Further screening process was based on content of articles like methodology used and conclusion given on the postural stability aspects in pregnant women. Duplicate articles were screened off minimizing the number to one hundred and two, out of which fifty-seven were found non relevant. Forty-five articles were checked under the eligibility criteria that the contents must be based on the postural stability under both the static and dynamic conditions respectively. Nineteen articles on pregnant women regarding lower back pain and lumber pelvic pain as well as based on medical aspects of pregnancy without any relevance to biomechanics were discarded. Hence after the application of exclusion and inclusion criteria, the final number of articles came out to be twenty-six in number (Fig. 1).

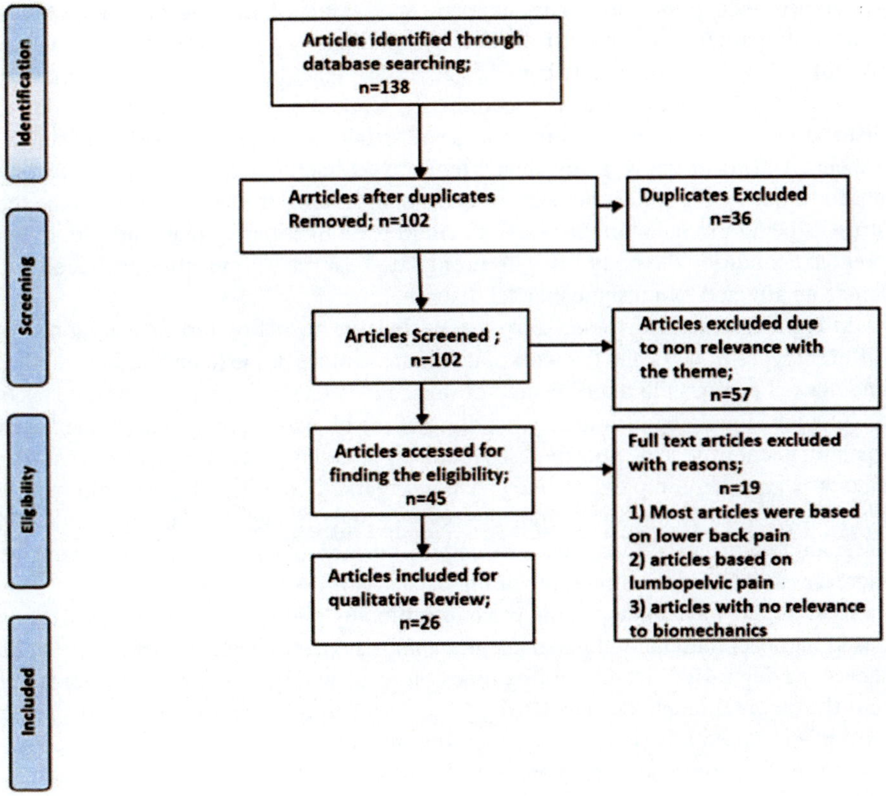

Fig. 1 Flowchart showing the process of screening and reviewing the articles as per (PRISMA) [1]

3 Literature Review

A systematic literature review has been done in exploring the findings done by various researchers, doctors, physiotherapist and gynaecologists. Some of the key references along with their findings are tabulated in Table 1.

4 Results and Discussions

Most of the researchers have focussed on anthropometrical and morphological changes. Spinal morphology comes to be first in light when the aforesaid aspects are being observed with the advent of pregnancy. The formation of Lordosis where the spines bend posteriorly. This happens as a result of shifting of the center of mass in the anterior direction followed by bending load on the spines due to the weight acting eccentrically. This makes adverse impact on the postures retained by the women to balance themselves accordingly. Upper body extremity changes due to distribution of mass of body by the release of certain hormones and postural changes are observed accordingly. Body always tends to get itself balanced in such a way that net force acting on it must get zero or be in equilibrium. In the lieu of the same the stress is being produced in the body at certain parts like joints, ligaments etc. In the pregnant condition the body loses its strength to bear the stresses thus produced and hence its affected by musculoskeletal disorders.

Literature review traversed across revealed the result of the investigation on different types of tasks and the workplaces that are more prone to postural in stability and hence becomes the major source of musculoskeletal disorders. Repetitive tasks and the jobs involving asymmetry handling of the objects by pregnant women leads to major instability in the posture. Jobs requiring standing and sitting for ling duration causes pain in certain parts of body. A non-pregnant condition body could handle the postural load as the strength of the body is high enough to resist it whereas in pregnant condition due to release of certain hormone, the laxity of the ligaments increases and hence the strength bearing capacity of the body decreases.

Researchers have made the major contribution by giving some recommendations based on observational and experimental analysis. Pelvic belts and maternity belts though having insignificant usability index but yet could be improvised further to get into the medical practices. The sitting jobs like computer workstations and similar workplaces could incorporate certain ergonomic interventions which could accommodate the anthropometrical changes to facilitate the ease of doing job by pregnant women.

Table 1 Literature review

Authors	Year	Findings
Joanne E. Bullock-Saxton	1991	The author has done investigation regarding postural imbalance postnatal period. She observed the changes in spinal morphology due to several factors arising during pregnancy and its persistence postpartum resulting in hampering the bodily posture [2]
Stephen J. Morrissey	1998	Author has identified certain risk factors for the pregnant women and its indirect effect on the fetus. A few ergonomic interventions have been recommended by the author which could open a wide scope of research with the proper ergonomic justification and hence leading to the decrease in risk factors and increase in workplace efficiency [3]
Yoji Hattori et al.	2000	The researchers have studied the changes in bodily posture as per the demand of the tasks and advent of stress on the different body parts with the nature of posture and plane of symmetry of vertical loading [4]
Wendy et al.	2002	Authors have investigated the alignment of the upper body in different postures and found that there has been week correlation between the anthropometrical changes in the upper portion of the body and the progression period of pregnancy however the spinal cord has been reported to get flattered postpartum [5]
P. L. Cheng et al.	2006	The authors have done a case study in determining the adverse effects of repetitive tasks on posture and its impact on the musculoskeletal disorders [6]
Yasuki Motozawa et al.	2008	The authors have studied the anthropometric characteristics of pregnant drivers in Japan and the effect of seating postures. They have done a comparative study with American drivers than with the Japanese ones and found a significant difference that could lead to the anthropometric design of car seats per the users requirement [7]

(continued)

Table 1 (continued)

Authors	Year	Findings
Genevieve et al.	2009	Researchers have made the ergonomic interventions in the form of desk board attachment for workstations. It improved the posture by reducing kyphosis and hence making them to sit in the upright posture but produced discomfort at the lower back and pelvis area. The impact of this intervention was found to be adverse in relevance to the upper part of the body [8]
Priya et al.	2010	The researcher has investigated the effects on lower extremities with the advent of pregnancy and found that the changes depend upon the anthropometrical characteristics of the body [9]
Erica Beaucage-Gauvreau et al.	2012	Authors have done case study on pregnant women in Benin, West Africa to find the cause of lower back pain due to different postural movements of trunk as per the workplace demand and found that the repetitive changes in the trunk posture are the main factors of fatigue in women causing lower back pain [10]
Yasuyo Sunaga et al.	2013	The authors have investigated the adverse effect of start of gait cycle immediately after standing off the chair and recommended for an ergonomic intervention that could avoid the same [11]
Ahmet et al.	2014	Investigation of changes in hormones during pregnancy and its effect on dynamic stability was studied. The risk of falls was quantified during all the three trimesters and probable factors causing the risks were observed. The critical stage of risk was found more in third trimester comparatively [12]
Gulcan Ozturk et al.	2015	Authors studied the correlation of lower back pain and postural and dynamic stability and found that with the advent of lower back pain the stability gets hampered [13]
Bulent et al.	2015	The authors have studied the various physiological changes occurring during pregnancy and the factors occurring due to these changes which are responsible for increasing the risk of fall [14]

(continued)

Table 1 (continued)

Authors	Year	Findings
Agnieszka Opala-Berdzik et al.	2015	The authors have made a comparative study on the postural stability in static conditions during various stages of pregnancy and postpartum and found that there is a significant change in the same [15]
Marco Branco et al.	2016	The researchers have studied the effect of certain bodily characteristics like body composition that change during the advent of pregnancy causing the change in gait pattern and inducing sway movement [16]
Marie et al.	2018	Author has investigated the usability index of maternity belts with the advent of pregnancy for the stabilization of pelvic girdle and its effect on musculoskeletal disorders. The effect of thc same has been reported satisfactorily in the first two trimesters, however with the increase in the mass of the body and shifting of center of gravity in third trimester, it lead to decrease in the effectiveness of maternity support belt [17]
Alessander et al.	2018	Authors have focused on certain strategies that would help the pregnant ladies to maintain an upright posture. The anthropometry of the body plays a major role in balance mechanism where the larger body would have more sway movement as compared to the average one. The risk of falls has been observed to be greater in the last two trimesters of pregnancy [18]
Ayumi et al.	2018	Author have studied the effect of lumbopelvic pain on gait pattern described by pregnant women. Asymmetry in the movement has been noticed in the form of enhancement of rotational and translational movements with the advent of lumbopelvic pain leading to the dynamic instability [19]
W. Forczek et al.	2019	Authors investigated the effect of gain in weight with the progression of pregnancy period which found to affect the gait stability to a larger extent. The authors have done a follow up study to establish a correlation between base of support in increasing the gait stability and opens a wide scope of research further [20]

(continued)

Table 1 (continued)

Authors	Year	Findings
Francisco et al.	2018	The researchers have done work on find the correlation of pain experienced and change in foot posture with the advent of pregnancy. There has been no significant relation thus found but postural stability could be accessed on the changes in foot posture during the period [21]
Jeanne Bertuit et al.	2018	Authors have done a comparative study on gait analysis with and without girdle pain and also compared the two types of pelvic belts. They found that pregnant women with pelvic girdle pain and with belt showed the same gait cycle as the pregnant women without pelvic girdle pain and hence emphasizing the usability of pelvic belts [22]
Kaname Takeda et al.	2018	The researchers have focused on the factors that would cause the risk of fall of pregnant women and found the inability of joints to maintain the balance and musculoskeletal changes of the body in the lead role for causing such risks [23]
Lyncyn L. R. Reliquias and Joy C. Kuebler	2019	The researchers have studied the effectiveness of ergonomic intervention in the form of sit-stand workstation especially at the workplaces requiring sitting for long hours and hence a rest period in the form of stand is being required on the employee part. The effectiveness of the sit to stand workstation has not been found significant but it has dormant impact which could be further modified to be thus exploited [24]
Lene et al.	2019	Researchers have focused on dynamic stability of the pregnant women in the progression period of 2nd trimester. The gait is observed to be affected significantly with the pain in pelvic girdle making them to describe swayed movements and slower pace [25]
Robert D. Catena	2019	Author conducted the study to investigate the changes in anthropometry succeeded by change in dislocation of center of mass and posture after pregnancy. The distribution of mass in the pregnant ladies was significantly studied in anterior, posterior and lateral directions [26]

(continued)

Table 1 (continued)

Authors	Year	Findings
A. G. Haddox et al.	2020	Author has done a tremendous work by doing the study on a biomechanical model of pregnant women to find the factors that could be responsible for the risk of falls. The generic model is said to be of a little use to predict the same as the anthropometry, musculoskeletal morphology as well as other bodily characteristics may vary from person to person. Author has studied the gait cycle on the model by varying the center of mass, distribution of mass and inertia. Such a model would help the researchers in biomechanical field to predict other dormant factors leading to dynamic instability [27]

5 Conclusions

A lot of work has been done in this area by physiotherapists, gynaecologist and doctors in the form of statistical analysis on the survey data and certain experimental results. A lot of work is yet to be done in finding the correct postures that could minimise the postural loads acting on the body. Further it would not help only by finding the correct postures but also the incorporations of ergonomic interventions is required at a large scale. Certain ergonomic interventions suggested by researchers are maternity belt, pelvic belts, slippers for pregnant women etc. that could minimise the postural load and assist in carting out the pregnancy. Work gap could be identified in terms of modifications in chairs that could be incorporated comprised of foot rest, floor mating, cushioned chairs, an adjustable back rest with lumber support. All angles of backrest are comfortable subjected to the time period of its exercise. Researchers can-do in-depth study in finding the optimised angle for minimising the postural loads. It has been observed that the pregnant women greatly require frequent breaks while in sitting and standing position. Researchers have devised stands like sit to stand chairs to facilitate the intervals of rest without hampering the productivity of the workplace. A good amount of work has been done to modify the workstation as per the anthropometric changes in pregnant women like adjusting the shoulder level, workbench etc. A simulation work is required to be carried in a suitable software for determining the level of stress on spines and pelvic girdle due to postural loads.

References

1. Moher D, Liberati A, Tetzlaff J, Altman DG (2009) The, P.G. Preferref Reporting items for systematic reviews and meta-analyses: the PRISMA statement. PLoS Med 6:e1000097. https://doi.org/10.1371/journal.pmed.1000097

2. Bullock-Saxton JE (1991) Changes in posture associated with pregnancy and the early post-nataal period measured in standing. Physiol Theory Pract 7:103–109
3. Morrissey SJ (1998) Work place design recommendations for the pregnant worker. Int J Ind Ergon 21:383–395
4. Hattori Y, Ono Y, Shimaoka M, Hiruta S, Shibata E, Ando S, Hori F, Takeuchi Y (2000) Effects of box weight, vertical location and symmetry on lifting capacities and ratings on category scales in Japanese female workers. Ergonomics 16(26):2031–2042
5. Gilleard WL, Crosbie J, Smith R (2002) Static trunk posture in sitting and standing during pregnancy and early postpartum. Arch Phys Med Rehabil 83:1739–1744
6. Cheng PL, Dumas GA, Smith JT, Leger AB, Plamondon A, McGrath MJ, Tranmer JE (2006) Analysis of self-reported problematic tasks for pregnant women. Ergonomics 49(3):282–292
7. Motozawa Y, Hitosugi M, Tokudome S (2008) Analysis of seating position and anthropometric parameters of pregnant Japanese drivers. Traffic Inj Prev 9(1):77–82
8. Dumas GA, Upjohn TR, Delisle A, Cahrpentier K, Leger A, Plamondon A, Salazar E, McGrath MJ (2009) Posture and muscle activity of pregnant women during computer work and effect of an ergonomic desk board attachment. Int J Ind Ergon 39:313–325
9. Ponnaupula P, Boberg JS (2010) Lower extremity changes experienced during pregnancy. J Foot Ankle Surg 49:452–458
10. Beaucage-Gauvreau E, Dumas GA, Lawani M (2012) Trunk postural demands of occupational activities of some merchant pregnant women in Benin, West Africa. Ergonomics 55(10):1218–1228
11. Anan YSM, Shinkoda K (2013) Biomechanics of rising from a chair and walking in pregnant women. Appl Ergon 44:792–798
12. Inanir A, Cakmak B, Hisim Y, Demirturk F (2014) Evaluation of postural equilibrium and fall risk during pregnancy. Gait Posture 39:1122–1125
13. Ozturk G, Kulucu DG, Aydog E, Kaspar C, Ugurel B (2015) Effects of lower back pain on postural equilibrium and fall risk during the third trimester of pregnancy. J Maternal-Fetal Neonatal Med. ISSN: 1476-7058
14. Cakmak B, Ribeiro AP, Inanir A (2015) Postural balance and the risk of fall during pregnancy. J Maternal-Fetal Neonatal Med. ISSN:1476-7058
15. Opala-Berdzik A, Blaszczyk JW, Bacik B, Cieslinska-Swider J, Swider D, Sobota G, Markiewicz A (2015) Static postural stability in women during and after pregnancy: a prospective longitudinal study. PLoS One 10(6):e0124207
16. Branco M, Santos-Rocha R, Vieira F, Silva M-R, Aguiar L, Veloso AP (2016) Influence of body composition on gait kinetics throughout pregnancy and postpartum period. Hindawi Publishing Corporation Scientifica, pp 1–12
17. Bey ME, Arampatizs A, Legerlotz K (2018) The effect of a maternity support belt on static stability and posture in pregnant and non-pregnant women. J Biomech 75:123–128
18. Danna-Dos-Santos A, Magalhaes AT, Silva BA, Duarte BS, Barros GL, Maria De Fatima C, Silva CS, Silva SM, Degani AM, Cardoso VS (2018) Upright balance control strategies during pregnancy. Gait Posture 66:7–12
19. Tanigawa A, Morino S, Aoyama T, Takahashi M (2018) Gait analysis of pregnant patients with lumbopelvic pain using inertial sensor. Gait Posture 65:176–181
20. Forczek W, Ivanenko Y, Curylo M, Fraczek B, Maslon A, Salamaga M, Suder A (2019) Progressive changes in walking kinematics throughout pregnancy—a follow up study. Gait Posture 68:518–524
21. Pardo FJV, del Amo AL, Rios MP, Gijon-Nogueron G, Yuste CC (2018) Changes in foot posture during pregnancy and their relation with musculoskeletal pain: a longitudinal cohort study. Women Birth 31:e84–e88
22. Bertuit J, Leyh C, Feipel V (2018) Pelvic belts and pregnancy-related pelvic girdle pain: influence on temporal and spatial gait parameters. Int Biomech 5(1):104–112
23. Tkeda K, Yoshikata H, Imura M (2018) Changes in posture control of women that fall during pregnancy. Int J Women's Health Reprod Sci 6(3):255–262

24. Reliquias LLR, Kuebler JC (2019) The behavior of pain in response to sit-stand workstations: a systematic review. Phys. Ther Rev 24(5):223–228
25. Christensen L, Veirod MB, Vollestad NK, Jakobsen VE, Stge B, Cabri J, Robinson HS (2019) Kinematic and spatiotemporal gait characteristics in pregnant women with pelvic girdle pain, asymptomatic pregnant and non-pregnant women. Clin Biomech 68:45–52
26. Catena RD, Campbell N, Wolcott WC, Rothwell SA (2019) Anthropometry, standing posture, and body center of mass changes upto 28 weeks postpartum in Caucasians in the United States. Gait Posture 70:196–202
27. Haddox AG, Hausselle J, Azoug A (2020) Changes in segmental mass and inertia during pregnancy: a musculoskeletal model of the pregnant woman. Gait Posture 76:389–395

Thermal Analysis of Discontinuity in Deposited Bead

Soumen Mandal, Manish Oraon, and Subrata Kumar

Abstract In this paper, the continuity and discontinuity analysis in deposited bead using Plasma Transferred Arc Welding (PTAW) process has been done through experiments and analytical approach. The calculated energies using experimental data points have been compared with the data points which are derived analytically using Rosenthal equations: line heat source and point heat source for different efficiencies. As a result, the continuous, discontinuous and partially continuous beads with respect to constant melt pool line have been identified. The energies for substrate melting per unit length and the sizes of the melt pool widths (bead widths) for the continuous beads are more compare to partially continuous and discontinuous beads.

Keywords PTAW · Continuous · Partially continuous · Discontinuous

1 Introduction

In PTAW process, the plasma beam, formed between the tungsten electrode (cathode) and the substrate (anode), melt the substrate and a melt pool is formed over the substrate. The mechanism of melt pool formation over the substrate depends on the magnitude and the distribution of heat. To understand the thermal behavior during welding, analysis of temperature distribution in welding has been done by many researchers [1–3]. However, most popular theory was developed by Rosenthal [4]. He solved conduction equation of heat transfer for fixed points for temperature distribution in welding analytically. He considered a line heat source moving with constant speed with respect to the fixed Cartesian coordinate system. For analytical solution, three types of heat sources namely, point source, line source and plane source are used. Eagar and Tsai [5] modified Rosenthal's theory by introducing a Gaussian distribution of heat source with constant distribution parameters and analytically

S. Mandal (✉) · M. Oraon
Birla Institute of Technology, Mesra, Ranchi, Patna Campus, Patna, India
e-mail: soumen.mandal@bitmesra.ac.in

S. Kumar
Indian Institute of Technology Patna, Patna, Bihar, India

R. Agrawal et al. (eds.), *Recent Advances in Industrial Production*, Lecture Notes in Mechanical Engineering, https://doi.org/10.1007/978-981-16-5281-3_20

solved it to predict the temperature distribution more accurately in the vicinity of the heat source regions. To model the broadening of the weld in the upper layer, Rosenthal's line source solution with point source were used on the top surface and the results were used to model experimental weld shapes [6]. The moving point source was used to derive a solution for a line source, where the source strength varies along the length of the line. The solution was used to model the shape of the fusion zone [7]. Resch and Kaplan [8] used Rosenthal's solution for a point source and integrated over a line of point source. Hilton solved analytical solution to the heat conduction equation and calculated the temperature distribution and finally determined the weld pool profile from the shape of the melting isotherm [9]. Kim [3] solved heat conduction equation analytically to find out the temperature distribution around a rectangular shape source and proposed a formula to predict the cooling rate and time. After Rosenthal, many researchers developed [10, 11] analytical model to find out the temperature distributions around the moving heat source.

With the advancement of powerful computer, the numerical methods have been more popular to solve it. The numerical methods can be used to solve very complex geometry more accurately compared to analytical methods. However, the analytical solutions have been continued to be developed for the inherent simplicity.

In this analysis, Rosenthal equations i.e. line heat source and point heat source have been used to find out the energy required for constant melt pool widths by varying scanning speed. The total input energy, energy available for substrate melting and powder melting have also been calculated for all deposited beads considering 22% loss of energy. Finally, comparisons have been done between the experimental data points of all the deposited beads with the constant melt pool width data calculated using line and point heat source considering different efficiencies.

2 Methodology

For heat transfer analysis of discontinuity in deposited layer, the PTAW experimental set-up has been used for material bead depositions. The schematic diagram of Plasma Transferred Arc Welding Process has been shown in Fig. 1. Here, SS316 and SS304L material have been used as a substrate and as a powder material respectively for experiments. For the experiments L27 design matrix has been used. For the experiments, Argon has been used as plasma gas as well as shielding gas. The Plasma gas flow rate and shielding gas flow rate are kept at 1.5 l/min and 7.5 l/min respectively during experiment. The stand-off distance is kept at 8 mm. The powder particle size varies in between 45 and 105 μ. Total 27 beads have been deposited during experiment. The control factors and their labels for the experiments are given in Table 1.

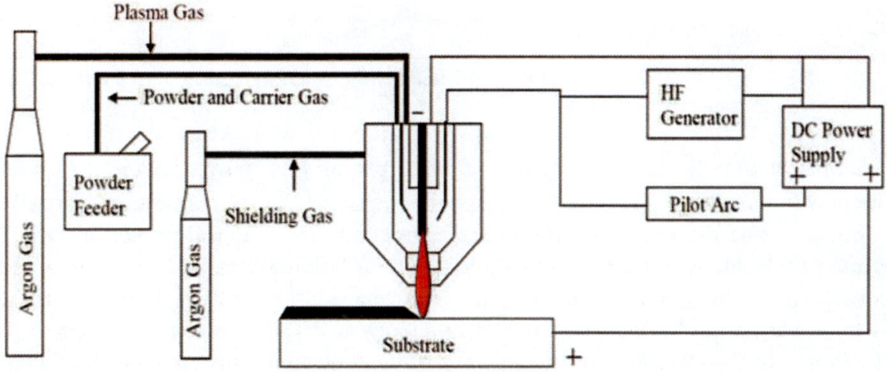

Fig. 1 Plasma transferred arc welding experimental set-up

Table 1 Control factors and their levels for L27 design matrix

Factors	Level-1	Level-2	Level-3
Current (A)	70	80	90
Scanning Speed (mm/min)	300	400	500
Powder feed rate (g/min)	12.64	13.85	16.09

3 Thermal Energy Analysis

Among the deposited beads, three different types have been identified, namely: continuous, partially continuous and discontinuous beads. The continuous, partially continuous and discontinuous beads are shown in Figs. 2, 3 and 4 respectively.

Fig. 2 Continuous bead pattern

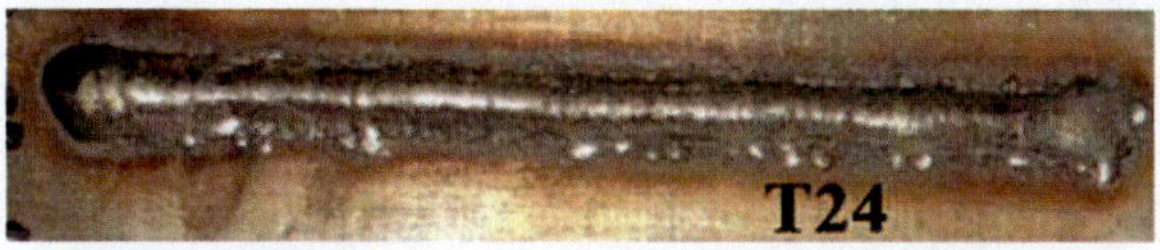

Fig. 3 Partially continuous bead pattern

Fig. 4 Discontinuous bead pattern

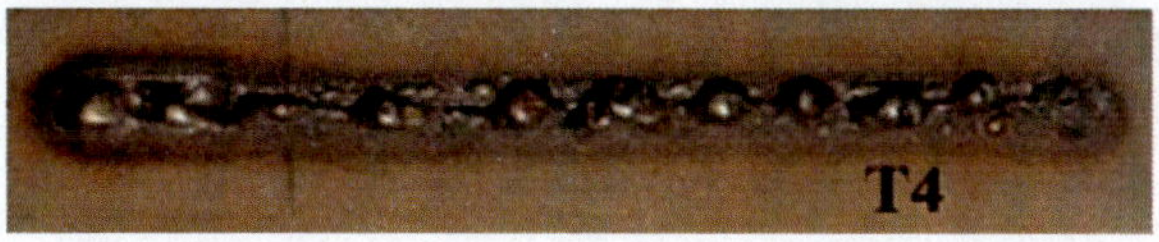

Table 2 Thermal properties of SS316 material [12]

T_m	ρ	C_p	k
1723 K	7269 kg/m^3	720 J/kg/K	32 W/m K

For heat transfer analysis, values of input power $(\dot{E})$, power required to melt the powder $(\dot{E}_p)$ and power left for substrate melting $(\dot{E}_b)$ for continuous, partially continuous and discontinuous tracks are calculated. The material properties which are used for calculation, shown in Table 2. The following data have been obtained considering a loss of 22% of the total input energy, i.e., 78% power efficiency. A large variation of considered power efficiency in PTAW process is observed in the literature [12–14]. So, for the analysis of bead discontinuity in deposited layer, different values of efficiency have been considered for the analysis of discontinuity i.e., 78% and further 60% and 85%.

Total power brought to the substrate,

$$\dot{E} = \eta \text{ IV W} \tag{1}$$

where η is the efficiency parameter, I is the current in A, and V is voltage difference between the electrode and the base material. Out of this total power, the power needed to melt the substrate and powder materials are $\dot{E}_b$ and $\dot{E}_p$ respectively.

$$\text{Where, } \dot{E}_b = \dot{E} - \dot{E}_p \tag{2}$$

$$\text{And } \dot{E}_p = \dot{m}_p\left[C_p(T_m - T_0) + L\right] \text{ W} \tag{3}$$

where $\dot{m}_p$ is the powder feed rate in kg/s, C_p is the specific heat in J/kg K, T_m and T_o are melting and ambient temperatures respectively in 0 K, and L is the latent heat of melting in J/kg.

$$\text{Mass of powder, } \dot{m}_p = \frac{\dot{E}_p}{C_p(T_m - T_0) + L} \text{ kg/s} \tag{4}$$

If S is scanning speed in m/s, then the compound parameter "energy deposition per unit travel length" and "powder deposition per unit travel length" are defined as:

Energy deposition per unit travel length $= \frac{\dot{E}}{S}$ J/m
Powder mass deposition per unit travel length $= \frac{\dot{m}_p}{S}$ kg/m

The direct plasma energy left to melt the surface of the base material per unit length is:

$$\overline{E}_b = \frac{\dot{E} - \dot{E}_p}{S} \text{ in J/m} \tag{5}$$

4 Analytical Solution of Temperature Field During Depositions

Two equations i.e. line heat source and point heat source have been used to find out the power required for constant melt pool width by varying scanning speed. Two equations are shown in below:

$$T = T_0 + \frac{\dot{E}_b}{2\pi\,kh}\exp\left(-\frac{S\varepsilon}{2\alpha}\right)K_0\left(\frac{Sr}{2\alpha}\right) \tag{6}$$

Equation (6) is known as Line heat source model of Rosenthal's equation [15]. Where T_0 (K) is the initial temperature of the substrate, k (W/m K) is the thermal conductivity, h (m) is the thickness of the plate, S (m/s) is the scanning speed, α (m^2/s) is the thermal diffusivity, $K_0\left(\frac{Sr}{2\alpha}\right)$ is the modified Bessel function, r is the distance from the moving heat source.

$$T = T_0 + \frac{\dot{E}_b}{4\pi kr}\exp\left(-\frac{S(r+\xi)}{2\alpha}\right) \tag{7}$$

Equation (7) is known as point heat source model of Rosenthal's equations [15]. The above equations i.e. line and point heat source have been solved analytically to find out the temperature field. Using these equations, required powers have been calculated for constant bead widths (or melt pool width) by varying scanning speed from 6 to 540 mm/min. The melt pool widths of 0.5, 1, 2, 3 and 4 mm have been considered for line and point heat source. The models have been validated with published numerical results [16, 17]. The flow chart of present solution method is represented in Fig. 5.

5 Results and Discussions

The compound parameters, i.e., energy deposition per unit length and powder deposition per unit length for the constant bead widths and experimental data points have been plotted for bead discontinuity analysis. The variations of power efficiency are very large in case of PTAW deposition process. So, for the analysis, the three different values of arc efficiency, i.e. 60, 78 and 85% is considered. A typical temperature field for constant bead width of 1 mm when the plasma torch moves with scanning speed 300 mm/min and input power to the substrate 1682 W, is shown in Fig. 6.

Many details can be extracted from the temperature field i.e. shown in Fig. 6. Constant bead width (i.e., melt pool width) lines are plotted for energy deposition per unit length and powder deposition per unit length. Figure 7a shows the comparison between the energy values required for constant bead width lines using the line

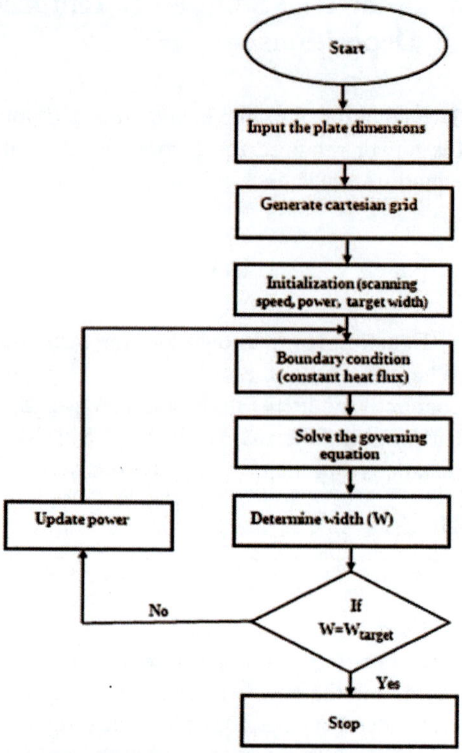

Fig. 5 Flow chart of solution method

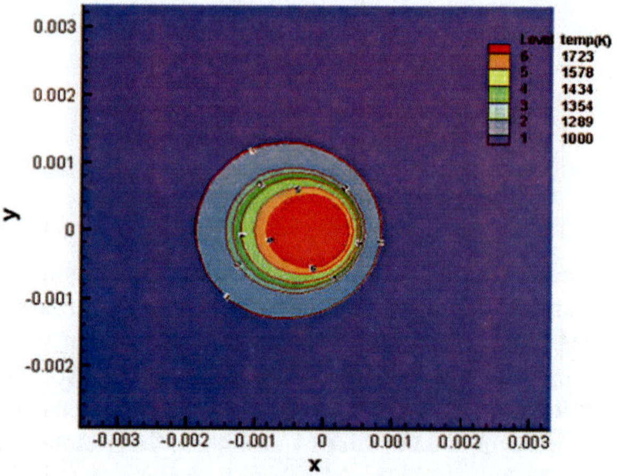

Fig. 6 Temperature contour plot for S = 300 mm/min and $\dot{E}_b = 1682$ W

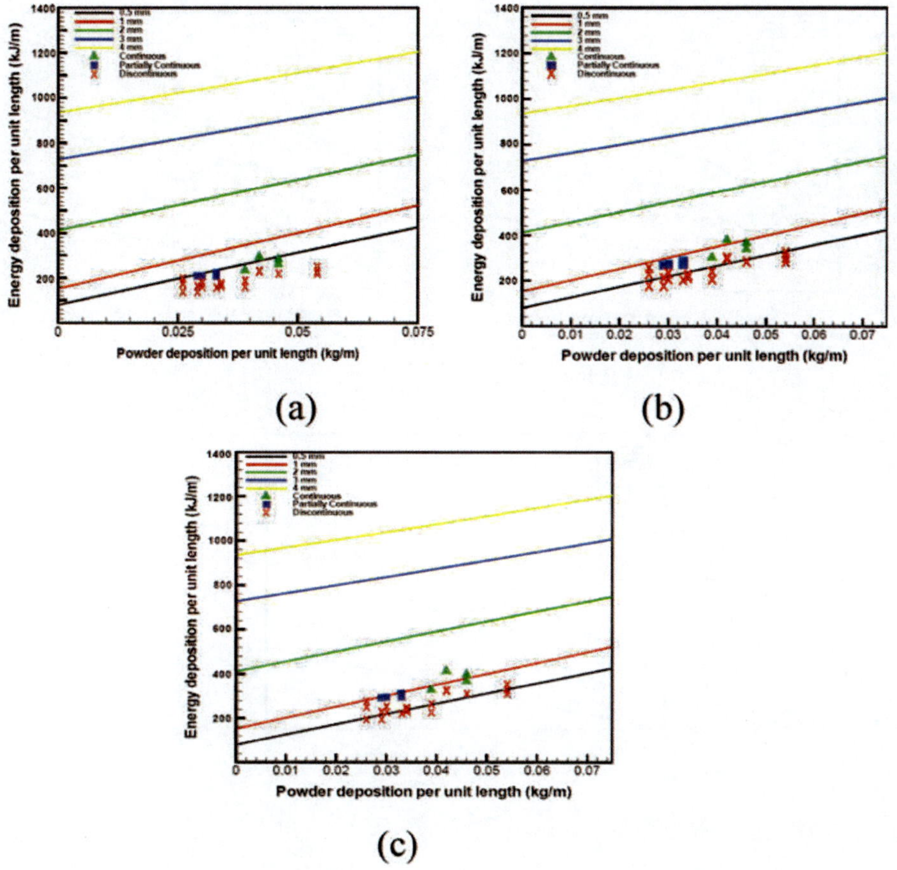

Fig. 7 Bead pattern analysis taking line heat source and efficiency **a** 60% **b** 78% **c** 85%

heat source and experimental values for continuous, partially continuous and discontinuous beads taking efficiency 60%. Similarly, Fig. 7b and c shows the same, but the efficiency is considered for the experimental data points as 78 and 85% respectively. From Fig. 7a, it has been observed that if the melt pool width is around 0.5 mm and above then bead becomes continuous. The bead may be continuous, partially continuous and discontinuous if bead width is below 0.5 mm. From Fig. 7b and c, it has been observed that if the melt pool width is 1 mm and above the bead becomes continuous. If the melt pool width is below 1 mm, then bead may be continuous, partially continuous and discontinuous.

Figure 8a–c show the comparison between the analytical results for constant bead width lines using point heat source and experimental values for continuous, partially continuous and discontinuous beads taking efficiency 60, 78 and 85% respectively. It has been observed that in case of continuous bead, the plasma energy reaches to substrate per unit length is higher than the energy reaches per unit length for partial

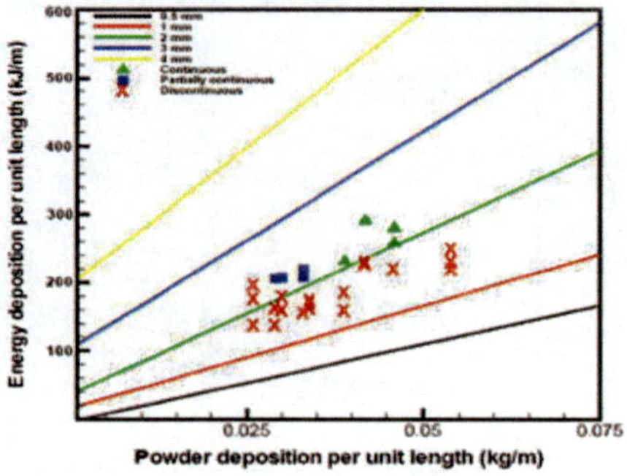

(a)

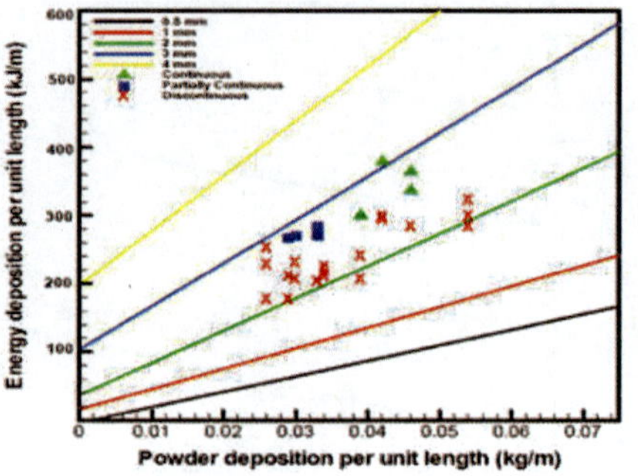

(b)

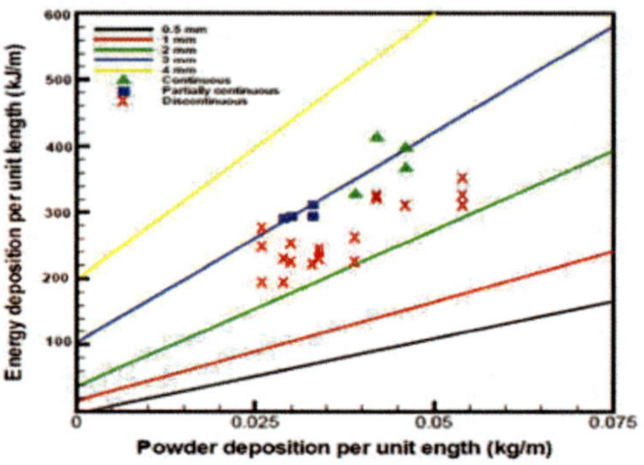

(c)

Fig. 8 Bead pattern analysis taking point heat source and efficiency **a** 60% **b** 78% **c** 85%

continuous and discontinuous beads. The energies per unit length on the substrate for continuous beads are 290.20, 336.43, 318.28 and 275.18 kJ/m. The melt pool size (width of bead) and melting of powder per unit length depends upon energy input per unit length. The energies per unit length on the substrate for partially continuous beads are 264.45, 252.33, 249.24 and 228.88 kJ/m. The energies per unit length on the substrate for discontinuous beads are less compare to energy per unit length required for continuous and partial continuous beads. Mandal et al. [18] experimentally observed that within the conducted experimental range for discontinuity analysis, the measured bead width of continuous bead is in between 4 and 5 mm. Therefore, the graph between energy deposition per unit length and powder deposition per unit length of constant bead widths are taken point heat source and experimental data points considering efficiency 85% shows more accurate results compared to others results. From the constant melt pool width line using point heat source and the experimental data points considering efficiency at 85% in Fig. 8c, it has been observed that when the melt pool width is above 3 mm, then the bead is continuous. If the melt pool widths are around between 2.8 mm and 3 m then the bead may be continuous, partially continuous and discontinuous. If the melt pool width is below 2.8 mm then beads are discontinuous.

The normal trends of this model's (line source and point source) results are over predicted [16]. The error is very large in case of line heat source compared to point heat source, so that, the line heat source model is not suitable for melt pool size or bead width prediction. Here, point heat source model has been accepted for prediction of bead width or melt pool size.

6 Conclusions

Higher thermal energies per unit length on the substrate show continuous beads. For partial continuous and discontinuous beads energies per unit length are less compare to energies per unit length required for continuous beads. If the melt pool width is above 3 mm, then the deposited bead is continuous. If the melt pool width is around in between 2.8 and 3 mm, then the deposited bead may be continuous, partially continuous and discontinuous. If the melt pool width is below 2.8 mm, then the deposited beads are discontinuous.

References

1. Hou ZB, Komanduri R (2000) General solutions for stationary/moving plane heat source problems in manufacturing and tribology. Int J Heat Mass Transf 43:1679–1698
2. Kwon Y, Weckman DC (2008) Analytical thermal model of conduction mode double sided arc welding. Sci Technol Weld Joining 13(6):539–549
3. Kim CK (2011) An analytical solution to heat conduction with a moving heat source. J Mech Sci Technol 25(4):895–899
4. Rosenthal D (1941) Mathematical theory of heat distribution during welding and cutting. Weld J 20(5):220s–234s
5. Eagar TW, Tsai N (1983) Temperature fields produced by travelling distributed heat sources. Weld J 62(12):346s–355s
6. Steen W, Dowden J, Kapadia P (1988) A point and line source model of laser keyhole welding. J Phys D Appl Phys 21(8):1255
7. Akhter R, Davis M, Dowden J, Kapadia P, Ley M, Steen W (1989) A method for calculating the fused zone profile of laser keyhole welds. J Phys D Appl Phys 22(1):23
8. Resch M, Kaplan A (1998) Heat conduction modelling of laser welding. Lasers Eng 7(3–4):229–240
9. Hilton P (1995) Eu194 in the United Kingdom. Opt Quant Electron 27(12):1127–1147
10. Ali Y, Zhang L (2005) Relativistic moving heat source. Int J Heat Mass Transf 48:2741–2758
11. Majumdar P, Xia H (2007) A green's function model for the analysis of laser heating of materials. Appl Math Model 31:1186–1200
12. Mills KC (2002) Recommended values of thermophysical properties for selected commercial alloys. Woodhead Publishing Limited and ASM International
13. Messler RW (2004) Principles of welding processes, physics, chemistry and metallurgy. Wiley-VCH

14. Meillot E, Guenadou D (2004) Thermal plasma flow modeling: a simple model for gas heating and acceleration. Plasma Chem Plasma Process 24(2):217–238
15. Elijah Kannatey-Asibu J (2009) Principles of laser materials processing. Wiley, Hoboken
16. Pinkerton AJ, Li L (2004) Modelling the geometry of a moving laser melt pool and deposition track via energy and mass balances. J Phys D Appl Phys 37:1885–1895
17. Bharathi RS, Shanmugam NS, Kannan RM, Vedan SA (2018) Studies on the parametric effects of plasma arc welding of 2205 duplex stainless steel. High Temp Mater Processes (London) 37(3):219–232
18. Mandal S, Kumar S, Bharagava P, Premsingh CH, Paul CP, Kukreja LM (2015) An experimental investigation and analysis of PTAW process. Mater Manuf Processes 30(9):1131–1137

A Case Study on Survey Plan for Digital Merchandising System and Consumer Association Management

Sundeep Kumar, Vikram Singh Rathore, and Alok Mathur

Abstract This exploration plans to investigate business measures and what the variables affect digital promoting and consumer relationship management frame works which information should be broke down and incorporated into the framework, execute it. How powerful for mix of digital showcasing and Consumer Relationship Management along huge information empowered by help Merchandising and Consumer Relation tasks. Exploration dependent on contextual investigations at ABC Companies (WVPS) Worldwide Vernacular Pedagogy Service. Examination are contemplating auxiliary information for upheld through subjective exploration strategies. Utilizing a purposive inspecting procedure with perception and meeting a few respondents who need framework reconciliation. The documentation of the meeting is coded to keep the classification of the source. Technique for broadening interest, divided into three dimension of information origins, conversations, with sufficiency for hypothesis is utilized for approve information. These models are provided by Huberman and Miles are utilized for examine the information meet. The consequences of the exploration are required to turn into an all encompassing way to deal with completely incorporate with digital Promoting Consumer Relationship Management framework with utilizing Big data analytics.

Keywords Digital merchandising · Consumer relationship management · Huge amount of data analytics · Three dimension of information origins

S. Kumar (✉) · V. S. Rathore · A. Mathur
Department of Technical Education, Centre for Electronic Governance, Government of Rajasthan, Jaipur 302004, India

R. Agrawal et al. (eds.), *Recent Advances in Industrial Production*, Lecture Notes in Mechanical Engineering, https://doi.org/10.1007/978-981-16-5281-3_21

1 Introduction to Digital Merchandising and Consumer Relationship

This exploration plans to investigate business measures and what the variables affect digital promoting and Consumer Relationship Management frameworks Which information should be broke down and incorporated into the framework, utilized with powerful of mix of digital showcasing and Consumer Relationship Management with huge information empowered to help Merchandising and Consumer Relation tasks. Exploration dependent with contextual investigations at ABC Companies Worldwide Vernacular Pedagogy Service. Examination is contemplating auxiliary information which is upheld by subjective exploration strategies. Utilizing a purposive inspecting procedure with perception and meeting a few respondents who need framework reconciliation. The documentation of the meeting is coded to keep the classification of the source. Technique for broadening interest, three dimension of information origins of information sectors, conversations, with sufficiency through hypothesis is utilized to approve information. According to Huberman's and Miles models are utilized by examine the information meet. The consequences of the exploration are required to turn into an all encompassing way to deal with completely incorporate with Consumer Relationship Management frameworks using Big Data Analytics program with digital promoting. Li [1]: the consequences of advanced information examination can improve showcasing techniques, locate the likely business sectors, and can address the aftereffects of reasoning and human investigation (advertisers). It is imperative to join information examination with a top to bottom comprehension of the market is deciphering the client venture. Business objectivity factors internet promoting and web based client commitment according to Fotaki et al. [2] increment latest clients, increment consumer loyalty and reliability, lessen beat rates, increment deals. Beloin [3]: the cooperation or association of under studies and clients assumes a significant function in the accomplishment of a consumer relationship management framework. By planning a framework that can include clients, for example, correspondence input, recommendations, grumblings including the clients following history will expand client maintenance.

2 Review of Literature

Organizations that can measure and use information dissemination will get numerous advantages for the way toward creating and deal their items or administrations, have competency in keeping up great connections and client reliability. Organized information that have information types, companies, characterized by Sirait [4: 118]. Information like as value-based information, Online Analytical Processing information, conventional Relational Database Management Systems, Comma Separated Values records, basic spreadsheets. While unstructured information is printed information with a sporadic configuration with nothing characteristic formation, so to put required information in a organized manner to used to attempt timely with instruments.

2.1 Getting Information from Big Data Sources

According to McKinsey Global (2011), Information from Big Data source used to characterized through information converted into plate magnitude, dissemination rate, the extremely huge assortment, or potentially suffering, in this way requiring the utilization of specialized engineering and creative scientific techniques to pick up knowledge into who can give new business esteem significant data. Also, in its turn of events, some such as seven V like Visualization, Variability, Value, Velocity, Veracity, Volume and Variety. Huge information are a term for a huge and composite arrangement for information that cannot be taken care of with customary PC innovation frameworks by Hurwitz (2013). While as indicated by Sirait [4: 114] information has a significant part in key dynamic. Along these lines, parties who can measure and use accessible information in enormous volumes, change in variety, high intricacy, and rapid of information augmentations can make huge benefits.

2.2 Digital Merchandising

Promoting Digital Marketing or electronic showcasing for utilization of information innovation by advertising examples or exercises is concept provided by Strauss and Frost [5: 23]. Toward this making, imparting, convey, and market provide an incentive for clients, Consumers, accomplices, and the more extensive network. Streamlined E-Merchandising is the consequence of data innovation that is received for conventional promoting strategies. E-Merchandising is essential for E-Business that utilizes digital media to direct promoting exercises to accomplish advertising objectives as per Hermawan and Ahmadi (2013: 186). Various types of promoting Electronic-Marketing, for example, web showcasing, intuitive promoting, and versatile advertising. In view of the phrasing above, it tends to be conclude that digital advertising (Digital-Marketing) or digital showcasing is otherwise called Web based and digital marketing or we can say that online marketing.

According to Kalyanam and McIntyre [6: 11], the advertising blends on digital promoting, digital showcasing capacities are depicted as an advertising blend component. The arrangement of costing, help, with a reduction of an object to one essential detail which has struck us as more important than the others called symbolization are essential capacity of digital advertising that is planned with items, and advancements as independent digital promoting components. Nine out of eleven are delegated fundamental capacities, while seven capacities moderate different impacts covering. The non-covering capacity is set on the outside of the three Dimension shape. Capacities that moderate different capacities are put at the lower part of the 3D shape to outline that the majority of these capacities work by directing capacities on a superficial level to add control to one another. The consequences through digital promoting blend possibly planned as per following.

- Four P as Price, Place, Product, Promotion.
- Two P as Privacy, Personalization.
- Two C as Community, Consumer Service.
- Four S as Sales Promotion, Security, Site.

According to McIntyre and Kalyanam likewise distinguished digital promoting devices are assembled by the digital showcasing blend. There are four areas of internet promotion as indicated Business to Business, Business to Consumer, Consumer to Business and Consumer to Consumer (C2C) as per Kotler and Armstrong [7: 526]. Through Digital media such as internet promoting, advertisers using websites, digital mail, web-based media, and many other online assets to discover brand-new corporal client occupation openings, offer with exist clients, and distribute clients successfully and productively in today's business scenario. By understanding the area of web-based showcasing, organizations and advertisers can without much of a stretch create key anticipating digital promoting frameworks.

As per McIntyre and Kalyanam also defined Digital Merchandising tools. Mainly there are four types of online Merchandising according to Kotler and Armstrong [7: 526]. Such as B2C, B2B, C2C, and C2B. Web-based Merchandising B2B, retailers used websites, digital-mail, social-media, and Consumers successfully with proficiently. Consumers contribution is required by companies in planned and making a special substance.

1. Developed websites.
2. Web Publishing and Advertising.
3. Developed Communication skills companies or Web servers services.
4. Information conversation through e-mails
5. SMS and Youtube, Radio channels.

2.3 *Consumer Relationship Management*

In this phase the Consumer Relationship Management or business companies used techniques that are related to inward cycles through works and outer companies. According to Buttle and Maklan [8: 16]. It improve development with incredibility on an incentive to clients. It indicate the fundamental of much capacity of consumer relationship expert and recognize information with the revolution frameworks. In this concept used three kind of strategic methods for consumer relationship management.

2.3.1 Strategic Consumer Relationship Management

Zeroed in at length creating items/administrations following the organization's business culture that is arranged to the requirements and wants of clients, to keep up client faithfulness by making and conveying more worth contrasted with contenders.

2.3.2 Operational Consumer Relationship Management

Centers around business computerization measures wanted by clients. Through a promoting robotization framework, deals power computerization, and administration mechanization.

2.3.3 Analytical Consumer Relationship Management

Zeroed in on the way toward acquiring, putting away, handling, incorporating, circulating, and utilizing client information, used to expand an incentive from between client with organization.

As per models of consumer relationship management used to develop on a specific date that is called IDIC have been created to date, such as Identify, Differentiate, Interact, Customize Models, Model developed by Adrian Payne the model identifies five core processes in consumer relationship management such as the strategy development process, value creation process, multichannel integration process, performance assessment process and the information management process and Value Chain method in reference of consumer relationship management.

According to Payne and Ford the IDIC model most suitable for proposed research paper. In this life there are four phases for consumer relationship management with retailer or customer.

Identify: What is the relationship with consumer with company.

Differentiate: Check whether the consumers can longer relationship with company.

Interact: In this phase the consumer is satisfy with company consumer orders, consumer relationship and understand the demand of consumer with satisfaction with branded items etc.

Customize Models: In this phase the retailer and company make sure that the consumer is satisfy with product according to offer and methods with guarantee that consumer is fully satisfy.

Through planning communications among clients and companies, an unmistakable business cycle and cooperation will be acquired to do a fitting consumer relationship management framework arranging examination. Logical consumer relationship management by using great client information organized in inner server farms and unstructured information, to be prepared and broke down will have the option to deliver an arrangement of connections among clients and organizations that are more custom and explicit.

3 Methodology and Techniques

Coming up next are a portion of the central matters of theory research dependent on hypothetical, Digital Merchandising, Data Source and Consumer Relationship Management Systems (Fig. 1):

3.1 Data Source Used in CRM as (Big Data Analytics)

A. Techniques and strategies: Cloudera, Microsoft, Power BI, Oracle Analytics Cloud.
B. Information Managed: In this phase data and type of data such as structured and unstructured example like Web based examination, Youtube channels, facebooks, instagram, chat analytics etc. In this method we discuss on digital promoting various analytical reports related to particular company business.
C. Information from the investigation cycle: Pattern information and examination with viability to promoting and consumer relationship management frameworks for complete review of company.

Fig. 1 Research methodology

3.2 Digital Merchandising

3.2.1 Merchandising Devices and Techniques

Websites, online transactions, SMS, electronic mail promotions impacts.

3.2.2 Prepared Informations

Actual Reports as output: factual detail from clients which get advancements through the framework.

3.3 Consumer Relationship Management

3.3.1 Procedure

Systematic consumer relationship Management, Functioning consumer relationship management.

3.3.2 Information Generation

Organized interior information data such as structured and unstructured example like Web based examination, Youtube channels, facebooks, instagram, chat analytics, WhatsApps etc.

3.3.3 Outcomes

In this phase the businessman or a companies required a customized details of sales or promotions of items in a particular field using consumer relationship management frameworks. It can also get from digital emails, SMS, WhatsApps groups etc. for business requirements satisfactions.

4 Research Design

In research study we examination utilizes subjective exploration techniques. In every framework improvement plan consistently requires elucidating information as organized meetings through gathering with exploration specific, for acquire genuine information extra profundity. Straight perception, perception, survey and conversation for likewise expected for additionally advance the exploration information on an issue. According to Suwendra [9: 8], a advantages for subjective examination with enhancement for training, contextual investigation exploration will follow the arranging, measures, and development of a program with the goal with entirely significant for development and executions. Basic fundamental for issue with authentic issue for beginning stage through examination (Table 1).

Issue recognizable proof is completed in week after week gatherings. Given the information classification factor, the gathering notes can't be completely portrayed in this investigation. To keep up the legitimacy through information with examination matters, are intelligent factors have been completed for lessen via issue that the study stays zeroed in on its goals:

A. Starting meet with Ms. Sita Narayan for top of the Digital Merchandising Promotion division: "As per market and companies requirements are fulfil by

Table 1 List of informations

S. No.	Conditions/Requirement/Information	Name	Post
1	Completely devoted to improvement and development of company	John	Managing Director
2	Considerable scholastic with expert knowledge between association's promoting framework and have abilities in advanced promoting arranging	Ms. Sita Narayan	Head of Merchandising and Digital Content
3	Have capability in keeping up great associations with understudies and guardians	Mr. SNS	Head of Consumer Section
4	Consumer relationship and advertisement	SSh. KD Singh Sh. PDS	Consumer Relationship Employee Management
5	Collecting data from market or open source from emails, SMS, whatsapps YouTube channels, radio channels, open query request, order lists analysis consumer complete information gathering etc.	Mr. RSR	Technical and Information System Officer
6	A person having good communication skills to improve and trained the company employees as well as communicate with out sides consumers	Mr. JJB	Director of Study
7	A Person have a complete knowledge of Business Study. Collaboration with companies	Dr. David	Assistant Director

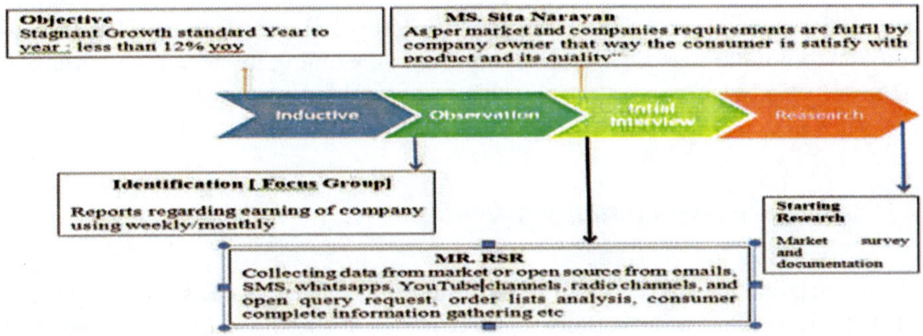

Fig. 2 Research stages

company owner that way the consumer is satisfy with product and its quality" for consumer relationship management frameworks.

B. Starting meet with Mr. RSR as Technical System Software Developer or System Officer "Collecting data from market or open source from emails, SMS, whatsapps, YouTube channels, radio channels, and open query request, order lists analysis, consumer complete information gathering etc." (Fig. 2).

Issue distinguishing proof is completed in week by week gatherings. Given the information secrecy factor, the gathering notes can't be completely depicted in this examination. To keep up the legitimacy and information with exploration matters, are an intelligent the work that is complete for diminish issue, i.e. research stays zeroed in destinations:

A. Starting meet with Ms. Sita Narayan for top of the Digital Merchandising Promotion division: "As per market and companies requirements are fulfil by company owner that way the consumer is satisfy with product and its quality" for consumer relationship management frameworks.

B. Starting meet with Mr. RSR as Technical System Software Developer or System Officer "Collecting data from market or open source from emails, SMS, whatsapps, YouTube channels, radio channels, and open query request, order lists analysis, consumer complete information gathering etc.".

Casual meetings are directed, trailed by finishing up the aftereffects of the week by week meeting, particularly on focuses identifying with this exploration research Digital Merchandise, Consumer Relations, and Program Development. These are rundown from inquiries accommodate with organized meeting convey, discussion were directed with every witness.

Shows information on development and number of understudies in the course of recent years.

Two principle matters influence to nature of exploration outcomes, in particular, the nature for instrument and the nature of information assortment. In subjective examination, information assortment is completed in characteristic settings, essential information sources, and more information assortment methods in member perception, top to bottom meetings, and documentation. Coming up next are a portion of the information assortment strategies this completed for examination.

4.1 Absolute Association Perception

Many scientists are engaged every day exercises related matters considered and utilized exploration information origin. Towards gathering information, analysts are completely engaged with finished through information origin air is characteristic, with specialists not interested in investigate. Towards exercises specialists workers continually have great connection with related matters, perceptions outcomes straightforwardly phases chosen perception smaller than expected visit. Analysts can portray the center found with the goal that the information is more essential.

4.2 Structured Meetings

Analysts utilize this meeting procedure as a technique for information assortment since they definitely know for sure about the data to be acquired. Respondents were given a similar inquiry, the specialist recorded and reported. Exploration instruments, for example, corporate optional information, pictures, and advanced voice recorders are utilized to enable the meeting to deal with run easily.

4.3 Documentation

Information as photograph documentation, and digital records recorded, at that point accomplices provided in quantity record along programming codes keep up classification for exploration research information.

5 Information as Analytical Work

Concerns with subjective exploration, primary standards of study information legitimate, dependable, impartial. Results are legitimacy information with examination utilizes validity examine are incorporates:

Supplement of Perception and Reinforce
As per examine the information acquired one of two straightforwardly primary witness and extra sources as indicated by the snowball technique. An endorsement legitimacy to the result expansion appended to the connection with exploration research.

Three Dimension of Information Origins
Utilizing origin three dimension of information origins technique, through investigate the accuracy of the information got through different origins are applicable to the concern with similar sections. Zero in with advanced advertising, consumer relationship management, with anticipated advantages along with a Large Source of Information coordination.

Employ Sufficient Recommendations
According to Sugiyono [10: 245], investigation for subjective information empirical to the world, in particular, examination dependent upon information got, at that point formed into a theory. Given the theories defined dependent on the information, at that point the information is looked again consistently so it would then be able to be closed whether the theory is acknowledged or dismissed dependent on extra information gathered. At the point when dependent on information that can be gathered over and again with Three dimension of information origins methods, it turns out the speculation is acknowledged, at that point the theory forms into a hypothesis. Information investigation is done at the hour of information assortment, and after finishing in a specific period.

As per above diagram activity are categories into three parts such as information compaction, Information output, and get conclusion using referred as result/verification/output. Straight-thinking Fig. 3.

In Fig. 4 we can understand that information is first collected in a automated form or manual survey then after it converted into compact by using special software's the information is a collection of fields and related records regarding data representations in the form of output or we can say that result or verification. The diagram is related to research matters.

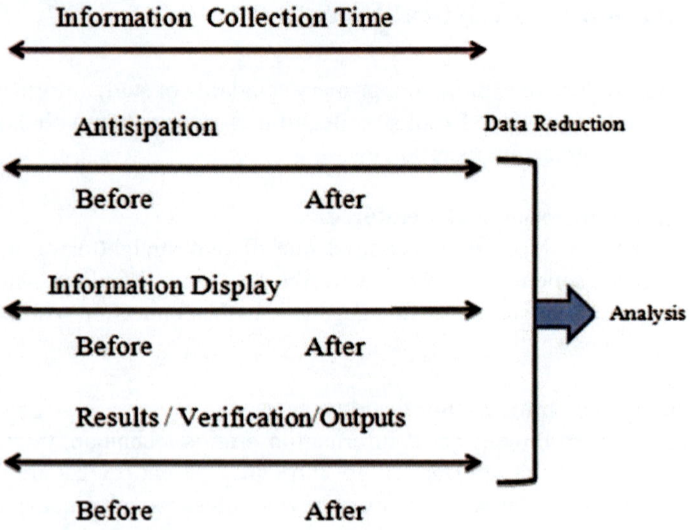

Fig. 3 Information/Data collection component model

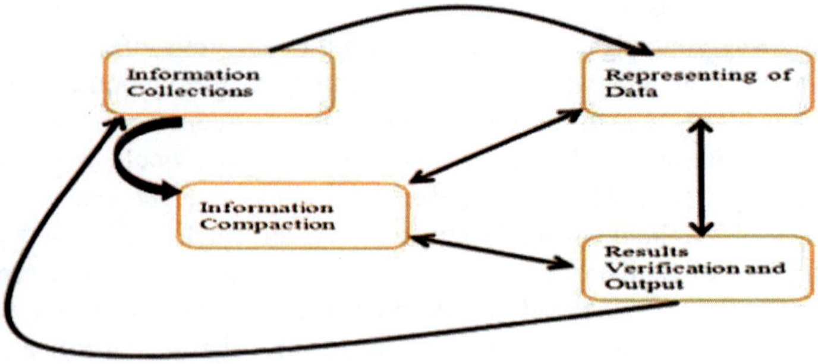

Fig. 4 Information collection model or interactive model

6 Result

Expanding investment, three dimension of information origins, conversations, according to Huberman and Miles models are accustomed to investigating the information meet. In this section we saw the result of the research or study.

6.1 Standard of Unification Through Big Data

Simple information approach, receptive mix to Digital Merchandising and consumer relationship management frameworks, with simplicity of getting information from central data origin. Other than frequently fields for information approach and reliability significant, moreover.

6.2 Digital Merchandising and Consumer Relationship Features

In concern with formation few data representing tools of correspondence with clients, showcasing content match to patterns, assessment, and development of correspondence data representing tools, with appearance of substance modernize highlights were found as an examination concentrate with Digital Merchandising and consumer relationship management Features gathering.

6.3 Investigation of Information Using Factors Related to Consumer Association

Consumer loyalty with grumbling investigation, examination for possible client information, business portion assessment with pattern examination is discoveries due to this address the inquiry “This is a different things should be broke down in the incorporation of this framework”.

7 Conclusion

To incorporating a large amount of data that is predicate from the Digital Merchandising System and Consumer Relationship Management framework, the study is concern with consumer and digital marketing commitment to make promoting understanding patterns. Likewise, the market cycles make consumer requirement have fundamentally influenced Digital Merchandising and consumer relationship management system execute viably. Simple information and approach, reliability, making a few data representing tools correspondences through web-based media and online visit are other significant focuses that were found from this examination study. Each one of those focuses were referenced by respondents while talk with, found in the information decrease from cycles of broadening interest. On other hand defendant concurred the Digital Merchandising frameworks and consumer relationship management with huge information empowered expanding effectively technically of the Merchandising and consumer relationship management division.

About examination the research has been directing in the Worldwide Vernacular Pedagogy Service division and population analysis. The subjective examination strategy should be directed to discover how the effectiveness of the framework by any means.

References

1. Li W (2016) Digital media data and market intelligence. Thesis management for international business not published, Erasmus University Rotterdam
2. Fotaki G, Spruit M, Brinkkemper S, Meijer D (2013) Exploring big data opportunities for online consumer segmentation. Technical report UU-CS2013-021. Utrecht University Netherland
3. Beloin CA (2018) A study of consumer relationship management and undergraduate degree seeking student retention. Ed.D. dissertations not published. Concordia University Portland
4. Sirait ERE (2016) Implement as technology big data di Lembaga Pemerintahan Indonesia. J Penelitian Posdan Informatika 6(2):113–136
5. Strauss J, Frost R (2014) E-marketing, 7th edn. Pearson Education Limited, England
6. Kalyanam K, McIntyre SH (2002) The e-merchandising mix: a contribution of the e-tailing wars. J Acad Merchandising Sci 30(4):483–495. (2008) 11:1621–1635
7. Kotler PT, Armstrong G (2014) Principles of marketing, 15th edn. Pearson Education Limited, England
8. Buttle FA, Maklan S (2015) Consumer relationship management concept and technologies, 3rd edn. Routledge, New York
9. Suwendra IW (2018) Metodologi penelitian kualitatif dalam ilmu sosial, pendidikan, kebudayaan dan keagamaan. IW Suwendra. Nilacakra 107
10. Sugiyono (2016) Metode Penelitian Kuantitatif, Kualitatif, dan R&D tahun terbitan
11. McNulty E (2014) Understanding big data: the ecosystem (online). https://dataconomy.com/2014/06/understanding-big-data-ecosystem. Accessed at 23 Sept 2018
12. Setiyaningrum A, Udaya J, Efendi (2015) Prinsip-Prinsip Pemasaran Plus Tren Terkini. Andi, Yogyakarta

Medical Applications of Rapid Prototyping Technology

Rakesh Chaudhari, Praveen Kumar Loharkar, and Asha Ingle

Abstract Rapid Prototyping (RP) has turned out to be an evolutionary technology in the medical and healthcare field. Technology is constantly getting evolved resulting in advanced applications in various domains of the healthcare industry. The use of RP and medical advancements are complementing each other in offering better services to the patients. In this context, this paper presents an overview of various rapid prototyping techniques which are collectively called additive manufacturing processes. The wide spread potential of RP in various medical applications such as prosthesis and implant development, surgical planning, modern drug delivery systems, tissue engineering and medical models and devices is also presented. Moreover, the case studies discussed in the paper show the use of RP in overcoming the complexities and anomalies in human anatomy. The aim is to present the capabilities of the RP technique with the help of reviewed literature and discuss the associated challenges.

Keywords Drug delivery system · Fused deposition modeling · Prosthesis · Rapid prototyping · Stereolithography · 3D Printing

1 Introduction

Rapid prototyping is being use in medical and healthcare field in many ways. RP potentially decreases product development time and cost with the use of CAD data to produce tangible prototypes of an object. These prototypes are used in number of medical applications such as diagnosis, implants designs, external prosthesis, artificial organs production and planning of complicated surgery [1–3]. The diverse features of RP to convert CAD data into 3D object provides new approaches to develop prototypes with better mechanical properties and good biocompatibility

R. Chaudhari (✉) · P. K. Loharkar
Department of Mechanical Engineering, SVKM's NMIMS, MPSTME Shirpur Campus, Dhule, Maharashtra 425405, India

A. Ingle
Department of Mechanical Engineering, SVKM's NMIMS, MPSTME Mumbai, Mumbai, Maharashtra 400056, India

R. Agrawal et al. (eds.), *Recent Advances in Industrial Production*, Lecture Notes in Mechanical Engineering, https://doi.org/10.1007/978-981-16-5281-3_22

[4]. One of the most important applications of RP in medical arena is designing, development and manufacturing of medical devices and instrumentation [4].

Rapid prototyping is also being applied in the context of drug delivery due to its capability of producing highly precise intricate forms of solid dosage. It is observed that these drug dosage forms show excellent uniformity and control of dosage compare to existing mixing and pressing techniques [5].

RP can be competently used in complex cases of surgery. The patient's musculoskeletal model can be generated using data information obtained from various sources. These musculoskeletal models are then referred to study the deformity, fracture or intricacies in any part of body [6, 7]. Following sub-section presents the procedure and techniques of RP technology.

1.1 Rapid Prototyping Procedure

The Rapid Prototyping (RP) is primarily an additive technique which produces the prototypes of components by adding volume elements using 3D computer aided design (CAD) data. The basic procedure of RP as generative manufacturing technique consists of (i) construction of a CAD model based on the original object or geometrical data, (ii) conversion of CAD model to STL format, (iii) Laying down the successive thin layers, (iv) fabrication of prototype by layered deposition (v) Post processing/cleaning [8].

In the literature, several RP techniques are presented based on the principle or mechanism involved in the layer by layer deposition such as Stereo-Lithography (SLA), Fused Deposition Modeling (FDM), Selective Laser Sintering (SLS), 3 Dimensional Printing (3DP), Laminated Object manufacturing (LOM) [9].

Even though there are advantages of RP technology, its usage in the medical field has not reached its potential. There are several reasons attributed to this situation such as biological mismatch of materials, surface finish of product, anatomical accuracy of generated model and high cost for model fabrication. These factors restrict broader use of RP in medical applications. In the subsequent sections, the use of RP in medical applications is elaborated which is followed by the discussion on case studies with an aim to understand the state of art and related challenges. The aim is to present a consolidated review to comprehend the scope for subsequent research so that the RP technology can be used at a larger scale in medicine and healthcare.

2 RP in Medical Applications

Use of RP in medical application has made an impressive progress in recent years. Figure 1 shows the key applications of RP in the medical field. It is widely used in various medical applications such as prosthesis and implants, modern drugs delivery systems, medical devices and instruments. In addition, surgery planning, dentistry,

Fig. 1 RP in medical applications

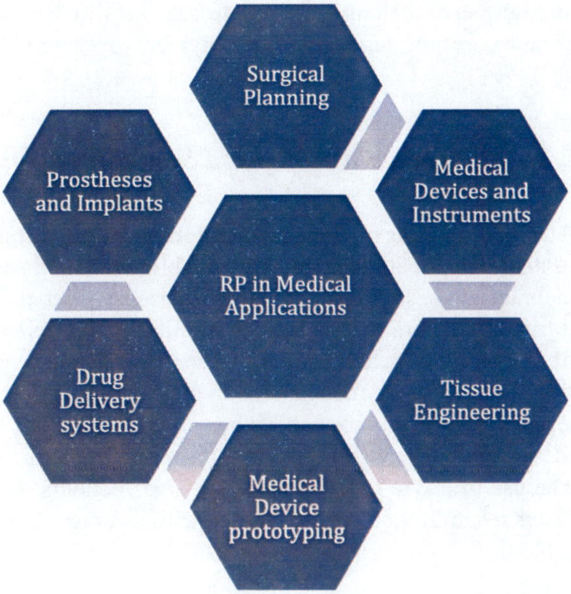

tissue engineering etc. are applications where RP is being used. These applications of RP in medical field are described in the subsequent sections.

2.1 *Prostheses and Implants*

Rapid prototyping is substantially used in the fields of prosthetics and implantation. The techniques help to manufacture a custom prosthesis of unique dimensional accuracy to precisely fit into patient. RP has been used in making hip sockets, knee joints and some other implants [10]. While making prosthesis, patient's data is a collected using different measurement technique such as Computed Tomography (CT) or Magnetic Resonance Imaging (MRI). The data is then used to fabricate the replica of patient's anatomy to replace deformed or fractured body part of patient.

In some cases, printed prosthetic components are found to be relatively weak and fragile [10]. A 3D scanner and with selective laser sintering can also be used to develop prototype of ankle and foot orthoses (AFO) [11]. The study to explore the feasibility of RP application in manufacturing AFO shows that selective laser sintering (SLS) is most suitable technique. The detailed study on use of RP in fabrication of AFO express that replica of carbon fiber AFO (CF-AFO) has shown better bending stiffness characteristics [12]. There is still a wide scope for fabrication of orthoses from scan data of human anatomy and evaluation of the mechanical effect.

The recent research shows that biocompatible materials are now being used for production of medical implants It is observed that biocompatible materials such as high-purity calcium phosphate ceramics hydroxyapatite, (HA or HAp) and its composites with biodegradable polymers gives good biocompatibility [13, 14].

2.2 *Surgical Planning and Scientific Applications*

Advance planning mitigates risk of complications during a surgery. RP helps surgeons to plan complicated surgeries by enabling detailed study of model produced by RP technique. These prototypes then are used for detailed study and understanding the intricacies before surgery. The Stereolithography (SL) is suitably used in these types of modelling as generated models are translucent to revealed internal details. Selective coloring in the models aids in clear visualization of special anatomical characteristics such as tumors and other anomalies within the bone or tissue. Moreover, these models are also used for study and clinical diagnosis of patient's disease [6]. The use of RP in surgical planning gives benefits of minimum complications with faster rehabilitation, enhanced surgical accuracy, lesser injury to healthy tissue etc. [15, 16].

2.3 *Biologically Active Implants and Tissue Engineering*

Tissue engineering can be used in femur bone implants, implants of hard and soft tissues of human body, reconstructive surgery, root canal treatments, etc. It involves development of scaffold: a porous artificial extracellular matrix essential to accommodate and progress the growth of mammalian cells with generation of new tissue [9]. The loss of a human organ or tissue is very costly affair in medical field. Rapid prototyping significantly contributes in tissue engineering field as it can manufacture bioactive implants due to its ability to use biomaterials. Tissue engineering is very beneficial to repair damaged tissue or organs due to accident or surgery with the biological substitutes where present medical treatment is not sufficient. The replacement is done by transplanting the cell on biological scaffolds. These scaffolds work similar to natural cells of the patient and allow its growth by providing provisional mould [17].

RP techniques such as fused deposition modelling (FDM), 3D printing (3DP) and selective laser sintering (SLS) are used to fabricate scaffolds. FDM, 3-DP and SLS are mostly used to fabricate the mould for tissue engineering scaffolds though these techniques can fabricate a functional scaffold directly [18]. Biodegradable scaffolds fabricated using FDM are good carriers of osteoblasts therefore proved successful to replace hard tissue. Number of biocompatible polymers such as, polyesters, poly-anhydrides, polysaccharides, and hydrogels are finding applications in tissue engineering applications [19, 20].

2.4 Modern Drug Delivery System

Drug delivery is process of delivering a pharmaceutical active ingredient (API) at the required location of body where desired therapeutic effect is to be obtained. The considerable inventions have been carried out in drug delivery with conventional dosage forms used in modern drug delivery systems [21]. In recent years, development of controlled-release systems has progressed though different multistep manufacturing technologies. Due to shorter biological life and limited therapeutic window, effect of current drug delivery system, a controlled release (CR) dosage are at the center of attraction of healthcare innovations.

Rapid prototyping is revolutionary technique in the field of drug delivery as it can produce dosage with complex drug release profiles and gives precise control of droplet dose and size [22]. The 3D printing is quite helpful to develop simple, accurate, cost effective and tailored drug delivery systems [23, 24]. The technique is able to produce solid dosage forms with different densities and diffusivities, intricate geometries and multiple drugs. The challenges of delivery of poor water soluble drugs, potent drugs those tends to produce violent effects, peptides and release of more than one drug can be resolve by 3D printing technique.

3D printing has also acquired the capability of fabrication of drug-loaded tablets with sustained drug release profiles changed by a cautious selection of printing variables [25]. However, the selection of appropriate excipients, binders and the properties of the formed product are the factors that limit the application of 3D printing commercially to fabricate novel drug delivery system.

2.5 Medical Models and Device

Medical modelling process is used to re-create specific human body part in as a prototype. These models are instrumental in preoperative surgery planning, fabrication of implants, prosthetics, and study of human anatomy [26, 27]. RP has found innovative use in medical applications to generate medical devices and models, physical prototypes of human anatomical parts etc. RP can produce hand-held appliances, anatomical models and large-scale treatment units [28]. The 3D models fabricated using RP are used to characterization of vital anatomy, simulation of procedure and testing of surgical tools. The pre-study of patient's anatomy with an availability of life size replica of structures is more advantageous than observing images on a computer monitor [29]. These models make treatment cost effective and faster for complicated diseases and surgeries. These models are generated according to patient's requirement with high accuracy. Medical models of hearts, bones, and joints allow surgeons to study and understand the medical problem and plan the treatment accordingly. These models are very beneficial to plan skulls, shoulders, jaw, hips, and knees surgeries more proficiently [30].

RP finds its use in prototyping of stents, catheters, syringes, fluid administration systems, cardio-vascular devices, vascular access devices etc. Besides medical device prototyping, few RP techniques such as FDM and 3D printing are turning out to be important tools for developing new surgical instruments and devices [31].

3 Case Studies on RP in Healthcare

In this section, few case studies on the implementation of RP in the field of healthcare have been discussed.

Herbert et al. [32] has developed prosthetic socket for transtibial and transradial patients using 3D printing technique at Rapid Design and Manufacture (RDM)-RP research Centre, Glasgow. The residual limbs of transtibial and transradial patients were scanned using the TracerCad Premier Prosthetic system. The digital data of patients' residual limb were then transferred to form 3-D surface. The surface data was then converted into three dimensional solid model. The researchers used data of both sets to construct 3D model which further was converted in STL files using some software. The sockets for transtibial as well as transradial residual limb were fabricated using 3D printing technique. These socket were found to be more comfortable than those fabricated by traditional method as per the patients' experience.

RP based biomedical models were also used to separate conjoined Guatemalan twins' head at University of California's Mattel Children Hospital U.S.A [33]. The twins have conjoined head with common membrane and separate brains and arteries. The medical model of the conjoined head was fabricated using 3D printing technique. The scan images of twins' brain and intersection of two skulls was used to design a 3D surface model. The sets of physical model with high accuracy were then fabricated using 3D printing. These models helped surgeons to study inside the model and plan the rerouting of existing blood vessels during the head separation. Few more case-studies on the use of RP in healthcare is presented in Table 1.

4 Conclusion

The paper presents the progress in the application of rapid prototyping (RP) techniques in medical and healthcare field. It is inferred that the bio-models designed and fabricated using various RP techniques play a vital role for planning of complicated surgery. It helps to reduce operating time and risk to patient's health. RP has also demonstrated significant capabilities in developing customized implants. Moreover, FDM and 3D printing have been instrumental in repairing damaged tissues and biological scaffolds.

RP techniques are making progress in fabrication of novel drug delivery system with various features such as optimized drug release, multiple drug release, dosage forms etc. RP has also proved to be a training tool in medical education for students

Table 1 Case studies on use of RP in healthcare

Authors	Year of publication	Objectives	Outcomes
Vijayaragavan et al. [34]	2014	Design of orthosis club-foot	With the help of rapid prototyping technology, an orthosis was designed that could replace the conventional casting used in the treatment of club-foot. It helps in speeding up the protyping process and improves quality as the data used for modelling can be captured through computer tomography scan
Dawood et al. [35]	2015	Discussion on the use of 3D printing in the applications of dentistry	It was inferred that the 3D printing technology which is one of RP techniques can be used in orthodontics, creating models for surgical planning, metal based restoration of teeth etc. Challenges such as post processing, high cost of materials and maintenance was also discussed
Mothes et al. [36]	2018	Creating 3D models for better understanding of shoulder bones to minimize the operability risks	RP can by successfully implemented in planning an operation for the treatment of deformities in the shoulder bone. This has helped in minimizing the risks during the operations
Mendonca et al. [37]	2016	Use of 3D printing technology to assist in the surgery of craniofacial deformity	RP was used for the craniofacial surgery. It was demonstrated that this technology helps in understanding the true defect, correct planning and execution of the surgery, training etc. within an affordable pricing

(continued)

Table 1 (continued)

Authors	Year of publication	Objectives	Outcomes
Zuniga et al. [38]	2015	Development of low cost prosthetic hand	The study resulted in an alternative prosthetic hand which was 3D printed and affordable. Although the complete functionality was to be tested

and budding practitioners. RP can produce multi-colored models that help students in better understanding of the human anatomy before surgery. Although RP is faster and flexible, its high cost linked to the technology, materials and maintenance along with the need for specialized expertise has restricted the use of rapid prototyping in medical applications. Future research in the field of RP can focus on overcoming the stated challenges for extensive use of this technology.

References

1. Gopakumar S (2004) RP in medicine: a case study in cranial reconstructive surgery. Rapid Prototyping J 10(3):207–211
2. Cruz F, Coole T, Bocking C, Simoes J (2003) Selective laser sintering of customized medical implants using biocomposite materials. Tehnicki Vjesnik 10(2):23–27
3. Raos P, Stoić A, Lucić M (2005) Rapid prototyping and rapid machining of medical implants. In: 4th DAAAM International conference on advanced technologies for developing countries, ATDC'05
4. Javaid M, Haleem A (2017) Additive manufacturing applications in medical cases: a literature based review. Alexandria J Med
5. Shreepad S, Ravi W (2015) New revolutionary ideas of material processing—a path to biomaterial fabrication by rapid prototyping. Procedia Soc Behav Sci 195:2761–2768
6. Katstra WE, Palazzolo RD, Rowe CW, Giritlioglu B, Teung P, Cima MJ (2000) Oral dosage forms fabricated by three dimensional printing™. J Controlled Release 66(1):1–9
7. Wu D, Hu Q, Lu Q, Xu G (2008) Study on the application of rapid prototyping in assistant surgical planning. In: 7th Asian-Pacific conference on medical and biological engineering, pp 729–732
8. Gibson I, Srinath A (2015) Simplifying medical additive manufacturing: making the surgeon the designer. Procedia Technol 20:237–242
9. Stanek M, Manas D, Manas M, Navratil J, Kyas K, Senkerik V, Skrobak A (2012) Comparison of different rapid prototyping methods. Int J Math Comput Simul 6:550–557
10. Tukuru N, Gowda SKP, Ahmed SM, Badami S (2008) Rapid prototype technique in medical field. Res J Pharm Technol 1(4):341–344
11. Herbert N, Simpson D, Spence WD, Ion W (2005) A preliminary investigation into the development of 3-D printing of prosthetic sockets. J Rehabil Res Dev 42(2):141–145
12. Milusheva S, Tochev D, Stefanova L, Toshev Y (2005) Virtual models and prototype of individual ankle foot orthosis. In: ISB XXth Congress—ASB 29th annual meeting, Cleveland, Ohio
13. Faustini MC, Neptune RR, Crawford RH, Stanhope SJ (2008) Manufacture of passive dynamic ankle–foot orthoses using selective laser sintering. IEEE Trans Biomed Eng 55(2):7840–7909

14. Awad A, Fina F, Goyanes A, Gaisford S, Basit AW (2020) 3D printing: principles and pharmaceutical applications of selective laser sintering. Int J Pharm 586:119594
15. Ravaglioli A, Krajewski A (2012) Bioceramics: materials properties applications. Springer Science & Business Media
16. Liu Q, Leu MC, Schmitt SM (2006) Rapid prototyping in dentistry: technology and application. Int J Adv Manuf Technol 29(3–4):317–335
17. Mavroidis C, Ranky RG, Sivak ML, Patritti BL, Di Pisa J, Caddle A, Gilhooly K et al (2011) Patient specific ankle-foot orthoses using rapid prototyping. J Neuroeng Rehabil 8(1):1
18. Hollister SJ (2005) Porous scaffold design for tissue engineering. Nat Mater 4(7):518
19. Butscher A, Bohner M, Hofmann S, Gauckler L, Müller R (2011) Structural and material approaches to bone tissue engineering in powder-based three-dimensional printing. Acta Biomater 7(3):907–920
20. Williams JM, Adewunmi A, Schek RM, Flanagan CL, Krebsbach PH, Feinberg SE, Hollister SJ, Das S (2005) Bone tissue engineering using polycaprolactone scaffolds fabricated via selective laser sintering. Biomaterials 26(23):4817–4827
21. Ratner BD, Hoffman AS, Schoen FJ, Lemons JE (2004) Biomaterials science: an introduction to materials in medicine. Elsevier, Amsterdam
22. Porter SC (2001) Novel drug delivery: review of recent trends with oral solid dosage forms. Am Pharm Rev 4:28–36
23. Schubert C, Van Langeveld MC, Donoso LA (2014) Innovations in 3D printing: a 3D overview from optics to organs. Br J Ophthalmol 98(2):159–161
24. Kolakovic R, Viitala T, Ihalainen P, Genina N, Peltonen J, Sandler N (2012) Printing technologies in fabrication of drug delivery systems. Expert Opin Drug Delivery 10(12):1711–1723
25. Preis M, Breitkreutz J, Sandler N (2015) Perspective: concepts of printing technologies for oral film formulations. Int J Pharm 494(2):578–584
26. Goyanes A, Buanz ABM, Hatton GB, Gaisford S, Basit AW (2015) 3D printing of modified-release aminosalicylate (4-ASA and 5-ASA) tablets. Eur J Pharm Biopharm 89:157–162
27. Chelule KL, Coole T, Cheshire DG (2000) Fabrication of medical models from scan data via rapid prototyping techniques. In: Proceedings of the 2000 conference on time compression technologies. Cardiff International Arena
28. Chee Kai C, Meng CS, Ching LS, Teik LS, Aung SC (2000) Facial prosthetic model fabrication using rapid prototyping tools. Integr Manuf Syst 11(1):42–53
29. Hieu LC, Zlatov N, Vander Sloten J, Bohez E, Khanh L, Binh PH, Oris P, Toshev Y (2005) Medical rapid prototyping applications and methods. Assembly Autom 25(4):284–292
30. Hoang D, Perrault D, Stevanovic M, Ghiassi A (2016) Surgical applications of three-dimensional printing: a review of the current literature & how to get started. Ann Transl Med 4(23)
31. McGurk M, Amis AA, Potamianos P, Goodger NM (1997) Rapid prototyping techniques for anatomical modelling in medicine. Ann R Coll Surg Engl 79(3):169
32. Herbert N, Simpson D, Spence WD, Ion W (2005) A preliminary investigation into the development of 3-D printing of prosthetic sockets. J Rehabil Res Dev 42(2):141
33. Conjoined twins benefit from rapid prototyping. Retrieved 08 Nov 2020, from https://www.theengineer.co.uk/conjoined-twins-benefit-from-rapid-prototyping/
34. Vijayaragavan E, Kurian LM, Sulayman H, Gopal TV (2014) Application of rapid prototyping in the treatment of clubfoot in children. Procedia Eng 97:2298–2305
35. Dawood A, Marti BM, Sauret-Jackson V, Darwood A (2015) 3D printing in dentistry. Br Dent J 219(11):521–529
36. Mothes FC, Britto A, Matsumoto F, Tonding M, Ruaro R (2018) Application of three-dimensional prototyping in planning the treatment of proximal humerus bone deformities. Rev Bras Ortop 53(5):595–601

37. Mendonca DA, Deraje V, Gujjalanavar RS, Gopal S (2016) Case series of three-dimensional printing technology applied in complex craniofacial deformity surgery. J Cleft Lip Palate Craniofac Anomalies 3(2):88
38. Zuniga J, Katsavelis D, Peck J, Stollberg J, Petrykowski M, Carson A, Fernandez C (2015) Cyborg beast: a low-cost 3D-printed prosthetic hand for children with upper-limb differences. BMC Res Notes 8(1):1–9

Prediction of Temperature During Susceptor-Assisted Microwave Heating of Aluminum Using Parametric Simulation

Praveen Kumar Loharkar, Asha Ingle, and Himanshu Singh

Abstract Numerous materials processing applications such as powder metallurgy, surface modification, additive manufacturing, joining, etc. require thermal energy. In most of these applications, heat is derived primarily by induction process, fossil fuel combustion, etc. which have their own limitations. Materials processing using microwave heating has distinct merits in terms of reduction in the process time and energy. However, there are significant challenges in metal heating using microwave energy. The use of numerical simulation can serve the purpose of improved understanding of microwave heating of metals. Owing to multiphysics nature of microwave metal heating, its numerical studies are still not extensively reported. In this work, multiphysics numerical simulation based on finite element analysis has been used to investigate the effect of the microwave heating parameters on the temperature during susceptor-assisted microwave heating of the aluminum specimen. The parameters chosen for the parametric study are load position, input power and time. The maximum temperature achieved for each parametric combination was used to develop a prediction model. It has been revealed that the input power and irradiation time have a synergistic effect while an increase in the distance of the specimen from the base has a negative impact on the temperature rise. The study provides the basis for utilizing the merits of the numerical simulation in parametric analysis and can serve as a reference model for predicting temperature rise during susceptor-assisted heating of metals using microwave.

Keywords Additive · Comsol · Load · Microwave · Parametric · Susceptor

---

P. K. Loharkar (✉)
Department of Mechanical Engineering, SVKM's NMIMS, MPSTME, Shirpur Campus, Dhule, Maharashtra 425405, India

A. Ingle
Department of Mechanical Engineering, SVKM's NMIMS, MPSTME, Mumbai, Maharashtra 400056, India

H. Singh
Department of Mechanical Engineering, Visvesvaraya National Institute of Technology, Nagpur, Maharashtra 440010, India

R. Agrawal et al. (eds.), *Recent Advances in Industrial Production*, Lecture Notes in Mechanical Engineering, https://doi.org/10.1007/978-981-16-5281-3_23

1 Introduction

Heating is a critical step in various materials processing applications involving metals in either bulk or powdered form. Few such applications are powder metallurgy, surface engineering, additive manufacturing, joining, etc. One of the objectives to impart thermal energy to a metal specimen or a metal powder specimen is to pre-heat them for removing moisture as it results in extended processing times, the formation of undesired compounds such as hydroxides and oxides [1], develops hydrogen porosity [2] and leads to inferior properties of the processed material. To avoid moisture-related issues, heating is carried out to dry the material. Drying is an age-old technique and involves a "transient transport" phenomenon [3]. It can be carried out using different heating technologies. In the case of metals, the heating is usually carried out in a furnace based on induction phenomenon or fossil fuel combustion. These are either high energy-consuming method with lesser efficiency and/or lead to longer processing times. In fact, Leung et al. [4] pointed out that for overcoming the disadvantages of pre-heating in conventional mode, a suitable preheating strategy is required to be developed. In processes dependent on high energy laser such as laser cladding, preheating of the substrate is must to avoid high thermal stresses during the cladding process [5]. The major challenge is to control the temperature gradient and the rate of temperature rise during the preheating process. Even in selective laser melting (SLM) powder bed is to be preheated for minimizing thermal residual stresses [6, 7]. Papadakis et al. [7] presented that around 40% energy of the total required in the whole process is consumed during the preheating stage.

An energy-efficient alternative to the extensively used existing mode of heating is to use microwave energy [8]. Microwave is a very important constituent of the electromagnetic radiation spectrum and characterized by longer wavelengths that range from 1 mm to 1 m [9]. It has large scale applications in communication engineering domain and a significant amount of utility in applications that require heating owing to its unique characteristics.

Quite a few applications such as food processing, drying of wood, rubber processing, medical therapy, polymer industries, ceramic and metal sintering, metal joining, metal cladding, metal casting, etc. have been developed which utilize microwave energy [8–16]. Of all these, metal related microwave-based processing applications are an advent of the past two decades when it was established that microwaves can be used to process metal-based systems [12, 17].

Microwave heating is predominately a result of polarization and conduction [18, 19]. Polarization is the back and forth movement of dipoles under the action of an external electric field inherent with the electromagnetic radiations. On the other hand, conduction is due to the movement of free charge and the ions under the same field. The latter type is known as ohmic heating [20]. The heat generated through the former mechanism is known as dielectric heating, which is essentially a case of volumetric heating, as seen in Fig. 1 [13]. This leads to several advantages such as rapid and uniform heating, superior mechanical properties and reduced processing times attributed to the phenomenon of volumetric heating [10].

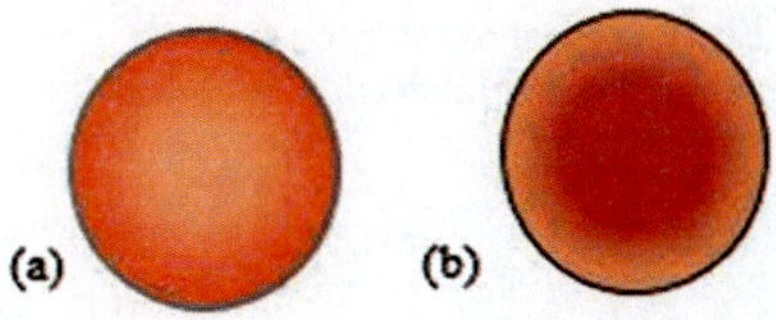

Fig. 1 **a** Conventional heating **b** microwave heating

In addition, an indirect microwave heating technique can be incorporated for materials with poor microwave interaction. In this technique, materials with good interaction with microwaves are used which readily converts microwave energy into thermal energy. These materials are known as susceptors. The heat is then transferred by the means of conventional heat transfer modes [21]. This method of microwave heating is called as "microwave hybrid heating" [22]. There are various parameters such as input power, load position, applicator type, etc. that have an effect on the efficiency of microwave materials processing [8]. In the reviewed work, prediction of temperature rise has seldom been done in conjunction with parametric numerical simulation.

Hence, in this work, numerical simulation of susceptor-assisted microwave heating of metal specimen is carried out to understand the effect of different parametric conditions such as load position, input power and time on the rise in temperature of the aluminum specimen. The objective was to develop a reference regression model as a function of the selected process parameters for the prediction of maximum temperature attained during the microwave heating process.

2 Numerical Modeling of Microwave Heating

A susceptor-assisted heating model was developed to simulate hybrid heating which combines heating through material-microwave interaction and conventional mode of heat transfer. The model includes a specimen enclosed inside a susceptor block of silicon carbide (SiC), an effective susceptor due to its favorable (with respect to microwave interaction) dielectric properties [23]. The steps required in numerical simulation of microwave heating is presented by Loharkar and Ingle [24]. Detailed discussion on the implementation of these steps is presented in the subsequent subsections.

2.1 Modeling Aluminum Specimen with Susceptor

The specimen has been modeled using a spherical geometry while the susceptor has been modeled as a block enclosing the specimen. Figure 2 depicts the model of the whole set-up. The rectangular port at the top right-hand side will take the power input

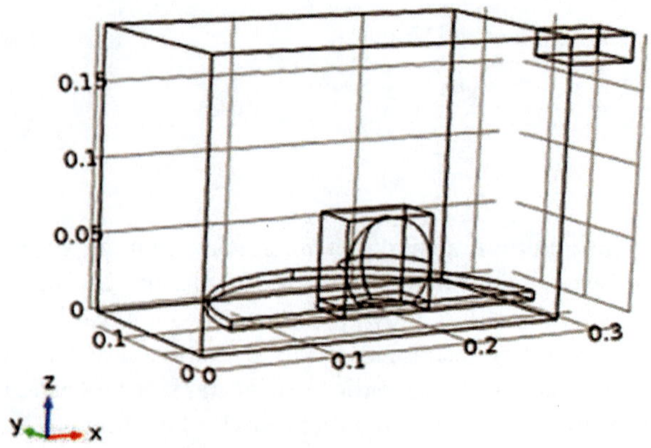

Fig. 2 Model of the aluminum specimen with the susceptor

at the frequency of 2.45 GHz. The rectangular port operates at TE10 mode which means that there is no electric field component in the direction of propagation.

2.2 *Material Properties and Simulation Conditions*

The analysis in COMSOL multiphysics is based on finite element analysis and as an important step in preprocessing, material properties are to be defined. The properties of aluminum and susceptor (SiC) were obtained from published literature [25, 26]. Table 1 comprises the properties of aluminum. The relative permittivity of SiC was taken as 11.45–2 * j, density as 3210 kg/m^3, electrical conductivity as 1.745 S/m while heat capacity and thermal conductivity were taken to be 690 J/kg K and 450 W/m K respectively. The material properties were kept constant over the time range of analysis.

The microwave cavity, microwave oven walls were modeled using built-in materials, air and copper respectively while properties of glass plate were derived from COMSOL application library.

Microwave heating is a multiphysics problem involving equations governing electromagnetic wave propagation (Maxwell's equation) and equations governing heat

Table 1 Properties of aluminum specimen

Name	Value
ε^*	18.44–0.30 * j
μ_r'	1
σ	3.7e7 S m^{-1}
k	205 W m^{-1} K^{-1}
ρ	2710 kg m^{-3}
C_p	900 J kg^{-1} K^{-1}

Table 2 Simulation parameters (factors) and levels

Parameters	Levels					
P_{input} (W)	250	500			750	
Height of the sample, H (mm)	15	45			90	
Time (s)	0	10	20	30	40	50	60

transfer (Fourier's law of heat conduction). Frequency-domain study step was chosen for the electromagnetic part of the study and the time-dependent study step was used for heat transfer part of the study.

Due to the symmetry of the problem, only half of the set-up was modeled. Hence, a boundary condition, $nxH = 0$ was introduced to incorporate this effect. A tetrahedral mesh was used to discretize the aluminum specimen and the SiC block. Since specimen is the key part of the study, hence its meshing was done with finer elements.

2.3 *Design of Experiments (Simulation) and Regression Modeling*

Table 2 presents the three parameters (factors) and their levels used for evaluating different simulation conditions.

The analysis was carried out using parametric sweep for investigating the effect of these parameters on temperature rise of the specimen. The parametric simulation was carried out after pre-validating the model based on the results presented for a similar model in the COMSOL multiphysics library.

3 Results and Discussion

The electric field distribution inside the cavity and across the aluminum specimen is an important criterion to identify the optimum positioning of the specimen.

Since the microwave furnace is a multimode cavity, the electric field distribution for the given conditions is as presented in Fig. 3. It is evident that the higher values of intensity are depicted by colors varying from yellow to red. As discussed earlier, dielectric properties have a major role in heating and higher intensity of electric field would result in greater dielectric losses and better heat conversion (from electromagnetic energy to thermal energy) efficiency.

Figure 4 shows the amount of temperature rise in the aluminum specimen for the input power of 750 W when the specimen is kept at the height of 90 mm from the base and the radiation is allowed to fall on the susceptor for 20 s.

It can be observed from Fig. 4 that the right hand side of the susceptor block has the maximum temperature which can be attributed to the larger electric field

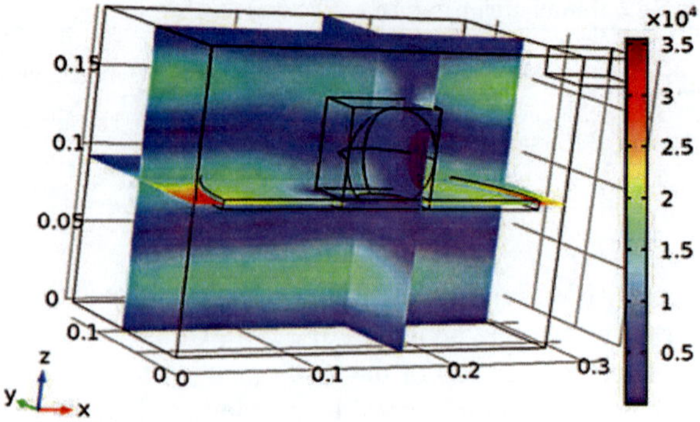

Fig. 3 Distribution of electric field (V/m) (P_{input} = 250 W, H = 90 mm, t = 60 s)

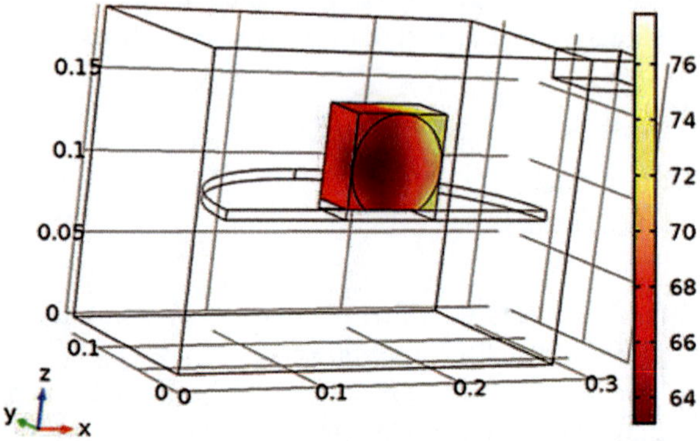

Fig. 4 Temperature rise (°C) in the sample (P_{input} = 750 W, H = 90 mm, Time = 20 s)

intensity causing higher amount of electric field losses around this location. For each parametric combination, the maximum temperature obtained was tabulated. On the basis of these results, a contour plot has been drawn as shown in Fig. 5. It is clearly evident that the maximum temperature attained is directly influenced by the magnitude of power input and increases with time.

It has also been revealed that the maximum value of temperature attained reduced with the increase in the distance of the specimen from the base as can be seen in Fig. 6. This can be attributed to the reduced dielectric losses in the absence of a high-intensity electric field in the vicinity of the aluminum specimen and SiC block.

Regression analysis was carried out with the help of the obtained simulation results. The p-value for all the coefficients of the regression model was found to be less than 0.05, which signifies that the model can be used for predicting the temperature rise. At the same time, the R-square value is also 88.83% that proves that the temperature is fairly influenced by all the input parameters. Equation 1 obtained through regression modeling, represents the maximum temperature attained by the aluminum specimen as the function of input parameters.

$$T(^{\circ}\mathrm{C}) = -18.14 + 0.1159 * P_{input}(\mathrm{W}) - 0.1927 * \mathrm{H(mm)} + 1.756 * \mathrm{t(s)} \quad (1)$$

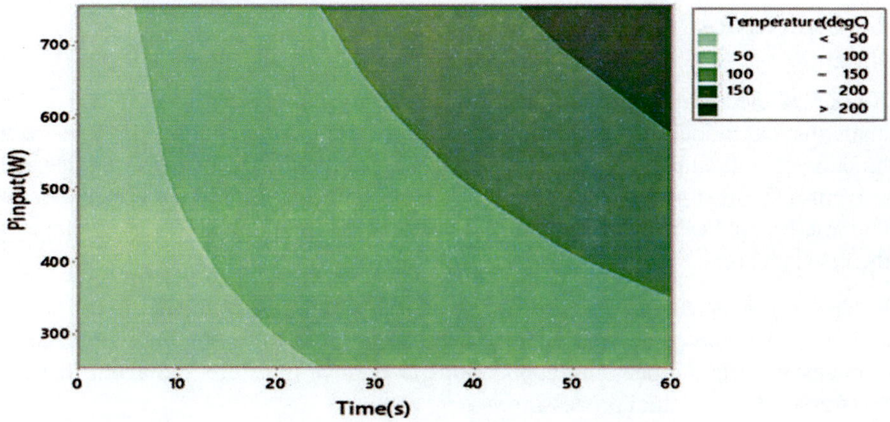

Fig. 5 Temperature rise (°C) versus P_{input} (W) and t (s)

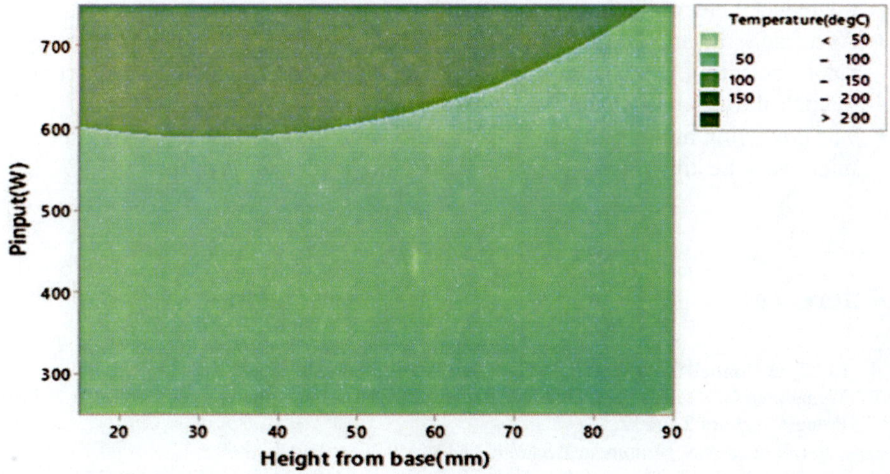

Fig. 6 Temperature rise (°C) versus P_{input} (W) and H (mm)

The sign of the coefficients reveal that input power and time have a synergistic effect due to increased availability of microwave energy for conversion into thermal energy. On the other hand, the increase in the distance of the specimen from the base has a negative impact on the temperature rise. This can be attributed to the lesser strength of the electric field around the specimen.

4 Conclusion

A coupled electromagnetic and heat transfer analysis has been used to derive the mathematical model for maximum temperature rise in an aluminum specimen as a function of critical parameters. This can be used to predict the temperature rise and to control the heating process at the desired rate. The results have fostered the point that microwave heating is faster than conventional heating methods. Key findings of the research are:

- The specimen reaches the temperature of 100 °C in a quick time for all the power inputs. This outcome would aid in preheating process design, a prerequisite in many materials processing applications to remove moisture and avoid thermal stresses during actual processing.
- Another important observation with respect to load position in the applicator is that the value of maximum temperature rise drops with the increase in the distance of the specimen position inside the microwave cavity. It is mainly attributed to the reduced electric field intensities and corresponding losses.
- As a result, this model can be used as a reference for designing microwave applicators, selecting the values of key input parameters and choosing the susceptors to control the maximum temperature rise in a metallic specimen.
- The prediction model can also help in addressing the issue of developing reliable microwave heating-based materials processing applications in future.

References

1. Li XP, O'Donnell KM, Sercombe TB (2016) Addit Manuf 10:10
2. Weingarten C, Buchbinder D, Pirch N, Meiners W, Wissenbach K, Poprawe R (2015) J Mater Process Technol 221:112
3. Lee DJ, Jangam S, Mujumdar AS (2012) KONA Powder Part J 30:69
4. Leung CLA, Tosi R, Muzangaza E, Nonni S, Withers PJ, Lee PD (2019) Mater Des 174:107792
5. Ding C, Cui X, Jiao J, Zhu P (2018) Materials (Basel) 11
6. Mertens R, Dadbakhsh S, Van Humbeeck J, Kruth JP (2018) Procedia CIRP 74:5
7. Papadakis L, Chantzis D, Salonitis K (2018) Int J Adv Manuf Technol 95:1325
8. Loharkar PK, Ingle A, Jhavar S (2019) J Mater Res Technol 8:3306
9. Chandrasekaran S, Ramanathan S, Basak T (2012) AIChE J 58:330
10. Thostenson ET, Chou T-W (1999) Compos Part A Appl Sci Manuf 30:1055
11. Kim J, Mun SC, Ko HU, Kim KB, Khondoker MAH, Zhai L (2012) Int J Precis Eng Manuf 13:2263
12. Mishra RR, Sharma AK (2016) Crit Rev Solid State Mater Sci 41:217
13. Singh S, Gupta D, Jain V, Sharma AK (2015) Mater Manuf Process 30:1
14. Kharissova OV, Kharisov BI, Valdés JJR (2010) Ind Eng Chem Res 49:1457
15. Ku HS, Siores E, Ball JAR (2001) HKIE Trans 8:31
16. El Khaled D, Novas N, Gazquez JA, Manzano-Agugliaro F (2018) Renew Sustain Energy Rev 82:2880
17. Roy R, Agrawal D, Cheng J, Gedevanlshvili S (1999) Nature 399:668
18. Sun J, Wang W, Yue Q (2016) Materials (Basel) 9

19. Mishra RR, Sharma AK (2016) Compos Part A Appl Sci Manuf 81:78
20. Reeja-Jayan B, Harrison KL, Yang K, Wang CL, Yilmaz AE, Manthiram A (2012) Sci Rep 2:1
21. Loharkar PK, Ingle A (2021) Mater Today Proc. https://doi.org/10.1016/j.matpr.2020.11.839
22. Gupta D, Sharma AK (2011) Surf Coatings Technol 205:5147
23. Bansal A, Sharma AK (2014) Int J Mech Eng Robot Res 1:7
24. Loharkar PK, Ingle A (2020) IOP Conf Ser Mater Sci Eng 810:012061
25. Fayos-Fernandez J, Pérez-Conesa I, Monzó-Cabrera J, Del Pino-De León S, Carlos Albaladejo-González J (2018) AMPERE Newsl 1
26. Saremi-Yarahmadi S, Whittow W, Vaidhyanathan B (2013) Appl Surf Sci 275:65

Tool Path Generation for Layer Specific Infill Density in Fused Filament Fabrication (FFF)

Krishnanand, Ankit Nayak, Shivam Soni, and Mohammad Taufik

Abstract Fused Filament Fabrication (FFF) is well known technology in 3D printing domain due to its economic and open-source availability. But parts printed using this technology have poor surface quality and low mechanical strength. Literature shows that infill density and infill pattern are important parameter to influence the mechanical properties of FFF printed part. Infill density is one of the important parameters on which strength, printing time and material requirement are depended. The novelty of this work is to present a new algorithm and tool path generation approach for providing layer specific infill density. This algorithm will provide two different infill densities in alternate layers. A code for tool path generation has been made using MATLAB. Feasibility study of this newly developed algorithm was checked by printing the part using a commercially available FFF 3D printer.

Keywords Fused filament fabrication · Tool path · STL files · Slicing

1 Introduction

Fused Filament Fabrication (FFF) is well technology in 3D printing domain. 3D printing is a method in which products are manufactured using layer over layer deposition of material, using a digital design data obtained from computerized solid model of the part to be manufactured. Initially, this method of manufacturing was used mainly for the development of prototypes that is why it is also called Rapid prototyping. Another name of Additive Manufacturing is 3D Printing. This technology can be utilized for product design as well as customization of products [1]. There is "design freedom" in 3D printing technology. It is a big supporting tool for researchers to develop the innovative products that can be economically manufactured [2]. There are various methods of 3D printing, all these methods could be divided in three types

Krishnanand · S. Soni · M. Taufik (✉)
Maulana Azad National Institute of Technology (MANIT), Bhopal 462003, India
e-mail: mohammad.taufik@manit.ac.in

A. Nayak
Banasthali Vidyapith, Vanasthali, Rajasthan 304022, India

R. Agrawal et al. (eds.), *Recent Advances in Industrial Production*, Lecture Notes in Mechanical Engineering, https://doi.org/10.1007/978-981-16-5281-3_24

based on the type of feed material used—liquid-based, powder-based and solid-based [3]. In liquid based, build material is in liquid form, like stereo-lithography (SLA) [4, 5]. In powder-based additive manufacturing material used is in powdered form. Similarly, laminated object manufacturing (LOM) and fused filament fabrications (FFF) are examples of solid based 3D printing, because build material used is in solid form [6, 7]. In subtractive manufacturing, parts with complex geometry cannot be manufactured, but additive manufacturing has advantage of manufacturing most complex geometry part in single run. AM technologies also reduces the overall lead time of manufacturing. Aerospace, medical, automobile are the main sectors where AM technologies are utilized to produce parts economically and, in less time, as compared to subtractive manufacturing process.

Fused filament fabrication is most popular among all the AM technologies due its economic availability, but high surface roughness and lower strength are main inherited properties of the parts manufactured using this technology. Orientation of part, layer height, road width, infill density and infill pattern are some of the process parameters used in FFF 3D printing [7, 8]. These process parameters could be optimized for better quality of printed part. Study of characteristics of parts manufactured using FFF used to be done mainly on three aspects—surface quality, dimensional accuracy and mechanical behavior. Fernandez-Vicente et al. [9] studied the influence of infill density tensile strength in FFF Technology. They concluded that mechanical properties are more affected by raster angle and air gap in comparison to other factors. Samykano et al. [8] also noted three factors i.e. raster angle, infill density and layer height and identified their effects on mechanical properties using ABS FFF 3D printed parts. They found the optimized parameters for best mechanical properties of ABS printed part are at 0.5 mm, 55° and 80% as layer height, raster angle and infill density respectively. Yadav et al. identified the effects on mechanical properties of FFF 3D printed part by varying the infill density and extrusion temperature. They noted that extrusion temperature has more effect on tensile strength of the FFF 3D printed parts, than the infill density [10].

Infill density is one of the important process parameters to alter the mechanical properties of FFF printed part. In this study a code to generate the tool path for FFF process has been developed using MATLAB. This code will generate a tool path which will be used to impart two different infill densities in alternate sliced layer of part to be manufactured.

2 Methodology

Generally, Fused Filament Fabrication (FFF) process divided in three stages—(i) CAD Modeling and STL file preparation, (ii) Process planning and (iii) Fabrication of part (see Fig. 1).

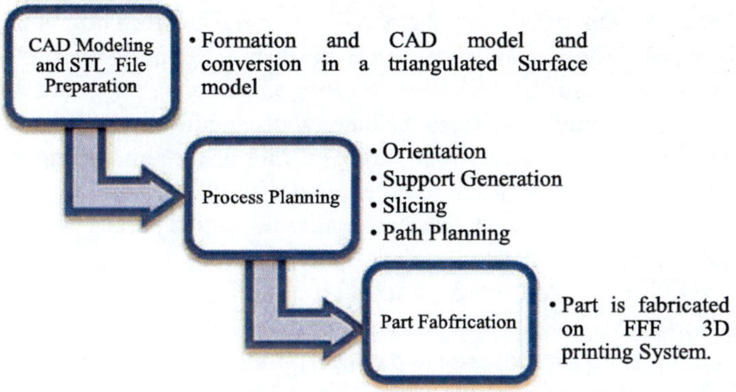

Fig. 1 FFF 3D printing process

2.1 CAD Model to STL File Preparation

Computer Aided Designing (CAD) tools like Solidworks, CATIA, AutoCAD etc. are used to prepare the 3D model of the part to be printed using FFF technology. STL (Stereo lithography) file format is the bridge between CAD modeling tools and additive manufacturing process planning tool. STL file is the data of CAD model in form of tessellated (triangulated) surface model [11, 12]. It is de facto industry standard as the raw data for any 3D printing system. However, there are various other file formats like—IGES, STEP etc. but STL file is simple to generate and having capability of defining any shape with any number of edges [13]. During tessellation CAD model is converted in triangulated surface model. These triangles are called facets. These triangles are represented by its vertices coordinates and normal vector.

2.2 Tool Path Generation Using MATLAB

After preparation of STL file, it is transferred to some process planning tools like Slicer and Cura. Where process planning is performed mainly in two domains—(i) Model domain and (ii) Layer domain [14]. In model domain, orientation and support generation are decided and in layer domain slicing and path planning is performed. Process planning is a critical phase, it will decide the physical and mechanical characteristics of the part to be printed. An adequate tool path planning enhances the precision, physical and mechanical properties of part also saves material and decrease the build period [15]. Build orientation and support generation are decided in such a way that build time should be less surface quality should be good and less material should be used as support [16, 17]. First step after build orientation and support generation is slicing. Slicing may be uniform or adaptive. If the distance between successive sliced planes is not constant than the slicing is adaptive. An algorithm is followed for slicing and out of that algorithm is intersection points of facet with slicing plane [18]. These point from same facets are joined together to form a line segment. Another algorithm uses these line segments to form a contour. After getting contour tool path planning is to form the outer walls and fill that contour with lines of extruded material. To fill the contour various filling pattern and filling densities are used. Raster, spirals, hatching are some examples of filling patterns. Filling pattern and filling densities will affect the characteristics of printed part as well as build time and amount of material extrusion.

In this study, to handle this phase an algorithm has been developed and implemented on MATLAB make a computer program (see Fig. 2), which will import the STL file and generate the tool path and convert it into G-Code. This G-code will be used in FFF 3D printer to print the part.

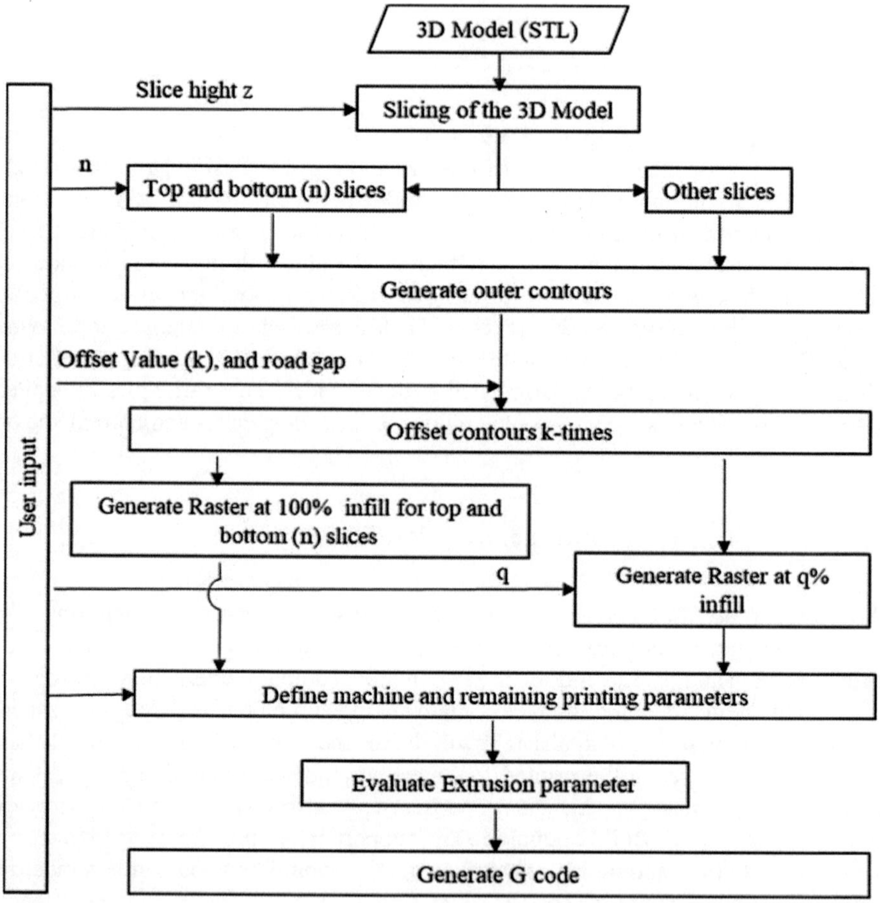

Fig. 2 Flow diagram for tool path generation using MATLAB

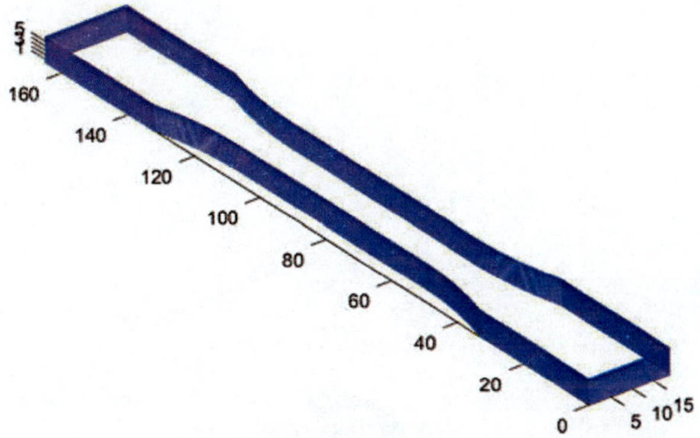

Fig. 3 Contour formed at each slice (z-height)

Tool path generation algorithm can be divided in three major steps i.e., STL processing and slicing, infill formation and G-code writing.

STL processing and slicing. Tool path for additive manufacturing is subjected to the outer boundary or the shell of 3D geometry. In proposed method of tool path generation, the STL file in ASCII format was supplied to computer program which has been written in the MATLAB. In first step coordinates of each facet was stored in a matrix having m × 3 × 3 dimension, where m is the total number of facets. Then according to the slice height contours for each z height was created followed by generation inwards offset contours. These offset contours are separated with the distance known as road gap (see Fig. 3).

Infill formation. Top and bottom layers of any part required to be solidly filled while other intermediate layers of fabricated part can be filled at varying fill density (see Fig. 4). So, raster fill at 100% filling density was generated for top and bottom layers and for intermediate layers, raster was generated according to the predefined filling density. Respective coordinates of raster have been stored in a matrix.

G-code writing. G-code for FFF based machine requires some machine and fabrication related parameters like nozzle diameter, slice height, printing temperature, nozzle temperature, bed temperature, road gap, nozzle velocity etc. According to the machine specification feed rate for nozzle travel and printing, speed has been calculated. On the bases of filament diameter, slice height and nozzle diameter the extrusion parameter E has been calculated. Then these parameters have stored in a directory. Initial machine parameters like the bed, nozzle temperature and printing speed have been defined in the initial part of the code after that the nozzle travel coordinates, feed rate and extrusion parameters have been defined in each line of the code. So that nozzle will follow these points sequentially and deposit the material for part fabrication.

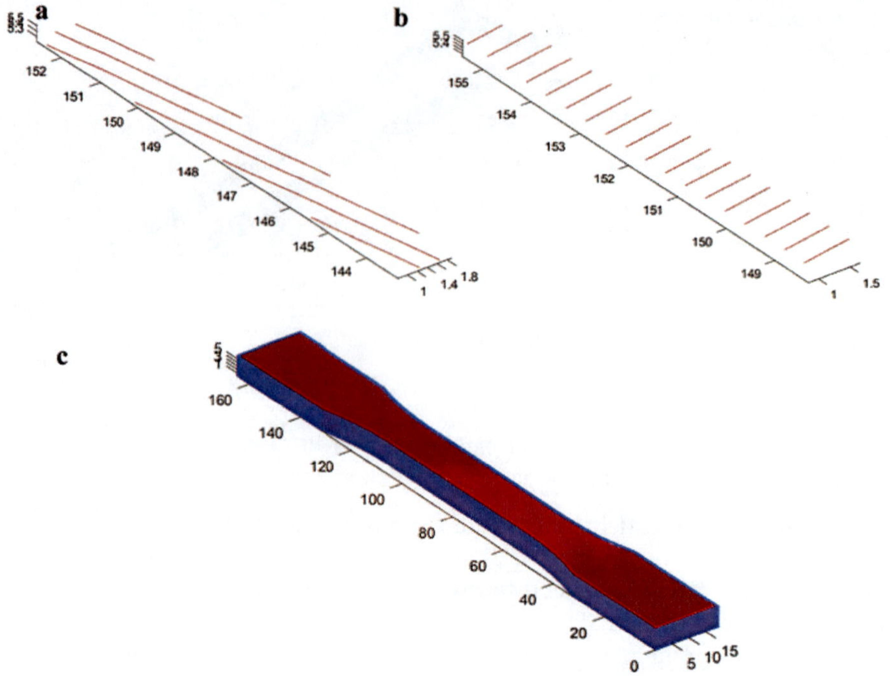

Fig. 4 Infill density and pattern (I) (**a**); infill density and patter (II) (**b**) and representation of tool path for each layer (**c**)

3 Feasibility Study

Developed MATLAB code is accepting the STL file as input and generating tool path for implementing different in fill densities in alternate slicing layer. Final output of this code is G-code file in ASCII format. To check the feasibility of that tool path, a STL file of dog-bone shape part was imported in this MATLAB code and it generated the suitable tool path with predefined infill densities in alternate layers, in form of G-code. This G-code file was transported to a commercial FFF 3D printer for printing of part to check the feasibility of code. Bed size, nozzle size and other required parameters of FFF 3D printer were already implemented in the code.

On the basis of observation during the printing process (see Fig. 5a), it has been found that each alternate layer was printing with predefined infill density as per the tool path generated using MATLAB (see Fig. 5b). Final printed part (see Fig. 5c) is having total 20 layers. Top and bottom layers are having 100% in-fill density.

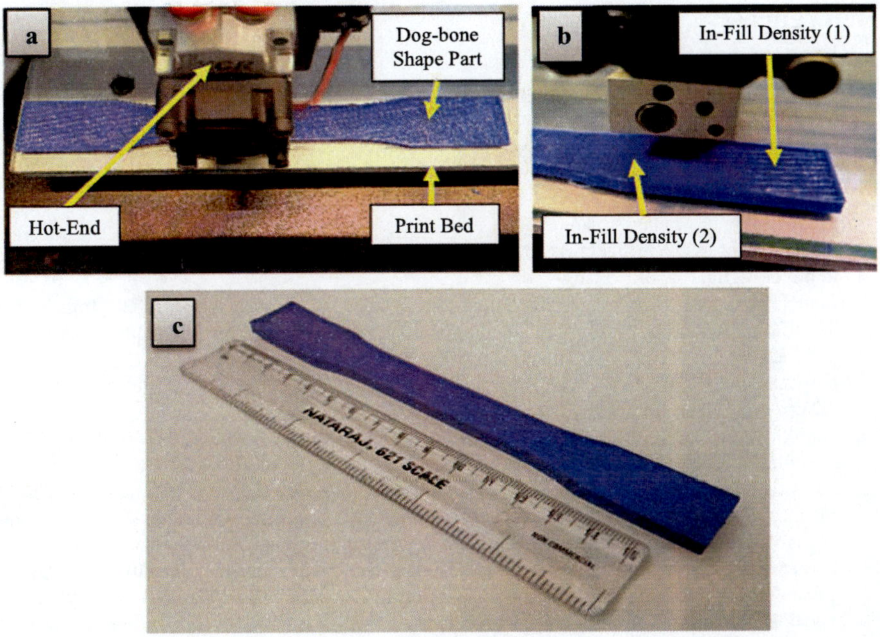

Fig. 5 Printing of dog-bone shape part using developed algorithm (**a**); two different infill densities in alternate layers (**b**) and Final FFF 3D printed part (**c**)

4 Conclusion

To enhance the physical and mechanical properties of FFF printed part, some initial attempt has been made by developing a new adaptive infill slicing procedure. In this study an algorithm has been presented and along with its feasibility study. MATLAB has been used to implement that algorithm and to generate tool path for providing layer specific infill density. Feasibility of algorithm and generated tool path has been verified successfully by printing the part on fused filament fabrication 3D printer. This tool path could be utilized for providing layer specific infill density to improve mechanical strength, save printing time and material. This algorithm could have future scope to be developed for providing more infill density at the locations where more stress is generated. So, this approach could help to provide better mechanical strength to printed part as well as it will save material and printing time.

Acknowledgements This work was supported by the Science and Engineering Research Board (SERB)—DST under its Start-up Research Grant (SRG) scheme [Grant number: SRG/2019/000943].

References

1. Shahrubudin N, Lee TC, Ramlan R (2019) An overview on 3D printing technology: technological, materials, and applications. Procedia Manuf 35:1286–1296
2. Bikas H, Stavropoulos P, Chryssolouris G (2016) Additive manufacturing methods and modeling approaches: a critical review. Int J Adv Manuf Technol 83:389–405
3. Jain MT, Jain PK (2013) Role of build orientation in layered manufacturing: a review. Int J Manuf Technol Manag 27:47–73
4. Gibson I, Rosen DW, Stucker B (2010) Additive manufacturing technologies: rapid prototyping to direct digital manufacturing
5. Francis V, Ukey P, Nayak A, Taufik M, Jain PK, Mankar SH, Srivastava SS (2020) Influence of 3D printing technology on biomedical applications: a study on surgical planning, procedures, and training. In: Advances in material processing, pp 269–278
6. Krishnanand SS, Taufik M (2020) Design and assembly of fused filament fabrication (FFF) 3D printers. Mater Today Proc
7. Taufik M, Jain PK (2020) Part surface quality improvement studies in fused deposition modelling process: a review
8. Samykano M, Selvamani SK, Kadirgama K, Ngui WK, Kanagaraj G, Sudhakar K (2019) Mechanical property of FDM printed ABS: influence of printing parameters. Int J Adv Manuf Technol 102:2779–2796
9. Fernandez-Vicente M, Calle W, Ferrandiz S, Conejero A (2016) Effect of infill parameters on tensile mechanical behavior in desktop 3D printing. 3D print. Addit Manuf 3:183–192
10. Yadav D, Chhabra D, Kumar Garg R, Ahlawat A, Phogat A (2020) Optimization of FDM 3D printing process parameters for multi-material using artificial neural network. Mater Today Proc 21:1583–1591
11. Ding DH, Pan ZX, Dominic C, Li HJ (2015) Process planning strategy for wire and ARC additive manufacturing. Adv Intell Syst Comput 363:437–450
12. Taufik M, Jain PK (2017) Laser assisted finishing process for improved surface finish of fused deposition modelled parts. J Manuf Process
13. Eragubi M (2013) Slicing 3D CAD model in STL format and laser path generation 4:410–413
14. Kulkarni P, Marsan A, Dutta D (2000) Review of process planning techniques in layered manufacturing. Rapid Prototyping J 6:18–35
15. Jin Y, He Y, Fu J, Gan W, Lin Z (2014) Optimization of tool-path generation for material extrusion-based additive manufacturing technology. Addit Manuf 1:32–47
16. Taufik M, Jain PK (2016) CNC-assisted selective melting for improved surface finish of FDM parts. Virtual Phys Prototyping 11:319–341
17. Taufik M, Jain PK (2016) Computer aided visualization tool for part quality analysis of additive manufacturing process. In: Proceedings of the ASME design engineering technical conference
18. Vatani M, Rahimi AR, Brazandeh F, Sanati A. An enhanced slicing algorithm using nearest distance analysis for layer manufacturing, pp 669–674

Cost Minimization in a Scheduling Problem with Unrestricted and Restricted Common Due Date

Prasad Bari and Prasad Karande

Abstract A scheduling problem is considered where the sequencing of multiple jobs in a single machine shop is to be done. In this paper, fundamental scheduling models are considered in search of understandings and interactions that suggests procedures for minimizing the total cost considering the due-date which is common for all the jobs, and the cost of completing the jobs early and late to be the same. Basic results from scheduling theory are considered involving the single machine model. An algorithm is developed which initially checks if the problem is of unrestricted or restricted version. Later it finds the optimal sequence of the jobs for processing which has minimum cost. The algorithm gives the solution of minimum cost by delaying the jobs in the unrestricted version. Further, in a restricted version the algorithm also gives solutions to some special cases of problems where the cost can be reduced further. Therefore, a single algorithm is applicable for both unrestricted and restricted versions. A numerical illustration is also solved with respect to the given algorithm considering all the stated cases.

Keywords Scheduling · Earliness · Tardiness · Common due date

1 Introduction

Scheduling is the allotting of resources to jobs so as to confirm completion of these jobs in an appropriate time period. In manufacturing industries, jobs are to be assigned to machine and that too in a sequence which reduces the overall execution period which is termed as sequencing. Most of the scheduling literature deals with the performance metrics like mean lateness, flow-time and tardiness. However, meeting

P. Bari (✉) · P. Karande
Veermata Jijabai Technological Institute, Matunga, Mumbai 400019, India
e-mail: prasad.bari@fcrit.ac.in

P. Bari
Fr. C. Rodrigues Institute of Technology, Vashi, Navi Mumbai 400703, India

R. Agrawal et al. (eds.), *Recent Advances in Industrial Production*, Lecture Notes in Mechanical Engineering, https://doi.org/10.1007/978-981-16-5281-3_25

due dates is also one of the vital objectives. In old scheduling techniques, due-dates are assumed to be given as exogenous conclusions. Still, they are found by considering the system's ability to match the given delivery dates. Therefore, in many research it has been noticed that due-date allotment is a part of the scheduling process. Mainly two types of ways can be considered which include performance metrics associated to due-dates. In the first case, due-dates are known, and a performance metrics like total tardiness, represents the usefulness of schedule at meeting the known due-dates. Here, some jobs can be rejected to enhance performance. In another case, due-dates are internal choices. These due- dates are frequently settled with clients or selected by production control department to monitor or speed up the development of work, and the performance metrics may comprise measures of due-date tightness.

A usual way of assessing conformance to due-date is considered to be mean tardiness; however, it overlooks the effect of jobs finishing early. The concept of Just-In-Time production promotes the idea that not only tardiness but also earliness should be dejected. Earliness and Tardiness (E/T) costs are few of the general measures considered to trace the performance of production. Finishing of the jobs early effects inventory carrying costs like storing and insurance costs. In contrast, jobs that finish later to their due-dates effect in fines like late-dues, harm to client concern and damage to trades.

In sequencing considering due-dates, the main intent is generally to finish entire jobs on time. If due-dates are unrestricted, obviously, this intent may be achieved by permitting the loose due-dates. Still, in a situation wherein due-dates can be carefully chosen, it is intended to allot due-dates to be tight as probable, which is restricted one. Due-dates which are tight, invite more clients than due-dates which are loose in a market full of competition and indicate improved client facility. Tighter due-dates also have a tendency to yield lesser inventory levels hence they are necessary in scheduling.

2 Literature Review

An algorithm was developed that finds minimum cost schedules whenever the tasks either all have the same length or are required to be executed in a given fixed sequence [1]. The researcher reviewed by considering the problem of scheduling 'n' jobs to reduce cost of earliness and tardiness [2]. The earliness plus tardiness scheduling problems are solved by taking into account the weighted deviance of end times about a common due-date [3]. Stochastic scheduling problem was formulated with E/T costs on a machine for static nature wherein the processing time of jobs are stochastic and due-dates are different and deterministic [4]. Here a heuristic approach was developed where the intent was to find an optimum sequence which reduces the entire E/T costs. Multiple machine scheduling problems were surveyed considering a common due-date mainly focusing on shop scheduling problems [5]. Authors introduced the idea in problems of E/T scheduling wherein non-execution of the task is allowed and accordingly are fined for every non executed job [6]. They developed an algorithm considering the non-execution penalties. The problems were solved on 2 single machine scheduling with tardiness fines and due-date allotment [7]. In the first case the objective was to reduce cost function which comprised fines suffered due to due-date allotment, earliness and tardiness. In the next case the objective was to reduce cost which included fines incurred due to due-date allotment and number of jobs that are tardy. Researchers considered one machine problem with uncertain processing time [8]. They proposed a heuristic method for sequencing of jobs and due-date allotment which utilizes average and variances of processing time of jobs.

3 E/T Model and the Heuristic Procedure

A job shop yields parts for succeeding assembly into final products. The due-dates for parts are dependent on the assembly schedule of the final product. If parts order is delayed, then the assembly of a product will be late. It impacts in terms of loss of assembly productivity and client goodwill. If a part order is finished before time it has to wait in stores until the assembly period of the product where it is required. The bad impact is the build-up of part inventory. Hence, an ultimate solution is one where all jobs completed just on their due-dates.

Let, E_j and T_j represent earliness and tardiness respectively, while $\propto_j$ and β_j earliness cost and tardiness cost respectively. Considering cost functions to be linear, the basic objective function for E/T of schedule S for n jobs is as follows:

$$f(S) = \sum_{j=1}^{n} \propto_j E_j + \beta_j T_j \tag{1}$$

In E/T problem, consider $\propto_j = 1, \beta_j = 1$ and due-date which is common, the objective function can be given as:

$$f(S) = \sum_{j=1}^{n} E_j + T_j \tag{2}$$

Ideally, a schedule should be constructed such that due-date d is at the mid of jobs. If d is very tight, it is not feasible to have adequate jobs ahead of d. This becomes restricted version. If d is not very tight it becomes unrestricted version.

Considering the unrestricted version for E/T problem, following properties are proposed [9],

(1) Schedules not having idle time among consecutive jobs consists a dominant set.
(2) Jobs that finishes ahead the due-date can be arranged in longest processing time, first (LPT) order and jobs that begins later the due-date can be arranged in shortest processing time, first (SPT) order.
(3) In an optimum schedule where job ends just at the due-date.

This job can be considered as job $[b]$, where $b = [n/2]$

The following steps are considered while developing an algorithm for the unrestricted version,

(1) Divide the jobs in two sets, list A and list B, starting with assigning the longest job to list B.
(2) Give the following longest job to list A and then the longest job to list B.
(3) Iterate the above step 2 until no job is left, or if one job is left assign the same to list A.
(4) Arrange the jobs in list B by LPT and the jobs in list A by SPT.

The complete processing time (PT) in list B is as follows:

$$\Delta = PT_n + PT_{n-2} + PT_{n-4} + \cdots \tag{3}$$

The begin time of the jobs is the due date minus total processing time of jobs in list B. Importance of Δ tells the definition of the restricted and unrestricted version. If $d \geq \Delta$ then it is unrestricted else, it represents restricted version.

Considering the restricted version of the E/T problem, the properties 1 and 2 hold but not property 3, as the result may have straddling job. An effective heuristic is available. Let, L indicate the period present afore the due-date, and let R indicate the period present later the due-date which can be mathematically represented as:

$$L = d \tag{4}$$

$$R = \sum_{j=1}^{n} PT - d \tag{5}$$

The jobs are arranged in LPT rule initially and the following rules are applied,

If $L > R$, allot the job to the beginning vacant place in the order and after doing so subtract PT_j of the job from L.

If $L \leq R$, allot the job to the ending vacant place in the order and after doing so subtract PT_j of the job from R.

From the above discussion of unrestricted and restricted versions it is clear that the significant outcome for the common due-date ordering is the V-shaped optimal schedule. For example, if x is the job having the least processing time among all the jobs to be organized, a schedule is V-shaped if the jobs located afore job x are in decreasing manner of processing period and the jobs located later the job x are in increasing manner of processing period.

In the restricted version where the due-date is very tight it is logical that the schedule should begin at zero time to get the optimal sequence which will give minimum cost. But this may not be true in all the cases. Let y be the number of jobs that complete prior to the due-date. Then the number of jobs that completes on or later the due-date will be $(n - y)$. If the start of the schedule is delayed by a small value, Δt then the cost of y jobs will get decreased by Δt, and cost of $(n - y)$ jobs will increased by Δt. The total cost is decreased due to this delay if,

$$y\Delta t > (n - y)\Delta t \quad (6)$$

Mathematically, if $y > n/2$ i.e. if over and above half of the jobs can finish before the due- date the schedule should be delayed so that the early job which is last should finish correctly on the due-date. The sequence will be an alternative to the sequence obtained from the procedure of restricted version but in this case also the sequence will be V-shape solution as well.

The algorithm **Optimal_Sequence_Minimum_Cost** for the unrestricted and restricted version for finding the minimum cost with optimal sequence is given below. Python programming is used to implement this algorithm.

Algorithm Optimal_Sequence_Minimum_Cost

```
Get n - Number of jobs processing in a machine
Get d – same due-date for the jobs
Set i ← 1
while i <= n
        Get PT1, PT2, PT3 ... PTn - Processing Time of job one by one
        sequence ← PT1, PT2, PT3 ... PTn
end while
//Sort Sequence in descending order
sort ← sort(Sequence)
//Solution of unrestricted version
first ← sort.pop(first)
B ← list of jobs
A ← list of jobs
B ← first
while i<=n-1
        A ← Find the next longest job. Assign it to last empty place in A
        B ← Find the next longest job. Append it to B
end while
finalsequence ← B jobs + A jobs
//Find Processing time of list B
Btime = Sum(processing time of jobs in list B)
if Btime < d
        Print "Unresticted Version"
        delay ← d - Btime
        Call procedure_mincost to find Cost of sequence
        Print "Cost for the sequence"
else
        Print "Resticted Version"
        left ← d
        process_time ← 0
        Bres ← list of jobs
```

```
A_res ← list of jobs
for t in seq:
        process_time ← sum(PT)
right ← process_time - d
for j in range(len(sort)):
        if left > right
        B_res.append(sort[j])
        left ← left - sort[j][1]
else
        A_res.append(sort[j])
        right ← right - sort[j][1]
A_res.reverse()
Finalsequence ← B_res + A_res
Print "Final Sequence"
Call procedure_ mincost to find Cost of sequence
Print "Cost1"
Updatesequence ← Find the other sequence which is a V-shaped
y ← number of jobs finished before due-date i.e. early jobs
if y > n/2 then
        delay ← d - sum (PT of y jobs)
        Call mincost procedure to find Cost of sequence
Print "CostUpdated"
if Cost1<=CostUpdated
  Process the jobs with Finalsequence
  Print "Finalsequence as optimum sequence"
else
  Process the jobs with Updatesequence
  Print "Updatesequence as optimum sequence"
Procedure mincost(sequence,d,delay):
  cos ← 0
```

```
for j in range(len(sequence)):
  delay ← delay + PT(j)
  cost ← cost + abs (d - delay)
return(cost)
```

4 Numerical Illustration

A following numerical illustration is being solved with respect to above E/T model and algorithm using input data given in Table 1. Processing time and common due date in days are given for six jobs. The intent is to find the optimal sequence of jobs which gives the minimum cost.

Arrange the jobs in the descending sequence of their processing time, 6-5-4-3-2-1.

Check whether the problem is unrestricted by applying the procedure of unrestricted version shown in Table 2.

Since time for jobs in list B is less than the due date, this is unrestricted version and the jobs can be delayed by 15 days shown in Fig. 1.

Cost = 50 + 5 + 0 + 3 + 11 + 59 = 128.

So the optimal sequence of jobs is 6-4-2-1-3-5 which gives minimum cost of 128 by delaying the jobs by 15 days.

For simplicity let us continue with the same numerical but with common due date as 90 as shown in Fig. 2. Now it is checked whether the problem is unrestricted or not. In this case the problem is of restricted version since time for jobs in list B is more than the due date. Now solving by restricted version procedure will produce the Table 3.

Cost = 35 + 10 + 13 + 18 + 26 + 74 = 176.

So the optimal sequence of jobs is 6-4-1-2-3-5 which gives minimum cost of 176.

In some special cases further reduction in cost is possible by delaying the jobs. Consider another V-shape sequence 6-3-2-1-4-5 which have more number of jobs that finish prior to the due date as compared to the jobs finishing later to the due date.

Table 1 Input data

Jobs	1	2	3	4	5	6
Processing time	3	5	8	45	48	55
Due date (common)	120					

Table 2 Sequence for unrestricted version

List B jobs	List A jobs	Sequence	Time for List B	Start time
642	135	642,135	105	15

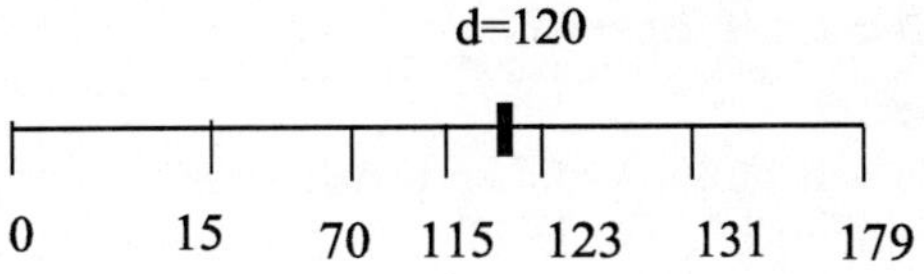

Fig. 1 Time line for unrestricted version

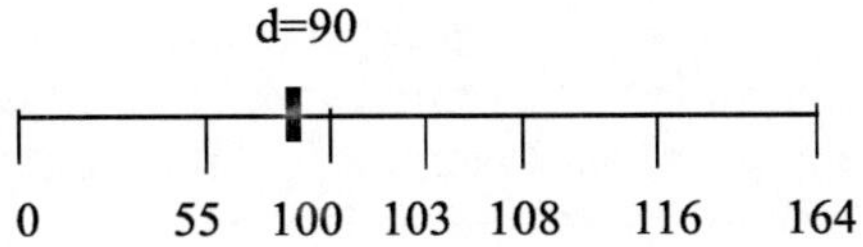

Fig. 2 Time line for restricted version

Table 3 Sequence for restricted version

L	R	Sequence
90	74	6 - - - - -
35	74	6 - - - - 5
35	26	6 4 - - - 5
– 10	26	6 4 - - 3 5
– 10	18	6 4 - 2 3 5
– 10	13	6 4 1 2 3 5

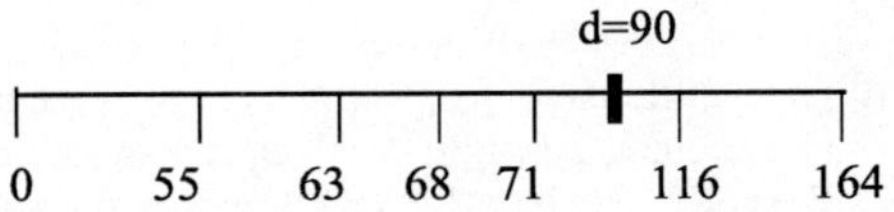

Fig. 3 Time line1 for restricted version

It may not guarantee cost lesser than the one obtained by the above procedure but for some special cases like this it may work.

Cost $= 35 + 27 + 22 + 19 + 26 + 74 = 203$.

From the above Fig. 3 it is noticed that the first four jobs in the sequence completes at 71 which is 19 days earlier than the due date, hence the jobs can be delayed by 19 days as shown in Fig. 4

Cost $= 16 + 8 + 3 + 0 + 45 + 93 = 165$.

By comparing the two costs that is 176 and 165 it is concluded that the optimum sequence is 6-3-2-1-4-5 and the jobs to be delayed by 19 days. It is to be noted that if this cost is not minimum then the sequence obtained by the procedure of the restricted version is considered to be the optimum one.

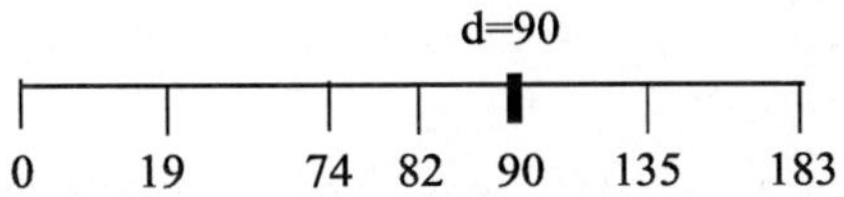

Fig. 4 Time line2 for restricted version

5 Conclusions

A scheduling problem is discussed in this paper, wherein number of jobs are to be scheduled on a machine. The processing times of the jobs are distinct with due date to be common. The intent of finding an optimum schedule of sequencing the jobs that incurs minimum cost considering the earliness and tardiness of jobs is achieved by constructing an algorithm that takes into account two different heuristics concepts of unrestricted and restricted versions. This algorithm single handily solves both type of problems. The algorithm is verified by executing various numerical through program which is developed in Python programming language and found to be correct. In unrestricted version the due date is considered to be loose while in restricted version the due date is too tight. Further, it is shown that how delaying the jobs to some extent can also reduce the cost further. Also, there are some special case problems in restricted version which if arranged in some other V-shape sequence of the jobs reduces the cost further but increases the completion time. In the numerical which is considered, the cost is reduced by 6.25%. The work can further be developed by considering the problems of distinct due dates and distinct earliness and tardiness cost.

References

1. Garey MR, Tarjan RE, Wilfong GT (1988) One-processor scheduling with symmetric earliness and tardiness penalties. Math Oper Res 13(2):330–348
2. Baker KR, Scudder GD (1990) Sequencing with earliness and tardiness penalties: a review. Oper Res 38(1):22–36
3. Hall NG, Posner ME (1991) Earliness-tardiness scheduling problems, I: weighted deviation of completion times about a common due date. Oper Res 39(5):836–846
4. Soroush HM, Fredendall LD (1991) The stochastic single machine scheduling problem with earliness and tardiness cost. Eur J Oper Res 77:287–302
5. Lauff V, Werner F (2004) Scheduling with common due date, earliness and tardiness penalties for multimachine problems: a survey. Math Comput Model 40:637–655
6. Hassin R, Shani M (2004) Machine scheduling with earliness, tardiness and non-execution penalties. Comput Oper Res 32:683–705
7. Shabtay D (2008) Due date assignments and scheduling a single machine with a general earliness/tardiness cost function. Comput Oper Res 35:1539–1545
8. Xia Y, Chen B, Yue J (2008) Job sequencing and due date assignment in a single machine shop with uncertain processing times. Eur J Oper Res 184:63–75
9. Baker KR, Trietsch D (2009) Principles of sequencing and scheduling. Wiley, Hoboken, New Jersey

Prioritization of Sustainability Criteria of Service Only Supply Chain: A Case Study of Indian Hospitals

Ramji Nagariya, Divesh Kumar, and Ishwar Kumar

Abstract Unlike manufacturing supply chain the adoption of sustainable practices is less evident in Service only supply chain (SOSC). The purpose of this paper is to identify sustainable practices applicable in SOSC and prioritize these practices for the inclusion of sustainable practices in SOSC. These practices are identified from literature review and finalized with expert's opinion. Fuzzy TOPSIS method is used for finding the ranks of the sustainable practices. The case study of a Hospital is taken for testing the applicability of the research. This study identifies fifteen key sustainable practices related to SOSC and prioritizes these practices. Limited studies are found in the area of SOSC and this study not only identifies applicable sustainable practices for SOSC but also prioritize these practices in order of their importance for the inclusion of sustainability in SOSC. This study will enable the working professional in identifying the important sustainable practices related to SOSC and better knowledge of these practices will enable them in managing these practices efficiently and effectively.

Keywords Service only supply chain (SOSC) · Sustainable service only supply chain (SSOSC) · Service supply chain (SSC) · Hospitals · India

1 Introduction

The growing contribution of services in the GDPs of many countries attracts the attention to carry out research in service supply chain [1]. Sustainability is the most relevant field in current world as the efforts to make the human life easy the mankind has exploited natural resources at fast pace and created different types of pollution

R. Nagariya (✉) · D. Kumar
Malaviya National Institute of Technology, Jaipur, India

I. Kumar
Entrepreneurship Development Institute of India, Ahmedabad, India
e-mail: ishwar@ediindia.org

R. Agrawal et al. (eds.), *Recent Advances in Industrial Production*, Lecture Notes in Mechanical Engineering, https://doi.org/10.1007/978-981-16-5281-3_26

and other things putting the society, environment at risk. Earlier studies of sustainability were limited to environmental issues but later [2] presented the comprehensive view of sustainability and since then the economic, environmental and social aspects became the three pillars of the sustainability. The difference of operations and processes executed in services and manufacturing industry also stresses the need to carry research in service supply chain separately. With in service supply chain, service only supply chain (SOSC) and product service supply chain exist. In SOSC products are pure services and there no presence of any physical product [3]. In literature, the most attention is given to PSSC and SOSC has not attracted much research. The fast development of service sector also necessitates the adoption of sustainable practices in SOSC [4]. This study focuses on the sustainable practices relevant to SOSC and prioritizes these practices for the inclusion of sustainability in SOSC.

This research work is aimed to find the answers of the following three research questions:

(i) What are different important factors of sustainable hospital service supply chain?
(ii) What is the order of importance of these factors?
(iii) What are critical factors managers need to give extra attention?

Based on these research questions following objectives were set for this study.

(i) To identify the important factors of sustainable hospital service supply chain.
(ii) To find the order of importance of these factors.
(iii) To find the critical factors requiring special attention.

The sustainable practices were identified from the literature review and were finalized after consulting with the experts. For the analysis of these practices and prioritization fuzzy TOPSIS method was used. The motive for using the fuzzy theory was to take care the human bias and vagueness in the human responses [5]. The case hospital was interested in improving the sustainability score of the hospital with the aim of cost effectiveness, better services and a wide reach.

The rest of the paper is organized as: the following section presents a literature review of the related articles. Section 3 presents the research methodology, Sect. 4 discusses the analysis using fuzzy TOPSIS and Sect. 5 presents the results and discussion.

2 Literature Review

Sustainability is defined as “it meets the needs of the present without compromising the ability of future generations to meet their own needs” [6]. In SOSC, the presence of customer is essential as the customer helps in producing, customizing the services of their choice and is the ultimate consumer of the services. The success of the SOSC depends on the suppliers, employees, customers and all its participants [7]. Therefore, the cooperation of all stakeholders is necessary to attain sustainability in the SOSC. The effective proliferation of sustainability can be achieved with the active participation of all the participants of SOSC [8]. The sustainable practices in SOSC are less evident and hence need further research in SOSC [9].

[10] was the first to define the SSC but later their definition was considered more suitable to manufacturing context only and a new definition of SSC was proposed by [11] as "The service supply chain is the network of suppliers, service providers, consumers and other supporting units that performs the functions of the transaction of resources required to produce services; transformation of these resources into supporting and core services; and the delivery of these services to customers." Wang et al. [3] was the first one to further categorize the SSC into PSSC and SOSC. In PSSC the physical product is present but the main aim is to provide timely delivery and good service experience. While in SOSC the physical product is missing and services becomes the product.

From the literature the common sustainable practices were identified. The authors have given importance to waste disposal, removal of unsanitary exercises [12]. In SOSC the involvement of customer stresses the need of customer involvement for the inclusion of sustainability [13]. The supplier relationship and participation [14, 15] and customer involvement was highlighted by [16]. The sustainable practices helps in achieving sustainability goals of the organization and the knowledge of such practices helps improving the sustainability score of the companies in long run. The identified and finalized sustainable practices are as- Understanding and managing demand (E1)[10, 11], Resource and capacity management (E2) [8, 17], Customer relationship management (E3) [8, 15, 18], Supplier relationship management (E4) [8, 10, 15, 18], Better service delivery performance (E5) [8, 10, 11, 19], Real-time information sharing through different channel (E6) [8, 14, 15], Financial performance (E7) [14, 15, 20, 21], Adoption of the environmental friendly activities (E8) [12, 20, 22], Reduce the contamination (E9) [8, 14], Reduce the energy consumption (E10) [8, 14], Meeting legal requirements (E11) [8], Understanding and maintaining relationship with Employees (E12) [8, 14], Responsibility towards Community and stakeholders (E13) [8, 14], Social accountability (E14) [8] and Business ethics (E15) [8, 14].

3 Research Methodology

In multiple-criteria decision making (MCDM) there are several methods which can be used to solve unstructured problems having many conflicting goals [23]. Several MCDM techniques have been extensively used in the literature to rank the alternatives based on the weights assigned by the experts. In this paper fuzzy TOPSIS [23] approach is employed for prioritization of the sustainability factors of service supply chain. TOPSIS uses many scalar values to compute the best and worst choices simultaneously and is easy to compute [24]. While, Analytical Hierarchy Process (AHP) uses the pair-wise comparison of the factors and is found more suitable for small number of factors [23]. The results of TOPSIS are more stable when an additional factor is added of eliminated. These reasons make the choice of Fuzzy TOPSIS for the prioritization of the factors more suitable. Crisp values are used in TOPSIS method but crisp values are not able to handle the real-life situations and hence the fuzzy concept is paired with TOPSIS in this study. The reason for using fuzzy concept is that it takes care of the ambiguity and imprecise information provided by the experts [25].

The steps taken for carrying out the research are shown in Fig. 1. The first step is to identify the important sustainability factors of service supply chain. The literature was reviewed critically and the identified factors were discussed with the working professionals of the case organization "XYZ" for the relatedness and completeness. The working professional also added two important factors in the list of identified sustainability factors.

For evaluating sustainability, six experts were requested to participate in this study. One of the authors of this research paper is in close contact with two experts

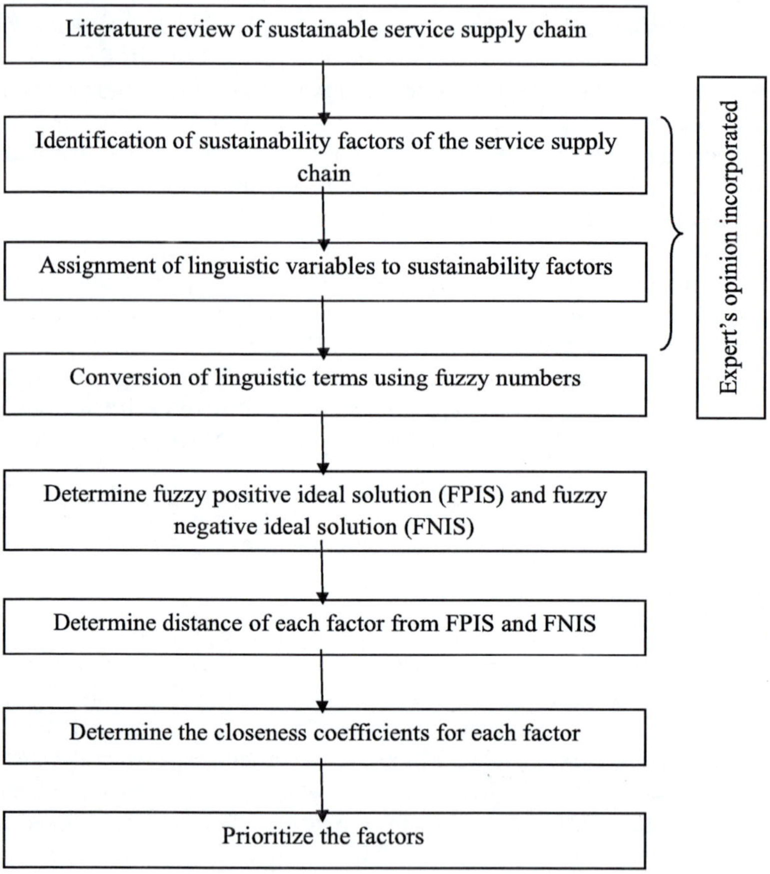

Fig. 1 Flow diagram of research method for prioritization of sustainability factors of SSC

Table 1 Linguistic terms and corresponding fuzzy numbers

Linguistic terms	Abbreviation	Fuzzy number
Low	L	(0.0, 0.1, 0.3)
Fairly low	FL	(0.1, 0.3, 0.5)
Medium	M	(0.3, 0.5, 0.7)
Fairly high	FH	(0.5, 0.7, 0.9)
High	H	(0.7, 0.9, 1.0)

of the case-organization, so the rest of the experts also agreed to participate in this study. The experts had a fair knowledge of sustainable practices and have a wide range of experience in handling operations, corporate social responsibility activities, environmental activities, and handling customers, etc. The purpose of the study was discussed with them. For making them familiar with the purpose of the study, the concept of sustainability was briefed to them. Linguistic terms (Table 1) are used for assigning the ratings and weights. Moreover, the experts also feel comfortable with linguistic terms as compared to crisp numbers. The experts were requested to mark their responses for the performance ratings and important weights for sustainable dimensions. The demographic details of the experts are provided in Table 2.

Certain steps suggested by [23, 24] for the implementation of fuzzy TOPSIS methods are as:

Step I Collect the expert's responses for the factors in linguistic terms then convert the responses into the fuzzy numbers as per the linguistics scales defined in Table 1.

Step II Calculate the aggregate fuzzy ratings of the factors

Table 2 Demographic details of six experts chosen from four different hospitals

S. No.	Designation of experts	Experience (in years)	Average patient visit per day in the hospitals (approximately)	Number of decision maker
Hospital 1	Medical superintend	22	500	2
	Operations manager	10		
Hospital 2	Quality control manager	8	250	2
	Human resource manager	10		
Hospital 3	Quality control manager	8	150	1
Hospital 4	Operations manager	7	150	1

The expert's responses converted in triangular fuzzy number $A_k = (l_k, m_k, u_k)$, where k = 1, 2, 3..., k are further converted into aggregate fuzzy ratings A = (l, m, u) k = 1,2, 3..., k where

$$l = \min_k\{lk\}, m = \frac{1}{k}\sum_{k=1}^{k} m_k, u = \max_k uk$$

Fuzzy ratings of Kth experts is $F_{ijk} = (l_{ijk}, m_{ijk}, u_{ijk})$ where i = 1, 2, 3,..., m, j = 1, 2, 3,...,n then fuzzy rating $F_{ij}(l_{ij}, m_{ij}, u_{ij})$.

where

$$l_{ij} = \min_k\{l_{ijk}\}, m_{ij} = \frac{1}{k}\sum_{k=1}^{k} m_{ijk}, u_{ij} = \max_k\{u_{ijk}\} \quad (1)$$

Step III Tabulate the fuzzy decision matrix for the key factors as

FD =		C1	C2	–	C_n
	R1	X11	X12	–	X1n
	R2	X21	X22	–	X2n
	–	–	–	–	–
	R_m	Xm1	Xm2	–	Xmn

I = 1, 2, 3,..., m and j = 1, 2, 3,..., n

Step IV Construct the normalized fuzzy decision matrix
The data of fuzzy decision matrix is normalized so that the various criteria scales can be compared. The normalized fuzzy decision matrix is presented as:

$N = [a_{ij}]$ m × n, i = 1, 2, 3, ... m and j = 1, 2, 3,..., n

where

$$a_{ij} = \left(\frac{l_{ij}}{n_j^*}, \frac{m_{ij}}{n_j^*}, \frac{u_{ij}}{n_j^*}\right) \text{ and } n_j^* = \max_i\{n_{ij}\} \text{ (benefit criteria)} \quad (2)$$

and

$$a_{ij} = \left(\frac{l_j^-}{l_{ij}}, \frac{l_j^-}{m_{ij}}, \frac{l_j^-}{u_{ij}}\right) \text{ and } l_j^- = \max_i\{l_{ij}\} \text{ (cost criteria)} \quad (3)$$

Step V calculate the weighted normalized matrix by multiplying the weight (b_j) of evaluated criteria with the a_{ij} to get weighted normalized matrix (wn).

$$\mathrm{wn} = [\mathrm{wn_{ij}}]\mathrm{m} \times \mathrm{n}\, \mathrm{i} = 1, 2, 3, \ldots, \mathrm{m} \quad \text{and} \quad \mathrm{j} = 1, 2, 3, \ldots, \mathrm{n} \tag{4}$$

where $\mathrm{wn_{ij}} = (\mathrm{a_{ij}}) \times (\mathrm{b_j})$.

$\mathrm{wn_{ij}}$ is a triangular fuzzy number represented by ($\mathrm{l_{ijk}}$, $\mathrm{m_{ijk}}$, $\mathrm{u_{ijk}}$).

Step VI Calculate the fuzzy positive ideal solution (FPIS) and fuzzy negative ideal solution by using the following equations

$$\text{Fuzzy positive ideal solution (FPIS)}\, P^+ = (p_1^+, p_2^+, p_3^+, \ldots, p_n^+) \tag{5}$$

where $p_j^+ = \left(u_j^+, u_j^+, u_j^+, \ldots, u_j^+\right)$ and $u_j^{\mp} = {}_i^{\max}\{u_{ij}\}$ and

$$\text{Fuzzy negative ideal solution (FNIS)}\, P^- = \left(p_1^-, p_2^-, p_3^-, \ldots, p_n^-\right) \tag{6}$$

where $p_j^- = \left(l_j^+, l_j^+, l_j^+\right)$ and $l_j^- = {}_i^{\min}\{l_{ij}\}$.

i = 1, 2, 3…., m and j = 1, 2, 3, …, n

Step VII Calculate the distances ($\mathrm{d_i}^+$, $\mathrm{d_i}^-$) of each factor from the FPIS and FNIS as per the Eqs. 7 and 8.

$$\text{Distance from FPIS}\, d_i^+ = \sum_{j=1}^{n} dp(wn_{ij}, p_j^+), \quad \mathrm{i} = 1, 2, 3, \ldots, \mathrm{m} \tag{7}$$

$$\text{Distance from FNIS}\, d_i^- = \sum_{j=1}^{n} dp(wn_{ij}, p_j^-), \quad \mathrm{i} = 1, 2, 3, \ldots, \mathrm{m} \tag{8}$$

Step VIII Calculate the closeness coefficient for each factor by Eq. 9

$$\text{Closeness coefficient}\, CC_i = \frac{d_i^-}{d_i^+ + d_i^-} \tag{9}$$

Step IX Rank the factors as per the decreasing values of CC_i.

4 Analysis of Key Success Factors Using Fuzzy TOPSIS

Fuzzy TOPSIS method was used in this paper to prioritize the key success factors of the sustainable service only supply chain. The expert's responses were analyzed in this section. The name of the hospitals and participants are kept confidential due

Table 3 Expert's responses in linguistic variables

S. No.	Important factors	Ex1	Ex2	Ex3	Ex4	Ex5	Ex6
SP1	Demand management	H	FH	FH	H	FH	FH
SP2	Customer relationship management	H	H	H	FH	H	FH
SP3	Resource management	FH	H	FH	H	FH	FH
SP4	Real-time information sharing with stakeholders	FH	FH	H	FH	FH	FH
SP5	Supplier relationship management	H	FH	H	FH	FH	FH
SP6	Commitment from top management	FH	H	FH	H	H	FH
SP7	Service delivery management	FL	M	FL	M	FL	FL
SP8	Reducing the energy consumption	L	M	FL	FL	M	FL
SP9	Social accountability	M	FH	M	M	FH	M
SP10	Designing new services	M	FH	FH	M	FH	M
SP11	Reducing the contamination	L	L	FL	L	FL	L
SP12	Meeting legal requirements	L	FL	L	L	L	M
SP13	Relationship with employees	M	M	FH	M	M	FL
SP14	Environmental friendly activities	FL	M	M	FL	M	M
SP15	Responsibility towards community and stakeholders	L	M	FL	FL	M	FL

to the confidentiality clause requested by the participants. All the experts were rich in experience, familiar with sustainable practices and understand the importance of sustainability for the next generation. The demographic details of the six experts are mentioned in Table 2. Two experts were familiar with the fuzzy TOPSIS approach and helped us in formulating the key success factors. All the experts were interested in knowing the hierarchy and importance level of the key success factors to implement these in their hospitals for enhancing the sustainability level of their hospital. The purpose of the study and the identified key success factors and their meaning were discussed with them. The expert's responses were collected in linguistics variables (Table 3) which were processed as per the steps discussed in Sect. 3 (Table 4).

5 Results and Discussion

The identified sustainable practices are ranked in their order of importance by fuzzy TOPSIS method. Six experts from the four different hospitals were consulted for prioritizing the sustainable practices related to SOSC. The hospitals from where the experts were consulted were accredited by National Accreditation Board for Hospitals and Healthcare Providers (NABH). The reason for selecting NABH hospitals was that these hospitals tend to adopt sustainable practices. The selected case hospital was also NABH accredited. The experts were requested to participant in this study

Table 4 Closeness coefficient values and prioritization of sustainability factors of SSC

S. No.	Important factors	d+	d−	CC_i	Ranking
SP1	Demand management	0.1234	0.8030	0.8668	3
SP2	Customer relationship management	0.3922	0.4941	0.5575	6
SP3	Resource management	0.1467	0.7558	0.8374	4
SP4	Real-time information sharing with stakeholders	0.0789	0.8344	0.9136	2
SP5	Supplier relationship management	0.2575	0.6987	0.7307	5
SP6	Commitment from top management	0.0173	0.8744	0.9806	1
SP7	Service delivery management	0.7142	0.1875	0.2079	15
SP8	Reducing the energy consumption	0.6651	0.2614	0.2821	14
SP9	Social accountability	0.4637	0.4506	0.4928	9
SP10	Designing new services	0.4093	0.4878	0.5438	7
SP11	Reducing the contamination	0.5646	0.3485	0.3817	12
SP12	Meeting legal requirements	0.4937	0.4653	0.4852	10
SP13	Relationship with employees	0.4880	0.4440	0.4764	11
SP14	Environmental friendly activities	0.5677	0.3281	0.3663	13
SP15	Responsibility towards community and stakeholders	0.4424	0.4928	0.5270	8

and after their consent the purpose of the research was discussed with them. To make familiar with the study, the identified sustainable practices and their meaning were discussed with the experts. The experts were provided the five-point linguistic scale (Table 1) and requested to mark their responses according to Table 1. The responses were collected for the fifteen sustainable practices on the quality patient care and financial performance. Due to space constraint the responses for the quality patient care are shown in Table 3.

For prioritizing the sustainable practices were further analyzed by the fuzzy TOPSIS method as per the steps discussed in Sect. 3. First the linguistic responses were converted into triangular fuzzy numbers using the Table 1. Further, combined decision matrix was developed and followed by normalized fuzzy decision matrix. Normalized fuzzy decision matrix was further used for obtaining weighted normalized fuzzy decision matrix. Fuzzy positive ideal solution and fuzzy negative ideal solutions were calculated. Equations 7 and 8 were used for obtaining the distance d+ and d− of each sustainable practice. Finally, CC_i values for each sustainable practice were calculated. The sustainable practices were ranked based on the values of CC_i. Greater the CC_i value, greater the importance of the sustainable practice. The results indicated 'commitment from top management' as the most important factor for the inclusion of sustainable practices in SOSC while taking care of quality patient care and aiming for enhanced financial performance. The order of importance of sustainable practices were identified as—SP6 > SP4 > SP1 > SP3 > SP5 > SP2 > SP10 > SP15 > SP9 > SP12 > SP13 > SP11 > SP14 > SP8 > SP7.

The findings of this study calls for the commitment from the top management for the inclusion of sustainable practices in SOSC. ‘Commitment from top management’ has been ranked first with the CC_i value 0.9806. If the top management is committed then they will implement the strategy and policies for the sustainability thus making a smoother way for achieving sustainability [8]. In SOSC information sharing with customers plays an important role and hence ranked second (CC_i value = 0.9136). This platform can be used to create awareness among the customers for the sustainability. This platform can be used for communicating the requirements to suppliers about the sustainable products and services and monitoring of suppliers can be executed to achieve sustainability [20]. ‘Demand management’ and ‘resource management’ were ranked third and fourth respectively. In SOSC context the demand management plays an important role specially in case of hospitals SOSC the ‘demand management’ plays an crucial role [10]. ‘Resource management’ motivates the organization for effective and efficient usage of manpower and other facilities therefore ensuring right practices in the organization and helping in achieving sustainability in SOSC [11]. Suppliers are the crucial part of a SOSC and hence maintaining relationships with suppliers helps in achieving the sustainability in SOSC. The role of customer in SOSC is prime as they are present from the generation of services to consumption of services therefore making customers aware about sustainability can help in achieving sustainability more effectively [5].

‘Designing new services’ is ranked at seventh place. The new services should be designed in such a manner that reduces the generated waste and energy consumption will lead the organization towards green practices. [14] also supported the result that new services should be designed in unique way so as to gain competitive advantage and unique services helps in gaining new market share. ‘Responsibility towards community and stakeholders’ is placed at eight positions. The stakeholders can be considered all associated individuals directly or indirectly being affected by the organization. If the organization reduces the energy consumption then ultimately it helps in keeping the environment clean. The organization must focus on reducing the energy consumption, contamination and should develop society. This can help in improving social sustainability of the SOSC [15]. ‘Meeting legal requirements’ will help in achieving sustainability in SSOC. Proper law should be framed and imposed so as to adopt such practices which focused on safe disposal of bio-waste, waste, recycle the water, focus on reuse of materials if possible, reduce the energy consumption, shift to solar panels, educate the society, develop facilities for the society, organizing free camp for the poor etc. will help the SOSC in achieving social and environmental sustainability [5, 14]. The role of employees in achieving sustainability cannot be neglected. Committed employees will focus on the proper execution of sustainability goals and will help in achieving sustainability to a next level therefore ‘relationship with employees’ is ranked at eleventh position.

‘Reducing contamination’, adoption of ‘environmental friendly activities’ and ‘reducing energy consumption’ are ranked as twelfth, thirteen and fourteen respectively. All these sustainable practices will help the organization in improving environmental sustainability [22]. ‘Service delivery management’ was found the least important factor in achieving sustainability in SOSC.

6 Conclusion

In present context, service only supply chain is the fastest growing supply chain and has started attracting the researchers in this area. The sustainable practices in SOSC are less evident and hence the inclusion of sustainable practices in SOSC will not only help in greening the environment but also will help the community and stakeholders in raising their living standard while being economical. The hospitals do not want to compromise with the quality of patient care while being profitable. Therefore the sustainable practices and their order of importance should be known to the managers of SOSC. This study makes an attempt to identify fifteen sustainable practices relevant to the SOSC and these practices were ranked by the fuzzy TOPSIS method. The identified fifteen practices are- 'Demand management', 'customer relationship management', 'resource management', 'real time information sharing with stakeholders', 'supplier relationship management', 'commitment from top management', 'service delivery management', 'reducing the energy consumption', 'social accountability', 'designing new services', 'reducing the contamination', 'meeting legal requirements', 'relationship with employees', 'environmental friendly activities', and 'responsibilities towards community and stakeholders'. The applicability of this research was tested in a hospital SOSC. Out of these fifteen practices 'commitment from top management', 'real-time information sharing with stakeholders', 'demand management' and 'resource management' were identified as the most crucial practices in SOSC context.

The role of healthcare in the economies of countries is growing significantly and hence the findings of this research will help this sector in achieving sustainability while being economic and providing quality patient care. The present research is limited to its applicability in health care sector. The study can be further analyzed with large sample size. The same study in future can conducted with new and relevant sustainable practices either by same methodology or by other multi criteria decision making technique.

References

1. Nagariya R, Kumar D, Kumar I (2020) Service supply chain: from bibliometric analysis to content analysis, current research trends and future research directions. BIJ. Ahead-of-print.https://doi.org/10.1108/BIJ-04-2020-0137
2. Elkington J (1998) Partnerships fromcannibals with forks: the triple bottom line of 21st-century business. Environ Qual Manage 8:37–51. https://doi.org/10.1002/tqem.3310080106
3. Wang Y, Wallace SW, Shen B, Choi T-M (2015) Service supply chain management: a review of operational models. Eur J Oper Res 247:685–698. https://doi.org/10.1016/j.ejor.2015.05.053
4. Nagariya R, Kumar D, Kumar I (2021) Enablers to implement sustainable practices in the service only supply chain: a case of an Indian hospital. Bus Process Manage J 27(5):1463–1495. https://doi.org/10.1108/BPMJ-10-2020-0469
5. Pourjavad E, Shahin A (2018) Hybrid performance evaluation of sustainable service and manufacturing supply chain management: an integrated approach of fuzzy dematel and fuzzy

inference system. Intell Syst Account Fin Manage 25:134–147. https://doi.org/10.1002/isaf.1431

6. WCED U (1987) Our common future. World Commission on Environment and Development, Oxford University Press, U.S.A.
7. Aliakbari Nouri F, Shafiei Nikabadi M, Olfat L (2019) Developing the framework of sustainable service supply chain balanced scorecard (SSSC BSC). Int J Product Perform Manage 68:148–170. https://doi.org/10.1108/IJPPM-04-2018-0149
8. Mirghafoori SH, Morovati Sharifabadi A, Karimi Takalo S (2018) Development of causal model of sustainable hospital supply chain management using the Intuitionistic Fuzzy Cognitive Map (IFCM) method. J Indust Eng Manage 11:588–605. https://doi.org/10.3926/jiem.2517
9. Nagariya R, Kumar D, Kumar I (2021) Sustainability evaluation of service supply chains: a case study of an Indian hospital. Int J Prod Performance Manage ahead-of-print(ahead-of-print). https://doi.org/10.1108/IJPPM-05-2020-0237
10. Ellram LM, Tate WL, Billington C (2004) Understanding and managing the services supply chain. J Supply Chain Manage 40:17–32. https://doi.org/10.1111/j.1745-493X.2004.tb00176.x
11. Baltacioglu T, Ada E, Kaplan MD, Yurt And O, Cem Kaplan Y (2007) A new framework for service supply chains. Serv Ind J 27:105–124. https://doi.org/10.1080/02642060601122629
12. Scavarda A, Daú GL, Scavarda LF, Korzenowski AL (2019) A proposed healthcare supply chain management framework in the emerging economies with the sustainable lenses: the theory, the practice, and the policy. Resour Conserv Recycl 141:418–430. https://doi.org/10.1016/j.resconrec.2018.10.027
13. Sallnäs U, Björklund M (2020) Consumers' influence on the greening of distribution—exploring the communication between logistics service providers, e-tailers and consumers. IJRDM. Ahead-of-print.https://doi.org/10.1108/IJRDM-07-2019-0213
14. Uribe D, Sarache G (2019) Sustainable supply chain management practices and sustainable performance in hospitals: a systematic review and integrative framework. Sustainability 11:5949. https://doi.org/10.3390/su11215949
15. Aliakbari Nouri F, Shafiei Nikabadi M, Olfat L (2019) Sustainable service supply chain practices (SSSCPs): a framework development. Int J Prod Performance Manage. Ahead-of-print. https://doi.org/10.1108/IJPPM-09-2018-0314
16. Sigala M (2014) Customer involvement in sustainable supply chain management: a research framework and implications in tourism. Cornell Hospitality Q 55:76–88. https://doi.org/10.1177/1938965513504030
17. Zailani SHM, Kumar KM (2011) Service supply chain (SSC): proposed SSC practices measurement items for empirical testing. J Syst Manage Sci 1:13–23
18. Haas DH, Hansen AP (2010) Facilities management in a service supply chain perspective. In: Proceedings of the 22nd annual NOFMA conference. Arlbjorn, J.S. (ED.), Denmark, Kolding, pp 631–645
19. Cho DW, Lee YH, Ahn SH, Hwang MK (2012) A framework for measuring the performance of service supply chain management. Comput Ind Eng 62:801–818. https://doi.org/10.1016/j.cie.2011.11.014
20. Valinejad F, Rahmani D (2018) Sustainability risk management in the supply chain of telecommunication companies: a case study. J Clean Prod 203:53–67. https://doi.org/10.1016/j.jclepro.2018.08.174
21. Liu W, Bai E, Liu L, Wei W (2017) A framework of sustainable service supply chain management: a literature review and research agenda. Sustainability 9:421. https://doi.org/10.3390/su9030421
22. Wu K-J, Zhu Y, Chen Q, Tseng M-L (2019) Building sustainable tourism hierarchical framework: coordinated triple bottom line approach in linguistic preferences. J Clean Prod 229:157–168. https://doi.org/10.1016/j.jclepro.2019.04.212
23. Gupta A, Singh RK, Suri PK (2018) Prioritizing critical success factors for sustainable service quality management by logistics service providers. Vis J Bus Perspect 22:295–305. https://doi.org/10.1177/0972262918786102

24. Bottani E, Rizzi A (2006) A fuzzy TOPSIS methodology to support outsourcing of logistics services. Surg Endosc Other Interv Tech 11:294–308
25. Kumar D, Rahman Z (2017) Analyzing enablers of sustainable supply chain: ISM and fuzzy AHP approach. J Model Manage 12:498–524. https://doi.org/10.1108/JM2-02-2016-0013

Incremental Sheet Metal Forming: The State of Art and Its Future Prospects

Bittu Toppo, Manish Oraon, and Manish Kr. Roy

Abstract Incremental sheet metal forming (ISMF) is an innovative sheet metal forming in which not dedicated die is required. The paper is a review of the ISMF for its future research potential. The present study aimed to enlighten the innovative strategies used in the ISMF and highlighted the complexities that is still present in the ISMF. The key benefits of the ISMF are customer-orientated process, flexible manufacturing, and the part formation at low cost. With few advantages, the several difficulties are observed. The achievement of exact dimension i.e. geometrical accuracy, required surface finish, springback effect, forming of complex shape, prevention of thin sheet failure, optimized tool path, tool wear etc. are the major challenges in the ISMF. The causes for aforesaid difficulties are due to the variation in the process, undefined input variability, tool size, tool path, machines efficiency, etc. Wide recognition of this technique enumerates its prospects and the acceptability in industries. The current study enumerated the ISMF strategies and optimization technique has been reported by the researchers and directed to the area in which more concentrated study is required for the optimization of the ISMF.

Keywords ISMF · Inputs parameters · Heat-assisted forming · Optimization

1 Introduction

A weighty mutation occurs in the industries related to sheet-metal part formation. It is crafted for the development of effective strategies in terms of flexibility and reduced production cost. Sheet-metal forming based industries attempt to complete the customer needs by producing the parts as per design specifications which is varied rapidly. The customers demand customized products at very low cost and for survival in the competitive market, production at low cost is strictly required for the margin of profit.

B. Toppo (✉) · M. Kr. Roy
Sikkim Manipal Institute of Technology, Majitar, East Sikkim 737136, India

M. Oraon
Birla Institute of Technology, Mesra, Ranchi 835215, India

R. Agrawal et al. (eds.), *Recent Advances in Industrial Production*, Lecture Notes in Mechanical Engineering, https://doi.org/10.1007/978-981-16-5281-3_27

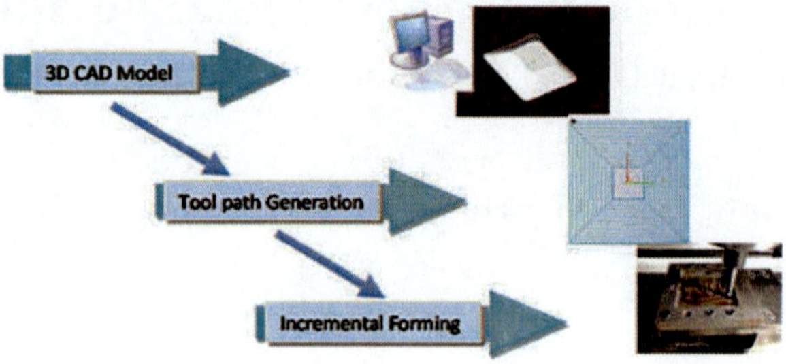

Fig. 1 Flowchart of the ISMF process [1]

The ISMF process is an emerging technique in which a simple tool imposed on the metal sheet and local deformation is consecutively made on the sheet. The hemispherical end shaped solid rod (forming tool) traverse over the sheet according to a predefined tool path. The metal-sheet is fixed at its edge through nut and bolts in a fixture. The geometry formation in ISMF included the incremental movement of the tool which punched the clamped sheet locally. The accurate incremental movement needed a computerized machining Centre. Sometimes, the supporting die e.g. full/partial is required behind the clamped sheet to prevent the early failure i.e. tearing, crack formation, excessive thinning of the blank, etc. The tool-paths are created by using CAD/CAM and import the file into the machining center for production. ISMF is more suitable for customized products, rapid prototyping, and low volume production. The processing step involved in the ISMF is quite similar to the deep drawing process. Initially the CAD/CAM model is developed for the recognition of the optimized tool-path. The STL file transferred into the machining centre and the local deformation process is conducted accordingly to the predefined program (see Fig. 1).

Concerning tool geometry, the ISMF tool is different from the milling cutter. A solid cylindrical rod of the hemispherical head is used for deformation despite the shaped die. Among the different chances, ISMF is satisfied to be a good solution in many distinct works done by Kim and Park [1]. They explored different shapes formed with an aluminum sheet under different ISMF forming conditions. They also discussed the forming limit curve (FLD) of the aluminum sheet [2]. The authors identified the different ISMF configurations. The results of analytical and the experimental investigation for the ISMF are discussed by the researchers on the basis of forming different shape like cones [4, 5] and pyramids [6]. The emulate figures of the pyramid formation through ISMF (see Fig. 2).

In the continuing ISMF investigation, the different shapes is produced with the help of partial and full die aiming to minimize the geometrical inaccuracy (see Fig. 3). The authors suggested that a larger tool end diameter along with low step down size in ISMF may reduce the springback effect [7]. The new method in the ISMF is introduced in which the slave tool is planned. The forming method is termed as 'Sculptor method'. In this technique, small and medium sculptor are engaged with two inclusively tools placed over and behind the clamped metal sheet that produced two independent flexible die systems (see Fig. 4) [8].

Fig. 2 ISMF processing of aluminum [3]

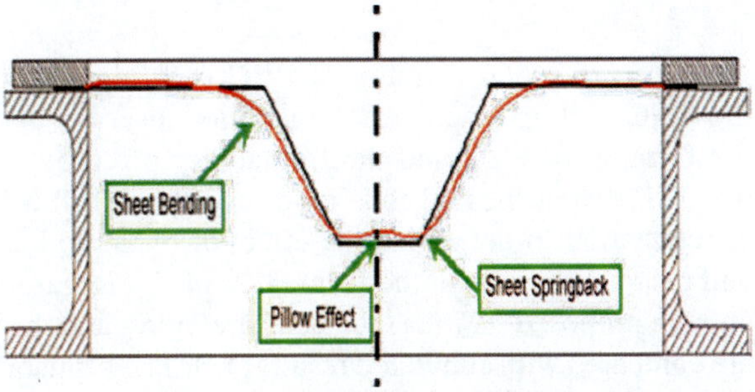

Fig. 3 Geometrical inaccuracies observed in the ISMF process [7]

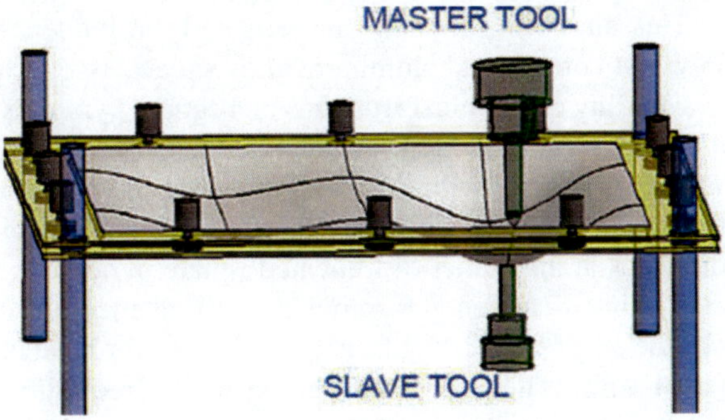

Fig. 4 Tool arrangement in sculptor technique used in ISMF [8]

The statistical analysis of ISMF of aluminum AA1050 is done with the variation of tool diameter, step size, wall angle and total depth. The authors reported that the geometrical error at the corners due to sheet thickness and the pillow effect is due to tool diameter [9]. The high-velocity water jet (WJIF) is suggested for ISMF of thin metal sheets. The advantage of this technique is that it eliminated the solid forming tool previously used by the researchers. The scheme of WJIF is compared with rigid tool incremental forming (RTIF) and (see Fig. 5). The authors also reported the significant parameters for WJIF that are jet pressure and nozzle diameter [10].

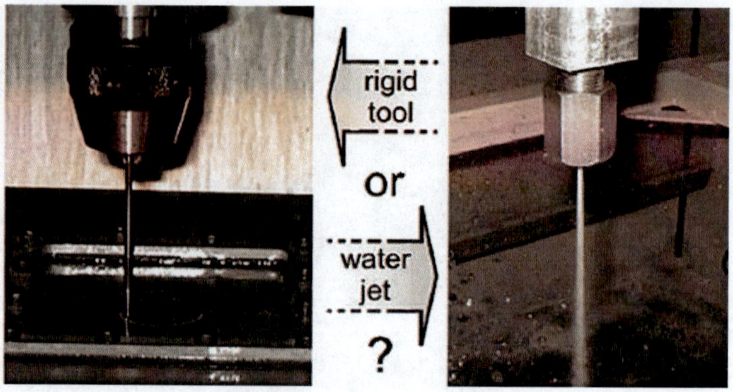

Fig. 5 The comparison of forming tool in RTIF and WJIF [10]

The incremental forming of PVC (polyvinylchloride) is done successfully on the CNC milling machine. The authors observed that the low force is required in the forming of PVC and the formability of curbs is a major operating specification [11]. The series of cone is formed through ISMF of aluminum alloy with the concentration to determine the effect of dynamic loading, local heating, accuracy, and residual stresses in the formed part [12]. Regarding tool path optimization, the authors proposed various tool-path strategies, and the resulting part geometries that are compared with simulated results [13]. The formability of metal sheet is increased by applying local heating during the ISMF. The authors reported the reduction in applied pressure during the forming [14] (Table 1).

Hussain et al. [15] have investigated the influences of wall angle (θ_{max}) in the ISMF of commercial aluminum alloy sheets. Two shapes were formed to assess the moldability e.g. conical frustums and square pyramids. A set of each is produced by a systematical range of the wall angle to explore θ_{max}. Table 2 shows that the conic section is formed with 70° wall angle but the pyramid is limited to only 66.5° due to complicated forming behavior of thin sheet at the of corners. After the said angle, the crack at the corner of truncated square pyramid is appeared (see Fig. 6).

The hot incremental forming is investigated for the easy forming of tungsten alloy. For the purpose, Nd-Yag laser beam is used for local heating of blank during process (see Fig. 7). The laser beams are synchronized with the forming tool. The authors reported the formability of tungsten is possible with controlled laser beam power for a specified time period but infeasible due to expensive process and initial setup cost [16]. Malwad et al. have revisited the ISMF process and confirmed the excessive shearing of aluminum sheet is possible with increasing wall angle which was earlier stated by Kim and Park. The authors noticed the maximum formability of aluminum sheet at 75° wall angle and a vertical step size of 0.1 mm [17]. Reddy et al. [18] presented the various configurations for the growth of the incremental forming of the thin metal sheet. In the continuing research, the authors reviewed the ISMF research till date and directed the research gap of forming process and a deep analysis is required for the optimization of the ISMF [19].

Table 1 ISMF at constant current @500 A with varying process parameters [14]

Formability test result of feed rate (tool diameter: 8 mm, step size: 0.2 mm)		
Feed rate	Wall angle	Remarks
700	Failed	Sheet burns
1000	64.3	No sheet burning
1300	53	No sheet burning
1300	43	No sheet burning
Formability test result of tool diameter (feed rate: 1000 mm/min, step size: 0.2 mm)		
Tool diameter	Wall angle	Remarks
6	Failed	Sheet burns
8	64.3	No sheet burning
12	44.3	No sheet burning
Formability test result of step size (tool diameter: 8 mm, feed rate: 1000 mm/min)		
Step size	Wall angle	Remarks
0.2	64.3	No sheet burning
0.3	49	No sheet burning
0.4	42	No sheet burning

Table 2 Results of varying wall angle [15]

Part type	Wall angle θ (°)	Outcome
Conical frustum	50	Formed
	55	Formed
	60	Formed
	65	Formed
	66	Formed
	66.3	Formed
	66.8	Formed
	68	Formed
	70	Formed
Truncated square pyramid	66	Formed
	66.3	Formed
	66.5	Formed

Soren et al. [20] formed the exterior body parts of the car e.g. mudguard, bumper, etc. through ISMF. For the purpose, the partial die is used to support the metal sheet (see Fig. 8). The statistical analysis of ISMF is done with varying the input variables e.g. sheet thickness, wall angle, tool diameter, and step down. The authors observed that step size and the tool diameter is significant [21]. The forming behavior of ISMF

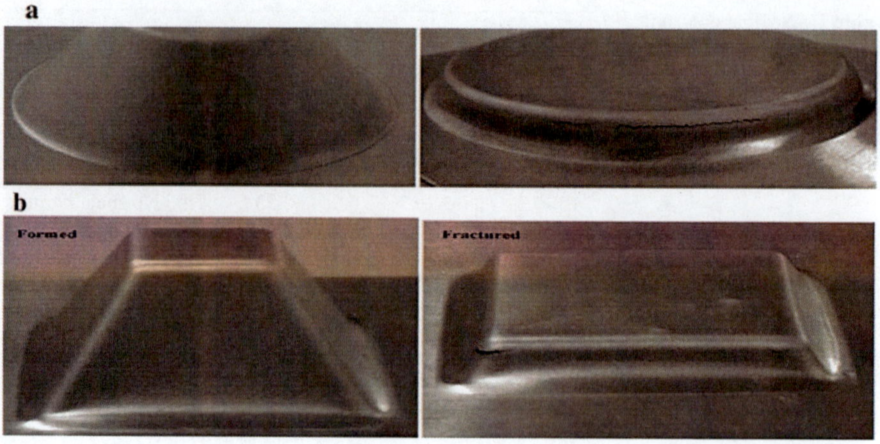

Fig. 6 Different shapes formed in the test-1: **a** conic frustums and **b** square pyramid [13]

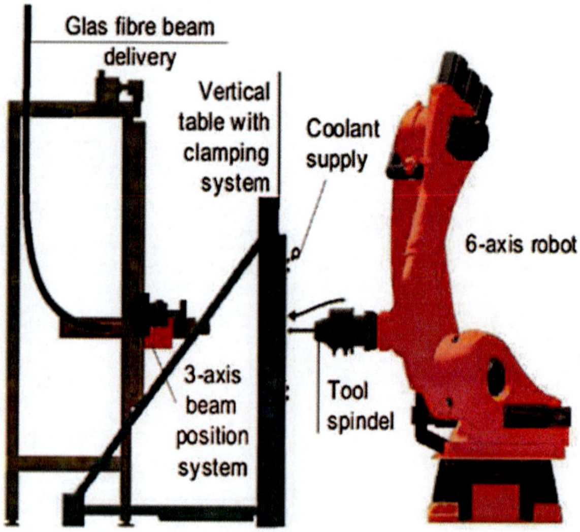

Fig. 7 The set up of laser-assisted incremental forming [16]

Fig. 8 3D visualization of the partial die used for car body parts in ISMF

is compared with two manufacturing processes i.e. deep drawing, hydro-forming. The authors also conducted the ISMF of aluminum alloys AW-1050 (H14-H24) and briefed the advantage of the emerging forming process over the others [22]. Hartmann et al. [23] proposed the automation of ISF process for the optimization of tool path with which the incremental forming of aluminum alloy is done successfully [24].

During the extensive review, it is observed that the processing technique of ISMF is quite simple but limited to shape the thin sheet only. Several advantages of ISMF arise some complex issues that should be taken care of. Some of these are noticed.

(a) The forming of any part even small in size taken longer processing time.
(b) The process is suitable i.e. up to small batch production only.
(c) Forming at corners is a complex issue due to tool shape.
(d) Achieving the required surface finish is difficult due to the tool feed mark and vertical step size.
(e) Reduction of Springback and pillow effect.
(f) Formation of complex profile,
(g) Optimization of tool-path for ISMF.

Sheet metal parts are widely utilized in general purpose machines and different application. Some of the industrial applications of the ISMF formed parts are as; Automobile sector:- Door panel (inner/outer), Hood panel, Engine cover, etc., Aerospace parts like Passenger seat cover Instrument panel, Body panel, etc. and customized products such as Denture plate, Metal helmet Ankle support, etc.

2 Future Research Potential

Several reports have been proposed but still not declared the optimized condition for ISMF; therefore lots of scopes are there for setting up the emerging technique. Since the formability characteristics of all metals and alloys are different, therefore it is necessary to know the forming limit (FL) and ductile fracture limit (DFL) of the blank in the ISMF. Even though with knowing FL and DFL, it is a complex issue to shape a thin metal sheet or foil without the use of partial or full die. The workability of the ISMF may be improved, if a optimized perform is utilized. On the other hand, it is observed that the tough metal is not deformed in the normal room temperature, therefore a system is to be designed for local heating during the ISMF by the use of heating equipment (halogen, electricity, laser etc.). The various strategies in the ISMF are proposed till date but the optimized condition of the input variables is not stated, therefore the study of sheet metal deformation mechanics in ISMF, fracture criteria for metal in ISMF and the ISMF process optimization is needed.

3 Conclusion

The ISMF is very simple technique and the product can be formed by controlling the input parameters like tool diameter, step size, wall angle, thickness of sheet, spindle speed, and tool feed rate.

The fast implementation of ISMF in sheet metal parts manufacturing industries required to promote the ISMF process mechanics and develop the workplace for it. This Process may act as the accelerator for industrial growth. The journey of ISMF is the started and research on their progression. Few features are discovered, identified, and described, but many complexities are still there.

References

1. Kim YH, Park JJ (2002) Effect of process parameters on formability in incremental forming of sheet metal. J Mater Process Technol 130–131:42–46
2. Kim YH, Park JJ (2003) Fundamental studies on the incremental sheet metal forming technique. J Mater Process Technol 140:447–453
3. Shim MS, Park JJ (2001) The formability of aluminum sheet in incremental forming. J Mater Process Technol 113:654–658
4. Ambrogio G, Filice L, Fratini L, Micari F (2003) Some relevant correlations between process parameters and process performance in incremental forming of metal sheets. In: 6th international proceedings on ESAFORM. Salerno, Italy, pp 175–178
5. Jadhav S, Goebel R, Homberg W, Kleiner M (2003) Process optimization and control for incremental sheet metal forming. In: Proceedings of the international deep drawing research group (IRRDG). Bled, Slovenia, pp 165–171
6. Kopac J, Kampus Z (2005) Incremental sheet metal forming on CNC milling machine tool. J Mater Process Technol 162–163:622–628
7. Micari F, Ambrogiob G, Filice L (2007) Shape and dimensional accuracy in single point incremental forming: state of the art and future trends. J Mater Process Technol 191:390–395
8. Maidagan E, Zettler J, Bambach M, Rodríguez P, Hirt G (2007) A new incremental sheet forming process based on a flexible supporting die system. Eng Mater 344:607–614
9. Ambrogio G, Cozza V, Filice L, Micari F (2007) An analytical model for improving precision in single point incremental forming. J Mater Process Technol 191:92–95
10. Petek A, Jurisevic B, Kuzman K, Junkar M (2009) Comparison of alternative approaches of single point incremental forming processes. J Mater Process Technol 209:1810–1815
11. Franzen V, Kwiatkowski L, Martins PAF, Tekkay AE (2009) Single point incremental forming of PVC. J Mater Process Technol 209:462–469
12. Duflou JR, Callebaut B, Verbert J, Baerdemaeker DH (2007) Laser assisted incremental forming: formability and accuracy improvement. Ann CIRP Ann 5(1)
13. Duflou JR, Verberta J, Belkassem B, Gu J, Sol H, Henrard C, Habraken AM (2008) Process window enhancement for single point incremental forming through multi-step toolpaths. CIRP Ann Manuf Technol 57:253–256
14. Fan G, Gao L, Hussain G, Haoli J, Wu B (2008) Electric hot incremental forming: a novel technique. Int J Mach Tools Manuf:1688–1692
15. Hussain G, Gao L, Dar NU (2007) An experimental study on some formability evaluation methods in negative incremental forming. J Mater Process Technol 186:45–53
16. Wulfsberg JP, Terzi M (2007) Investigation of laser heating in microforming applying sapphire tools. Ann CIRP 56(1)

17. Malwad DS, Nandedkar VM (2014) Deformation mechanism analysis of single point incremental sheet metal forming. J Proc Mater Sci:505–1510
18. Reddy NV, Lingam R, Cao J (2015) Incremental metal forming processes in manufacturing. Springer-Verlag, Berlin Heidelberg
19. Pathak J (2015) A brief review of incremental sheet metal forming. Int J Latest Eng Manage Res:35–43
20. Soren S, Alexander P, Peter S, Sebastian M, Dieter W, Zdenek R (2019) Incremental sheet metal forming on the example of car exterior skin part. Proc Manuf 29:105–111
21. Dwivedy M, Kalluri V (2019) The effect of process parameters on forming forces in single point incremental forming. Proc Manuf 29:120–128
22. Scheffler S, Pierer A, Scholz P, Melzer S, Weise D, Rambousek Z (2019) Incremental sheet metal forming on the example of car exterior skin parts. Proc Manuf 29:105–111
23. Hartmann C, Volk W (2019) Knowledge-based incremental sheet metal free-forming using probabilistic density functions and voronoi partitioning. Proc Manuf 29:4–11
24. Dabwan A, Ragab AE, Saleh MA, Anwar S, Ghaleb AM, Rehman AU (2020) Study of the effect of process parameters on surface profile accuracy in single-point incremental sheet forming of AA1050-H14 Aluminum Alloy. Int J Adv Mater Sci Eng 14

Lean Implementation Value in Automobile Sector

Arshit Kapoor, Krishna Mohan Agarwal, and Aaryan Sheokand

Abstract Lean manufacturing is the organized removal of waste from all phases of an organization's activities in its simplest form, in which waste is seen as the use or lack of resources which do not actually contribute to the production of a product or service a consumer needs. The research indicates the adoption of ergonomic circumstances in Lean production for the enhancement of the industry's organizational efficiency. The goal of ergonomics and research is to apply modern methods in an effective and healthy way to their function in order to optimize human health conditions and to increase the rate of production. By this research the practices and methodology followed by the leaders of lean manufacturing in automobile industry has been brought into notice so that it could be followed by the emerging automobile industries. In automotive part manufacturing, streamlined tools and approaches are very common, enhancing growth initiatives in the automotive industry over the last two decades. In this paper, the study is done related impact of lean techniques and improvements in selected automobile industries.

Keywords Lean manufacturing · Toyota production system · Ford production system · Chrysler operating system

1 Introduction

Lean Production has been a concern for many industrial firms. Development industries face global pressure by shifts in consumer demands, reduced order size, model selection, shorter product cycles and fast-track consumer production requirements. Lean Manufacturing is an important subject of management which focuses on the process reduction of waste, lower inventories, avoiding unnecessary expenditures and strengthening business as a whole by reducing non-value-adding activities [1]. Lean Manufacturing is capable of reducing processing time, lowering production costs, minimizing material handling and increasing efficiency. It also generates a area of minimized waste, decreases product distribution time and increases customer

A. Kapoor (✉) · K. M. Agarwal · A. Sheokand
Department of Mechanical Engineering, Amity University Uttar Pradesh, Noida 210303, India

R. Agrawal et al. (eds.), *Recent Advances in Industrial Production*, Lecture Notes in Mechanical Engineering, https://doi.org/10.1007/978-981-16-5281-3_28

loyalty, which lets businesses meet strategic targets. Lean Manufacturing's goal is to allow companies to give consumers the best values. It also deals basically with excess goods due to overproduction, transport and anticipation, etc. as an interview to investigate whether the implementation is to be handled with Malaysian automobile part manufacturers [2]. Kaizen can be interpreted loosely as a continuous development program and is appropriate for a manufacturing line. The lean manufacturing strategy also relies on model line systems such as the small space, bottlenecks and supply area. This also basically refers to the management of the inventory standard, and in the production of lean goods the following rules must be considered as benefit, flow benefit direction, push, drag, perfection etc. Present study helps the researchers to consider and use these techniques in the automotive industry for auto-components. The company uses the following methods, such as Kaizen, Total Production Management (TPM), Cellular Manufacturing, Just-in-time (JIT), 5S, Six Sigma, Pre-production planning (3P and continual changes in production line or assembly line as well as in quality control. In this study we have seen the realistic mechanism of these tools which helps to boost the profitability, quality and overall business growth by waste reduction. This also tends to prevent overproduction, idle and material excess wastage. It also reveals how the lean production model has really been applied and how the sector has changed [3]. Techniques of lean manufacturing is presented in Fig. 1.

Fig. 1 Techniques of lean manufacturing

2 Literature Review

Lean production is also literally "simple," a systemic form of waste management for a production system without losing efficiency. This ideology derives from the lean-identified Toyota development program (TPM) in the 1990s. Lean Manufacturing has already been in operation for over three decades, but everyone has not yet decided on a definition. Any plant that does not incorporate lean production would undoubtedly face output problems in comparison to plants that, given their age and the scale of the facility, have implemented lean production. In recent years, a lot of research has been performed in the field of advanced manufacturing, for example in lean manufacturing, overall quality control, overall plant maintenance and implementation in different manufacturing firms like automobile, electronics, aerospace etc. Most scientific studies have shown that lean production is in the twenty-first century the strongest production strategy [4, 5].

Flinchbaugh [6] performed the survey on various automobile companies revealed how the development rate and ergonomic requirements of humans are affected. The integration of Ergonomics during lean manufacturing can achieve significant productivity gains and improve working conditions at the same time. Arumugam et al. [7] designed the study on Comprehensive Dynamic Equation mathematical model to verify that lean social practices and human factors are essential in applying lean practices to improve the success rate of lean transition in small to medium-sized automotive part manufacturer OEM's. The results show that lean social practice is influenced positively by lean technical practices. Arunagiri and Gnanavelbabu (2014) conducted a survey in the automotive industry to find the most important lean tools on the basis of the ranking [8]. The analysis was performed using the weighted average technique and shows that use of these 5 tools (8 Step Problem Solving, (OEE) Overall Equipment effectiveness, Pareto analysis, waste management and 5S) have a positive impact on automotive industry productivity. Kumar et al. [9] research would help to define tools for the lean and sustainable market, and help to make choices for sustainable lean manufacturing (SLM) adoption to achieve better output, distribution on a timely basis and reduce costs. By using the systemic model of understanding the environmental conditions can be strengthened by the introduction of SLM. Wee and Simon [10] discussed the case study of Ford Motor Company and the value of the Values stream mapping (VSM) as a supply chain tool to implement lean production and described the Lean Supply Chain through (VSM). They discussed about both observable metrics that help to minimize prices, increase efficiency and decrease lead time. Jadhav et al. (2015) provided a roadmap to implementation of lean management in the Indian automobile industry using United Nations Industrial Development Organization (UNIDO)-ACMA and ISM Model for lean management. The successful implication of the lean management in company is focused by ISM Model. In a organization that aims to maintain sustainable lean activities across the entire cluster was focused by UNIDO-ACMA. For Lean to be effectively adopted, it is utterly necessary to promote Lean practice and consistency in-process [11]. Chaple et al. [3] reviewed the lean concepts and standards in

Indian Industrial manufacturing. The automotive sector in India has high lean implementation execution level, electronics and IT sector followed. Apart from automotive and electronics, other industries have mild to low lean management implementation distributed in India. Gaps that have been found in the literature review are that Lean manufacturing implementations in automotive industry have not been thoroughly investigated and little has been implemented in industries at local level. There were very less literatures available in Lean manufacturing implementation in automobile company. The manufacturing industries are not able to take the advantage of Lean Manufacturing implementation due to lack of planning and proper awareness.

3 Objectives of Current Study

The current study should act as the base for various scholars, researchers, automobile and other companies to follow up strategies and implementation of the lean manufacturing and its management in there industry or organization.

This study tries to address the following research questions for mapping the lean manufacturing in automotive sector.

RQ1: What are the lean manufacturing techniques mainly followed by automobile industries?
RQ2: What are the key outcomes if the lean manufacturing techniques are followed in automobile industries?
RQ3: Who are the leaders of lean manufacturing in automotive industry?

1. To identify lean manufacturing technique that can help the automobile industries to eliminate waste and enhance efficiency.
2. To study the impact of Lean manufacturing implementation in the top automobile manufacturers such as Toyota (TPS), Ford (FPS) and Chrysler
3. Comprehensive analysis with the help of Scopus database by using appropriate keywords. Further, time evolution of industry is presented in the Fig. 2.

4 Case Study Methodology

World is developing as a new destination for manufacturing and many companies are searching for ways to improve their productive potential by eliminating waste from their processes. Lean is a strategy that various automobile companies aim to make both local and global markets more competitive. The Automobile industries have a very high lean implementation level and the electronics/IT/engineering industry has a very moderate implementation level [1]. The medium-to-low level is seen in many industries. Even though Lean production is an instrument for enhancing organizational efficiency across a wide range of services and development industries in the world, there is still a significant gap in implementation in order to achieve

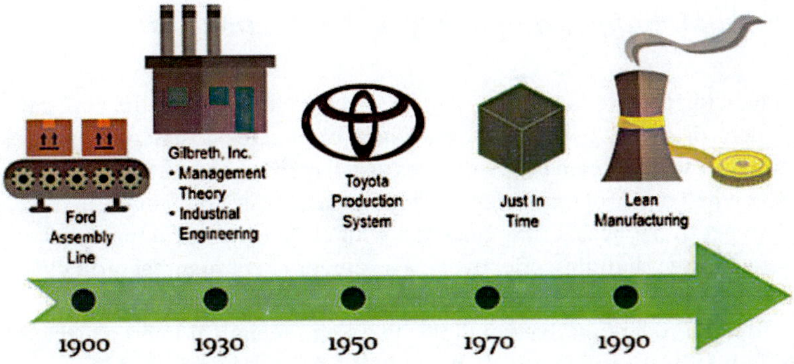

Fig. 2 Time evolution of industry [12]

all the benefits of Lean manufacturing. In the early twentieth century, the three major automobile firms-General Motors, Ford & Chrysler- known as Big 3 were virtually dominating the global industry. The firm replaced Chrysler in 1994 by Toyota and became the world's No. 2 automobile maker in 2003. Toyota is the biggest automaker worldwide followed by GM since 2008. All began with Kiichiro Toyoda, the first President of Toyota Motor Corporation, who set the target of the organization "to use small lot sizes for cheaper vehicles, to match American motor corporations' costs by continually reducing waste disposal costs." Eiji Toyoda, the second chief, has managed to develop the manufacturing cycle in Toyota. Since the Second World War in 1950, Toyota learnt from Henry Ford's book, "Today and Tomorrow" about continuous content movement, operation standardization, and waste disposal. Toyota also created the "Pull System" model, influenced by the American supermarket, The Pull System model. The ideals and the ideas above form the foundation of judoka and just-in-time (JIT) after decades of experience and development, rendering TPS more practical. In both the US and Europe, the common theory was that manufacturing by mass alone could minimize production costs. But Japan's automakers have accomplished low-cost, lower-volume production with higher complexity and shorter periods of delivery. Fuel-efficient and long-lasting or decent values for reselling Toyota goods are trustworthy. Comprehensive type of review approach has been used with the existing available literatures on lean manufacturing in automobile industry been reviewed. All the research papers with keywords as Lean Manufacturing and Automobile have been found from year 2000 to the latest. And mainly the impact of Lean manufacturing implementation in the top automobile manufacturers such as Toyota (TPS), Ford (FPS) and Chrysler has been focused upon. The study about the methodology followed by industry leaders in lean implementation and their key achievements after the lean implementation has been brought to notice through this research [13].

- Toyota Motor Company—Toyota Production System
- Ford Motor Company—Ford Production System
- Chrysler—Chrysler Operating System.

4.1 Lean Manufacturing in Toyota Production System

The manufacturing mechanism for Toyota was developed using the application and development of a very useful strategy, which eliminates costs and resources, thus questioning all value stream behavior. This uses a method known as "The Five whys." This asks whether an operation is carried out and also challenged if it is always feasible to get to the heart of the question with each answer. Comprehension of the root cause allows rebuilding effectively. The aim of every manufacturing business is value-added versus non-value-added activities.

The Toyota Production System (TPS) is a socio-technical interconnected framework built by Toyota to effectively manage the manufacturing and logistics sector, including cooperation with manufacturers and customers, to reduce costs and waste. Between 1948 and 1975, TPS was created. The two key persons credited to the creation of this system are the Japanese industrial engineers Taiichi Ohno and Eiji Toyoda. TPS is regarded as one of the main lean manufacturing processes, and all of the lean ideas that people currently recognize can be specifically related to the Toyota. The goal is to operate smartly to reduce waste such that only a small inventory is needed. It increases cash flow and decreases a physical space demand which makes it possible to produce the necessary products to the final client in a seamless manner via internal processes (single piece flow). Toyota's JIT (just in time) journey had started in 1934 when Toyota moves away from textiles for the production of its first car, founder of the Toyota Motor Corporation. By intensively researching every step of the cycle, he decided to avoid restoring low quality. When Toyota was given the first truck deal with the government in 1936, the systems were met with new obstacles and it produced a "healthy improvement." After visits, they understood that the job plan would not be dictated by demand or development targets, but instead by actual sales. In this time, overproduction needed to be avoided because of the financial situation and the idea of pull was backed by a production schedule (construct to order, instead of goal push) [14].

Major Principles followed by Toyota

1. Reducing single production lot: Commodity development in smaller lots eliminates storage costs, high initial supply expenses and much more.
2. Reduce Start-up times: The time it takes to set up a new job is always time consuming. TPS wants to make attempts to reduce the time it takes to launch a new manufacturing cycle.
3. Improve the consistency of the supply: Acquisitions of reliable source quality materials can help avoid errors during the whole manufacturing cycle.
4. Maintain machines correctly: This can never be interrupted when doing repairs on a machine.
5. Using pull manufacturing technology: The use of pull manufacturing will reduce the need to store large stocks.

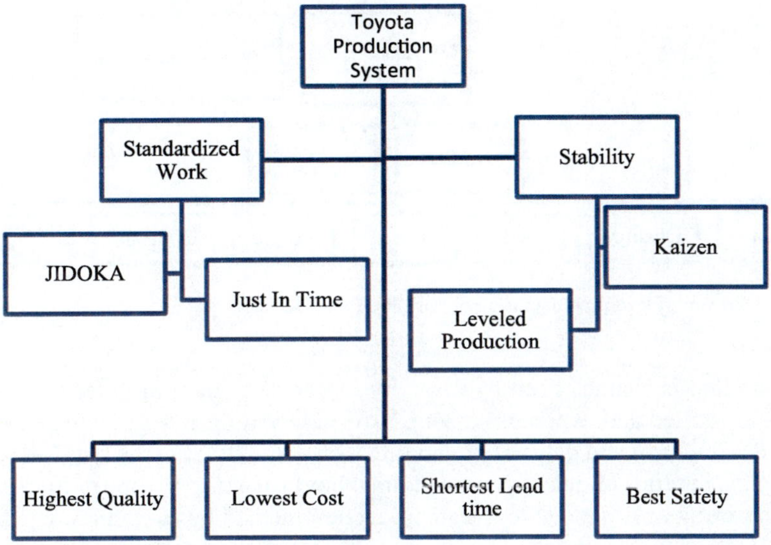

Fig. 3 Flow diagram of Toyota Production System

6. Partner suppliers: Parts manufacturers become associates and support is provided as to how their consistency and performance can be enhance. Flow diagram of Toyota Production System is shown in the Fig. 3.

4.2 Lean Manufacturing in Ford Production System

The American automotive corporation Ford Motor Company or simply Ford, founded in 1903. His headquarters are located in Michigan in Dearborn. Under the Ford brand name, it manufactures and markets various types of vehicles. Luxury vehicles are sold under the Lincoln brand name. These also used large-scale workforce training strategies through a hierarchical linear manufacturing method distinguished by mobile production lines. In the line-assembly and manufacturing process of Henry Ford, roots of TPS can be found. The Ford produces vehicles in vast numbers in similar prototypes, making the operation highly efficient and offering a low-cost vehicle. Ford's core concept is a generic vehicle, leading to uniform processes. Both workers have their own responsibilities and only have to carry out a minor part of the work, such as fixing a screw or oiling a part. Moving production line allowed Ford to manufacture in a manner that had to adapt to the conveyor speed. The personal dimension will be eliminated from the production line. The whole system was synchronized with the production line rhythm.

The Ford 2000 program was launched in January 1995 with the goal of designing and introducing a modern production system known as Ford Production System (FPS). The dream of FPS is a compact, versatile and robust joint manufacturing

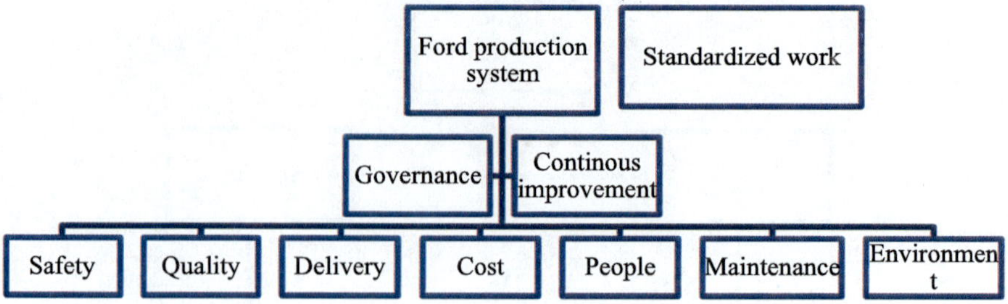

Fig. 4 Diagram of lean manufacturing in Ford Production System

structure that is characterized by a variety of concepts and processes that require groups of skilled and experienced people to study and operate peacefully together in the development and delivery of goods that consistently meet the standards of our consumers in terms of quality, cost and time. Ford sought to transform from a mass manufacturing environment to a lean production method by using FPS. By the fall of 2003, all of Ford's plants had taken on FPS. It was projected that Ford had saved $500 million a year since introduction of FPS. The workers at the factories were often educated on the Web as best practices were defined and exchanged. By mid-2003, Ford was standardizing the production practices of all its plants worldwide [10, 14, 15].

FPS Continuous Improvement Model processes—For all manufacturing practices, the FPS Concept of Continuous Development is the basis. It is a 10-step method intended to consistently develop procedures through standardization and appreciation. In the model, stabilization is generated for inputs, processes are optimized and output improvements are found (if necessary). The 10 performance improvement types of FPS are as following: Continuous Improvement Board, Startup Confirmation, Results Process, Support Process, Time and Data Management, Basic Administration, Kaizen, Standardized Work, Star Points. Diagram of lean manufacturing in Ford Production System is shown in Fig. 4.

4.3 Lean Manufacturing at Chrysler Production System

Chrysler launched the quest by 2000 to become the world's largest corporation for cars and trucks and started with the goal of becoming the premier car and truck manufacturer in North America in 1996. Still well behind Toyota and Honda, and behind domestic rivals Ford and GM even after making significant changes in quality. Recognized Toyota as the world leader of automotive manufacturing, if not in general manufacturing, marked a launch to figure out what Chrysler wanted to do. A comprehensive benchmarking and strategy building initiative has led to a decision to evolve the Chrysler operating system (COS) as the unique Chrysler version of the popular Toyota production system.

COS started by realizing that management was the greatest tool for any attempt to improve. There is no basis for COS not to be the subject of attempts to develop plants because the values and skills of COS are enshrined in plant managers and workers. The first 8 courses in COS Education, which began in May 1995, were performed by the Manufacturing Executive. The second key component of the COS method is that a learning laboratory line is required for each plant. Growing plant must assign an appropriate COS production line for a specific portion of the factory. The line must reflect regular output, but it must not be so necessary that there can be no failures. The line gives managers the ability to work in the sense of real life development and improve their COS competences. Most companies already engage in related training programs, based on the performance of their MLT practices. In reality, other non-automotive companies, with best practices are adopting for their own practices and are benchmarking the DaimlerChrysler Corp operating standards. Development efficiencies from quality training and the introduction of the lean production method of Chrysler Corporation in the two years from its adoption had an additional $300 million in positive results for the company [6, 16]. The four primary sub-systems of the Chrysler Operating System are as follows: Human Infrastructure; Level and Balânced Schedules; Value-Added Activities; Robust, Capable, and In-Control Processes.

5 Discussion of Findings

Starting in 1985, TPS consistently tried to achieve greater mass production and workplace savings through Jidoka (automation with a human touch) and the use of robots. In order to accomplish a general purpose and versatility, as well as the initiative to create revolutionary equipment and dies ("simple and slim") were also undertaken to improve accuracy and weight reductions. By the 1990s, Ford concentrated again on its manufacturing and financial services sectors and implemented lean manufacturing through FPS. Jaguar, a British luxury car company, was purchased by Ford in 1989–90. In 1993, Aston Martin was fully owned. Subsequent acquisitions included Hertz Corporation's car rental business in 1994, Volvo's vehicle group in 1999, and Land Rover's mobility sports car brand in 2000. Ford also bought the Mazda Motor Corporation's significant share. However, Ford continued to market these products in the early twenty-first century. Hertz was sold in 2005 by Ford and in 2007 by Aston Martin. In 2008, it sold to Tata Motors Ltd. of India Jaguar and Land Rover.

For an accredited TPS plant, all facets of the Toyota Manufacturing System have to be demonstrated over a time span by preparation and observance. According to the Shingo Prize the Ford Motor Company's Q1 Quality Systems Award was won by Textron Automotive Group, and McCord Winn of Manchester and New Hampshire, and are also TPS accredited, which marked a very significant accomplishment because Ford wanted its Tier 1 supplier base to deliver on terms superior to all other automotive components and manufacturing components. Many other automobile OEM's have been led by Ford guidelines to introduce similar systems. At the

Table 1 Achievements and awards of TPS and FPS based on innovation and its implementation

Year	Company	Type	Scope
1991	TPS	Production	Development of integrated body panel production system using laser-welding
1993	TPS	Production	Development of new automobile assembly line
1999	TPS	Production	Development of motoring-based, engine-function automatic inspection system
2000	TPS	Special production	Short lead time and low-cost development of stamping dies for automobile body steel sheets made possible by developing high-strength, tempered cast iron
2002	FPS	Quality	Tier 1 supplier components to be of best quality
2007	FPS	Quality Performance	Shingo prize for best quality performance
2007	FPS	Quantity	Platinum award for the highest production worldwide

height of FPS in 1996, the production of cars took an average of 37.59 work hours. (Assembly facilities: how they fit-Harbor research on the factory efficiency of the auto assembly in 2007). According to the new Harbor Survey (Harbour Survey: more effective GM and FORD, but losing money in every vehicles 2006), this rate is down to 35.82 h after Lean implementation. The 2008 study says the difference in the overall production work (measurement including assembly, engine and transmission operations) between the more and less efficient automakers has decreased to 3.5 working hours per car (or some $260 per car) in 2003 from 10.51 h (or $790 per vehicle). In the 2008 study, the Detroit Three, led by Chrysler demonstrated multiple performances. The 7.7% decline in Chrysler hours per vehicle to 30.37 contributed to Toyota's collapse, rendering Chrysler and Toyota the production leaders that year. Since 1998, FORD has been at the top spot for this award for the first time. In addition, FORD's Wixom Assembly Plant has also secured the Platinum Award for the highest production worldwide. FORD also won the leading position in the 2007 J. D. Power and Associates' Production Ranking. Table 1 shows some famous achievements and awards of TPS and FPS after lean implementation in their production.

6 Managerial Implications

From the current study we got to know that it was observed that because of lack of planning and sufficient training, the organisations do not enjoy the advantages of Lean Manufacturing practise in their organization. Because of not implementing lean management properly in their organization they face some drawbacks of its improper implementation. Inappropriate lean output is often seen with a decreased emphasis on cost control due to inadequate information. It raises the risk of inability to recognize the variety of investments that are the dominant indicator of any industry's performance. Smaller batch processing decreases the production time and lowers inventory costs, but the distribution of components in smaller quantities increases shipping costs. The vital assembly lines range from average to the peak level, producing an unstable working atmosphere with insufficient expertise of Lean manufacturing. Repeated and hasty operation on these lines raises the risk of health risks and causes operator safety problems [17].

7 Conclusion

From the findings of the current research it was concluded the Toyota Production system (TPS), Ford Production System (FPS) and Chrysler operating system (COS) suggest the bench mark for the implementation of Lean management and follow the practices of lean in their industry. Current study covers the basic aspects of lean implementation in top automobile companies and also suggests effective measures and drawbacks of sustainable lean manufacturing implementation. Due to lack of available previous research on lean implementation in automobile companies there was limited access to data for reference. The current study should act as the base for various scholars, researchers, automobile and other companies to follow up strategies and implementation of the lean manufacturing and its management in there industry or organization.

(1) Because of lack of planning and sufficient training the organisations do not enjoy the advantages of Lean Manufacturing practise.
(2) Inappropriate lean output is often seen with a decreased emphasis on cost control due to inadequate information.
(3) There is an increase in efficiency and effectiveness in Automotive manufacturing with proper lean implementation.
(4) In case of absence of a buffer stock that could serve as a stabilizer in case of change in demand and other unfavourable condition, lower inventories and repeated deliveries will raise the risk of stock shortage condition to clients.

References

1. Kumar R, Kumar V (2015) Lean manufacturing in Indian context: a survey. Manage Sci Lett 5(4):321–330. https://doi.org/10.5267/j.msl.2015.2.009
2. Nagar N, Tyagi M, Garg SK (2016) Lean manufacturing and its. Int J Adv Mech Eng 4(2):9–10 [Online]. Available http://www.ripublication.com/ijame.htm.
3. Chaple AP, Narkhede BE, Akarte MM (2014) Status of implementation of lean manufacturing principles in the context of Indian industry: a literature review. Feb 2014
4. Agrawal Pulak RT, Agarwal KM (2016) A review on quality management system in automotive sector and ISO/TS 16949. Int J Adv Eng Res Appl 2(8):525–536
5. Krishna Mohan Agarwal SKS, Kumar N (2018) Analysis of process characteristics for a batch production unit and controlling the variation for effective performances. Adv Power Gener Renew Energy Sources (APGRES 2017):291–298

6. Flinchbaugh BJW (1998) Implementing lean manufacturing through factory design
7. Arumugam V, Kannabiran G, Vinodh S (2020) Impact of technical and social lean practices on SMEs' performance in automobile industry: a structural equation modelling (SEM) analysis. Total Qual Manage Bus Excell:1–27.https://doi.org/10.1080/14783363.2020.1791067
8. Arunagiri P, Gnanavelbabu A (2014) Identification of major lean production waste in automobile industries using weighted average method. Procedia Eng 97:2167–2175
9. Kumar N, Mathiyazhagan K, Mathivathanan D (2020) Modelling the interrelationship between factors for adoption of sustainable lean manufacturing: a business case from the Indian automobile industry. Int J Sustain Eng 13(2):93–107. https://doi.org/10.1080/19397038.2019.1706662
10. Wee HM, Simon S (2009) Lean supply chain and its effect on product cost and quality: a case study on ford motor company. Supply Chain Manage Int J 14(5):335–341. https://doi.org/10.1108/13598540910980242
11. Jadhav JR, Mantha SS, Rane SB (2015) Roadmap for Lean implementation in Indian automotive component manufacturing industry: comparative study of UNIDO Model and ISM Model. J Ind Eng Int 11(2):179–198
12. Haseeb M, Hussain HI, Ślusarczyk B, Jermsittiparsert K (2019) Industry 4.0: a solution towards technology challenges of sustainable business performance. Soc Sci 8(5). https://doi.org/10.3390/socsci8050154
13. Sharma SL, Mudgal P, Jha AK, Kumar A, Singh GK (2018) Study of lean manufacturing for manufacturing of auto components. Apr 2018. https://doi.org/10.13140/RG.2.2.18963.17443
14. Yadav OP, Nepal B, Goel PS, Jain R, Mohanty RP (2010) Insights and learnings from lean manufacturing implementation practices. Int J Serv Oper Manage 6(4):398–422. https://doi.org/10.1504/IJSOM.2010.032916
15. Ryan KA (2006) The re-innovation of ford motor company to a sustainable lean enterprise. Masters Abstr Int [Online]. Available http://search.proquest.com/docview/33564683?accountid=14643
16. Kumar SS, Kumar MP (2014) Cycle time reduction of a truck body assembly in an automobile industry by lean principles. Proc Mater Sci 5:1853–1862. https://doi.org/10.1016/j.mspro.2014.07.493
17. Singh B, Garg SK, Sharma SK (2010) Scope for lean implementation: a survey of 127 Indian industries. Int J Rapid Manuf 1(3):323. https://doi.org/10.1504/ijrapidm.2010.034253

Optimization of Turning Process Parameters Using Entropy-Gra and Dear Methods

K. Srinivasulu Reddy, V. Venkata Reddy, and Ravi Kumar Mandava

Abstract The advantage of amalgamating particles reinforced with metal matrix composites is enhancing their physical and mechanical properties. The objective of this study is to conduct the machining operation on the developed composite using single point cutting tool on lathe machine and to investigate the influence of machining process parameters on surface roughness and material removal rate. To identify the best optimal machining parameters the authors implemented two approaches that is, multi-objective optimization of Grey relational analysis (GRA) and Data Envelopment Analysis based Ranking (DEAR). The methodology of GRA allows the problem solver to assign weight fractions for each output. Further, the accuracies of the solution differ with the fractions of weight allocated to each output. The aggregate values calculated by the method ENTROPY-GRA and DEAR were used to determine optimum levels of factor and their contributions. The optimum levels calculated by the DEAR method performed better to access machining quality characteristics.

Keywords Turning · Optimization · ENTROPY · GRA · DEAR

1 Introduction

Turning is one of the most common machining operations in which rotational parts are produced by removing material by obtaining a reduced size of the required diameter. Nowadays, most industries are using this operation for machining the various components and producing the required shapes [1, 2]. The research conducted around the world in the area of MMC's in the turning process, it has been found that the

K. Srinivasulu Reddy (✉)
Department of Mechanical Engineering, Sreenidhi Institute of Science and Technology, Hyderabad, India

V. Venkata Reddy
Department of Mechanical and Industrial Engineering, Dire Dawa University, Dire Dawa, Ethiopia

R. K. Mandava
Department of Mechanical Engineering, MANIT Bhopal, Bhopal 462003, India

R. Agrawal et al. (eds.), *Recent Advances in Industrial Production*, Lecture Notes in Mechanical Engineering, https://doi.org/10.1007/978-981-16-5281-3_29

main aspects affecting MMC turning operation predominantly depend on the type of material being used [3, 4]. According to the guidelines set by standard handbooks, based on the operators' experience, and knowledge, the machining parameters are selected. However, if the chosen machining parameters are not optimum, it may eventually increase the product's cost [5]. By selecting the best machining parameters, one can achieve high machining performance [6]. In [7] researchers used various optimization techniques to choose the best combination and identify the best machining parameters. So, an experimental study is required here [8]. Metal matrix composites (MMCs) find applications in aircraft components, automobile, marine, structural equipment, etc. They possess a combination of properties like superior hardness, enhanced strength, and better wear resistance. Reddy and Gopi [9] developed an MMC's by varying the weight percentage of reinforcements: 5% and 10% and they observed 10% wt shows better mechanical properties compared with 5% wt composite. Lin et al. [10] discussed taguchi based grey relational analysis to optimize turning operations with multiple performance characteristics. Further, Taguchi-DEAR approach is straightforward and effective. Under Taguchi-DEAR approach, Muthuramalingam et al. [11] analyzed the process parameters of abrasive flow orientation in AWJM. Seveal studies conducted by the authors [12] focusing on optimizing AWJM factors while processing Al7075 reinforced composites with TiB2 particles. In the present work, during the machining of AA7075 MMC, an attempt was made to address the above-mentioned problems. The current work focused on determining multi-response problems in the dry condition for optimum turning process parameters.

2 Experimentation

The composites were prepared by using Stir casting routing techniques. Al7075 Melting of A7075 ingots was performed in an electric furnace with graphite crucible. At 770 °C, molten metal pool was stirred at the centre of the crucible with the help of a mechanical stirrer that rotates at 500 rpm. SiC and flyash particulates were preheated and dropped in a uniform fashion into the melt. With the purpose of avoiding agglomeration at the time of stirring, the particles were ensured to have a smooth and continuous flow. Figures 1 and 2 show the stircasting process and specimens prepared for turning operation. Table 1 lists all the factors and their selected levels. The material removal rate (MRR) is calculated as the difference in weights of the work piece before and after machining over a period of machining time T seconds.

Fig. 1 Stir casting process

Fig. 2 Composite specimens after machining

Table 1 Factors and levels selected

S. No.	Factor	Unit	Levels of factors			
			1	2	3	4
1	Cutting speed, v	m/min	20	50	75	115
2	Feed, f	mm/rev	0.05	0.10	0.16	0.20
3	Depth of cut, d	mm	0.2	0.4	0.6	0.8

3 Methodology, Results and Discussion

Multi-objective optimization refers to the optimization of the process or product performance that involves two or more outputs simultaneously with or without conflicting responses. The current work aims to increase the rate of material removal simultaneously while reducing the surface roughness of the material. Every inexperienced user can incorporate the proposed offline optimization tools (ENTROPY-GRA and DEAR) to obtain the resulting benefits in industries.

3.1 Entropy Weight Method

The definition of entropy has been widely used in the social and physical sciences. Entropy can be regarded as a criterion for the degree of uncertainty of a discrete distribution of probabilities. In the process of decision making, theory of entropy can be used efficiently as it tests current comparisons between data sets and clarifies the average intrinsic information transmitted to decision makers.

The following method should be used to evaluate objective weight by Shannon entropy (Hwang and Yoon 1981):

Step 1 Decision matrix is normalized

$$K_{lm} = \frac{y_{lm}}{\sum_{i=1}^{n} y_{lm}} \tag{1}$$

Step 2 Entropy measurement of results calculation using the Eq. (2):

$$E_m = -p\sum_{i=1}^{n} y_{lm} \ln y_{lm} \tag{2}$$

where $p = {}^{1}/_{\ln(n)}$.

Step 3 Weights of the objective are computed. Computed weights are presented in Table2.

$$\omega_m = \frac{1 - E_m}{\sum_{m=1}^{q}(1 - E_m)} \tag{3}$$

3.2 Grey Relational Analysis

Prof. Deng introduced the Grey Theory to handle weak, incomplete, and uncertain information in 1982. Grey color is neither black nor white. Generally speaking, the system is described with color reflecting the amount of specific system details (i.e. internal characteristics or dynamic detailing mathematical formulations). Grey system refers to the information in between the known and the unknown. The current work is based on optimizing the turning process for the maximization of MRR while minimization of SR. The actions taken to improve the use of ENTROPY-GRA are listed below.

In this procedure, all actual data values of the sequence [(P a^*(b))] are converted between 0 and 1. Different formulae are used to convert the original sequence data into standardized data, depending on the response data requirement. Equation (4) is used to evaluate the larger-the-better and Eq. (5) for normalization values when the response data parameters are lower-the-better. In this analysis, surface roughness is

Table 2 Experimental values and ENTROPY weight method

Exp. No.	Experimental values		Normalization		K_{lm}	
	MRR	SR	MRR	SR	MRR	SR
E1	0.02	1.52	0.0097	0.0506	– 0.0451	– 0.1510
E2	0.038	1.78	0.0185	0.0593	– 0.0738	– 0.1675
E3	0.066	2.48	0.0321	0.0826	– 0.1105	– 0.2060
E4	0.17	2.8	0.0828	0.0932	– 0.2062	– 0.2212
E5	0.042	1.72	0.0204	0.0573	– 0.0795	– 0.1638
E6	0.038	1.5	0.0185	0.0500	– 0.0738	– 0.1497
E7	0.22	2.28	0.1071	0.0759	– 0.2393	– 0.1957
E8	0.174	2.32	0.0847	0.0773	– 0.2091	– 0.1978
E9	0.07	1.36	0.0341	0.0453	– 0.1152	– 0.1402
E10	0.21	1.48	0.1022	0.0493	– 0.2332	– 0.1484
E11	0.112	2.14	0.0545	0.0713	– 0.1586	– 0.1882
E12	0.162	2.24	0.0789	0.0746	– 0.2003	– 0.1936
E13	0.14	1.31	0.0682	0.0436	– 0.1831	– 0.1366
E14	0.22	1.39	0.1071	0.0463	– 0.2393	– 0.1422
E15	0.25	1.71	0.1217	0.0569	– 0.2563	– 0.1632
E16	0.122	2	0.0594	0.0666	– 0.1677	– 0.1804
Sum	2.054	30.03			– 2.5910	– 2.7456
				E_m	0.9345	0.9902
				$1 - E_m$	0.0655	0.0098
				W_m	0.87	0.13

a minimzation requirement, while MRR is a maximization requirement for productivity enhancement. Table 3 displays the measured normalization values for different responses.

Larger the better:

$$P_a^*(b) = \frac{P_a(b) - \text{Min } P_a(b)}{\text{Max } P_a(b) - \text{Min } P_a(b)} \tag{4}$$

Smaller the better:

$$P_a^*(b) = \frac{\text{Max } P_a(b) - P_a(b)}{\text{Max } P_a(b) - \text{Min } P_a(b)} \tag{5}$$

$P_a^*(b)$ Sequence after data processing.
$P_a(b)$ Original sequence.
Max $P_a(b)$ Maximum value of output response 'b'.

Table 3 Normalized data, Grey relational coefficients and WGRG values

Exp. No.	Normalized data		Δ (MRR)	Δ (SR)	GRC		WGRG	S/N ratio
E1	0.0000	0.8591	1.0000	0.1409	0.3333	0.7801	0.1957	– 14.1679
E2	0.0783	0.6846	0.9217	0.3154	0.3517	0.6132	0.1928	– 14.2962
E3	0.2000	0.2148	0.8000	0.7852	0.3846	0.3890	0.1926	– 14.3071
E4	0.6522	0.0000	0.3478	1.0000	0.5897	0.3333	0.2782	– 11.1127
E5	0.0957	0.7248	0.9043	0.2752	0.3560	0.6450	0.1968	– 14.1194
E6	0.0783	0.8725	0.9217	0.1275	0.3517	0.7968	0.2048	– 13.7745
E7	0.8696	0.3490	0.1304	0.6510	0.7931	0.4344	0.3732	– 8.5603
E8	0.6696	0.3221	0.3304	0.6779	0.6021	0.4245	0.2895	– 10.7669
E9	0.2174	0.9664	0.7826	0.0336	0.3898	0.9371	0.2305	– 12.7470
E10	0.8261	0.8859	0.1739	0.1141	0.7419	0.8142	0.3757	– 8.5040
E11	0.4000	0.4430	0.6000	0.5570	0.4545	0.4730	0.2285	– 12.8233
E12	0.6174	0.3758	0.3826	0.6242	0.5665	0.4448	0.2753	– 11.2026
E13	0.5217	1.0000	0.4783	0.0000	0.5111	1.0000	0.2873	– 10.8323
E14	0.8696	0.9463	0.1304	0.0537	0.7931	0.9030	0.4037	– 7.8789
E15	1.0000	0.7315	0.0000	0.2685	1.0000	0.6507	0.4773	– 6.4243
E16	0.4435	0.5369	0.5565	0.4631	0.4733	0.5192	0.2396	– 12.4099

Min $P_a(b)$ Minimum value of output response 'b'.
a Number of responses (1, 2).
b Number of experiments (1,2 … 0.16).

Step 1 GRC values are calculated for different outputs based on Eq. (6) and results are shown in Table 3. The GRC explain the relationship between the actual and desirable data.

$$\in_a(b) = \frac{\Delta_{Min} + \zeta\Delta_{Max}}{\Delta_{oa}(b) + \zeta\Delta_{Max}} \tag{6}$$

Here,

$$\Delta_{oa}(b) = \left|P_a^o(b) - P_a^*(b)\right| \tag{7}$$

$$\Delta_{Min} = Min_a Min_b\, \Delta_{oa}(b) \tag{8}$$

$$\Delta_{Max} = Max_a Max_b\, \Delta_{oa}(b) \tag{9}$$

Step 2 Calculation of weighted grey relational grade (WGRG) WGRG values can be used to evaluate multiple performance characteristics. This can be determined by taking different responses to the average WGRC values. Equation (10) used to evaluate the WGRG value and the results were summarized in Table 3.

$$\varphi_a(b) = \frac{1}{N}\sum_{i=0}^{N} W_a \in_a (b) \tag{10}$$

ENTROPY method supplies the required weights in the present work (see Table 2). Weight fractions calculated for MRR and SR are found to be 0.87 and 0.13 respectively.

3.3 Data Envelopment Analysis Based Ranking (DEAR)

In 1978 Charnes et al. proposed the definition of Data Envelopment Analysis (DEA) as the estimation of effectiveness of a mix of decision-making units that produces multiple outputs using multiple inputs. Recall that the DEAR approach used to solve the optimization of numerous responses does not allow calculation of weight fractions for individual output characteristics. Here, real output set is compared as a ratio with simple mathematical approximation so that the calculated values match the ratio ranks. These ranks are therefore used to estimate and optimize the optimal factor level.

The sequential steps for multiresponse performance index (MRPI) estimatio followed in DEAR are:

Step 1 Determine the weights (Performance measurement ratio at any trial to the sum of all performance measurements) for each test corresponding to all the experiments. The calculation is made using the following Eqs. (11) and (12) to calculate the weight fraction of each product.

$$\omega_{MRR} = \frac{MRR}{\sum MRR} \tag{11}$$

$$\omega_{SR} = \frac{\left(1/_{SR}\right)}{\sum\left(1/_{SR}\right)} \tag{12}$$

Step 2 Transforming the output data into weighted data by multiplying the output according to Eqs. (13) and (14) with their respective weight fractions.

$$M = \omega_{MRR} * MRR \tag{13}$$

$$S = \omega_{SR} * SR \tag{14}$$

Step 3 Calculate the MRPI by dividing the larger the better performance characteristics with smaller-the-better performance characteristics by using Eq. (15)

$$\text{MRPI} = \frac{\text{M}}{\text{S}} \tag{15}$$

3.4 Determination of Optimal Factor Levels for All Outputs

Multi-objective optimization methods (ENTROPY-GRA and DEAR) are used to evaluate the set of optimum factor levels of turning operation. The WGRG and the MRPI values represent the composite values that correspond to all the responses calculated using ENTROPY-GRA (Table 3) and DEAR (Table 4) respectively. The Taguchi method is now applied to the combined WGRG and MRPI and the S/N ratio values are calculated and the results are shown in Tables 5 and 6 respectively. Minitab 18.0 is used to measure the WGRG mean and MRPI mean for each stage of process parameters. The absolute mean of MRPI in DEAR method respective process parameters were taken as Taguchi predicted optimum process parameter settings. The optimum settings were predicted by Taguchi as v = 115 m/min, f = 0.2 mm/rev and d = 0.8 mm (v4-f4-d4) respectively as shown in Table 5.

Table 4 Weights, MRPI values

Exp. No.	MRR	SR	Weights for each output		M	S	MRPI	S/N ratio
E1	0.02	1.52	0.0097	0.0732	0.0002	0.1112	0.0018	– 55.1356
E2	0.038	1.78	0.0185	0.0625	0.0007	0.1112	0.0063	– 43.9855
E3	0.066	2.48	0.0321	0.0449	0.0021	0.1112	0.0191	– 34.3950
E4	0.17	2.8	0.0828	0.0397	0.0141	0.1112	0.1265	– 17.9588
E5	0.042	1.72	0.0204	0.0647	0.0009	0.1112	0.0077	– 42.2468
E6	0.038	1.5	0.0185	0.0742	0.0007	0.1112	0.0063	– 43.9855
E7	0.22	2.28	0.1071	0.0488	0.0236	0.1112	0.2118	– 13.4799
E8	0.174	2.32	0.0847	0.0479	0.0147	0.1112	0.1325	– 17.5548
E9	0.07	1.36	0.0341	0.0818	0.0024	0.1112	0.0214	– 33.3729
E10	0.21	1.48	0.1022	0.0752	0.0215	0.1112	0.1930	– 14.2880
E11	0.112	2.14	0.0545	0.0520	0.0061	0.1112	0.0549	– 25.2081
E12	0.162	2.24	0.0789	0.0497	0.0128	0.1112	0.1149	– 18.7962
E13	0.14	1.31	0.0682	0.0849	0.0095	0.1112	0.0858	– 21.3317
E14	0.22	1.39	0.1071	0.0800	0.0236	0.1112	0.2118	– 13.4799
E15	0.25	1.71	0.1217	0.0650	0.0304	0.1112	0.2736	– 11.2592
E16	0.122	2	0.0594	0.0556	0.0072	0.1112	0.0651	– 23.7224

Table 5 Response table of S/N ratio for WGRG

Level	V	f	D
1	– 13.471	– 12.967	– 13.294
2	– 11.805	– 11.113	– 11.511
3	– 11.319	– 10.529	– 11.425
4	– 9.386	– 11.373	– 9.752
Delta	4.085	2.438	3.542
Rank	1	3	2

Table 6 Response table of S/N ratio for MRPI

Level	V	f	D
1	– 37.87	– 38.02	– 37.01
2	– 29.32	– 28.93	– 29.07
3	– 22.92	– 21.09	– 24.7
4	– 17.45	– 19.51	– 16.76
Delta	20.42	18.51	20.25
Rank	1	3	2

Table 7 Comparison of confirmation test results

	Taguchi based WGRG (v4-f3-d4)	Taguchi based DEAR v4-f4-d4	% improvement
Surface roughness	1.72	1.62	5.81
Material removal rate	0.28	0.32	12.5

3.5 Confirmation Experiments

Confirmation tests were performed to validate the predictions of optimal techniques and select the best optimization method to improve the Turning process's multiple output features. It is observed that DEAR method outperformed compared with ENTROPY-GRA to determine the optimum levels resulting in desired high MRR values and low SR values. DEAR approach provided an increase of 12.5% in MRR, 5.81% in SR when compared with ENTROPY-GRA method (Table 7).

4 Conclusions

The following conclusions are drawn from the experimental study:

1. The weight fractions calculated by ENTROPY method for MRR and SR are 0.87 and 0.13, respectively. Remember that the summation of all weights must be held equal to one output.

2. For multi-objective optimization, GRA allow weight fractions to be allocated. Therefore, ENTROPY method supplies the weights ready to solve the problem.
3. The optimal combination of cutting conditions as v = 115 m/min, f = 0.16 mm/rev and d = 0.8 mm (v4-f3-d4) were determined from ENTROPY-GRA.
4. The optimum turning parameters were observed at v = 115 m/min, f = 0.2 mm/rev and d = 0.8 mm (v4-f4-d4), using the DEAR multi-objective optimization method.
5. For the optimal levels proposed by both the methods, the confirmation experiments are conducted. DEAR, method outperformed than the model (ENTROPY-GRA) to produce higher material removal rates with low surface roughness.

References

1. Kanta DD, Mishra PC, Singh S, Thakur RK (2015) Tool wear in turning ceramic reinforced aluminum matrix composites—a review. J Compos Mater 49(24):2949–2961
2. NMuthukrishnan MM, Rao KP (2008) Machinability issues in turning of Al-SiC (10p) metal matrix composites. Int J Adv Manuf Technol 39(3–4):211–218
3. Niknam SA, Kamalizadeh S, Asgari A, Balazinski M (2018) Turning titanium metal matrix composites (Ti-MMCs) with carbide and CBN inserts. Int J Adv Manuf Technol 97(1–4):253–265
4. Dandekar CR, Shin YC (2012) Modeling of machining of composite materials: a review. Int J Mach Tools Manuf 57:102–121
5. Thakur D, Ramamoorthy B, Vijayaraghavan L (2009) Optimization of high speed turning parameters of superalloy Inconel 718 material using Taguchi technique
6. Yang WP, Tarng YS (1998) Design optimization of cutting parameters for turning operations based on the Taguchi method. J Mater Process Technol 84(1–3):122–129
7. Zuperl U, Cus F, Milfelner M (2005) Fuzzy control strategy for an adaptive force control in end-milling. J Mater Process Technol 164:1472–1478
8. Kadirgama K, Noor MM, Rahman MM (2012) Optimization of surface roughness in end milling using potential support vector machine. Arab J Sci Eng 37(8):2269–2275
9. Reddy VV, Krishna MG, Kumar KP, Kishore BN, Rao JB, Bhargava NRMR (2018) Studies on microstructure and mechanical behaviour of A7075-Flyash/SiC hybrid metal matrix composites. IOP Conf Ser Mater Sci Eng 310(1):012047
10. Lin CL (2004) Use of the Taguchi method and grey relational analysis to optimize turning operations with multiple performance characteristics. Mater Manuf Processes 19(2):209–220
11. Muthuramalingam T, Vasanth S, Vinothkumar P, Geethapriyam T, Rabik MM (2018) Multi criteria decision making of abrasive flow oriented process parameters in abrasive water jet machining using Taguchi-DEAR methodology. Springer Science + siness Media B.V., Partof Springer Nature
12. Manoj M, Jinu GR, Muthuramalingam T (2018) Multi response optimization of AWJM process parameters on machining TiB2 particles reinforced Al7075 composite using Taguchi-DEAR methodology. Springer Science + Business Media B.V., Part of Springer Nature

An Impact of Internet Based Supply Chain Management Using IOT in Current Scenario

Sundeep Kumar, Vikram Singh Rathore, and Alok Mathur

Abstract Supply chain the executives has become an inexorably significant administration apparatus to assist associations with improving their business tasks. In spite of the fact that data and correspondence innovations have been utilized widely in Supply chains, there is an absence of deliberate proof with respect to the instruments through which IT makes esteem. This researches the impacts of the Supply chain aggregate information the board ability SCM execution. According to study and investigate advancement of SCM and the employments of Internet of Things innovations in current situation. Lead an assessment on the attributes of supplely chain framework in Internet of Things setting and dissect how to understand the advancement of Supply chain framework in Internet. The Data repository is the fundamental aspect of SCM utilized for reserve or stock items deal with stock amount. An impressive distribution center guideline framework for prompt price decrease with furthermore an enhance consumer loyalty. Predictable distribution center administration replica possess obtain smaller effective with inadmissible to current scenario of retail requirements. In this research study alteration authorize with relationship to common articles for convention in progress details with divide them the consecutive information used for to be applying to help automation active. In this research study suggest implementing the procedure for store the information the board and requirement of Internet of Think, providing ongoing deceivability of all data distribution center, faster with potency, and anticipate goods deficiency with duplicating.

Keywords IoT · Warehouse · SCM · Internet · Electronic-SCM

S. Kumar (✉) · V. S. Rathore · A. Mathur
Centre for Electronic Governance, Department of Technical Education, Government of Rajasthan, Jaipur 302004, India

R. Agrawal et al. (eds.), *Recent Advances in Industrial Production*, Lecture Notes in Mechanical Engineering, https://doi.org/10.1007/978-981-16-5281-3_30

1 Introduction to IoT for Supply Chain Management

In concern with a clear quality of modern public with monetary advancement opposition, existence additional serious, with advancement for innovative achievements, fast, inciting endeavors to produce a progression interchange with in the field of Information Technology coordination with buying some extra organizations for embrace to considerations with techniques for Supply Chain Management (SCM), there is exclusively could decrease to working expenses for ventures, increase the rate of reaction for advertise request, yet in addition can improve the intensity of undertakings in market rivalry. The innovative work in the field of IoT carries brand up to date occasions for advancement of Supply Chain Management. An existence of Personnel Computers with the facility of communication through Internet, with versatile correspondences, Internet of Think is advance progressive improvement for data companies. Since affective layer, application layer to network layer in concern with Internet of Think includes broad scope of various areas of principles, center advances, with items, just mix with cooperation together with different advances, frameworks, items, organizations, and applications. In this study, examination of IoT has been featured as of late, and its connected innovative work have additionally drawn unexpected consideration of an assortment of nations. Other unrest can be portrayed as a move in the assembling rationale towards an undeniably decentralized, automatic methodology of significant worth creation, empowered by ideas and advances, for example.

In concern with security cyber physical systems, (IoS) Internet of Services with (IoT) Internet of Things with Distributed Computer Network to added substance assembling to keen production lines. Introduce the ideas of Internet of Thinks with Cyber Physical System assembling climate own prompted to meaning of Companies or Industries is provide creative ideas for creating savvy creation, shrewd items and keen administrations. By moving towards mechanization and electronic developments, associations can accomplish more benefit Imran [1]. Individual of the normal meanings of the Supply Chain Management for contain to everything assembly incorporate, legitimately or traffic circle way, to fulfilling a client requirement. Through Supply Chain Management integrate the maker and distributers, still in addition carriers, administration offices, market or business clients individual. The study depicts an assortment for cycles with assets expected create to convey an item and support of client at beginning as far as possible. Supply Chain Management as (SCM) to interdependence of relationship that recognizes with each extra for demanding with next connection in the middle of the cycle it produces a stimulant to a conclusive buyer as commodity and management. At stage of primary success of achievement from Supply Chain for comprehend with perform clients requirement with heigher caliber items to schedule, due to this realistic for liquidation not about esteem Social Science-2019, included exercises, improving cycles with executive for Supply Chain Management additional dexterous. In the greater part of the cutting edge ventures, ordinary Supply chain measures are overseen by programming bundles, for example, enterprise resource management arranging ERP and arrangement ahead of time and booking Advanced planning and scheduling. On the other hand, these frameworks are insufficient to confront the expanding difficulties of the present Supply chains, for example, adaptability, responsiveness and readiness. Subsequently, new methodologies have been acquainted with address these difficulties. With the unavoidable worldwide move towards Industry 4.5 and keen associations, Internet of Thinks innovation is assuming an essential part in this progress. The center idea is that regular items can be outfitted with recognizing, detecting, systems management and preparing capacities that will permit them to speak with each other and with different gadgets and administrations over the Internet to accomplish some helpful target. Internet foundation depends on numerous advances, for example broadcasting Internet to accomplish some helpful target Radio-frequency identification as Bluetooth, Radio Frequency Identification, Sensors, wi-fi and Distributed Computer Networks. Internet of thinks is helpful to upgrade the exhibition entire supply chain with changing perceptive each instance very well may utilized to checking, following items, making a clever transportation framework, and request estimating. As per Ibrahim [2] the SCM is most important part of the critical regions where price decrease to accomplished supply chain management. In particular, IoT can lower stockpile expenses just ruler impact over the SCM. Main principle target to study for consider monetary with community effect to adjusting companies 4.5 with Internet of Think innovation regarding repository, for help in setting aside cash to modern association with how it can increase presentation to suggest in attribute a supposed structure actualizing internet distribution center. The study destinations consider effect of companies 4.5 for Supply Chain Management with influences capacities, with to observe principle segments those are based on application of Internet of Thing and uses of current scenario.

2 Literature Review

Submerged Arc welding is commonly used in industrial application due to good mechanical property and narrower heat affected zone. In Submerged Arc Welding Flux plays critical role in welding, approximately 50% cost is carried out by flux and in this research work flux preparation is the main objective before starting the submerged arc welding. In this research work the flux is prepared using the agglomeration techniques based on red mud with some alloying elements. The composition of red mud that was used to prepare a basic material as mentioned below.

The management area goes about as a key driver for the monetary development of the greater part of the nations. Company or Industry 4.5 turn on 4 keys innovations for Cyber Physical Systems, an incorporated ordering of correspondence, figuring and authority employ for connect the corporeal and essential infinity in many attributes for e.g., vehicle engine assembling, shipment, and co-ordinations. The successive revolution portable web and (IoT) advancements for manufacture communications among person with instrument, and between instrument to instrument make things ready for speak with certainly execute savvy ID, area, observing, following and control. The other alteration is distributed system alteration process a processing help for web services ease with superior and services provided by Internet Service Provider, such as, programming stages, equipment and innovation IT sector foundations. Next innovation enormous information and progressed examination procedures which are utilized for preparing assorted information types by utilizing new handling strategies to create concrete data rapidly, subsequently helping organizations in dynamic, improving cycles, improving operational efficiencies and lessening costs.

Information sharing has assumed a significant function at SCM the board. According to Sohal to Baihaqia and introduced experimental examination present effect data share SCM. They present the share data in middle of accomplices fundamental yet insufficient to accomplish a critical development. As per center zero in manufacturing SCM accomplices extra satisfying reinforce inward combination with the by accomplishing errands in conjunction with objective with relationship using found on confidence. Supervisors ought accept information divide best equipment to use it, target of enhancing every output through Supply Chain Management Execution. According to Choy [3] introduce a theoretical miniature using 7 speculations observe effect by squeeze IT application as SCM, for e.g. data and correspondence innovation Information and Communication Technologies, LIS and Business Information. Two components of administration execution were mulled over administration standard for business regarding and upper hand asset according to the sector. In this concern outcomes indicated greater part from co-ordinations specialist organizations Logistics Service Provider don't execute numerous strategies despite the fact that suggested by a few scientists, e.g., radio recurrence identification radio frequency indications. Suggested speculative technique a guide used as Logistics Service Provider for enhance their intensity. According to Grabara [4] showed the job and effect of data frameworks on transportation exercises in the venture, for example, enhance effectiveness shipment cycle, experienced motorist usage, extra skilled data trade, with monetary outcomes. In supply chain management contended great data frameworks the executives, associations won't have the option to settle on dependable choices shipment subsequently confront chances as concern with retailers necessities. According to Mattsson and Jonsson utilized recreation technique is comprehend worth with effect by distributed form by 4 kinds arranging data purpose deals information, supply chain information, client conjectures and arranged requests for stock earning to utilizing request tip strategies. The technique also discovered estimation common data relies upon if the interest is fixed when request is fixed the stock available information has high worth, while when the interest is not fixed the interest conjecture and arranged request information have high worth. Sharing purpose of deals has no worth if the interest is fixed, hence it is essential to choose how and when to share arranging data. Vnpoucke have built up an expository system for examine that exploit the combination Information technologies companies and supply chain management operations. Utilizing application of information technology strangely affects the operational presentation when utilized for upstream incorporation instead of the bringing together with clients. This can speed up and precision and can improve conveyance execution. In any case, as indicated by Azab, many Supply chains actually experience the ill effects of miscommunication between various partners and wasteful trade of data. Henceforth, new methodologies and strategies ought to be received to give more proficient data sharing. The Methodology and Technique in Internet of Things as Supply Chains Management.

3 Research Methodology Technique for IoT as Supply Chain

With the development of Internet of things, scientists began to investigate the capability of utilizing this innovation in a few fields. 2012 talked about the meaning of Internet of things, its difficulties and how it could be applied in various zones, for example, coordinations and Supply chain for following, following, checking and regulating. 2017 contemplated the effect of executing Internet of things over the supply chain, and contended that Internet of things adds to the development of the organization and in confronting the present difficulties, it likewise positively affected the organization's future economy. They delineated how to interpret the information gathered by Internet of things into important data to support the provider/maker in dealing with the Supply chain by picking between two systems. These techniques help the administrators to handily track, follow and assess items whenever and at any stage, which improves the realness and nature of the items and consequently expands the supply chain proficiency. 2015 audited a few uses of actualizing IoT in Supply chains, for example, employ on farming area, with underscored function overseeing item data, decreasing expenses related to supply chain management, enhance SCM proficiency. According to proposed a system by Li gave in this system represent the IoT Cloud Computing is used as supply chain management, particularly Supply Chain Intelligence by giving incorporated information on exercises, assets and cycles is enhanced the general exhibition proposed an evaluation model that indicated the effect of utilizing Radio-frequency identification on the issue of invalid data which happened the impact outcomes from data meandering using supply chain management.

4 Development of Supply Chain Framework in IoT Setting in Current Scenario

Development of Supply chain framework under the Internet of things setting primarily incorporates two angles: advancement of Supply chain capacity and advancement of Supply chain measure. Alongside the constant improvement of Internet of things innovation, the capacity of IoT based Supply chain framework has changed fundamentally. All in all, supply chain measures incorporate acquisition, creation and handling, coordinations, deals, and after deals management measure, which additionally should be improved in like manner.

4.1 The Objective of Supply Chain with Regards to IoT

To improve the degree of representation, straightforwardness, and steadiness of Supply chain, the arranging capacity must be changed. With regards to Internet of things, the utilization of detecting gadgets, for example, Radio Frequency Identification labels and remote sensors, the coordination and appropriation of wired and remote innovation and Internet, and the headway of mass information mining innovation can give data for arranging better exactness and speed, subsequently extraordinarily improving the precision of anticipating the other hand, the appearance of huge information bewilders the interior and outside climate of Supply chain framework, incredibly enhancing the trouble in making medium and long haul plan for leaders in Supply chain, diminishing the power of Supply chain. In any case, it is basic that significant leaders in Supply chain have great predictability for the improvement of IoT innovation and hold onto advancement inclination of Internet of things precisely. The crucial objective of Supply tie is to amplify the advantage of the whole chain. By and by, in Supply chains, there are consistently sure irreconcilable circumstances among individuals from Supply chain, blocking the acknowledgment of the essential objective of Supply chains. As per a few researchers, embracing proper component configuration to tackle issues, for example, deviated of data and mindlessness of collaboration instrument among Supply chain individuals, is huge in the acknowledgment of principal objective of Supply chain. IoT innovation offer specialized help for the amplification of the advantage of the whole Supply chain, acknowledging visual administration, and insightful administration all through the whole Supply chain and upgrading the straightforwardness of Supply chain, and empowering data sharing among individuals from Supply chain. As needs be, the goal of contentions among the individuals in Supply chain is significantly facilitated.

The fundamental mechanical empowering influences that drive profitability and increment perceivability along the assembling Supply chain are Radio Frequency Identification and IoT.

Radio Frequency Identification: Radio Frequency Identification frameworks comprise of three parts: Radio Frequency Identification labels, Radio Frequency Identification radio wires and Radio Frequency Identification per users. Radio Frequency Identification labels are CPUs conveying encoded advanced information about the items they are connected to for example, investigating the information from Radio Frequency Identification labels appended to the stock things and sensors introduced at a shrewd stockroom, a Supply chain the board framework can furnish distribution center specialists with continuous information about the area, the status and the state of each stock thing, tell laborers of inbound conveyances, give bits of knowledge into the utilization of the warehouse, and so Simultaneously, with cutting edge information science procedures, the Industrial Internet of things establishes a strong framework for the robotization of Supply chain measures (Fig. 1).

In the interim, the utilization of Internet of things innovation likewise prompts new issues and irreconcilable circumstances among the individuals from Supply chain. For instance, unbalanced venture cost on IoT from leaders in Supply chain, concealed

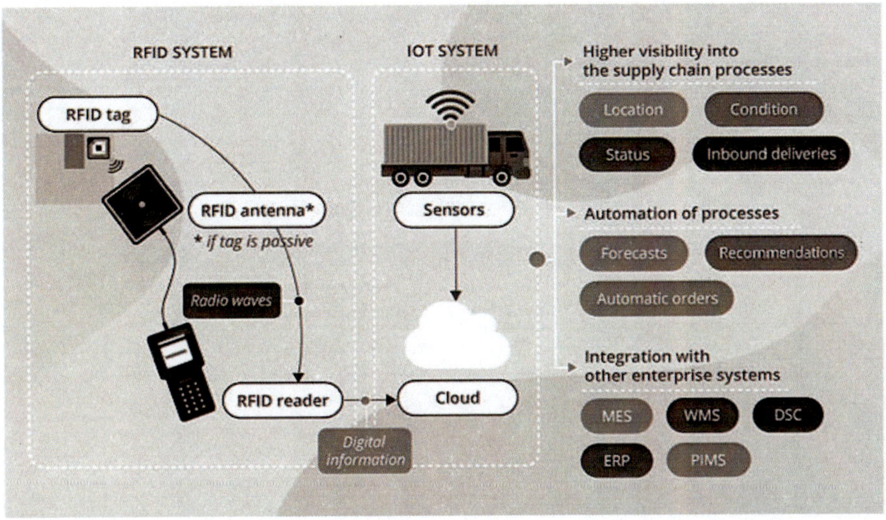

Fig. 1 Activity between RFID and IoT system

security issues, and issues related security insurance, which involve unending coordination among individuals in Supply chain and better instrument plan, so the new issue and irreconcilable circumstances showing up during the time spent Supply chain can be settled. This model demonstrates that IoT innovation has acknowledged outrageous observation and power over framework preparing, offering strong specialized help for smart control, programmed control, and exact control.

4.2 Development of Supply Chain Measure in IoT Setting

In this regards to the Internet of things with the use of IoT innovation, for example, Radio Frequency Identification, sensor, and so forth, exact data of the entire cycle including acquirement, coordinations, creation and preparing, deals, and after-deals administrations of products can be gotten. 160 Multi-Criteria Methods and Techniques Applied to Supply Chain Management Monitoring of warehousing, move, dissemination, and transportation is realized through IoT innovation. IoT advancements have the ability to associate your kin, cycles, information and merchandise by means of brilliant gadgets and sensors. With keen sensors which can catch and send information, you would now be able to get an away from of the condition, area and climate of your products as they are on the way (Fig. 2).

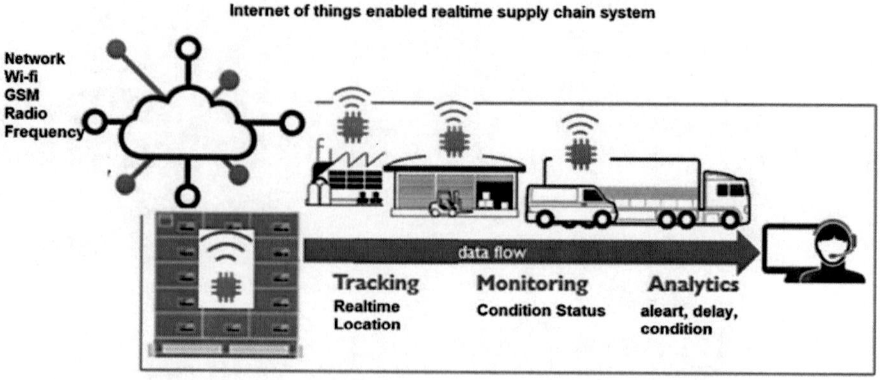

Fig. 2 Internet of thing enabled realtime supply chain system

5 Conclusion

Internet of things is viewed as important encouraging advances command and enhance exhibition for Supply Chain Management data repository is main pieces for SCM affix add achievement modern association, it is up to date innovations increasing huge consideration various data sources scope of undertakings for enhance execution, notoriety subsequently acquire clients and benefit. The research study an audit of Company 4.5 standards advances given the emerging technology of Cyber-Physical Systems and also internet of thing with distributed information technology, just data partaking as supply chain management, with past examination as executing internet of thing as SCM. And afterward, we lead a methodical assessment on the attributes of Supply chain framework in IoT.

References

1. Imran M, Hameed W, Haque A (2018) Influence of Industry 4.0 on the production and service sectors in Pakistan: evidence from textile and logistics industries. Soc Sci 7:246
2. Ibrahim S, Elayat HA, Khater MM, Mostafa NA (2011) Data analysis for inventory management in a multi-echelon supply chain. Int J Econ Resour 2:138–150
3. Choy KL, Gunasekaran A, Lam HY, Chow KH, Tsim YC, Ng TW, Tse YK, XA Lu (2014) Impact of information technology on the performance of logistics industry: the case of Hong Kong and Pearl Delta region. J Oper Res Soc 65:904–916
4. Grabara J, Kolcun M, Kot S (2014) The role of information systems in transport logistics. Int J Educ Res 2:1–8

An Analytical Study on Big Data Management for Supply Chain Analytics

Sundeep Kumar, Vikram Singh Rathore, and Alok Mathur

Abstract Progression in data and correspondence innovation Information and Communication Technologies (ICT) has offered ascend to a explore of information in each area by tasks. Functioning with colossal capacity of information in area of Big Information related Data, as it is famously familiar concerning the withdrawal of valuable data for help dynamic only the wellsprings for upper hand for associations nowadays. Ventures through utilizing for intensity for investigation with planning occupation system with every instance for their tasks with alleviate occupation hazard. An unpredictable worldwide market situation has constrained the associations to reclassify their Supply Chain the board SCM. Through this study, the main goal of outlined for pertinence with Huge Data with significance in overseeing start finish gracefully supply with accomplishing enterprises greatness. In big data engineering for Supply Chain Management is suggested misuses present status of the relationship innovation of information the executives, examination, and perception or forecasting. The protection and security prerequisites of a Big Data framework have likewise been featured and a few components have been talked about to actualize these highlights in a true Big Data framework organization with regards to Supply Chain Management. Some future extent of work has likewise been called attention.

Keywords Supply chain management · Analytics · Security · Cloud · Architecture · Protocols · Privacy · Big data

1 Introduction

As per SCM has gotten only key empowering agents for accomplishing the upper hand. Expanded client interest and assortment, strengthened rivalry, extracting multiple type and energetic, of countrywide venture, the tension for the growth for object with management, drive with alteration particular data with correspondence innovation as ICT have merged intricacy with organization with overseeing SCA.

S. Kumar (✉) · V. S. Rathore · A. Mathur
Centre for Electronic Governance, Department of Technical Education, Government of Rajasthan, Jaipur, Rajasthan 302004, India

R. Agrawal et al. (eds.), *Recent Advances in Industrial Production*, Lecture Notes in Mechanical Engineering, https://doi.org/10.1007/978-981-16-5281-3_31

Successful coordination and cooperation empower the individuals from a graceful chain to accomplish its worldwide destinations.

Nonetheless, the sharing of data assumes an irreplaceable function in Supply Chain Management joining. It sets and screens key execution markers to feature changes and failures and mitigates the bullwhip impact which is basically caused because of the contortion of interest data while moving from downstream to upstream Cheng [4]. How opportune and precisely an association can plan a successful and advanced methodology has become a basic issue with regards to present-day Supply Chain Management. Scientists have discovered that to accomplish consistent coordination or amicability among the individuals from a flexible chain for taking the correct choice at an opportune time, to convey assets ideally and channelize all exercises the correct way, to give the correct item to the clients at the perfect time, data goes about as an imperceptible string among the individuals. With the quick advancement and selection of ICT by the ventures, a goliath measure of information is being created all inescapably from each movement over a graceful chain. As indicated by the International Data Corporation for assessed development for computerized information is large as 46 thousands GB in year 2020 when contrasted with around 3.3 trillion gigabytes in 2013. This opens up a huge open door for business associations to successfully use such a huge measure of information for settling on judicious business choices. In Phase 1 for enterprises countrywide is advancing towards Industries 5.2, each article connected in a flexible fetters is currently going about constant alternator for information in an organized structure. Unnecessary to make reference to, overseeing such a colossal capacity of information big data, is famously understand an awesome undertaking. In Phase 3 comprehension and following of information age and afterward handling of information for inferring valuable data to work a keen gracefully chain remains as the way to progress. In accordance with our past work Sen and Biswas [1], propose engineering with a big data driven gracefully fetters with conformance through structural norms. In Phase III, a idea of big data with pertinence for big data investigation with regards to Supply Chain Management are called attention to. In Phase IV, work on portrayed unique big data engineering for writing and afterward to suggest a graceful fetters explicit big data framework. Segment V introduce different protection with secure gives to address for big data framework with furthermore talks about various components and conventions for planning a safe Big Data design.

2 Supply Chain Analytics with Its Collision on Supply Chain Production (Review Literature)

According to Christopher characterized gracefully chain as "The organization of associations that are included, through upstream and downstream linkages, in the various cycles and exercises that produce an incentive as items and administrations conveyed to a definitive shopper". Specialists accept that it is the "concordance" rather than "rivalry" which decides the reason for cutting edge Supply Chain Management procedure. Cooper and Lambert spoke to the graceful chain as an organization of business and cooperation seeing someone over the chain in quest for business greatness. As per the creators, main cycles for planned with level or the upright measurement for achieve elements for the flexibly fetters like client support with connection the executives, request the board, gracefully the board, assembling, and bring the executives back. Exploration has uncovered that the adequacy of asset—execution connections connected with worldwide destinations over the flexibly chain rely altogether upon the shared cooperation and coordination among the accomplices, which thusly encourages successful dynamic and plan of viable gracefully chain systems as per Dong at the end of the day, SCM is the key collaboration among for individuals from the gracefully chain to incorporate gracefully and request the board Stevenson. Basically as per the flexibly chain activities reference (Supply-chain operations reference) structure, the coordination of interest and gracefully the board over a gracefully chain happens through four expansive cycles like an arrangement, source, make, and conveyance, which includes the progression of material, data, and asset. Data with regards to Supply Chain Management comprehensively incorporates client data, deals data, market and contender data, item and administration level prerequisite, advancement/brand data, request anticipating, stock, limit use, measure arranging and control data, aptitude stock, human data, sourcing/merchant data, organizing data, coordination's, stockroom arranging, estimating and store stream/working capital data (Fig. 1; Table 1).

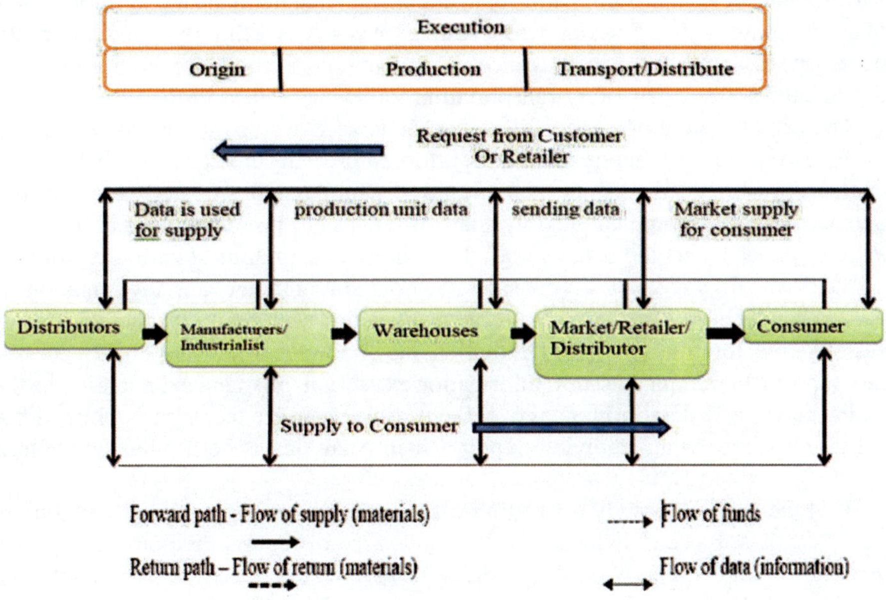

Fig. 1 Information and data used in supply chain structure

Table 1 In supply chain management various types of information collected in different stages

Type of vender	Related information
Distributor	Plan information, order status, stock level, schedule, shipment, and routing, example a/c receivable, charge, valuing, and so on
Manufacturer	Fundamental/activity information, design information, gauging information, prod. plan/plan, capacity arranging information, process information lot size, process duration, take time, throughput time, measure ability, and so forth, yield information, quality/reliability information FTR, percent dismissal, percent disappointment, and so on, goods raw materials/work in progress/financial graph, documentation, consumer criticism information, salesperson information, people information, finance information wage, transformation cost, and so forth, return/arrange
Market/retailer/warehouse/distributor/	Dispose, customer criticism, finance information valuing, installment and so forth /demand, stock level, schedule, shipment& routing, order, return etc
Consumer	Demand, product input, customer conclusions, payment, delivery, new item, promotion/recommendation, return/ purpose of deals POS, order status

Nonetheless, the developing universal environment of versatile with community correspondences with amplification converter achieved a flood the age of information of differentiated nature from numerous sources over a "wise" Supply chain. It impacts numerous essential operational choices like an appraisal of client stir choices, item sentiments, prescient support, transportation following, and so forth.

The essential goal of a gracefully affix tactician is to separate valuable data by investigating the humongous measure of information being created from all the items over the flexible chain. As per them, Supply Chain Analytics depicts an expansive perspective on the complete flexibly chain as it does not just display detail development of items, asset, and data over all the exercises being done at various phases of a graceful chain yet besides it considers key outcome markers and investigation to comprehend the cycle for taking restorative, preventive and prescient choices. In any case, the creators likewise thought that characterizing business and client necessities opposite hierarchical issues, information extraction, mix, and examination lastly taking commonly durable choices over the flexible chain are the three essential difficulties or desires while applying Supply Chain Analytics as per Ranjan and Sahay (Table 2).

That the analytical study for getting data for the process is going on start it used historical data, piece of evidence with various type of reports that introduction the analytics to known that what is happening now, how it is related to past, what may happen also known as Predictive analytics for finding what is the best or worst possible outcome, what is the easiest way to process and reduce the rules known as Prescriptive investigative 2010.

Table 2 Future applications of supply chain analytics

Phase	Role of application in supply chain analytics
Arrangement	Forecast of the market necessity of items and administrations, estimating, dynamic valuing
Obtained information	Provider determination and assessment, value arrangement
Development	Creation arranging and control, stock arranging
Conveyance	Coordination's the executives, area arranging, stockroom arranging, network arranging

3 Action on Information Analysis in Concern of SCM (Methodology and Techniques)

All of those demand should be supported by new data models supporting all data states and stages during the whole data lifecycle and new infrastructure services and tools that allow also obtaining and processing data from a variety of sources including sensor networks and delivering data in a variety of forms to different data and data customer and equipments.

According to Demchenko the structure acknowledge in Fig. 2 compare 6 elements:

A. Photoelectric cell and other Information Required Equipments.
B. Big Data Deployment.

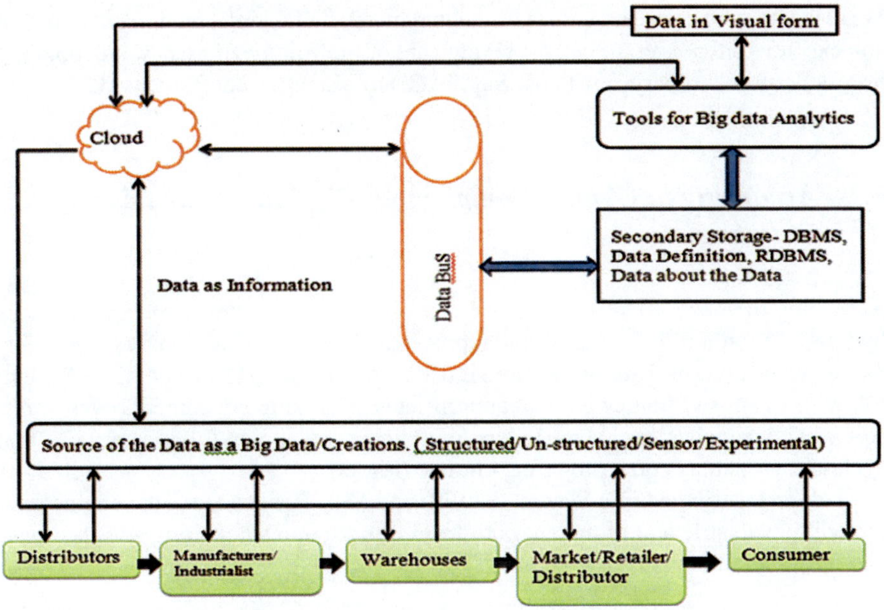

Fig. 2 Supply Chain Structure define next generation of Big Data

C. Information Registers.
D. Information Storage Devices and System Management.
E. Search Engine used for data analysis.
F. Graphical Representation of Data and Supply System.

The photoelectric devices and another required equipments for data include all methods and equipments that access unstructured data as input various scales in SCM. Information register are basically received the data into consideration the storage capacity requirement of the bus so that a large volume of data might be stored for real-time data processing and analytics. The Information Storage Devices and System Management receive information from the data bus for efficient storage and retrieval of data.

Synchronization of all the equipments is designed as per structure of real time system importance so that high availability, robustness, and interoperability of the sub-systems can be achieved. High availability of the system may be achieved by using replications of the subsystem of process. Software durability and communication through the components are used to improve the use of hardware as well as software in reliable cost.

4 Integrated Supply Chain with Big Data Architecture (BDA)

We at that point propose a big data framework used in SCM information examination and executives that is equipped for taking care of the volume, intricacy, and ongoing necessity of an ordinary certifiable Supply Chain Management framework.

4.1 According to Chan a Complicated Big Data Analytics Structure

According to above scientist recognizes the working, attributes with expected uses by Large Amount of Information and represented in a specific design by Huge Information investigation. This architecture is used while Hadoop Distributed File System fills in as a capacity framework that appropriates information documents over countless worker bunches, the Map-Reduce is a disseminated handling framework that cycles records in an equal preparing climate (Fig. 3).

Hadoop Distributed File System (HDFS) and Map-Reduce make the engineering incredibly versatile and proficient both from information stockpiling and the board and information handling viewpoints.

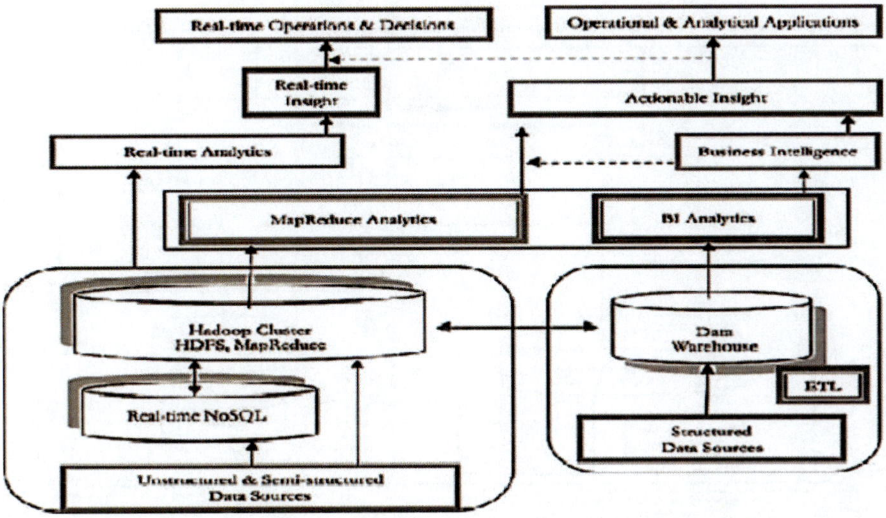

Fig. 3 Big data analytics architecture as per adapted from Chan

4.2 SCM Proposed Big Data Architecture

As per engineering design and an investigation system through SCM applications. The purposes behind getting the information from the source information distribution center into the Demand Driven Supply (DDS) and afterward questioning the Demand Driven Supply (DDS) as opposed to questioning the source information stockroom straightforwardly is that in a Demand Driven Supply (DDS) the information is masterminded in a dimensional arrangement that is more reasonable for examination motor. The metadata is an information base containing the rundown of the information in the information stockroom and incorporates data, for example, the information structure, the information meaning, the information use, the information quality guidelines, and other data about the information (Fig. 4).

Such systems are viable in Large amount information for medical services and focused on the forecasting of information is circulated remote sensor organizations. According to Membrey have tended to the test of establishing a confided in dispersed registering climate for handling touchy information in a Large Information from biological system. According to creators introduce a powerful foundation faith reset convention as Dynamic Infrastructure Trust Bootstrapping Protocol—DITBP that conveys believed figuring bunch reference engineering as (TCGRA) with confided in stage objects as TPM for setting up believe the registering elements [2].

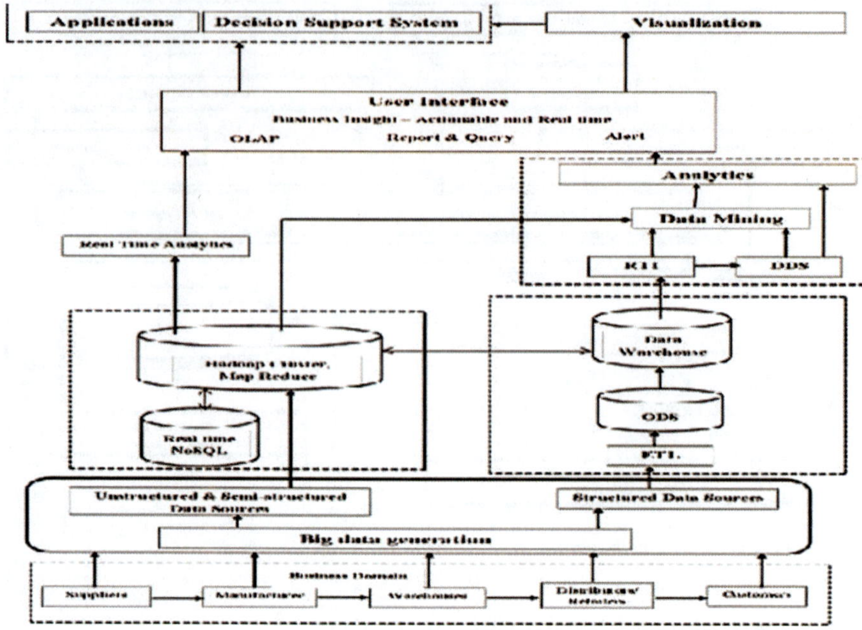

Fig. 4 Supply chain management proposed architecture model for big data analytics

5 Conclusion

This research paper has featured on Supply Chain Analytics for of Big Data examination with regards to Supply Chain Management. At that point, concerning the graceful chain activities reference as Supply-chain operations reference (SCOR) structure, the kinds of information produced over the average information-driven flexibly chain have been referenced. At that point, engineering through big data examination in SCM introduce working as nonexclusive big data design previously is writing. In this study likewise explained on the protection and secure data prerequisites in a big data framework with remembered a concise conversation for different conventions and instruments to authorize these necessities in true Big Data frameworks organizations.

References

1. Biswas S, Sen J (2016) a proposed framework of next generation supply chain management using big data analytics. In: Proceedings of national conference on emerging trends in business and management: issues and challenges, Kolkata, India. Available at SSRN http://ssrn.com/abstract=2755828

2. Brickell E, Camensich J, Chen L (2004) Direct anonymous attestation. In: Proceedings of the 11th ACM conference on trust and security in computer systems
3. Chae B, Olson DL (2013) Business analytics for supply chain: a dynamic-capabilities framework. Int J Inf Technol Decis Mak 12(1):9–26
4. Chen H, Chiang RHL, Storey VC (2012) Business intelligence and analytics: from big data to big impact. MIS Q 36(4):1165–1188
5. Clemons E, Reddi S, Row M (1993) The impact of information technology on the organization of economic activity: the "move to the middle" hypothesis. J Manag Inf Syst 10(2):9–35

Die Design and Its Parameters for Grain Refinement of AA6XXX Series Through Equal Channel Angular Pressing

Arshit Kapoor, Bhuwan Gupta, Abhishek Singhal, and Krishna Mohan Agarwal

Abstract Basic need of industry is to have a material with less weight and high strength. There are number of techniques to enhance mechanical properties and microstructure, among which equal channel angular pressing is found to be most effective technique. Equal Channel Angular Pressing (ECAP) is one of the famous grain refinement techniques of severe plastic deformation. For this process the die plays an important role. Factors like corner angle, channel angle, friction, number of passes, routes, back pressure and many more may affect the results of the process. In the current study all the parameters for the ECAP process influencing the mechanical properties for the AA6XXX series has been reviewed and analyzed for the better conclusion. Mainly ECAP passed materials have better mechanical, microstructure and physical properties. Two main alloying elements for the 6000 series are Si and Mg. AA6xxx is known for its ease in design, high strength to weight ratio, resistant to corrosion, plasticity and has various applications such as they are used in aircraft, automobile, defense, medical applications etc.

Keywords Equal channel angular pressing · Plastic deformation homogeneity · Die design · Fabrication · corner angle

1 Introduction

Severe plastic deformation is the technique which can enhance the material properties by decreasing grain size leading to refinement in microstructure and ultimately increasing the mechanical properties. Imposing high strain in material helps in changing the mechanical properties and microstructure. There are different techniques in SPD like high pressure torsion, multi axial forging, accumulative roll bonding, equal channel angular pressing etc. Among these, ECAP is best technique for enhancing the mechanical properties. Average grain size of material can be used to define its physical and mechanical properties. The relationship between yield stress and grain size d is given in equation. The relationship is mentioned in Eq. 1. Where

A. Kapoor (✉) · B. Gupta · A. Singhal · K. M. Agarwal
Department of Mechanical Engineering, Amity University, Uttar Pradesh, Noida 210303, India

R. Agrawal et al. (eds.), *Recent Advances in Industrial Production*, Lecture Notes in Mechanical Engineering, https://doi.org/10.1007/978-981-16-5281-3_32

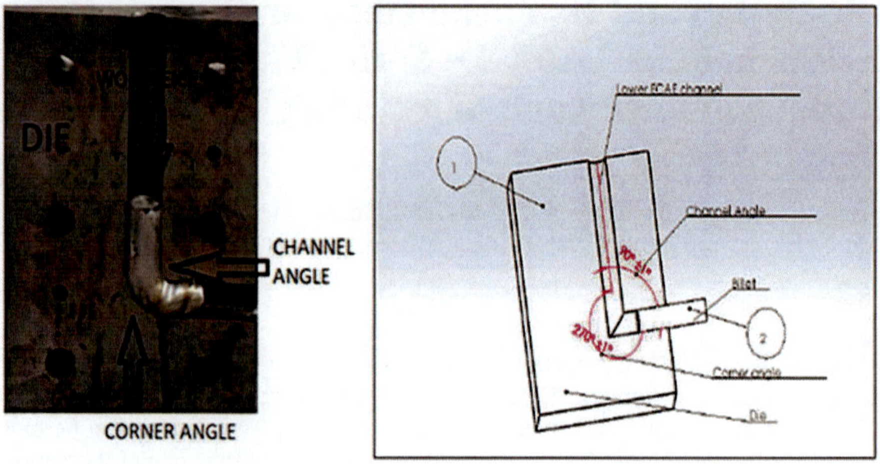

Fig. 1 Schematic diagram of ECAP

σ0 is the friction stress, and k is the yield constant [1, 2].

$$\sigma_y = \sigma_0 + k\, d^{-1/2} \tag{1}$$

A die with two intersection channels is a requirement for ECAP process, through which the workpiece is pressed to impose high stress at intersection. Two major factor ϕ and Ψ (channel angle and corner angle respectively), which affect the rate of strain and ultimately affecting the results of process. A schematic diagram of ECAP is shown in Fig. 1 [3]

ECAP is defined as a homogenous process. The homogeneity here means that there is uniform stress distribution in workpiece, but some factors affects its homogeneity. Die design plays an important role in ECAP. There can be internal and external factors affecting the results of die. External factors like back pressure, loading rate, ram speed, temperature, friction, geometry of die and passing routes. Internal factors majorly depend on material of workpiece and its properties. This paper studies various factors affecting homogeneity. Experimental study is made for mechanical properties to know the actual percentage enhancement in properties for AA6063 aluminum sample [4–7]. First application of aluminum came in 1894, Hartford railroads-built aluminum seat frames for lightweight car. Later its use started in logistic industry, railways and later to household applications. Light weight, high strength and corrosion resistant are the main reasons for its use. The benefits were highly efficient for the industry. This increased the use of aluminum in every field. Due to intensive research on ECAP and its applications, different Aluminum alloys are processed under ECAP for various applications. AA 6000 series is one of the most used series of aluminum which is extensively processed by ECAP. This series consist of Magnesium and Silicon as the major alloying elements. The alloys in this series possess moderate strength. Due to the combination of magnesium and silicon with aluminum, it leads to high machinability by heat treatment. It is widely used in welding applications, electrical hardware and many more important areas [8].

2 Literature Review

In Djavanroodi [9], investigated the effect on channel angle, coefficient of friction and back pressure on strain rate developed when a sample is passed through equal channel angular pressing. He reported through FEM analysis that reducing the channel angle will increase the strain in sample leading to high homogeneity and more refinement in microstructure. Smaller channel angle and high coefficient of friction eliminates the gap in corner which is formed by sample which helps in achieving high strains. Considering both the factors will increase the press pressure requirement also. Reducing the channel angle from 120° to 60° may increase the pressure requirement to 3 times. Applying back pressure will increase the homogeneity but the effect is less noticeable. Ebrahimi [10] studies about the parameters affecting the strain and homogeneity while processing in ECAP. The channel displacement and channel angle were analyzed. He reported that, channel angle has more effect on amount of strain developed in workpiece. A parallel die was used by him and he concludes that channel displacement has less effect on amount of strain but certainly, increasing the homogeneity with increase in displacement. Reduction in channel angle and decreasing the length of displacement will increase strain magnitude but will also give increment in requirement of pressure. Parallel die can be used to eliminate the corner gap which ultimately increases the effectiveness of ECAP process.

In Thakur [11], studies the past and new advancement in die design for ECAP. He investigated that apart split die, spring loaded die, rotary die, spiral die and T-shaped die are new advancement in die design. In rotary die, the main advantage is, for number of passes the material is not removed from the die because this die has 4 channels intersecting at 90°. The rotation of die helps to do number of passes without material removement from die. The spring-loaded die is basically used to decrease the plunger force requirement as spring used helps in force restoring. Now a days, along with ECAP there are other processes attached such as ultrasonic ECAP or torsional ECAP. Ultrasonic energy is used while process which helps in reduction of cracking of material and making process smoother. In torsion ECAP, at the outlet channel of ECAP, torsion is used for more enhancement in properties and microstructure. Mathieu [12], proposed a new design for ECAP die. He states that as friction plays a major role in the process so, to reduce the friction a new modified die with movable channel is made. He suggested that material of die can be Inconel 718 or stainless steel with certain modification like thermal insulation can be used. The material should be of high strength. It has two fixed part and one moving part with die channel. The sample is placed in moving part and then passed through the fixed part. The outlet channel is in fixed part. He investigated that this reduces the friction in die and helps in improving the properties and microstructure. In [13] Patricia Ponce-Peña studies about the optimized die design for ECAP. He investigated by a FEM analysis having 6 different die configuration and different coefficient of friction. He concludes that multiple passes are beneficial and with decrease in friction coefficient, there is increment in homogeneity of structure. The FEM analysis concluded that die with ϕ = 90°, Ψ = 16° and friction of coefficient = 0.05 are optimized die parameter for ECAP process. Dayal [14] studies about the die design for ECAP by minimizing the corner gap as corner gap is a consequence of corner angle and friction coefficient. He investigated through FEM analysis and concludes that for a circular die, corner angle more than 30° does not lead to formation of corner gap. He concludes that harder material is independent of friction whereas softer material has more homogeneity with respect to high friction. He also states that corner angle of 30° and fillet radius of 5 mm is best suited for homogenous structure with less concentration of strain.

3 Method and Material

Based on literature review and earlier studies, a split die is made in laboratory with channel angle ϕ = 90°, and corner angle Ψ = 20°. The outer dimension of die is 220 × 160 × 100 mm. The billet size is taken as 20 mm of diameter and the length of workpiece is used as 150 mm. The material used for die is H13 tool steel with carbon content 0.5% approximately because of its high strength and highly suitable for cold and hot work application. The two parts of split die are joined by 6 Allen bolts and 2 keys. Keys helps in better alignment of two parts and limit the movement.

3.1 Equal Channel Angular Pressing

Equal Channel Angular Pressing (ECAP) was first given by Segal [5] in the Minsk Institute, Soviet Union. In this process, a plunger is used to apply pressure on the material. It consists of a special die having two special channels meeting at an oblique angle known as the channel angle Φ. The angle made by the cross section of the two channels at the convergence is known as the corner angle Ψ. The strain tends to develop on the material at the meeting point of the two channels [8].

$$\epsilon_N = \frac{N}{\sqrt{3}}\left[2\cot\left(\frac{\emptyset}{2}+\frac{\varphi}{2}\right)+\varphi\operatorname{cosec}\left(\frac{\emptyset}{2}+\frac{\varphi}{2}\right)\right] \tag{2}$$

Above Eq. 2 gives the relation between strain developed and external factors. Here number of passes is N. There can be 4 routes for performing this process. Figure 2 is depicting all 4 routes. Route A is a simple route in which no change is orientation with number of pass. In route BA, the workpiece is twisted 90° alternatively in clockwise and anti-clockwise direction in next pass. Whereas in route Bc, the sample is rotated in only one direction by 90° in successive passes. In route C, the sample is rotated by 180°. All routes give different microstructures, correct route can be selected as per

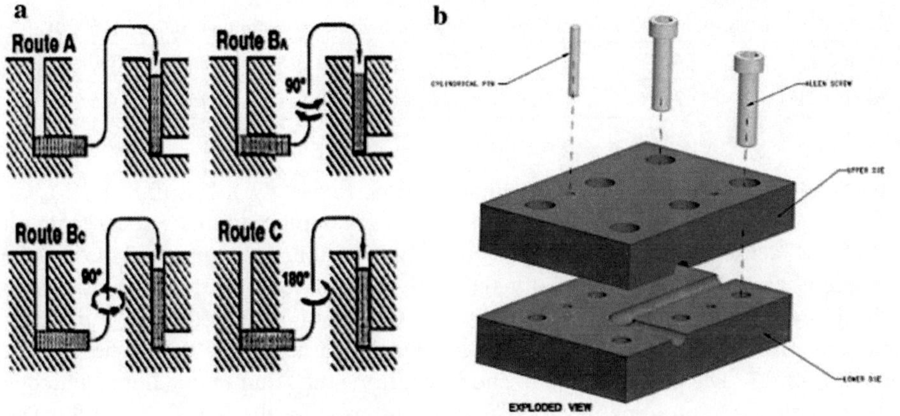

Fig. 2 **a** Different routes of ECAP, **b** exploded view of die

requirement. The availability of different routes in this process allows the possibility of obtaining grain refinement of different magnitudes which further allows flexible manufacturing of different components without much strain on the equipment and workpiece. As seen in Fig. 2a, the billet is also rotated in different orientations through different angles. The workpiece is rotated in different orientations in order to obtain grain refinement in different orientations and of different magnitudes. Due to high machinability, ECAP can be used for a high number of industrial applications like aircraft manufacturing, automobile manufacturing, sheet forming and many other industrial applications. The major advantage in ECAP process is that the dimensions of billet are not changed, and it is recognized as the best method for severe plastic deformation. The improvement in mechanical properties and getting UFG (ultra-fine grains) is exceptional with respect to other process [7, 15].

3.2 Advancement in AA6XXX

Two main alloying elements for the 6000 series are Si and Mg. AA6xxx is known for its ease in design, high strength to weight ratio, resistant to corrosion, plasticity and many more. Si and Mg are to be added in proper ratio (Mg/Si ratio 1.73) to form Mg2Si. Other alloying elements can also be added to get different properties as per requirement. If copper is added, it provides exceptional precipitation strengthening but it may increase the weight. Each alloying element plays a significant role. Addition of manganese can increase the strength by age hardening and enhances the resistance to corrosion. There can be many methods to improve the strength of AA6xxx such as precipitation strengthening, grain boundary strengthening and work hardening. By addition of Mn, Cr, Scand Zr helps in decreasing the grain size and ultimately refining the grain structure. This may reduce the grain size up

Table 1 Generalized composition of the experimental material Aluminum (grade AA-6000)

Element	Al	Si	Fe	Cu	Mn	Others
% composition	98.78	0.465	0.145	0.003	0.056	0.551

to (5–50 nm). AA6066 with the composition Si (0.9–1.8%), Mg (0.8–1.4%), Cu (0.7–1.2%), Mn (0.6–1.1%), Cr (0.40%) has the maximum ultimate shear strength of 152 MPa after annealing among its other composition. Annealing and the T6 (the solution heat treated and artificially aged) are the two-better process for increasing the strength. Table 1 shows the weight percentage composition of the 20 mm diameter experimental material commercial Aluminum (Grade AA-6063) as per ASTM E: 1251-2017. The presence of Mg and Si confirms the Aluminum Grade, which is responsible for reinforcing metals. The lubricant used for the process is MoS2. The coefficient of friction is assumed as 0.1 and ram speed is 1 mms^{-1} [16].

4 Results and Discussion

After an exhaustive study of different die design parameters and their effects, it is been easily concluded that channel angle, corner angle, type of route and number of passes are main 4 factors which are directly affecting the results of equal channel angular pressing. Other factors also play an important role but their effect can once be neglected in comparison to the above mentioned 4 factors. A reviewed data is being tabulated in Table 2, which shows that for AA6000 which all optimized parameters are used. The table is being constrained for 6000 series aluminum alloys and 4 major mechanical properties. The table shows an increase in mechanical properties such as Hardness and Ultimate tensile strength, yield strength and a decrease in percentage elongation to failure in the AA6000 series after performing ECAP. Table 2 increased has been denoted by IN and decreased by DE.

Above data evidently concludes that channel angle 90° is the optimized angle as maximum research and results are concluded on 90°. The variation in the use of different routes can be seen from the table. The reason for this can be concluded as the variation on type of application. For different applications, each route plays a different role. After reviewing several papers of AA6000 as shown in above table, analysis conclude that for AA6005 in which the major composition is of Silicon and Magnesium, for the optimized results the number of passes should be more than 4 and corner angle should be around 20.5°. Major research has been done on AA6061 due to its vast application capability and the chemical percentage of Magnesium is more than silicon in it.

The graph evidently depicts that rout Bc is used 13 times out of 21 papers. This proves that the optimized, majorly practiced and most researched route is Bc. After Bc, both A and C are used. As both are basically simplest routes. Figure 1b shows that earlier research has been done majorly on corner angle 20° and stated that 20° is the

Table 2 Variation in mechanical properties with parameters

Material	Pass	Channel angle	Corner angle	Number of passes	Hardness	UTS	Yield strength	%Elongation to failure	References
6005	BC	90°	20°	8	IN	–	–	–	[17]
	C	90°	0°	4	–	IN	IN	DE	[18]
	A	90°	20.6°	7	IN	IN	–	–	[16]
6012	BC	90°	37°	4	IN	IN	IN	DE	[19]
6013	BC	90°	20.6°	4	–	IN	IN	–	
6060	A	90°	20.6°	9	IN	IN	–	–	[16]
6061	BC	90°	–	6	–	IN	–	DE	[20]
	BC	90°	–	8	–	–	–	DE	
	C	90°	20°	4	IN	IN	IN	DE	[21]
	BC	90°	–	4	–	IN	IN	–	[22]
	BC	90°	20°	8	IN	–	–	–	[23]
	BC	90°	–	4	–	–	IN	DE	
	BC	90°, 105° and 120°	–	4	IN	–	–	–	[24]
	BC	90°	–	12	–	IN	IN	DE	
6061 (Powder route)	A	90°	20°	4	IN	IN	–	DE	[25]
	C	90°	20°	4	IN	IN	–	DE	[26]
6063	BC	90°	20°	9	IN	–	–	–	[27]
	BC	90°	37°	6	IN	IN	IN	DE	
6082	–	90° & 120°	–	1	IN	–	–	–	[16]
	A	90°	20.6°	6	IN	IN	–	–	
	BC	90°	37°	4	IN	IN	IN	DE	

(continued)

Table 2 (continued)

Material	Pass	Channel angle	Corner angle	Number of passes	Hardness	UTS	Yield strength	%Elongation to failure	References
	C	90°	37°	3	IN	IN	IN	DE	

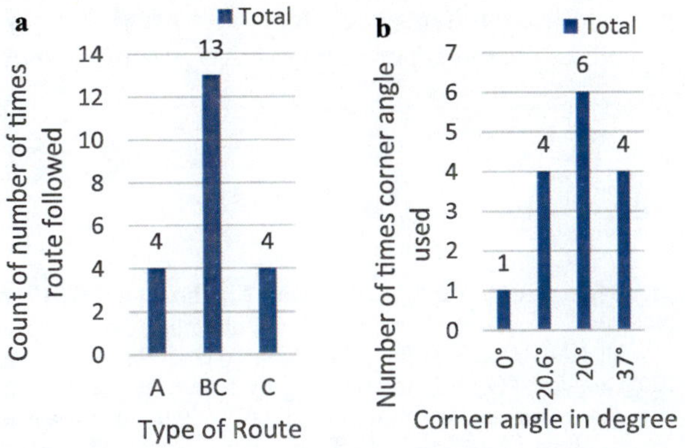

Fig. 3 **a** Type of route, **b** corner angle in degree

optimized corner angle for the equal channel angular pressing. The variation in graph is between 0° and 37°, 67% of sample size has used corner angle nearby 20°. It can be concluded from past research that more the corner angle less the force required to press the material and more the corner gap formation. Therefore, considering these factors and analyzing the work of other researchers, it can be easily concluded that 20°–22° is the range of corner angle for the best results of ECAP. 0° is the ideal corner angle but due to other factors like back pressure and coefficient of friction, its use is constrained (Fig. 3).

5 Conclusions

After conducting research and studying various research papers, we came to know about the impact of ECAP on the mechanical properties and thermal properties of metals and alloys. We plan on further conducting test runs for different channel angles of the die to compare the grain refinement obtained while performing ECAP.

- Thus, the preferred channel angle for the die during ECAP is 90° for obtaining a high amount of grain refinement without any changes in the dimensions of the sample.
- Preferred corner angle for performing the experimental tests is 20° and the most preferred route for performing ECAP on AA6000 series is route BC for number of passes to be 4 or more than that.
- After conducting literature review on 6000 series Aluminum alloys, we can say that there is an increase in mechanical properties such as Ultimate tensile strength, Hardness, Yield Strength, and decrease in percentage elongation of alloys with change in parameters such as channel angle, corner angle, number of passes, routes etc.
- More the corner angle less the force required to press the material and more the corner gap formation.

- Major research has been done on AA6061 due to its vast application capability and the chemical percentage of Magnesium is more than silicon.

References

1. Bagherpour E, Pardis N, Reihanian M, Ebrahimi R (2019) An overview on severe plastic deformation: research status, techniques classification, microstructure evolution, and applications. Int J Adv Manuf Technol 100(5–8):1647–1694. https://doi.org/10.1007/s00170-018-2652-z
2. Langdon TG (2011) Processing by severe plastic deformation: historical developments and current impact. Mater Sci Forum 667–669:9–14. www.scientific.net/MSF.667-669.9
3. K. Mohan Agarwal, R. K. Tyagi, and A. Dixit, "Theoretical analysis of equal channel angular pressing method for grain refinement of metals and alloys," *Mater. Today Proc.*, vol. 25, no. xxxx, pp. 668–673, 2020, doi: https://doi.org/10.1016/j.matpr.2019.08.026.
4. Wang S, Liang W, Wang Y, Bian L, Chen K (2009) A modified die for equal channel angular pressing. J Mater Process Technol 209(7):3182–3186. https://doi.org/10.1016/j.jmatprotec.2008.07.022
5. Segal VM (1995) Materials processing by simple shear. Mater Sci Eng A 197(2):157–164. https://doi.org/10.1016/0921-5093(95)09705-8
6. Langdon TG, Furukawa M, Nemoto M, Horita Z (2000) Using equal-channel angular pressing for refining grain size. Jom 52(4):30–33. https://doi.org/10.1007/s11837-000-0128-7
7. K. Mohan Agarwal, R. K. Tyagi, V. K. Chaubey, and A. Dixit, "Comparison of different methods of Severe Plastic Deformation for grain refinement," *IOP Conf. Ser. Mater. Sci. Eng.*, vol. 691, no. 1, 2019, doi: https://doi.org/10.1088/1757-899X/691/1/012074.
8. K. M. Agarwal, R. K. Tyagi, and A. Kapoor, "Deformation and strain analysis for grain refinement of materials processed through ECAP," *Mater. Today Proceeding*, no. 1, 2019.
9. D. M. Jafarlou, E. Zalnezhad, A. S. Hamouda, G. Faraji, N. A. Bin Mardi, and M. A. Hassan Mohamed, "Evaluation of the Mechanical Properties of AA 6063 Processed by Severe Plastic Deformation," *Metall. Mater. Trans. A Phys. Metall. Mater. Sci.*, vol. 46, no. 5, pp. 2172–2184, 2015, doi: https://doi.org/10.1007/s11661-015-2806-7.
10. Djavanroodi F, Ebrahimi M (2010) Effect of die parameters and material properties in ECAP with parallel channels. Mater Sci Eng A 527(29–30):7593–7599. https://doi.org/10.1016/j.msea.2010.08.022
11. P. Thakur, P. Surve, and S. Sanas, "Advancement in Die Design of Equi-Channel Angular Pressing (ECAP) Process: A Review," *Int. J. Sci. Eng. Res.*, vol. 5, no. 12, pp. 93–96, 2014, [Online]. Available: http://www.ijser.org.
12. Mathieu JP, Suwas S, Eberhardt A, Tóth LS, Moll P (2006) A new design for equal channel angular extrusion. J Mater Process Technol 173(1):29–33. https://doi.org/10.1016/j.jmatprotec.2005.11.007
13. P. Ponce-Peña, E. López-Chipres, E. García-Sánchez, M. A. Escobedo-Bretado, B. X. Ochoa-Salazar, and M. A. González-Lozano, "Optimized design of an ECAP die using the finite element method for obtaining nanostructured materials," *Adv. Mater. Sci. Eng.*, vol. 2015, 2015, doi: https://doi.org/10.1155/2015/702548.
14. A. Dayal, K. Hans Raj, and R. S. Sharma, "ECAP Die Design for Minimising Corner Gap," *Mater. Today Proc.*, vol. 5, no. 1, pp. 1686–1690, 2018, doi: https://doi.org/10.1016/j.matpr.2017.11.264.
15. Djavanroodi F, Ebrahimi M (2010) Effect of die channel angle, friction and back pressure in the equal channel angular pressing using 3D finite element simulation. Mater Sci Eng A 527(4–5):1230–1235. https://doi.org/10.1016/j.msea.2009.09.052
16. Roven HJ, Nesboe H, Werenskiold JC, Seibert T (2005) Mechanical properties of aluminium alloys processed by SPD: Comparison of different alloy systems and possible product areas. Mater Sci Eng A 410–411:426–429. https://doi.org/10.1016/j.msea.2005.08.112

17. Veveçka A, Cabibbo M, Langdon TG (2013) A characterization of microstructure and microhardness on longitudinal planes of an Al-Mg-Si alloy processed by ECAP. Mater Charact 84:126–133. https://doi.org/10.1016/j.matchar.2013.07.016
18. Chang JY, Shan A (2003) Microstructure and mechanical properties of AlMgSi alloys after equal channel angular pressing at room temperature. Mater Sci Eng A 347(1–2):165–170. https://doi.org/10.1016/S0921-5093(02)00577-4
19. M. P. Liu *et al.*, "Aging behavior and mechanical properties of 6013 aluminum alloy processed by severe plastic deformation," *Trans. Nonferrous Met. Soc. China (English Ed.*, vol. 24, no. 12, pp. 3858–3865, 2014, doi: https://doi.org/10.1016/S1003-6326(14)63543-3.
20. Z. Horita, T. Fujinami, M. Nemoto, and T. G. Langdon, "Equal-channel angular pressing of commercial aluminum alloys: Grain refinement, thermal stability and tensile properties," *Metall. Mater. Trans. A Phys. Metall. Mater. Sci.*, vol. 31, no. 3, pp. 691–701, 2000, doi: https://doi.org/10.1007/s11661-000-0011-8.
21. Sahai A, Raj KH, Gupta NK (2017) Mechanical Behaviour and Surface Profile Analysis of Al6061 alloy Processed by Equal Channel Angular Extrusion. Procedia Eng. 173:956–963. https://doi.org/10.1016/j.proeng.2016.12.155
22. Chung CS, Kim JK, Kim HK, Kim WJ (2002) Improvement of high-cycle fatigue life in a 6061 Al alloy produced by equal channel angular pressing. Mater Sci Eng A 337(1–2):39–44. https://doi.org/10.1016/S0921-5093(02)00010-2
23. Xu C, Furukawa M, Horita Z, Langdon TG (2005) The evolution of homogeneity and grain refinement during equal-channel angular pressing: A model for grain refinement in ECAP. Mater Sci Eng A 398(1–2):66–76. https://doi.org/10.1016/j.msea.2005.03.083
24. Chaudhury PK, Cherukuri B, Srinivasan R (2005) Scaling up of equal-channel angular pressing and its effect on mechanical properties, microstructure, and hot workability of AA 6061. Mater Sci Eng A 410–411:316–318. https://doi.org/10.1016/j.msea.2005.08.023
25. Chang SY, Lee KS, Choi SH, Shin DH (2003) Effect of ECAP on microstructure and mechanical properties of a commercial 6061 Al alloy produced by powder metallurgy. J Alloys Compd 354(1–2):216–220. https://doi.org/10.1016/S0925-8388(03)00008-2
26. Majzoobi GH, Nemati J, Pipelzadeh MK, Sulaiman S (2016) Characterization of mechanical properties of Al-6063 deformed by ECAE. Int J Adv Manuf Technol 84(1–4):663–672. https://doi.org/10.1007/s00170-015-7709-7
27. El-Danaf EA (2011) Mechanical properties, microstructure and texture of single pass equal channel angular pressed 1050, 5083, 6082 and 7010 aluminum alloys with different dies. Mater Des 32(7):3838–3853. https://doi.org/10.1016/j.matdes.2011.03.006

Integrating the Challenges of Cloud Computing in Supply Chain Management

Subhodeep Mukherjee, Venkataiah Chittipaka, Manish Mohan Baral, and Sharad Chandra Srivastava

Abstract Cloud Computing is an advanced technology in todays' scenario. The purpose of this paper is to identify and evaluate the challenges of cloud computing adoption in the SMEs of the supply chain management. SMEs supply chain management play a vital role in achieving operational excellence and in achieving its objectives. The research will give cloud service providers knowledge about the various challenges that the firms are facing while adopting this technology in their systems. Literature review was utilized to identify the components of challenges for the cloud computing adoption. Data was collected utilizing questionnaire method from various firms across the nation. Factor analysis was utilized for the analysis. The major influencing factors that was identified while adopting cloud computing were lock-in, loss of governance, compliance challenges, malicious insiders, resource and service management, service level agreement, data loss or leakage, legal jurisdiction, data privacy and protection, licensing risk.

Keywords Cloud computing · Supply chain management · Challenges · Factor analysis · SMEs

1 Introduction

Small and medium-sized enterprises (SMEs) assume a significant part in the economy of the non-industrial nation. It helps in producing work in the nation, sends out from the nation, and mechanical yield. SMEs in non-industrial nations particularly like India produce a ton of incomes and give heaps of work openings both in metropolitan and country zones. SMEs are being characterized based on interest in plant and apparatus. Micro enterprises for manufacturing firms are those whose interest in

S. Mukherjee (✉) · V. Chittipaka · M. M. Baral
Department of Operations, GITAM Institute of Management, GITAM (Deemed to be University), Visakhapatnam, Andhra Pradesh, India

S. C. Srivastava
Department of Industrial and Production Engineering, Guru Ghasidas Vishwavidyalaya (A Central University), Bilaspur, Chhattisgarh, India

R. Agrawal et al. (eds.), *Recent Advances in Industrial Production*, Lecture Notes in Mechanical Engineering, https://doi.org/10.1007/978-981-16-5281-3_33

plant and hardware doesn't surpass 25 lakh rupees. Micro firms for administration enterprises are those whose interest in hardware ought not to be in excess of ten lakh. Small enterprises for assembling firms are those whose interest in the hardware ought to be more than 25 lakh and ought not to surpass five crores. Small enterprises for administration areas are those whose interest in hardware ought not to be in excess of ten lakh and ought not to surpass two crores. Medium enterprises for assembling firms are those whose interest in the gear ought to be in excess of five crores and ought not to surpass ten crores. Medium enterprises for administration areas are those whose interest in machines ought not to be in excess of two crore rupees and ought not to surpass five crore rupees.

The changing business environment and increasing competition has made an urge for the companies to adopt latest information technology in their process so that excellence can be achieved [1, 2]. A disruptive technology evolved in the market is Cloud computing (Cc) which has the capability of changing the business functions in and around the world [3]. Cc model groupings are based on registering prerequisites of the clients; essentially three sorts are there: Software as a Service (SaaS), Infrastructure as a Service (IaaS), and Platform as a Service (PaaS) [4]. The evolvement of Cc in the market has a great deal of potential in changing the general market circumstance by changing the registering experience [5].

Most of the SMEs in developing countries face lots of problems like lack of finance, low production ability, less knowledge in marketing products, absence of skilled workers, accessibility to old technologies and lack of knowledge for use of newer technologies [6]. It has been observed that the adoption rate of Cc in the supply chain (SC) for the large enterprises is faster in comparison to smaller enterprises [7]. There is no doubt that Cc can change the overall functioning structure of SMEs SC but still there is a less focus on adoption of Cc at the organizational levels [7, 8]. Despite the numerous advantages of Cc in the SC, its adoption in developing countries is very much less in comparison to developed countries. Main challenges are security, reliability, cost, integration and many more [5]. The aim of this paper is to exploring the challenges that the Indian SMEs are facing towards adopting Cc.

2 Literature Review

2.1 Overview of Cloud Computing

Cc has become an arising pattern in innovations and business presently [9]. As per the National Institute of Standard and Technology (NSIT), Cc has been characterized as a "... a model for empowering helpful on-request network admittance to a shared pool of configurable registering assets (e.g., networks, workers, stockpiling, applications, and administrations) that can be quickly checked and delivered with insignificant administration exertion or specialist co-op collaboration" [10]. SaaS gives programming arrangements and applications to clients to utilize cloud administrations running in the cloud framework through a customer interface, similar to an internet browser [11]. PaaS conveys a processing stage like equipment, working frameworks, or storerooms through a web association. IaaS virtual foundation of processing assets is as workers, equipment, and storerooms [12, 13]. Additionally, there are four sending models for cloud i.e. private cloud (PC), community cloud (COC), public cloud (PUC), and hybrid cloud (HC). PC is a sort of Cc model which are made sure about and just indicated clients can just work. COC is another kind of Cc model that can be constrained by various associations yet can be upheld by a comparable network that has normal strategies, mission, and security networks [10]. PUC is another sort of Cc model that can be worked by various customers all at once through the assistance of shared organizations. HC is the mix of at least two different kinds of cloud models (PC, COC, or PUC) that are comparative in innovation and permit perfection in information applications.

2.2 Cloud Computing Challenges in the Supply Chain Management of the SMEs

Cc can has some significant effects on SMEs running in India in comparison of ICT as the cost of ICT infrastructure are more in Cc adoption. Various software's or its application can be accessed with the help of Cc like Enterprise Resource Planning (EPR) Customer Relationship Management (CRM) can be made available to the SMEs. Another feature of Cc is pay per usage which means customers will pay according to their usage only makes it high adaptable [14]. The advantages of Cc adoption for SMEs are easy installations, easier way of access, automatic updates, back up facilities, scalability and flexibility. SMEs SC play a vital role in achieving operational excellence [15]. But with the advantages several concerns are also associated with Cc in the SC. Issues like data lost, infrastructure of information technologies, lock-in, loss governance, multi-tenancy, service level agreement, loss of backups, data privacy and many more [9, 16]. Here in this paper we have identified ten potential challenges.

Lock-In (LI) refers to the capability issues a cloud customer faces during the movement of the data's and programs away from the cloud service provider (CSP) to another [17]. LI can create several problems for the customers. Loss of Governance (LG) happens during the Cc adoption; the customer sometimes denies giving proper access to the CSP leading to some dispute. Compliance challenges (COMC) are the security accountability of the organizations, which needs to meet by the CSP before the installation process begins. In some cases CSP does not meet the security protocols of the company [18].

Malicious insiders (MI) in cloud might get access to important information and produce damage to cloud user [19]. When the organisations ask for the share platforms of the services, there may a risk arises of threats due to sharing of infrastructure, platforms and applications [20]. Resource and service management (RSM) means the capacity of CSP to fulfil the demand of the customers in providing the resources. The target of the administration provider in the present circumstance is to distribute and de-assign assets from the cloud. A service level arrangement (SLA) is the understanding that gives the clients ensures from the sellers on the administration gave by them. Yet, the issue lies in the assurance cycle of SLA particularization. Data loss or leakage (DLL) can occur for the two clients and organizations in a circumstance where malignant assailants gain admittance to the information put away in the cloud and it gets lost. Information can likewise be taken out by CSP incidentally or an actual spillage can prompt the lasting spillage of the customer's information [21]. Legal jurisdiction (LJ) implies cloud clients and CSP should know the laws and guidelines of the nation where it's being utilized [22]. Data privacy and protection (DPP) is one of the fundamental things which should be taken locked at by both CSP and the customers from their sides, so it would be beneficial for both the CSP and the customers to know the laws in regards to data privacy [23]. Licensing risk (LR) can lead to some serious problems in which customers may have to pay more to the CSP [24].

3 Research Methodology

The objective of the study was to validate the challenges that are being faced by SC in the SMEs in adopting Cc. From the review of literature challenges were identified for this study. A questionnaire was developed for the survey of the research.

3.1 Instrument Development

Exhaustive literature review was done to identify the challenges affecting the CC adoption in SC of SMEs. The questionnaire had been divided into two parts, first part comprises of the SMEs profile, type of business they are into it. The second part comprises of the challenges affecting the CC adoption in SMEs. The constructs were measured using a seven point Likert scale ranging from "strongly disagree to strongly agree".

3.2 Data Collection

The sampling technique that has been used here is simple random sampling as it is an accurate representation of the larger population and it eliminates bias by giving all individuals SMEs an equal chance to be chosen. Target population were mainly owner, director, and plant manager. Correctly filled questionnaires were obtained

Table 1 Demographics of the survey

S. No.	Characteristics	Percentage
I	*Total number of employees*	
A	1–9 employees	8
B	10–25 employees	16
C	26–50 employees	9
D	51–100 employees	20
E	101–150 employees	27
F	151–250 employees	15
G	251 and above	5
II	*Respondents current position*	
A	Owner	24
B	Director	30
C	Plant manager	46
III	*Type of organizations*	
A	Micro enterprises	33
B	Small enterprises	24
C	Medium enterprises	43
IV	*Big data adoption status*	
A	Already adopted	38
B	Intend to adopt in next 1 year	33
C	Intend to adopt in next 3 year	18
D	Do not intend to adopt	11

from the SMEs through mail after making follow up and rest of the questionnaire were either incomplete or were not completely filled. The questionnaire was sent to 600 SMEs in India out of which a total of 366 (61% response rate) correctly filled questionnaire were obtained (Table 1).

The SMEs in which survey was done were from manufacturing sectors mainly suppliers of the OEMs.

4 Results

Exploratory factor analysis (EFA) had been used to get the desired results of the study. The reliability test was conducted in the SPSS using Cronbach's alpha and the result obtained was 0.822.

4.1 Factor Analysis

Factor analysis, a multivariate procedure investigation, is utilized to bunch a more modest arrangement of factors [25]. It helps in acquiring a gathering of a more modest arrangement of uncorrelated factors [25]. The KMO esteem is 0.794 which is more prominent than 0.70 showing that there are adequate things for every segment. The essentialness level is 0.00 (under 0.05), showing that the relationship framework is fundamentally not the same as a character grid, where connections between factors are every one of the zero.

Table 2 shows the absolute difference clarified i.e. how the fluctuation is split between the 10 potential components. It was discovered that three parts have eigenvalues more noteworthy than one. Orthogonal rotation (varimax) had been utilized. This implies that the last factors will be at right points with one another. Subsequently, we can expect that the data clarified by one segment is autonomous of the data in the other part. The combined level of the multitude of three components separated was 66.883%. The percentage of total variance explained by component 1 (26.880%), component 2 (20.149%), component 3 (19.853%).

The components whose Eigen values are greater than 1 explain the total variance. Here all the 10 factors were clubbed into three components. Component 1 contains technical factors like MI, RSM, SLA, and DLL. Component 2 contains policy and organizational factors like LI, LG, and COMC. Component 3 contains all the legal factors like LJ, DPP, and LR (Table 3).

5 Managerial Implications

This study highlights the challenges that are being faced by the SMEs in adopting the CC. One of the major finding was that 38% of SMEs are using some of the features of SaaS platform while 33% of the SMEs are using PaaS platform. Major challenges that SMEs are facing while adopting CC were LI, LG, COMC, MI, RSM, SLA, DLL, LJ, DPP, and LR. These are serious issues which need to be addressed by the CSPs so that SMEs can use CC services in a smooth way without any hurdles. Cloud CSPs and government institutes which are responsible for promoting or providing cloud services needs to create a lot of awareness for CC adoption in the country. After doing factor analysis 10 factors or the challenges were clubbed together into three principal components factors. Component 1 contains technical factors like MI, RSM, SLA; DLL which needs to be addressed by CSP and the technical problems needs to be solved. Component 2 contains policy and organizational factors like LI, LG, and COMC which need to be addressed by the top management of organizations or its employees for a smooth running. Component 3 contains all the legal factors like LJ, DPP, and LR which needs to be addressed by the legal department of the organization or the cloud CSP so that there is legal compliance while using Cc.

Table 2 Total variance explained

Total variance explained									
Component	Initial eigenvalues			Extraction sums of squared loadings			Rotation sums of squared loadings		
	Total	% of variance	Cumulative %	Total	% of variance	Cumulative %	Total	% of variance	Cumulative %
1	3.953	39.526	39.526	3.953	39.526	39.526	2.688	26.880	26.880
2	1.558	15.578	55.105	1.558	15.578	55.105	2.015	20.149	47.030
3	1.178	11.778	66.883	1.178	11.778	66.883	1.985	19.853	66.883

Table 3 Rotated component matrix; extraction method: principal component analysis; rotation method: varimax with Kaiser normalization

Rotated component matrix			
	Component		
	1	2	3
LI		0.846	
LG		0.800	
COMC		0.704	
MI	0.770		
RSM	0.831		
SLA	0.841		
DLL	0.716		
LJ			0.836
DPP			0.834
LR			0.692

6 Conclusion

The potential and significance of SMEs is very much perceived in the creating and created nation. In any case, this area faces numerous difficulties in their framework particularly in the agricultural nations as the asset are restricted in contrast with created countries. Proper utilization of Cc technologies could change the face of these SMEs. Owners and employees know the potential benefits of Cc adoption. But SMEs are facing a lot of challenges in Cc adoption. And these challenges need to address by cloud CSP so that these problems gets solved. The study offers some of the major challenges SMEs are facing nowadays, which will be a help for the cloud CSPs to solve these challenges in a long run.

References

1. Pan MJ, Jang WY (2008) Determinants of the adoption of enterprise resource planning within the technology-organization-environment framework: Taiwan's communications industry. J Comput Inf Syst 48(3):94–102
2. Sultan N (2010) Cloud computing for education: a new dawn? Int J Inf Manage 30(2):109–116
3. Al-Hujran O, Al-Lozi EM, Al-Debei MM, Maqableh M (2018) Challenges of cloud computing adoption from the TOE framework perspective. Int J E-Bus Res (IJEBR) 14(3):77–94
4. Alshamaila Y, Papagiannidis S, Li F (2013) Cloud computing adoption by SMEs in the north east of England. J Enterp Inf Manage
5. Gangwar H, Date H, Ramaswamy R (2015) Understanding determinants of cloud computing adoption using an integrated TAM-TOE model. J Enterp Inf Manage
6. Shiralashetti AS (2012) Prospects and problems of MSMEs in India—a study. Int J Multidiscip Acad Res 1(2):1–7
7. Gupta P, Seetharaman A, Raj JR (2013) The usage and adoption of cloud computing by small and medium businesses. Int J Inf Manage 33(5):861–874

8. Alharbi F, Atkins A, Stanier C (2016) Understanding the determinants of cloud computing adoption in Saudi healthcare organisations. Complex Intell Syst 2(3):155–171
9. M'rhaoaurh I, Okar C, Namir A, Chafiq N (2018) Challenges of cloud computing use: a systematic literature review. MATEC Web Conf 200:00007
10. Mell P, Grance T (2011) The NIST definition of cloud computing
11. Modi C, Patel D, Borisaniya B, Patel A, Rajarajan M (2013) A survey on security issues and solutions at different layers of cloud computing. J Supercomput 63(2):561–592
12. Ibrahim S, He B, Jin H (2011) Towards pay-as-you-consume cloud computing. In: 2011 IEEE International conference on services computing. IEEE. pp 370–377
13. Qi H, Shiraz M, Liu JY, Gani A, Rahman ZA, Altameem TA (2014) Data center network architecture in cloud computing: review, taxonomy, and open research issues. J Zhejiang Univ Sci C 15(9):776–793
14. Rad BB, Diaby T, Rana ME (2017) Cloud computing adoption: a short review of issues and challenges. In: Proceedings of the 2017 international conference on e-commerce, e-business and e-government, pp 51–55
15. Kumar D, Samalia HV (2016) Investigating factors affecting cloud computing adoption by SMEs in Himachal Pradesh. In: 2016 IEEE international conference on cloud computing in emerging markets (CCEM). IEEE, pp 9–16
16. Sultan N (2011) Reaching for the cloud: how SMEs can manage. Int J Inf Manage 31(3):272–278
17. Sahandi R, Alkhalil A, Opara-Martins J (2013) Cloud computing from SMEs perspective: a survey based investigation. J Inf Technol Manag 24(1):1–12
18. Hamouda S (2012) Security and privacy in cloud computing. In: 2012 international conference on cloud computing technologies, applications and management (ICCCTAM). IEEE, pp 241–245
19. Iqbal S, Kiah MLM, Anuar NB, Daghighi B, Wahab AWA, Khan S (2016) Service delivery models of cloud computing: security issues and open challenges. Secur Commun Netw 9(17):4726–4750
20. Caldarelli A, Ferri L, Maffei M (2017) Expected benefits and perceived risks of cloud computing: an investigation within an Italian setting. Technol Anal Strateg Manag 29(2):167–180
21. Ramachandra G, Iftikhar M, Khan FA (2017) A comprehensive survey on security in cloud computing. Procedia Comput Sci 110:465–472
22. Senarathna I, Warren M, Yeoh W, Salzman S (2015) A conceptual model for cloud computing adoption by SMEs in Australia. In: Delivery and adoption of cloud computing services in contemporary organizations. IGI Global, pp 100–128
23. Attaran M, Woods J (2019) Cloud computing technology: improving small business performance using the Internet. J Small Bus Entrep 31(6):495–519
24. Alsharari NM, Al-Shboul M, Alteneiji S (2020) Implementation of cloud ERP in the SME: evidence from UAE. J Small Bus Enterp Dev
25. Hair JF, Black WC, Babin BJ, Anderson RE (2010) Multivariate data analysis: a global perspective. Pearson Education Inc.

Finite Element Analysis of Infill Density on the Compressive Strength of 3D Printed Parts by Fused Deposition Modelling

Anurag Kumar Mishra, Abhishek Kaushal, Rabesh Kumar Singh, and Anuj Kumar Sharma

Abstract 3D printing, one of the subsets of Additive Manufacturing (AM), has gained popularity not only at prototyping but also for commercial applications because of its capability to prototype the complex parts in one go. At present, the 3D printed parts provide solutions starting from some common requirements to high-end ones, such as aerospace, medical, and engineering applications. 3D printed parts that are used for such high-end applications are usually subjected tensile, compressive bending and torsional loading and may lead the failure in the component. To minimize such failure there should be understanding of the effect of process parameters over the mechanical behaviour of the 3D printed parts. In this paper, the compressive FEA analysis has been conducted to examine the effect of infill density over the compressive strength of 3D printed parts and the results are validated with experimental results of available literature. The standard specimen design as per ASTM-D695 has been drafted using SOLIDWORKS 2018 and the honeycomb infill structure has been created using ANSYS SPACECLAIM 19.1. Finite Elements Modelling (FEM) has been done with infill densities 0, 20, 30, 40 and 100% using ANSYS STRUCTURE 19.1. Compressive strength increases with the increase of infill density the sample with 0 and 100% infill densities shows the minimum and maximum compressive strength. Results are validated with the experimental results and maximum deviation between experimental and finite element analysis has been recorded at 5.94%.

Keywords 3D printing · Finite element analysis · FDM · Compressive strength · Honeycomb

1 Introduction

The recent technological advancement in 3D Printing technologies made it a versatile tool for prototyping of complex parts. Manufacturing of complex parts via conventional, subtractive manufacturing processes usually takes several steps leading to

A. K. Mishra · A. Kaushal · R. K. Singh (✉) · A. K. Sharma
Centre for Advanced Studies, Dr. A.P.J. Abdul Kalam Technical University, Lucknow, Uttar Pradesh 226031, India

R. Agrawal et al. (eds.), *Recent Advances in Industrial Production*, Lecture Notes in Mechanical Engineering, https://doi.org/10.1007/978-981-16-5281-3_34

wastage of material and increment in the production time. Additive manufacturing (AM) technologies can prototype complex geometrical bodies just from CAD file by layering mechanism reducing material wastage and intermediate manufacturing sub-steps resulting in reduced production time and cost [1]. AM technologies have become popular among engineers and researcher because of their versatile capabilities like customizability, cost-effectiveness, single-step prototyping resulting in higher production efficiency [2]. Because of such capabilities, AM technologies provide solutions for some basic household requirements to high-end requirements of biomedical and aeronautical industries [3, 4].

At present based on layering mechanism seven types of the AM technologies are available in the market namely binder jetting, material jetting, direct energy deposition, material extrusion, powder bed fusion, sheet lamination and vat polymerization [5]. Based on materials, built area, printing speed, layer fusion, print finish each prototyping technique has its limits and capabilities. Fused depositing modelling (FDM) based on material extrusion is one of the most accepted and popular AM technique among available prototyping techniques because of its simplicity and affordability but having the problem of anisotropic and weak bonding between layers that may lead failure of the printed part under mechanical loading environment [6]. Therefore, there is a need for a comprehensive study of printing parameters affecting mechanical behavior of the printed parts. There are several print parameters that affect the mechanical behavior of print objects such as type of infill, infill density, print orientation, raster angle, print bed temperature and printing speed [7]. It is a very exhaustive task to study the effect of all these parameters as there will be a large number of combination by taking all these parameters. Therefore, FEM analysis may be an efficient tool to perform studies to optimize the printing parameters, but there may be differences between actual and simulated results as FEM model consider the material as isotropic while FDM printed parts are highly anisotropic [8].

The axial loads which try to compress the body can be termed as the compressive loads. Most of the time components are subjected to compressive loads that can cause failure, so the number of researchers investigated the effect of process parameters over the compressive strength of the 3D printed parts. Domínguez-Rodríguez et al. [9] has investigated the effect of infill pattern, infill density and infill orientation of FDM printed ABS specimen and concluded that the specimen which was printed along longitudinal direction showed higher stuffiness than those were printed along the transverse direction, also concluded that honeycomb structure showed better response than rectilinear under compressive loading also with an increment of infill density the stiffness also increases. Abbot et al. [10] has performed experimental as well as FE analysis to find the effect of layering orientation and density and found the compressive strength is highly dependent on layering orientation and the FEA results show a greater deviation than experimental results. Divyathej et al. [11] has researched to investigate the effect of layer thickness and also performed the comparative study between FDM printing and injection moulding concluding that FDM printed parts exhibits better properties than moulded one. Hernandez et al. [12] conducted the study to examine the effect of print orientation at compressive strength of print orientation. For such study, they printed specimen at (0°, 45° and

90°) and observed maximum compressive strength at 90° orientations. Abeykoon et al. [13] performed a study to examine the effect of infill density and infill structure and concluded that the specimen with 100% infill density gives maximum strength and linear infill pattern shows best response among diamond, cat fill, Hilbert and hexagonal. Dev and Srivastava [14] has researched to optimize the print orientation, type of infill, and infill density and concluded that gyroid with 80% fill density printed at 45° has better print quality. Further, study has performed to optimize the infill pattern and found that triangular infill pattern has the highest strength in comparison of Grid and cubic. FEA is the mathematical tools which virtually analyses the behaviour of engineering components under different conditions. FEA tools are capable of simulating the problems such as static structural, fluid behaviour, thermal behaviour or combination of more than two analyses. FEA techniques divide the whole component into large numbers of finite units and then investigates the response of these bodies under applied boundary conditions. Thus FEA techniques can provide an alternative solution to the exhaustive, time consuming experimental works.

In this paper finite element modelling has been performed to evaluate the effect of infill Density on the Compressive Strength of 3D printed parts. The Simulated Results are validated with the experimental literature available.

2 Materials and Methodology

The study of the effect of infill density on compressive strength of FDM printed acrylonitrile butadiene styrene (ABS) material, a total of five models has been studied taking honeycomb infill pattern with densities ranging from 0, 20, 30, 40 and 100% infill densities into consideration. The CAD files as per ASTM-D695 standards have been generated using SOLIDWORKS 2018 (Fig. 2a). The specified infill structure has been generated with the help of shelling operation using ANSYS-Spaceclaim the sliced body and cross-sectional view of the sliced body (Fig. 2b, c) respectively. FEM modelling has been done using ANSYS 19.1 structure with similar boundary condition applied by Domínguez-Rodríguez et al. [9]. The simulation results have been validated with experimental results presented by Domínguez-Rodríguez et al. [9] (Fig. 6). The present study has been conducted into three major parts (Fig. 1).

Acrylonitrile butadiene styrene commonly known as ABS is an amorphous thermoplastic polymer of Acrylonitrile, butadiene and styrene. It is capable of providing solutions for the applications where mechanical properties are a prior requirement because of its durability, and resistance to the higher temperature. Flexibility and electrical insulation properties are some other properties that make it a good choice as 3D Printing material. Mechanical properties of the ABS are listed in Table 1.

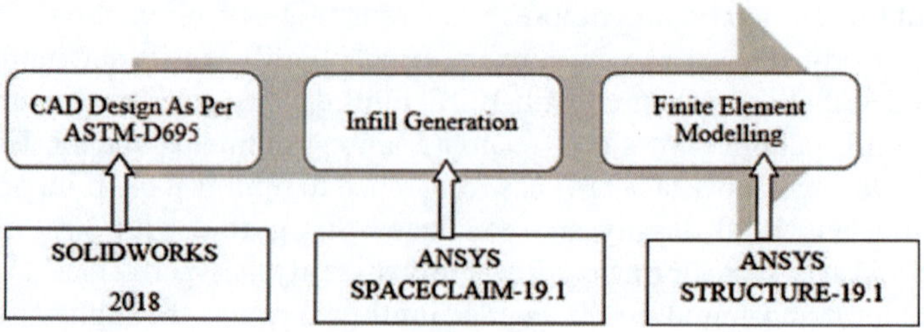

Fig. 1 Scheme of design, modelling and simulation of 3D printed parts

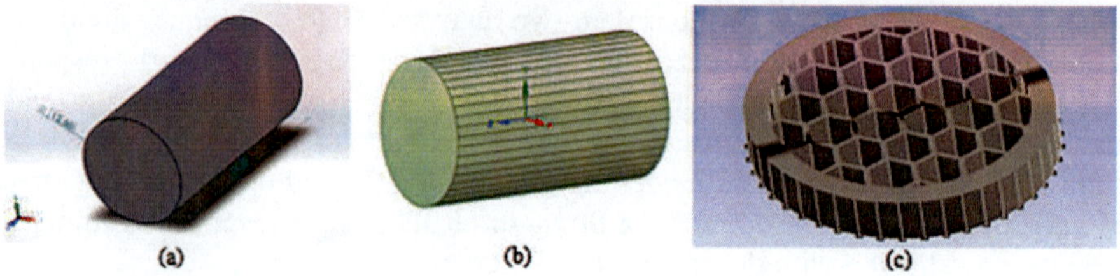

Fig. 2 **a** CAD design, **b** sliced CAD design, **c** cross sectional view of sliced body

Table 1 Mechanical properties of ABS

Property	Value	Unit
Density	2900	kg/m^3
Young's modulus	59.2	MPa
Poison ratio	0.197	Unit less
Bulk modulus	32.56	MPa
Shear modulus	24.72	MPa
Tensile yield strength	13.6	MPa
Tensile ultimate strength	13.6	MPa

3 Results and Discussion

Finite element analysis is performed on the ASTM D695 standard specimen and the meshing performed during the analysis (Fig. 3a), while the applied boundary conditions (Fig. 3b).

One of the surfaces is kept fixed and displacement has been applied at another face along the z-axis perpendicular to the infill which has been generated along the x–y plane. The reference deformation applied to each specimen taken from the literature available [9].

The stress-induced (Fig. 4b) along the body, the maximum simulated stress 16.33 MPa induced is quite similar to the first peak stress 16.33 MPa at 4.4% of strain

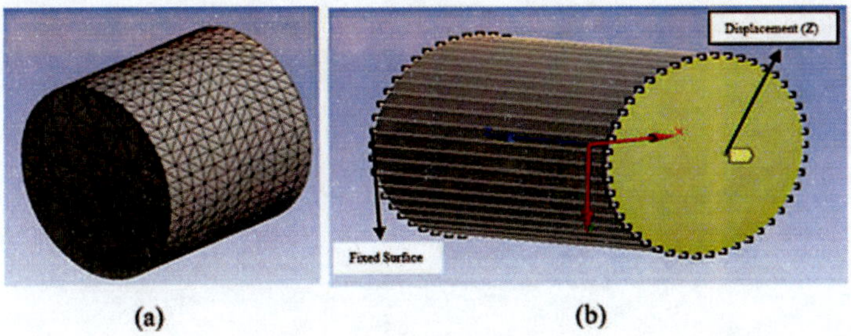

Fig. 3 **a** Meshing diagram, **b** boundary conditions

from the experiments performed by Domínguez-Rodríguez et al. [9]. Other specimens showed the response very similar to first. The response of the hallow specimen at 0% infill density (Fig. 4a) against the compressive strain of 3% and the peek stress has been observed 8.74 MPa. The stress behavior in the body (Fig. 4c) having 30% infill density against the strain of 4.47% with the highest stress of 23.75 MPa. Stress distribution in the body (Fig. 4d) for 40% infill density at 4.9% strain with the highest stress of 30.28 MPa. The stress in the completely solid specimen that is at 100% infill density (Fig. 4e) against 5% strain and the maximum induced compressive stress has been observed 46.49 MPa.

The maximum compressive stresses developed (Fig. 5) at mentioned infill densities. As it can be observed from Fig. 6 stress increases linearly with increase in percentage infill densities until 40% but after 40% infill density there is a sudden increase in induced stress.

The comparison between maximum induced stresses (Fig. 6) computed by Finite element analysis and experimental results by available literature [9]. In the article axial compressive testing has been performed to analyze the effect of print orientation, infill pattern and infill density. As per standards ASTM-D695 testing procedures, he printed the specimens having 12.5 mm diameter and 25 mm length then axial loading with crosshead speeds of 1.33 mm/m using Introns 8802 universal testing machine, and recorded the peek stresses in different specimens at a different percentage of strain. In this article similar conditions were applied in finite element modelling. The simulated results are very close to experimental results the percentage error between experimental and simulated results is observed in the range of 5.94 to −5.87% while maximum error occurred for hallow specimen.

4 Conclusion and Future Scope

In the current work effect of infill density over compressive strength has been studied using finite element analysis and the following points concluded from the work:

- The increment in the infill density resulted in the increment in induced stress.

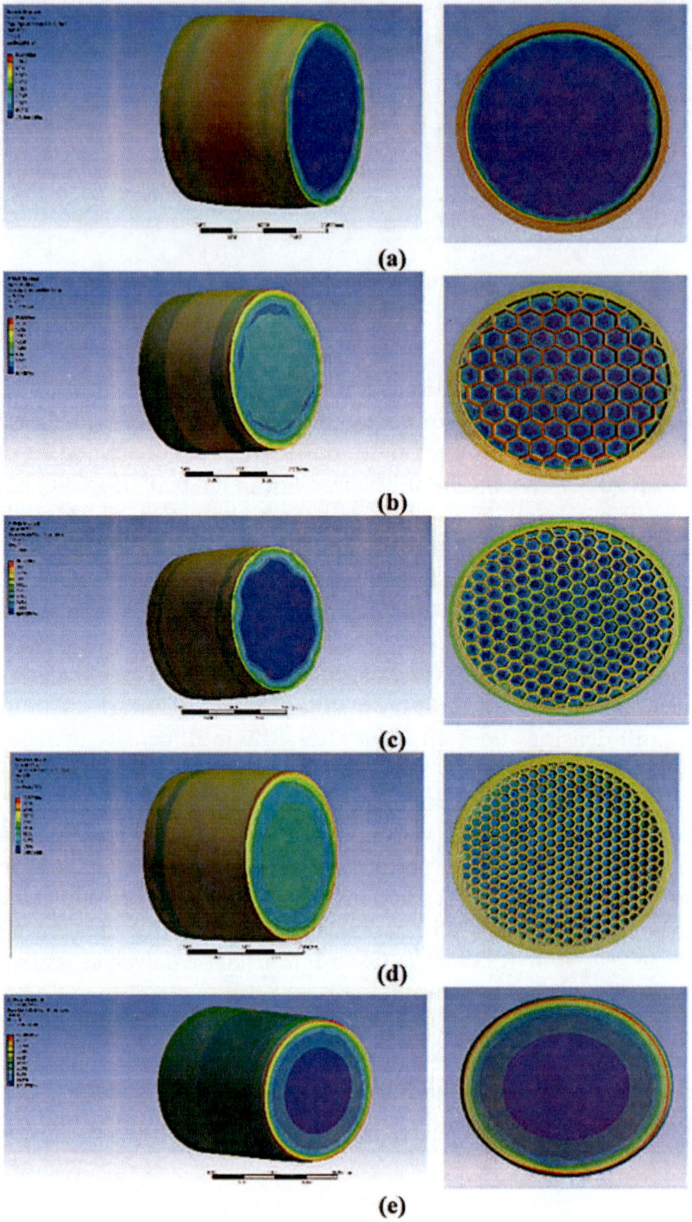

Fig. 4 Stress distribution **a** honeycomb (0%), **b** honeycomb (20%), **c** honeycomb (30%), **d** honeycomb (40%), **e** honeycomb (100%)

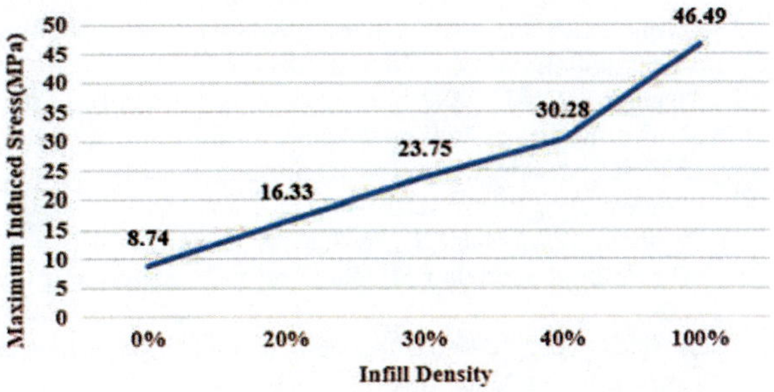

Fig. 5 Maximum stress induced (simulated) for 0, 20, 30, 40 and 100%

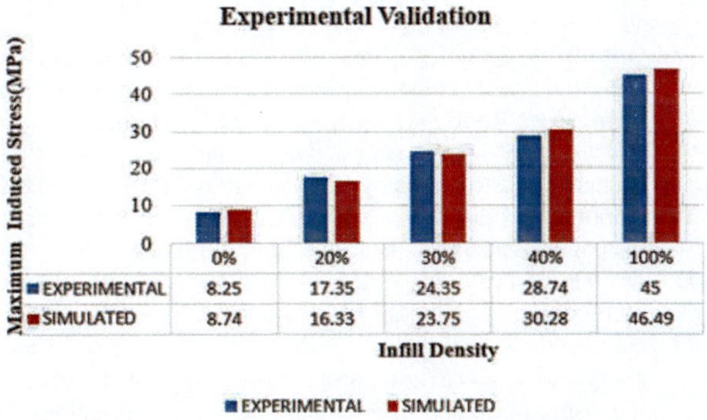

Fig. 6 Experimental versus simulated stress

- The hollow body (0% infill density) and complete solid body (100% infill density) showed minimum and maximum compressive strength respectively.
- The Strength increases proportionally till the 40% infill density and after that a sudden increment in the strength of the body observed.
- The results computed from Finite Element analysis are well close to experimental results. From the validation 5.94%, −5.87%, −2.46%, 5.36% and 3.31% of deviation of in compressive strength has been recorded at 0%, 20%, 30%, 40% and 100% of infill densities respectively.

The study can be further extended in FEM analysis of 3D printed parts to study the effect of various process parameters such as infill structure with various infill densities under tensile, flexural, torsional and fatigue loading environment. Multi-physics FEM analysis can be used to study the cumulative effect of layer fusion and layer orientation. The same studies can be performed for other printing technologies.

References

1. Srinivasan R, Pridhar T, Ramprasath LS, Sree Charan N, Ruban W (2020) Prediction of tensile strength in FDM printed ABS parts using response surface methodology (RSM). Mater Today Proc 27:1827–1832

2. Dizon JRC, Espera AH, Chen Q, Advincula RC (2018) Mechanical characterization of 3D-printed polymers. Addit Manuf 20:44–67
3. Ahangar P, Cooke ME, Weber MH, Rosenzweig DH (2019) Current biomedical applications of 3D printing and additive manufacturing. Appl Sci 9
4. Joshi SC, Sheikh AA (2015) 3D printing in aerospace and its long-term sustainability. Virtual Phys Prototyp 10:175–185
5. Swetham T, Madhana K, Reddy M, Huggi A, Kumar MN (2017) A critical review on of 3D printing materials and details of materials used in FDM. Int J Sci Res Sci Eng Technol 2:353–361
6. Es-Said OS et al (2000) Effect of layer orientation on mechanical properties of rapid prototyped samples. Mater Manuf Process 15:107–122
7. Kuznetsov VE, Tavitov AG, Urzhumtsev OD, Mikhalin MV, Moiseev AI (2019) Hardware factors influencing strength of parts obtained by fused filament fabrication. Polymers (Basel) 11
8. Gopsill JA, Shindler J, Hicks BJ (2018) Using finite element analysis to influence the infill design of fused deposition modelled parts. Prog Addit Manuf 3:145–163
9. Domínguez-Rodríguez G, Ku-Herrera JJ, Hernández-Pérez A (2018) An assessment of the effect of printing orientation, density, and filler pattern on the compressive performance of 3D printed ABS structures by fuse deposition. Int J Adv Manuf Technol 95:1685–1695
10. Abbot DW, Kallon DVV, Anghel C, Dube P (2019) Finite element analysis of 3D printed model via compression tests. Procedia Manuf 35:164–173
11. Divyathej MV, Varun M, Rajeev P (2016) Analysis of mechanical behavior of 3D printed ABS parts by experiments. Int J Sci Eng Res 7:116–124
12. Hernandez R, Slaughter D, Whaley D, Tate J, Asiabanpour B (2016) Analyzing the tensile, compressive, and flexural properties of 3D printed ABS P430 plastic based on printing orientation using fused deposition modeling. In: Solid freeform fabrication 2016: proceedings of the 27th annual international solid freeform fabrication symposium—an additive manufacturing conference SFF 2016, pp 939–950
13. Abeykoon C, Sri-Amphorn P, Fernando A (2020) Optimization of fused deposition modeling parameters for improved PLA and ABS 3D printed structures. Int J Light Mater Manuf 3:284–297
14. Dev S, Srivastava R (2019) Experimental investigation and optimization of FDM process parameters for material and mechanical strength. Mater Today Proc 26:1995–1999

An Experimental Investigation of Ribbed Solar Air Heater—Free Convection

Niraj Kumar, Manoj Kumar Singh, Vinod Singh Yadav, Vineet Singh, and Anurag Maheswari

Abstract In this study, we have designed and developed a ribbed type solar air heater on the basis of theoretical studies. After the successful development of solar air heater, we have carried out some experimental result like outlet temperature and velocity of air. On the basis of experimental study, we have found that the temperature improvement is about 30–40 °C at free convection. The impact of temperature on mass flow rate, heat transfer rate and mass flow rate on pressure drop, friction factor, Reynolds number and Nusselt number is investigated. A significant enhancement is found in Thermal efficiency due to the ribs.

Keywords Solar air heater · Ribs · Mass flow rate · Thermometer

Nomenclature

A	Surface area (m^2)
M	Mass flow rate (kg/s)
T	Temperature (°C)
V	Velocity (m/s)
d_e	Equivalent diameter (m)
L	Length of absorber plate (m)
Q	Heat transfer rate (W)

N. Kumar (✉) · M. K. Singh · V. Singh · A. Maheswari
Department of Mechanical Engineering, F.E.T., M.J.P. Rohilkhand University, Bareilly, Uttar Pradesh, India

V. S. Yadav
Department of Mechanical Engineering, National Institute of Technology, Uttarakhand, Srinagar, India

R. Agrawal et al. (eds.), *Recent Advances in Industrial Production*, Lecture Notes in Mechanical Engineering, https://doi.org/10.1007/978-981-16-5281-3_35

Greek

Υ	Kinematic viscosity (m^2/s)
F	Friction coefficient
δP	Pressure loss (N/m^2)
P	Density (kg/m^3)
R_e	Reynolds number
N_u	Nusselt number
C_p	Specific heat of air (kJ/kg K)

Indices

In	Inlet (T_{in} = temperature of absorber plate at inlet)
Out	Outlet (T_{out} = temperature of absorber plate at outlet)

Abbreviation

RSAH	Ribbed solar air heater
MFR	Mass flow rate

1 Introduction

Now-a-days the whole world is suffering from the high demand of energy in every area such as Industries, agriculture, refineries etc. The maximum demand of energy is fulfilled by the fossil fuels (Petrol and Diesel) and these fuels are responsible for environment pollution and dangerous for human also. The source and process of fossil fuels are limited and costly respectively. So, to counter the requirement of fossil fuels and to minimize the pollution level some researchers are working on the renewable energy resources from decades. In renewable energy, solar energy is easily available source and this source generates hot air and this air can be used for various required applications. Some researchers have been worked on air heater based on solar energy are given below:

Forson et al. [1] studied the single-chamber air heater and double-chamber air heater and calculated the performance of those heaters with the assistance of air mass stream rate and number of absorber plates. Mittal et al. [2] compared the effectiveness of smooth surface solar based heater with rough surface solar heater and found by using inclined ribs inside the pips the efficacy of the system increases. Ramadan et al. [3] studied the twofold pass system solar based heater using limestone and gravel and contrasted with packed bed FCSAH execution and found that the air mass stream rate was progressed to 0.05 kg/s and high thermohydraulic proficiency was accomplished. Saini and Verma [4] examined the effect of different parameters on solar air heater such as, roughness stature and circular segment point on N_u and friction coefficient at different R_e varies from 2000 to 17,000. Esen [5] examined the proficiency of a FCSAH using obstacles upon the absorber plate. The efficiency of FCSAH was expanded from 38 to 45%. Kumar et al. [6] investigated friction and heat transfer characteristics of a sun-oriented heater with separated W-formed roughness on wall of a solar heater at a perspective proportion 1:8. In this study three parameters were considered namely incident angle, roughness of absorber plate and Reynolds Number. Lanjewar et al. [7] studied the upstream and downstream flow of air on w-molded ribs rectangular channel. It was seen that, the effect of downstream flow was high than the up-stream flow on the performance of rectangular channel. Chabane et al. [8] reported that heat transfer was found to increase by using fins on the absorber plate in different shape and orientation. Nwosu [9] studied FCSAH utilizing pin-balance absorber plate, pin-blades type absorber plate were having impressive impact on heat transfer rate the flat a plate. Tanda [10] reported that the performance of ribbed based roughened channels was found high as compared to smooth channel for low to medium Reynolds numbers. Chabane et al. [11] studied the effectiveness of solar air heater using genetic algorithm at different parameters such as, glass plate and plate emissivity. Bouadila et al. [12] experimentally found that the efficiency of a FCSAH outfitted with paraffin wax as heat stockpiling materials. On the basis of results, it was found that the energy efficiency shifted from 32 to 45% with a normal estimation of 40%. While, the exergy proficiency fluctuated somewhere in the range of 13 and 25% with a normal estimation of 22%. Saxena et al. [13] studied the effect of 300-W halogen lights inside the tubes of solar heater and found the efficiency of solar heated was increased. Abed et al. [14] reported the effectiveness of a hybrid solar collector system and found that the system was highly efficient under the two conditions namely: (i) High flow rate of air, (ii) clear sky (No clouds/obstructions). Ansari and Bazargan [15] studied the impact of ribs on flat plate SAH found that the efficiency was increased by 9% at lower mass flow rate [16]. The exhibition of new plan of the SAH absorber comprising of nearby tubes (TSAH) was explored experimentally at different air speeds and found that heat loss of TSAH were lowest. Saleh et al. [17] examined experimentally the exhibition of double pass SAH of new planned absorber plate (TSAH) at various channel air Mass Flow Rates through the SAH, and demonstrated that TSAH has more air temperature, more prominent power, higher efficiency and lower heat misfortunes as contrasted with FSAH.

2 Analysis of Solar Air Heater

Some mathematical equations and empirical relations are used to find out the mass flow rate, R_e, N_u, (f) and pressure drop (δP).

Reynolds number is given by

$$Re = (Vd_e/\upsilon) \tag{1}$$

Nusselt number is given by (Kays equation)

$$N_u = 0.0158R_e^{0.8} \tag{2}$$

Friction coefficient is given by (Blasius Equation)

$$f = 0.079R_e^{-0.25} \tag{3}$$

Pressure drop is given by

$$\delta P = 2f\rho LV^2/d_e \tag{4}$$

Heat Transfer Rate

$$Q = mC_p(T_{out} - T_{in}) \tag{5}$$

3 Experimental Setup

A RSAH is designed for the experimental findings of some important parameters such as; outlet temperature and velocity of air. The length of the RSAH is 1.8 m and the width of the heater is 0.92 m. The total area is about 1.656 m^2. The total 94 ribs are used to improve the temperature of exit air. A black painted tin sheet of 1 mm thick is used in this set up as a absorber plate. A sheet of glass is used above the absorber plate with thickness 5 mm to improve the temperature of exit air. Experimental set up is shown in Fig. 1. The height of the ribs are 2.54 cm and width are 5 cm.

4 Results and Discussions

In this section we have drawn some graph on the basis of experimental reading and mathematical equations. The following parameters are considered for draw the graphs.

(a) Outlet temperature (b) Velocity of air (c) Mass Flow Rate (d) R_e (e) N_u (f) f and (g) δP

In Fig. 2: X-axis shows outlet temperature and Y-axis shows mass flow rate. From the graph it is concluded that MFR of air increases as the outlet temperature increases.

In Figs. 3, 4 and 5: X-axis shows R_e, N_u and pressure drop and Y-axis shows MFR. It can be seen that all the parameters are increases as the MFR increases.

Fig. 1 Experimental set up

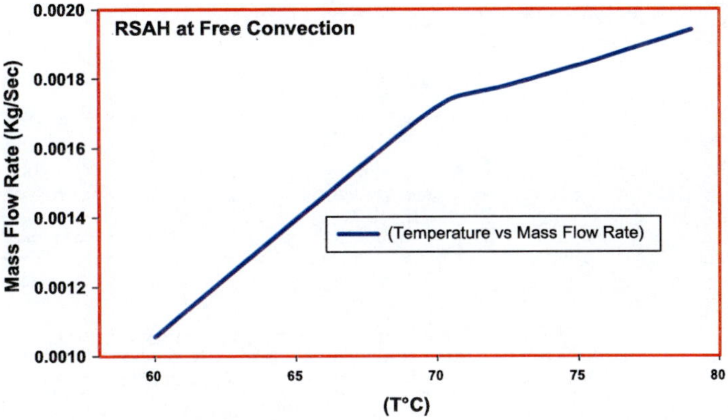

Fig. 2 Temperature versus mass flow rate

In Fig. 6: X-axis indicates friction factor and Y-axis indicates MFR of air. It can be clearly seen that the friction factor decease as the MFR increases.

In Fig. 7: X-axis indicates temperature and Y-axis indicates heat transfer rate. Heat transfer rate increase linearly as the temperature increases.

In Fig. 8: X-axis indicates time and Y-axis indicates solar flux. In this graph shows variation of solar flux with time.

In Fig. 9: Shows the variation of thermal efficiency w.r.t. various time of the day. From the plot it can be clearly seen the thermal efficiency is increasing as the temperature increasing in the following day. It is also observed from the trend after 1:00 p.m. temperature is decreasing the thermal efficiency still continue to increase this shows that the ribs incorporated on the plate store some solar energy in the form of internal energy.

Fig. 3 Reynolds number versus mass flow rate

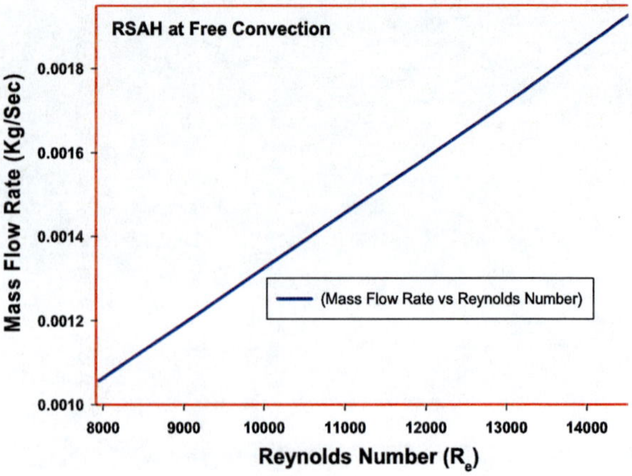

Fig. 4 Nusselt number versus mass flow rate

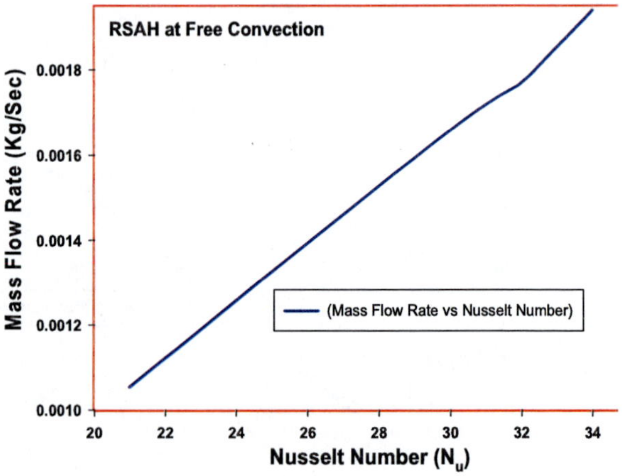

Fig. 5 Pressure drop versus mass flow rate

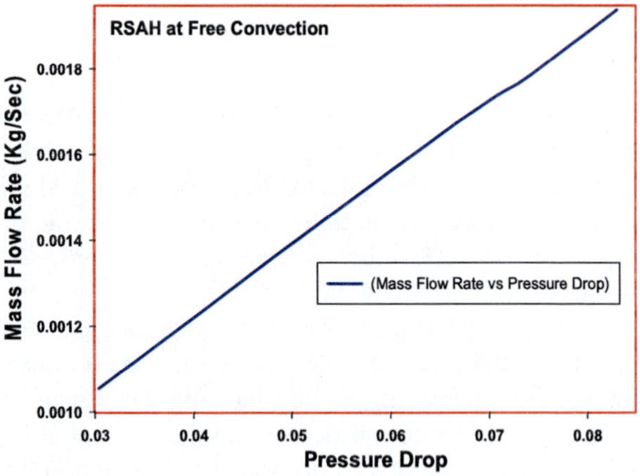

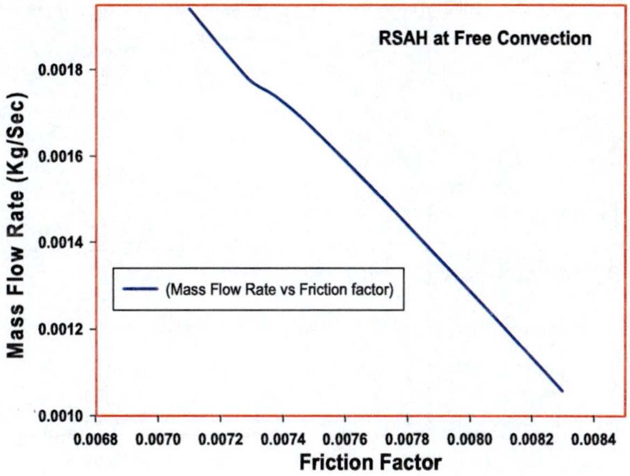

Fig. 6 Friction factor versus mass flow rate

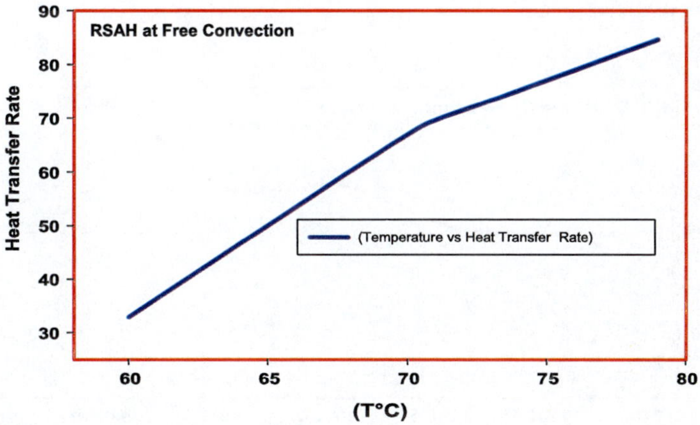

Fig. 7 Temperature versus heat transfer rate

5 Conclusion

An experimental study is carried out and some conclusions are made:

(i) MFR is directly dependent to the outlet temperature in RSAH at free convection.
(ii) It is concluded that MFR influences the Reynolds number, Nusselt number and Pressure drop in RSAH at free convection.
(iii) MFR is inversely proportional to the friction factor in RSAH at free convection.
(iv) Heat transfer rate majorly depends upon temperature in RSAH at free convection.
(v) Thermal efficiency is increasing due to the ribs provides on the absorber plate in RSAH at free convection.

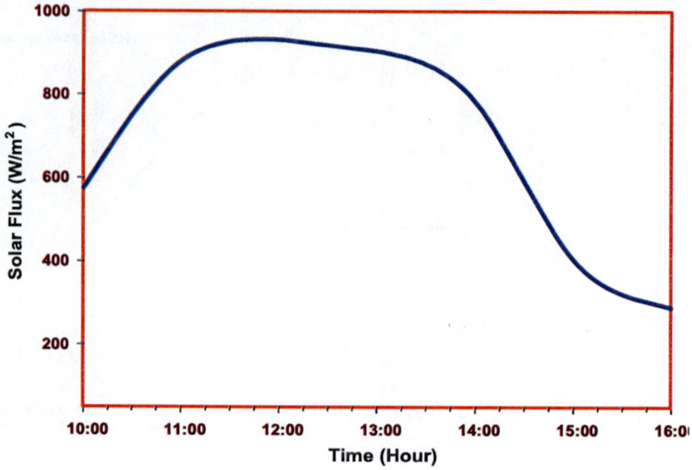

Fig. 8 Time versus solar flux

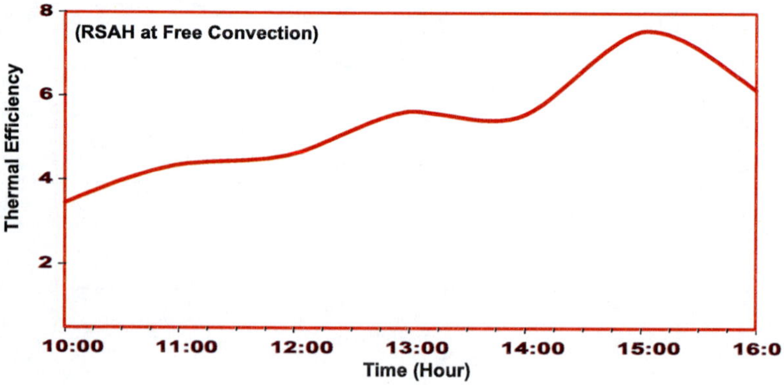

Fig. 9 Time versus thermal efficiency

Acknowledgements This work was supported by the collaborative project scheme (CRS) fund under NATIONAL PROJECT IMPLEMENTATION UNIT (NPIU) (A Unit of MHRD, Govt. of India for Implementation of World Bank Assisted Projects in Technical Education) [CRS Project ID: 1-5728003471].

References

1. Forson FK, Nazha MA, Rajakaruna H (2003) Experimental and simulation studies on a single pass, double duct solar air heater. Energy Convers Manage 44(8):1209–1227

2. Mittal MK, Varun, Saini RP, Singhal SK (2007) Effective efficiency of solar air heaters having different types of roughness elements on the absorber plate. Energy 32:739–745
3. Ramadan MRI, El-Sebaii AA, Aboul-Enein S, El-Bialy E (2007) Thermal performance of a packed bed double-pass solar air heater. Energy 32:1524–1535
4. Saini RP, Verma J (2008) Heat transfer and friction factor correlations for a duct having dimple-shape artificial roughness for solar air heaters. Energy 33:1277–1287
5. Esen H (2008) Experimental energy and exergy analysis of a double flow solar air heater having different obstacles on absorber plates. Build Environ 43:1046–1054
6. Kumar A, Bhagoria JL, Sarviya RM (2009) Heat transfer and friction correlations for artificially roughened solar air heater duct with discrete W-shaped ribs. Energy Convers Manage 50:2106–2117
7. Lanjewar AM, Bhagoria JL, Sarviya RM (2010) Heat transfer enhancement in solar air heater. Indian J Sci Technol 3(8):908–910
8. Chabane F, Moummi N, Benramache S, Bensahal D, Belahassen O, Lemmadi Z (2010) Thermal performance optimization of a flat plate solar air heater using genetic algorithm. Appl Energy 87:1793–1799
9. Nwosu NP (2010) Employing exergy-optimized pin-fins in the design of an absorber in a solar air heater. Energy 35:571–575
10. Tanda G (2011) Performance of solar air heater duct with different types of ribs on the absorber plate. Energy 36:6651–6660
11. Chabane F, Moummi N, Benramache S (2014) Experimental study of heat transfer and thermal performance with longitudinal fins of solar air heater. J Adv Res 5:183–192
12. Bouadila S, Lazaar M, Skouri S, Kooli S, Farhat A (2014) Energy and exergy analysis of a new solar air heater with latent storage energy. Int J Hydrogen Energy 39:15266–15274
13. Saxena A, Srivastava G, Tirth V (2015) Design and thermal performance evaluation of a novel solar air heater. Renew Energy 77:501–511
14. Abed QA, Badescu V, Soriga I (2017) Performance of a hybrid solar collector system in days with stable and less stable radiative regime. Int J Sustain Eng. https://doi.org/10.1080/19397038.2017.1333542
15. Ansari M, Bazargan M (2018) Optimization of flat plate solar air heater with ribbed surfaces. Appl Therm Energy. https://doi.org/10.1016/j.applthermaleng.2018.02.099
16. Hassan H, Saleh AE, El-Dosoky MF (2019) An experimental investigation of the performance of new design of solar air heater (tubular). Renew Energy. https://doi.org/10.1016/j.renene.2019.11.112
17. Saleh AE, Hassan H, El-Dosoky MF (2020) Study of the performance of double pass solar air heater of a new designed absorber: an experimental work. Sol Energy 198:479–489

Insolation Effect on Solar Photovoltaic Performance Parameters

Navneet, Neha Khuran, and Smita Pareek

Abstract Solar photovoltaic converts sunlight into electrical energy. Now days, it is quite popular for domestic purpose due to availability of sun insolation in abundance. But the atmospheric conditions vary depending upon depending upon the location. Therefore, it is important to know and predict the behaviour of the solar photovoltaic under different atmospheric condition. This paper studies the effect of insolation, one of the atmospheric parameters, on solar photovoltaic output. Two modules, each contains of 54 cells in series and parallel, is considered. The sun insolation is varied on these modules and solar photovoltaic characteristics i.e. current–voltage (I–V) and power–voltage (P–V) are observed. Preliminary observed results show that series connected modules are less sensitive to current and power if the different insolation is applied on both modules compare to parallel connected modules. But if the same insolation is applied on both the modules then current and power increases with the insolation intensity on the solar panel both in series and parallel. Also, the short circuit current varies in series and parallel connected modules with insolation variation, but open circuit voltage is less affected.

Keywords Cell · Modules · Solar photovoltaic · Insolation · Series · Parallel

1 Introduction

Electricity can be generated by numerous ways such as thermal, hydro, nuclear etc. But out of these numerous ways, solar energy is considered to be one of the best alternate due to adverse environment effect and it is green energy [1]. Solar energy uses solar photovoltaic cells which converts solar light energy into electrical energy [2]. In current times, solar energy is used in various fields such as transportation, domestic and agriculture usage, streetlights etc. Moreover, solar energy can be a standalone source, or it can be grid-connected [3–5]. The photovoltaic power usage

Navneet (✉) · N. Khuran
Department of Electrical Engineering, MDU, Rohtak, India

S. Pareek
Department of Electrical and Electronics Engineering, BKBIET, Pilani, Rajasthan, India

R. Agrawal et al. (eds.), *Recent Advances in Industrial Production*, Lecture Notes in Mechanical Engineering, https://doi.org/10.1007/978-981-16-5281-3_36

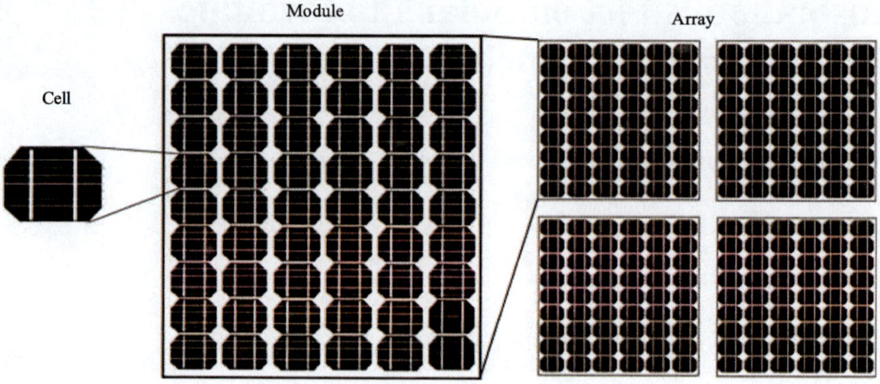

Fig. 1 Generic image of solar panel showing the photovoltaic hierarchy

is increasing day by day mainly due to increasing solar cells efficiency, and improved manufacturing technology. Also, different kinds of materials are used to manufacture the solar panels which are mainly divided into three categories [6] i.e. crystalline silicon, thin film, and multijunction cells. Cell is the smallest unit of a photovoltaic system and the power generated by this cell is not enough to meet the high-power requirements. Therefore, these cells are connected either in series, parallel or series–parallel to meet the high-power requirement. Once, these cells are interconnected then it is termed as module. Further, modules are connected to form an array. Figure 1 shows hierarchy of a PV cell with mono-crystalline structure.

There are many factors that affecting the solar photovoltaic performance like air mass effect [7], dust and dirt [8–10], soil [11], temperature [12] etc. But the sun insolation is the only source and most affecting the solar photovoltaic performance. Therefore, it is important to know and predict the solar power output based on the insolation [13, 14]. A simulation study is performed with sun insolation variation and the effect on characteristic parameters such as opens circuit voltage (V_{oc}), short circuit current (I_{sc}), and maximum power (P_{max}) is observed.

2 Modeling of Photovoltaic Cell/Module/Array

The equivalent circuit diagram of a solar photovoltaic is shown in Fig. 2.

Solar photovoltaic cells generate current due to light. This light current is linearly dependent on the solar insolation and has the influence of temperature, which is mentioned in Eq. (1).

$$I_{pv} = \left(I_{pv,n} + K_I \Delta T\right)\frac{G}{G_n} \tag{1}$$

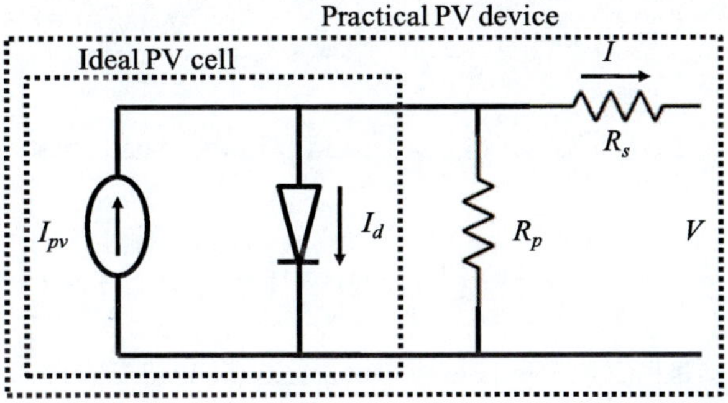

Fig. 2 Equivalent circuit of practical photovoltaic cell

$$\Delta T = T - T_n$$

Here I_{pv} is the current due to light, $I_{pv,n}$ is the current at nominal condition due to light (in A). 'K_I' is current coefficient (A/K), ΔT is change in temperature where T is the actual temperatures [in K] and T_n is nominal temperatures [in K], G is the insolation and G_n is the nominal insolation on the surface of the device [W/m^2]. Also, temperature dependent saturation current, I_o, is defined below by Eq. (2):

$$I_o = (I_{SC,n} + K_I \Delta T) / \left(\exp\left(\frac{(V_{OC,n} + K_v \Delta T)}{aV_t} \right) - 1 \right) \quad (2)$$

where $I_{SC,n}$—short circuit current at nominal condition (in A),
K_I—Current co-efficient,
ΔT—Change in temperature,
$V_{OC,n}$—Open circuit voltage at nominal condition,
K_v—Voltage coefficient,
a—ideality constant of diode,
V_t—The thermal voltage is given by equation:

$$V_t = \frac{N_s kT}{q}$$

N_s is series connected cells, Boltzmann constant (k) is 1.381×10^{-23} J/K, temperature (T) in K, charge of electron (q) is 1.602×10^{-19} °C.

Therefore, the mathematical I–V characteristic equation of a solar PV cell is defined by Eq. (3) from the circuit diagram is

$$I = I_{pv} - I_d - I_p$$

$$I_d = I_o \left[\exp\left(\frac{(V + R_s I)}{aV_t} \right) - 1 \right]$$

$$I_p = \frac{(V + R_s I)}{R_p}$$

Now substituting I_{pv}, I_d, and I_p in the characteristic equation modifies as defined below:

$$I = \left(I_{pv} - I_o\left[\exp\left(\frac{(V + R_s I)}{aV_t}\right) - 1\right] - \frac{(V + R_s I)}{R_p}\right) \quad (3)$$

V is load voltage, I is current across the load, R_s and R_p are the equivalent series and parallel resistance of the array. The other terminologies in the characteristic equation are explained earlier in this paper.

If Rs is not zero in Eq. (3) then it cannot be solved directly. Therefore, Lambert W function can be used to solve Eq. (3).

3 Simulation and Results of Solar Photovoltaic Array

A simple model can be built if the three basic important parameters of a solar PV cell are known viz. open circuit voltage (V_{OC}), short circuit current (I_{SC}) and maximum power point (P_{max}). But for accurate model, some additional information may be required. The maximum power that can be extracted from a PV cell is at the maximum power points. The simplified equivalent circuit diagram shown in Fig. 2 can be used to simulate the PV array. This circuit diagram can be simulated either by using script language of simulator or simulation block function generator. This paper presents the simulation results generated from a two modules M1 and M2 of 54 cells in each, connected in series and parallel by MATLAB SIMULINK simulator. This is shown pictorial in Figs. 3 and 4 and insolation is varied on the modules. First 1000 W/m^2 of insolation is given on both modules M1 and M2. Further, the insolation is reduced on modules in stepwise manner which is shown in Case-B and Case-C. In Case-B, insolation of 1000 and 500 W/m^2 is applied whereas in Case-C it is 500 W/m^2 on each module.

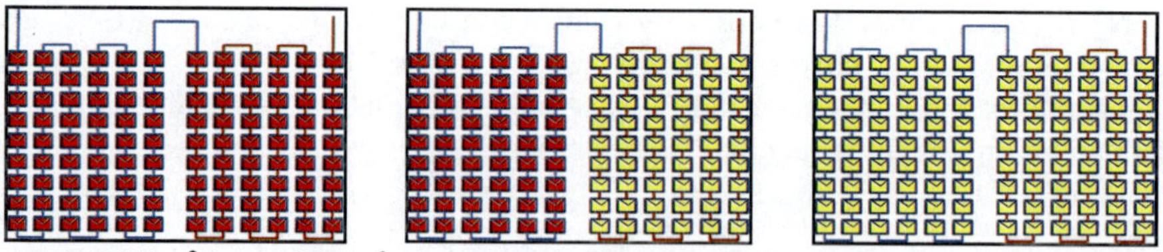

a) 1000W/m^2& 1000W/m^2 b) 1000W/m^2& 500W/m^2 c) 500W/m^2& 500W/m^2

Fig. 3 Insolation variation on series connected modules M1 and M2 of 54 cells in each

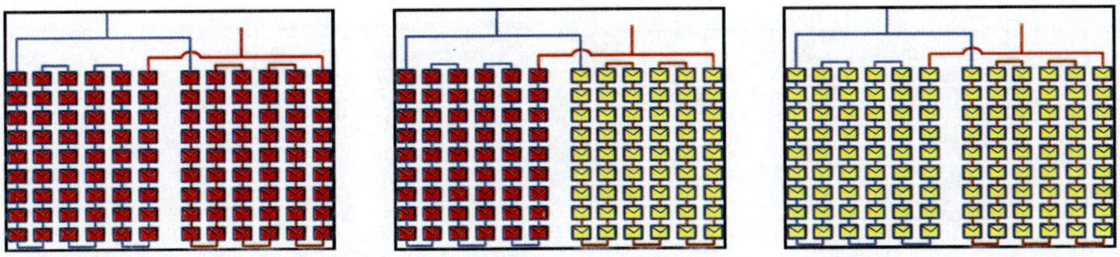

Fig. 4 Insolation variation on parallel connected modules M1 and M2 of 54 cells in each

Table 1 Characteristic parameters of solar PV modules, containing 54 cells in each

Cases	Connection and insolation (W/m^2)	I_{SC} (A)	V_{OC} (V)	P_{max} (W)
	Modules M1 and M2 in series with insolation variation			
Case-A	(M1: 1000 and M2: 1000)	7.34	64.76	366.8
Case-B	(M1: 500 and M2: 1000)	3.67	63.31	194.8
Case-C	(M1: 500 and M2: 500)	3.67	61.85	173.6
	Modules M1 and M2 in parallel with insolation variation			
Case-D	(M1: 1000 and M2: 1000)	14.68	32.4	366.8
Case-E	(M1: 500 and M2: 1000)	11.01	31.8	269
Case-F	(M1: 500 and M2: 500)	7.34	30.95	173.6

Similar variation of insolation is studied for parallel connected modules which are shown in Case-D–F of Fig. 4. Further, the same information is listed in Table 1 for simplification, which explains six different load cases. First three cases are for series connected modules from Fig. 3 and last three cases are for parallel connected modules from Fig. 4.

The characteristic curves are generated for both series and parallel connected modules. Figure 5 shows the characteristic parameters for series connected modules and Fig. 6 shows the characteristics parameters variation for parallel connected modules.

When both modules are connected in series and depending upon the insolation intensity from Case-A to Case-C, then the open circuit voltage (V_{OC}) is less/minimal affected and its variation is between ~62 and ~65 V. But short circuit current (I_{SC}) is same (3.67 A) for Case-B and Case-C and it is almost double (7.34 A) for Case-A. Therefore, it is observed from Fig. 5 and Table 1 that the current is more affected in case of same insolation on both modules than different on both the modules for series connect modules. The maximum power (P_{max}) in these three cases viz. Case-A, Case-B and Case-C is 366.8 W, 194.8 W and 173.6 W respectively.

Now both modules are connected in parallel and the insolation variation is studied, as done earlier for series connected modules. Here Case-D (M1: 1000 and M2: 1000) denotes the insolation of 1000 W/m^2 on each module. Similar notation is followed for other load cases viz. Case-E and Case-F. Figure 6 and Table 1 shows that open circuit

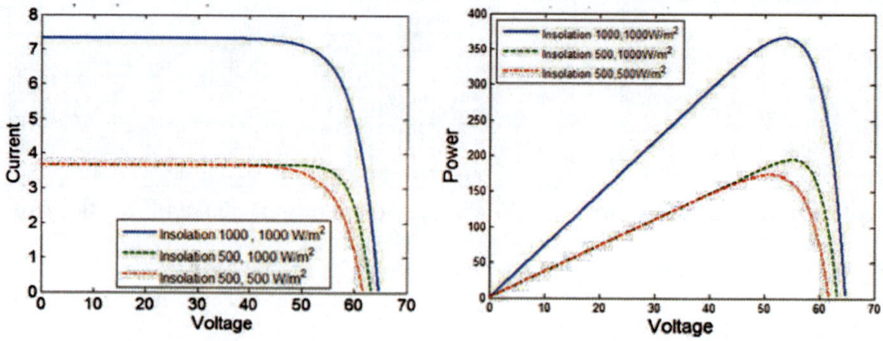

Fig. 5 I–V and P–V characteristic curves of two modules connected in series with insolation variation

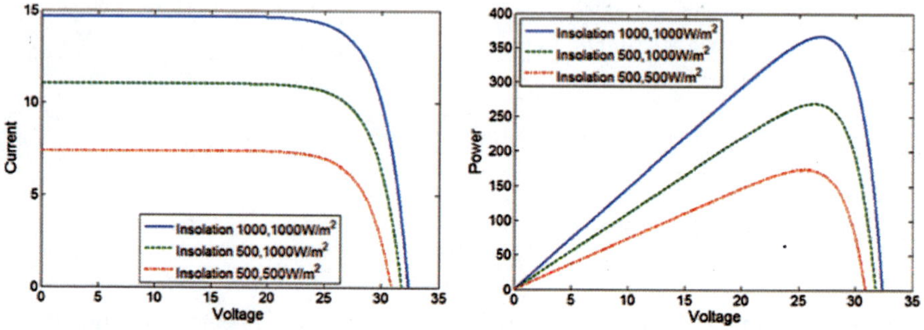

Fig. 6 I–V and P–V characteristic curves of two modules connected in parallel with insolation variation

voltage (V_{OC}) (~31 to ~33 V) has minimum or no impact of insolation variation for these three cases of parallel connected modules. Whereas, insolation variation affects the short circuit current irrespective of same or different insolation on both modules unlike series connected modules. The short circuit current for parallel connected modules can be interpolated based on the insolation variation. Maximum power (P_{max}) in these three cases (Case-D, Case-E and Case-F) is 366.8 W, 269 W and 173.6 W respectively.

Therefore, short circuit current (I_{SC}) is double and open circuit voltage (V_{OC}) is half for parallel connected modules (Case-D to Case-F) than series connected modules (Case-A to Case-C) for similar insolation on both modules. The maximum power (P_{max}) is same for both types of connection but at different voltage. If insolation is different on both modules, then parallel connected modules gives more power than series connected modules.

4 Conclusion and Future Scope

The open circuit voltage (V_{OC}) of the modules when connected in series is ~65 V whereas it is ~32 V for parallel connected modules with ideal P–V curve. The open circuit voltage (V_{OC}) for series connected modules is ~half of that connected in parallel. Similarly, ideal I–V characteristic curve shows that short circuit current (I_{SC}) for modules connected in series have minimum or negligible impact on the current and power with different insolation on each module. But if the same insolation either 500 or 1000 W/m^2 is applied on both modules then the short circuit current is almost half in series connected to the parallel connected modules.

The maximum power output is almost similar but at different point for both series/parallel connected modules when insolation is identical on both modules. But if different insolation on both modules is applied then parallel connected modules are having higher power output than series connected. Therefore, power output can be interpolated based on the insolation intensity in parallel connected modules.

Further, the shading effect of various kinds of shape pattern with insolation variation will be studied and presented in coming conferences.

References

1. Rahman S (2003) Green power: what is it and where can we find it? IEEE Power Energy Mag 1(1):30–37
2. Jager-Waldau A (2007) Photovoltaics and renewable energies in Europe. Renew Sustain Energy Rev 11(7):1414–1437
3. Bialasiewicz JT (2008) Renewable energy systems with photovoltaic power generators: operation and modeling. IEEE Trans Ind Electron 55(7):2752–2758
4. Bouhafs A, Lokmane BM, Mohamed D (2015) Grid connected photovoltaic system, for a 800 W. In: International conference on technologies and materials for renewable energy, environment and sustainability, TMREES15, pp 414–422
5. Belmili H, Haddadi M, Bacha S, Almi MF, Bendib B (2014) Sizing stand-alone photovoltaic–wind hybrid system: techno-economic analysis and optimization. Renew Sustain Energy Rev 30:821–832
6. Irena working paper (2012) Renewable energy technologies: cost analysis series. IRENA 1(4/5)
7. Rida KS, Al-Waeli AAK, Al-Asadi KAH (2016) The impact of air mass on photovoltaic panel performance. https://doi.org/10.18282/ser.v1.i1.41
8. Sulaiman SA, Singh AK, Mokhtar MMM, Bou-Rabee MA (2014) Influence of dirt accumulation on performance of PV panels. In: International conference on technologies and materials for renewable energy, environment and sustainability, TMREES14. Energy Procedia 50:50–56
9. Ketjoy N, Konyu M (2014) Study of dust effect on photovoltaic module for photovoltaic power plant. Energy Procedia 52:431–437
10. Mejia F, Kleissl J, Bosch JL (2014) The effect of dust on solar photovoltaic systems. Energy Procedia 49:2370–2376
11. Maghami MR, Hizam H, Gomes C, Radzi MA, Rezadad MI, Hajighorbani S (2016) Power loss due to soiling on solar panel: a review. Renew Sustain Energy Rev 59:1307–1316
12. Hasan MA, Parida SK (2015) Temperature dependency of partial shading effect and corresponding electrical characterization of PV panel. 978-1-4673-8040-9/15/$31.00 © IEEE

13. Pareek S, Runthala R, Dahiya R (2013) Mismatch losses in SPV systems subjected to partial shading conditions. In: International conference on advanced electronic systems (ICAES), pp 343–345
14. Pareek S, Dahiya R. Environmental effect consideration on photovoltaic module. RTESC-13-260

Comparing Theoretical and Practical Aspects of Process Management Practices for Competitive Potential in SMEs

Satyajit Mahato, Amit Rai Dixit, and Rajeev Agrawal

Abstract Process management practices (PMPs) are applied to gain 'Competitive-potential' (CP). However, their diffusion in SMEs is not very widespread yet. The relationships between the constructs of PMPs and CP have mainly been ignored. There are two steps in this research, i.e., identifying PMPs' constructs by factor analysis then establishing their empirical relationships by structural equation model (SEM). Data for the deployment scores of PMPs were collected through a survey questionnaire. The findings of this research provide unique insights into PMPs' practical aspects in SMEs, such as; improving 'Consistency' and 'Product innovation' doesn't improve CP unless mediated through other practices. This research advances the theory by deciphering the relation between PMPs. More focus is required to make such procedures more compatible with SMEs.

Keywords Process management practices · Competitive potential · SMEs · Structural equation modeling

1 Introduction

CP of an organization is gauged by parameters such as cost, profit, revenue, responsiveness, product quality [10], and its capability to develop better products in a shorter time [5]. Prester [14] reported that PMPs' effective deployment is the essential factor in achieving CP PMPs referred to approaches, systems, tools, and techniques in-built with the operations in the present context. A significant difference in cost, quality, and responsiveness were reported in a survey. These sources of the differences were identified to be the usage of PMPs. Authors have identified those practices where

---

S. Mahato (✉) · A. R. Dixit
Department of Mechanical Engineering, Indian Institute of Technology (ISM), Dhanbad, Dhanbad, Jharkhand 826004, India

R. Agrawal
Department of Mechanical Engineering, Malaviya National Institute of Technology Jaipur, Jaipur, Rajasthan 302017, India

R. Agrawal et al. (eds.), *Recent Advances in Industrial Production*, Lecture Notes in Mechanical Engineering, https://doi.org/10.1007/978-981-16-5281-3_37

high performers outperformed others, such as shorter time to delivery, high responsiveness, product quality, design performance, raw material meeting specification at low cost, low rejection rate etc. These differentiating factors to have higher CP is attributed to the usage of PMPs [2]. Past research on manufacturing strategy has reported the trade-offs between different types of practices. Still, their interaction with each other, and the overall impact on CP has mostly remained unexplored [6].

2 Literature Review

PMPs are intended to make processes more effective and efficient, maintain a specific quality, enhance supply capabilities, etc. Eventually, to improve CP. However, PMPs often fail due to a lack of understanding of strategically linking them with the CP [8]. There are practical challenges in framing an implementation strategy if their relationship pattern is unknown. According to Bello-Pintado and Merino-Díaz-de-Cerio since the interrelation between PMPs is unknown, their implementation cannot ensure any tangible benefit. Most companies try to take a shot in the dark [7]. A few researchers have attempted to study their qualitative relationship patterns and concluded to explore more on the various strands of PMPs, such as; Marodin and Saurin [11] pointed that a practice in isolation cannot improve CP. This view has been supported by Radnor and Johnston [16], who observed that exploratory research on PMPs and CP has not received enough attention. Perez-Arostegui et al. [13] concludes that predominantly the focus was on the qualitative assessment of 'Consistency' and 'Efficiency' related practices, such as Six-Sigma, Lean, but 'Supply-chain' and 'Product innovation' arena have received little attention in the context of gaining CP. Similar views were echoed by Prester [14] on 'Conformance-quality' practices, such as, TQM. The authors further noted that a firm could not achieve a competitive edge if methods are implemented without assessing their impact on CP.

Contemporary literature on business process management indicates a persistent debate on the trade-offs between different sets of practices. The prevailing confusion on implementing practices doesn't guarantee an increase in CP [12, 18]. Hence, there is a need to ascertain PMPs' relationship aspects for devising a robust strategy for their deployment [16]. According to Prester [14], large enterprises deploy many practices to achieve higher 'CP'. SMEs don't have adequate resources to do so, therefore, they tend to make cost economy a priority against performance. However, in the process, they end up spending more on manufacturing and material costs. The absence of relationship models has led to the adoption of an ad hoc approach towards process management. In this situation, the decision to implement a practice depends solely on the qualitative assessment of its potential to impact CP [21]. So the need to have a quantitative approach for the deployment of PMPs, based on their relationship pattern with CP, is even more significant in SMEs [3].

The present research developed a set of hypotheses on the relationship pattern of PMPs from the theoretical literature. An empirical model between PMPs and CP was created, followed by a comparative assessment with the research hypothesis.

The comparison provided significant insights into the practical aspects of PMPs when implemented in SMEs.

3 Research Methodology

It has a three-prong approach: (i) a survey through a fixed questionnaire, (ii) consultation with focused groups, and, (iii) semi-structured interviews. An inventory of PMPs was created from the prior research and also included contemporary practices, such as; Prester [14] and Epstein [5]. A comprehensive 37 PMPs with six indicators of CPs were shortlisted that are extensively used by SMEs. A total of 317 responses were received on the diffusion of practices on a Likert scale of 1–5. With the literature review's help, the PMPs' constructs were identified, followed by a set of research hypotheses. The collected data were subjected to reliability and construct validity. A structural equation model was developed on the collected data to decipher the relationship pattern between PMPs.

4 Key Performance Constructs of the Process Management Practices

This research attempted to examine the constructs of PMPs based on the underlying purpose of their deployment. Extant literature on practices was reviewed to decipher the underlying reason for their deployment and then classified under a construct.

Table 1 provides an inventory of the PMPs, their description, literature evidence, their deployment's underlying purpose, and their Key performance constructs through columns one to five. Cronbach's alpha for each construct was more than 0.8 implies the construct validity of the measurement instrument.

5 Research Hypothesis

The literature infers that PMPs exhibit specific relationship patterns with each other. This section outlines these patterns, as evident in the framework of literature. The accepted view of the literature has been formulated as the research hypothesis:

H1: 'Consistency' of the process output has a positive impact on the 'Efficiency' (H1a), 'Competitive potential' (H1b) and, on 'Conformance-quality' of the process (H1c).

H2: 'Conformance-quality' has a positive impact on 'Competitive potential.'

H3: 'Supplier-efficacy' practices positively impacts 'Competitive-potential' (H3a) and, 'Conformance-quality' (H3b).

Table 1 Key performance constructs of the process management practices (PMPs)

PMP_{ij}	Description of the practice	Literature references	The key performance construct of the practices	Key performance construct (P_i)	Mean (p_{ij})	Standard deviation	Cronbach's alpha
PMP_{11}	Run Six-Sigma projects	[15, 17]	To reduce variation and defects in a process	Consistency (P_1)	3.43	0.84	0.931
PMP_{12}	Quality Circle				3.38	0.89	
PMP_{13}	Statistical process control				3.43	0.79	
PMP_{14}	Cost, quality, delivery targets for suppliers				3.65	0.77	
PMP_{15}	Quality function deployment (QFD)				3.25	0.89	
PMP_{16}	Benchmarking competitors				3.53	0.78	
PMP_{17}	Failure mode effect analysis (FMEA)				3.65	0.76	
PMP_{21}	Optimize transportation	[9, 19]	To improve the efficiency of the supply chain	Supplier efficacy (P_2)	2.51	1.05	0.944
PMP_{22}	Outsource non-core functions				2.47	1.09	
PMP_{23}	Reduce the supply base				2.25	0.95	
PMP_{24}	Use common, standardized parts				2.34	0.99	
PMP_{25}	Build long term relationships with key suppliers				2.36	1.01	

(continued)

Table 1 (continued)

PMP_{ij}	Description of the practice	Literature references	The key performance construct of the practices	Key performance construct (P_i)	Mean (p_{ij})	Standard deviation	Cronbach's alpha
PMP_{31}	Value Stream Mapping (VSM)	[1, 18]	To increase efficiency by eliminating waste and making the processes Lean	Efficiency (P_3)	2.87	0.91	0.891
PMP_{32}	Collaborative planning				2.76	0.86	
PMP_{33}	Visual control				3.14	0.88	
PMP_{34}	Low-cost automation				2.92	0.88	
PMP_{35}	Poka-yoke				2.98	0.92	
PMP_{36}	JIT				2.99	0.86	
PMP_{41}	Cross functional design/development teams	[4, 14]	To channelize creativity for improving product and process	Product innovation (P_4)	3.99	0.62	0.920
PMP_{42}	Involving suppliers and customers in design process				3.92	0.64	
PMP_{43}	Study feedback from customers				3.93	0.65	
PMP_{44}	Designing based on commonality of parts				3.91	0.66	

(continued)

Table 1 (continued)

PMP_{ij}	Description of the practice	Literature references	The key performance construct of the practices	Key performance construct (P_i)	Mean (p_{ij})	Standard deviation	Cronbach's alpha
PMP_{45}	Design based on product needs				4.01	0.64	
PMP_{46}	Design meets specific customer requirement				3.95	0.70	
PMP_{47}	Design includes profit considerations				3.97	0.65	
PMP_{51}	Manufacturing to meet product design	Flynn et al.	To conform to the set quality standard	Conformance quality (P_5)	3.82	0.69	0.828
PMP_{52}	Ensuring the incoming material meets order				4.04	0.57	
PMP_{53}	Product reaches right place at right time				3.92	0.67	
PMP_{54}	No product damages during material flow				4.08	0.54	
PMP_{55}	Incoming material meets order specifications				3.91	0.66	

(continued)

Table 1 (continued)

PMP_{ij}	Description of the practice	Literature references	The key performance construct of the practices	Key performance construct (P_i)	Mean (p_{ij})	Standard deviation	Cronbach's alpha
PMP_{56}	Material rejection rate is low				3.68	0.80	
PMP_{61}	Responsiveness	[5, 10, 14]	To gauge the organization's performance on its competitive potential	Competitive potential (P_6)	3.58	0.67	0.856
PMP_{62}	On-time delivery				3.75	0.69	
PMP_{63}	Return on assets				3.78	0.66	
PMP_{64}	Cost-saving				3.61	0.75	
PMP_{65}	Product quality				3.78	0.53	

Responses were received on implementing PMPs on a Likert scale of 1–5 [1—not implemented. 5—fully implemented]

H4: 'Product innovation' practices have a positive impact on 'Competitive potential' (H4a) and also on 'Conformance-quality' (H4b).
H5: Increase in 'Efficiency' of the process improves the 'Competitive potential.'

6 Structural Equation Modelling (SEM)

The six constructs, validated through the factor analysis, revealed a variation of 74.8%. As a result of the factor analysis, the latent constructs of PMPs, identified from the literature, were empirically validated. A structural equation model depicting the empirical relationship between the constructs was developed. AMOS 16.0 software was used to build an SEM.

Figure 1 depicts the SEM output. The standardized regression weight is written alongside the arrow, indicating an independent variable on a dependent variable. The number of stars inscribed as superscript depicts the statistical significance level of the regression weight. For example, 'Product innovation' has an effect of 0.51^{***} on 'Conformance quality', and its significance level is 0.001, as indicated by the three stars inscribed on it. Each construct's practices in the final model are also provided in it, such as 'Product innovation', which has its defining practices from PMP_{41} through PMP_{47}. The description of these items is provided in Table 1.

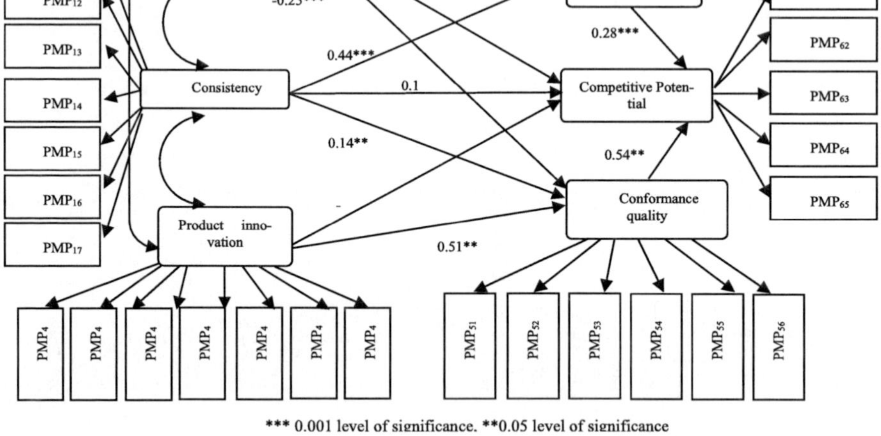

Fig. 1 Structural equation model of the key performance constructs of PMPs

As far as model fit indices of the SEM are concerned, Chi-square/degrees of freedom was found to be 1.258, which was less than 5. All the model fit indices were higher than 0.9, viz., Comparative Fit Index (CFI) was 0.992, Goodness of Fit Index (GFI) was 0.985, and Normed Fit Index (NFI) was 0.966. Root Mean Square Error of Approximation (RMSEA) was 0.05 at a *p*-value of 0.412, within acceptable limits. Root Mean Square Residual (RMR) was 0.022, which was less than 0.08. All the critical model fit indices were in the acceptable range. Hence, the overall SEM model fitted the data well.

7 Findings and Managerial Implications

The empirical model depicted in Fig. 1 was compared with the research hypothesis. The deviations exhibited by the model vis-à-vis the hypothesis highlight the nuances of PMPs in an SME environment. These are noted below as the findings of this study:

i. The exploratory factor analysis indicated a factor loading of more than 0.7 for each factor and there is no cross loading more than 0.3. Hence the existence of the six constructs of PMPs—'Consistency', 'Supplier-efficacy', 'Product innovation', 'Efficiency', 'Conformance quality', and 'Competitive-potential (CP)' was validated in an SME environment. This finding is in accordance to the theoretical literature.
ii. Contrary to the theoretical narrative of literature, a 'Consistency' practice in isolation is inadequate to improve 'CP', but does so when mediated by 'Efficiency' and 'Conformance-quality'.
iii. In contrast to the theory, 'Supplier-efficacy' deteriorates the 'Conformance-quality' and because of this does not carry enough potential to improve 'CP'. This contrasting relation is attributed to the fact that such practices in SMEs are implemented to meet the regulatory requirements [20] without considering their effect on other functions.
iv. Opposite to the literature perspective, 'Product innovation' initiatives don't improve 'CP'. This is attributed to the high cost and low success rate of SMEs' research initiatives [4]. But its relation with 'Conformance-quality' as per the theory, therefore, has the potential to improve it.
v. 'Efficiency' and 'Conformance-quality' demonstrate their relationship pattern precisely as per the theoretical model. Therefore, potentially they improve the 'CP'.

8 Conclusion and Future Work

To sum up, this research has aimed to achieve the objective of advancing the theory of process management by proposing an empirical model of CP for SMEs as a function of PMPs, which is also a novel scientific contribution of this paper. To meet this objective, six performance constructs were listed from the literature. A set of research hypotheses on their relationship was also derived. The SEM developed by using the empirical data was compared with the hypotheses. The deviations shown by the model depicts the practical aspects of PMPs in SMEs vis-à-vis the theoretical literature. One example is 'Consistency', and 'Product innovation' practices do not have a significant impact on 'CP' unless mediated by other practices, such as; 'Conformance-quality'. The future research shall focus on making 'Consistency' and 'Supplier efficacy' compatible with SME environment.

References

1. Bititci U, Garengo P, Dörfler V, Nudurupati S (2012) Performance measurement: challenges for tomorrow. Int J Manag Rev 14(3):305–327
2. Bloom N, Van Reenen J (2007) Measuring and explaining management practices across firms and countries. Q J Econ CXII(4):1–55
3. Boon Sin A, Zailani S, Iranmanesh M, Ramayah T (2015) Structural equation modelling on knowledge creation in Six Sigma DMAIC project and its impact on organizational performance. Int J Prod Econ 168:105–117
4. Dowlatshahi S (2010) The role of product design in designer-buyer-supplier interface. Prod Plan Control 8(6):522–532
5. Epstein MJ (2010) Thinking straight about sustainability. Stanford Soc Innov Rev 8(3):51–55
6. Giacosa E, Mazzoleni A, Usai A (2018) Business process management (BPM). Bus Process Manag J 24(5):1145–1162
7. Kuei C, Lu MH (2013) Integrating quality management principles into sustainability management. Total Qual Manag Bus Excell 24(1–2):62–78
8. Kumar M, Antony J, Tiwari MK (2011) Six Sigma implementation framework for SMEs—a roadmap to manage and sustain the change. Int J Prod Res 49(18):5449–5467
9. Li L, Su Q, Chen X (2011) Ensuring supply chain quality performance through applying the SCOR model. Int J Prod Res 49(1):33–57
10. Madu CN, Kuei C (2012) Introduction to sustainability management. In: Handbook of sustainability management. World Scientific, pp 1–22
11. Marodin GA, Saurin TA (2013) Implementing lean production systems: research areas and opportunities for future studies. Int J Prod Res 51(22):6663–6680
12. Narasimhan R, Swink M, Wook Kim S (2005) An exploratory study of manufacturing practice and performance interrelationships. Int J Oper Prod Manag 25(10):1013–1033
13. Perez-Arostegui MN, Benitez-Amado J, Huertas-Perez J-F (2012) In search of loyalty: an analysis of the determinants of buyer–supplier relationship stability under a quality management approach. Total Qual Manag Bus Excell 23(5–6):703–717
14. Prester J (2012) Competitive priorities, capabilities and practices of best performers: evidence from GMRG 4 data. Total Qual Manag Bus Excell. https://doi.org/10.1080/14783363.2012.704275
15. Radnor Z (2010) Transferring Lean into government. J Manuf Technol Manag 21(3):411–428
16. Radnor Z, Johnston R (2013) Lean in UK government: internal efficiency or customer service. Prod Plan Control 24(10–11):903–915
17. Radnor Z, McGuire M (2004) Performance management in the public sector: fact or fiction? Int J Product Perform Manag 53(3):245–260
18. Sila I (2007) Examining the effects of contextual factors on TQM and performance through the lens of organizational theories: an empirical study. J Oper Manag 25(1):83–109

19. Sweeney MT, Szwejczewski M (2010) A study of practices and performance, Porto, Portugal. In: Proceedings of the 17th EUROMA conference, manufacturing excellence: a study of practices and performance, Porto
20. Westphal P, Sohal A (2016) Outsourcing decision-making: does the process. Prod Plan Control. https://doi.org/10.1080/09537287.2016.1159350
21. Zhang C, Viswanathan S, Henke JW (2011) The boundary spanning capabilities of purchasing agents in buyer-supplier trust development. J Oper Manag 29(4):318–328

Development of a Mathematical Model for the Software Defect Rework Process to Optimize Defect Rework—A Six-Sigma Case Study

Satyajit Mahato, Amit Rai Dixit, and Rajeev Agrawal

Abstract The software development industry spends a staggering amount of resources on reworking the defects. The SMEs have limited resources; therefore, such enterprises look for cost-effective solutions. Past Six-Sigma studies in the software domain are limited to reducing defect generation opportunities. There is limited research that has dealt with the longer rework time and high cost during the rework phase. This work has attempted to fill into this niche by developing a mathematical model of defect rework-phase. This model is integrated with the DMAIC framework of Six-Sigma to formulate the defect rework process as a multi-objective problem between rework time and cost. Genetic algorithm was applied to obtain the non-dominated optimal solution. The proposed solution reduced the rework time by 31% at an optimum cost.

Keywords Mathematical model · Genetic algorithm · Multi-objective optimization · Defect rework · Software SMEs

1 Introduction

Resolving software defects after the project delivery is very expensive, which is why the defect rework-phase is an integral part of the software development life-cycle [6]. Defect rework is a non-value add activity [3]. Therefore, the project managers face the dual challenge of quickly completing the defect rework and reducing rework costs. These are conflicting objectives, thus to reduce rework time, most companies spend a massive amount of resources on managing the rework phase. But comparing with the large enterprises (LEs) this approach becomes unsustainable for the SMEs [7, 10, 14]. This paper has attempted to make a novel contribution to the theory

S. Mahato (✉)
Department of Mechanical Engineering, Indian Institute of Technology (ISM), Dhanbad, Dhanbad, Jharkhand 826004, India

A. R. Dixit · R. Agrawal
Department of Mechanical Engineering, Malaviya National Institute of Technology Jaipur, Jaipur, Rajasthan 302017, India

R. Agrawal et al. (eds.), *Recent Advances in Industrial Production*, Lecture Notes in Mechanical Engineering, https://doi.org/10.1007/978-981-16-5281-3_38

by proposing a mathematical model for the software defect rework process. It is implemented through a Six-Sigma case study conducted in a software SMEs rework process. Its application allowed fewer developers in the rework phase and reduced the rework 'time' by 31% at an optimum 'cost'.

2 Literature Review

Although the defect rework-phase chaos has been underscored by many, organizations still make terrible decisions on managing it. An ill-managed rework-phase aggravates the 'time' and 'cost' of rework, which leads to schedule deviation [9, 11].

These best practices to handle such disruptions have been implemented through the Six-Sigma framework in the manufacturing sector [12]. But the chaotic mismanagement of the rework phase is still an unattained issue [3]. Grant and Mergen [6] found that the Six-Sigma projects in the software domain put a lot of emphasis on reducing defects during product development and called it a mere replication of manufacturing studies. A framework like DMAIC will be more effective in the software domain if customized in conjunction with other analytical techniques. Suitable customization is needed to make it useful in the software industry [12].

The lack of a practical and systematic approach to dealing with the perpetual defect rework problem creates a vast disparity between SMEs and LEs. The LEs spend a colossal amount on DMAIC projects to eliminate defects without considering the 'cost' [16]. SMEs, however, are short of resources and sensitive towards the cost [2]. Therefore, they look for improvement framework to obtain cost-optimized solutions, particularly to address the high rework 'time' [5]. Integrating new methods and techniques within the Six-Sigma framework, such as; Machine learning, Artificial intelligence, Big data, and Analytics, has been recommended to make managers responsive rather than reactive [8]. This research has attempted to fill this niche by proposing a novel mathematical model of the defect rework process. This model is used in conjunction with the DMAIC Six Sigma framework and then implemented in a software firm's real-time case study. The obtained solution reduced the rework 'time' by 31% at an optimum 'cost'.

3 Research Methodology

The present work aims to fill into this niche by defining the following novel objectives,

- Propose a mathematical model of the defect rework process and then integrate with the proposed framework in a case study Six-Sigma DMAIC framework.
- Implement the integrated framework to derive defect the rework process as a multi-objective problem between rework 'time' and 'cost' and then solve for minimum rework time at optimum cost.

4 The Mathematical Model for Defect Rework-Phase

Software project development is managed into multiple modules. Every module is a set of its line of codes. Based on the skill and cost, developers' pool can be divided into 'n' quantiles [13].

As a function of the software development parameters, the rework time and cost is expressed by Eqs. (1) and (2),

$$T_R = \frac{\sum_{i=1}^{m} d_i \cdot \sum_{j=1}^{k}(\xi_j \times (\bar{t}_r)_j)}{(\xi_j)^2} \tag{1}$$

$$Z = \sum_{j=1}^{k} \xi_j \cdot C_j \cdot \frac{\sum_{i=1}^{m} d_i \cdot \left(\sum_{j=1}^{k} \xi_j \times (\bar{t}_r)_j\right)}{(\xi_j)^2} \tag{2}$$

where d is the number of predicted defects, ξ is allocated developer in rework, $\bar{t}$ is weighted rework time. The indices *i*, *j*, *m*, and *k*, represent the module, total modules, quantile, total quantile, and *r* stands for rework (Fig. 1).

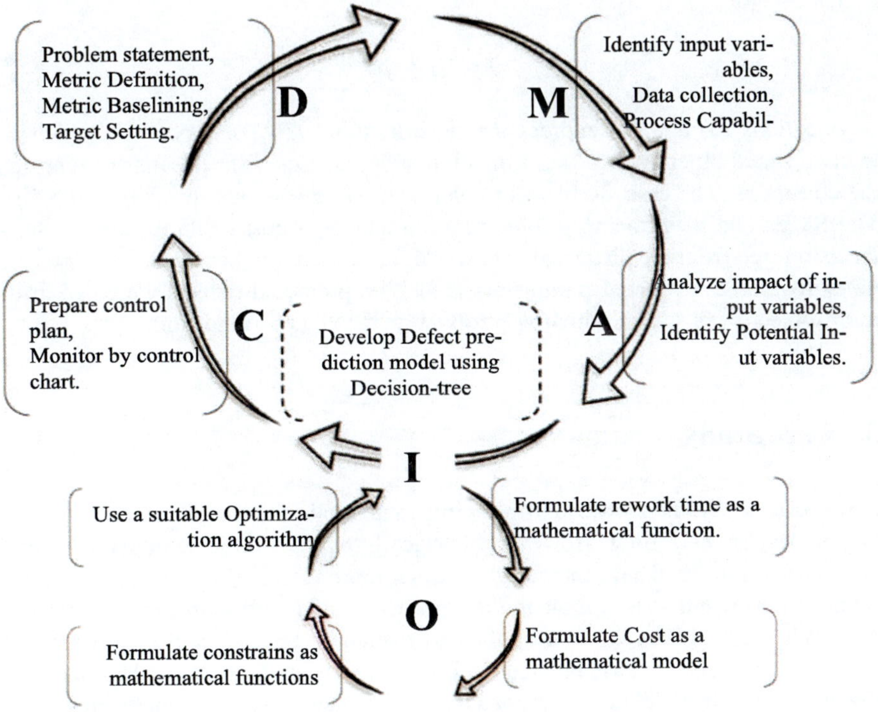

Fig. 1 Integrated framework of Six-Sigma

The constraints are also based on the software development process's parameters. The sum of the defects resolved by each quantile of developers shall be at least equal to the sum of errors from all modules. Therefore,

$$\left(\sum_{j=1}^{k}\frac{\xi_j}{(\bar{t}_r)_j}\right)\frac{\left(\sum_{j=1}^{k}\xi_j\times(\bar{t}_r)_j\right)}{(\xi_j)^2}\geq 1 \tag{3}$$

The case study aims to reduce the 'rework time' to be maintained below a predefined percentage (ϑ%) of the development time. Therefore,

$$T_R\leq\vartheta\cdot\left\{\sum_{j=1}^{k}(\bar{t}_c)_j(l_c)_j+\sum_{j=1}^{k}(\bar{t}_{rev})_j(l_{REV})_j+\sum_{j=1}^{k}(\bar{t}_t)_j(l_t)_j\right\} \tag{4}$$

where p is the number of developers, index j represents the jth quantile of the developer's pool, $\bar{t}$ *and* l represent time and lines of code. The indices *c*, *rev*, *r*, and *t* represent coding, review, rework, and testing phases. The historical data was collected to calculate empirically $(\bar{t}_c)_j$, $(\bar{t}_{rev})_j$, $(\bar{t}_t)_j$ *and* $(\bar{t}_r)_j$. Besides, the number of allocated developers from any category can only be a subset of the total number of developers in that category

$$0\leq\xi_j\leq p_j \tag{5}$$

Equations (5) and (6) express the rework 'time'. T_R, and 'cost' Z, which are the two stated objectives for minimization, whereas Eqs. (7), (8), and (9) represent the constraint. The case study was implemented in a software development SMEs. The SMEs had been facing a long duration and high cost in its rework phase. A Decision-tree from the historical data was developed to predict the defects by using the software development parameters [15]. The predicted defects were fed in the multi-objective problem, which was solved by the Genetic algorithm.

5 Case Study

In this case study, the software firm aspires to maintain the rework time within 15% of the Development time. However, historically 67% of projects delivered during Oct'2018 to Jan'2020 have their rework 'time' more than 15% of development time. Management wants to minimize the rework 'time' at an optimum cost. The conventional phases of DMAIC were applied to formulate the problem in a Six-Sigma project. The 'Define' phase created the project charter with a goal of less than 15% rework time. 'Measure' phase mapped the process steps and then concluded with the variables affecting the number of defects by referring to relevant literature [15]. In the 'Analyze' section, root cause analysis (RCA) was done for the rework phase's long duration. The RCA concluded that developers of different modules and new developers are allocated to fix defects that take a longer duration. This finding was also validated by a statistical test from historical data of developer allocation and rework time. Significant work on the implementation of DMAIC has been reported in the literature [1]. Therefore, in this work outcome of these conventional steps are not repeated. A Decision-tree model was developed in the 'Improve' phase to predict defects using the input variables. Data from more than 14,500 modules were collected on the independent variables and defects count. The SPSS20.0 output of the decision tree model. The number of defects in a module can be predicted by locating its terminal node based on independent variables' values. The suitability of the defect prediction model was validated by applying it to the test cases, where it reported an accuracy of 92.8%. The predicted defects count of a project were used in formulating the defect rework process as a multi-objective problem.

5.1 *Optimize Phase-Formulation of the Multi-objective Problem for the Case Study*

The exact formulation of this case study's problem was done in the 'Optimize' phase of the integrated Six Sigma framework. It was done by using the mathematical model proposed in Sect. 4. The software development project under consideration of this study had four modules, and the developer pool was also divided into four quantiles. By inputting the predicted defects count and empirically calculated average rework time for each quantile, we obtained the,

$$Rework\ \text{'time'}\ (T_R) = 497\left\{\frac{(17\xi_i + 20\xi_{i+1} + 29\xi_{i+2} + 41\xi_{i+3})}{\left(\sum_{i=1}^{4}\xi_i\right)^2}\right\} \tag{6}$$

$$Rework\ \text{'Cost'}\ (Z)$$

$$= 497\left\{\frac{(121\xi_i + 82\xi_{i+1} + 60\xi_{i+2} + 40\xi_{i+3})(17\xi_i + 20\xi_{i+1} + 29\xi_{i+2} + 41\xi_{i+3})}{\left(\sum_{i=1}^{4}\xi_i\right)^2}\right\} \tag{7}$$

Subjected to constraints,

$$497\left\{\frac{(17\xi_i + 20\xi_{i+1} + 29\xi_{i+2} + 41\xi_{i+3})}{\left(\sum_{i=1}^{4}\xi_i\right)^2}\right\} - 18{,}967 \le 0 \tag{8}$$

$$
\begin{aligned}
&0 \le \xi_j \le 14 \\
&0 \le \xi_{j+1} \le 19 \\
&0 \le \xi_{j+2} \le 35 \\
&0 \le \xi_{j+3} \le 75 \\
&For\ j = 1
\end{aligned}
\tag{9}
$$

The next section describes the application of GA to solve the multi-objected problem formulated in this section.

5.2 *Optimize Phase-Applying Genetic Algorithm to Solve the Formulated Problem*

The 'Optimize' phase continues to solve the formulated problem in the previous section by applying the Genetic Algorithm (GA). In this study, we used the controlled elitist non-dominated sorting algorithm, a variant of NSGA-II [4]. The Pareto optimal non-dominated solutions are shown in Fig. 2, only for some of the trials for the sake of brevity. Single point crossover, uniform mutation function, and tournament selection method size of 2 was adopted. Every time 18 solutions were obtained, but the first trial (Fig. 2a) converged in 800 iterations and had a relatively small average distance. In Fig. 2b–e, the population size was increased until 200. Simultaneously, crossover

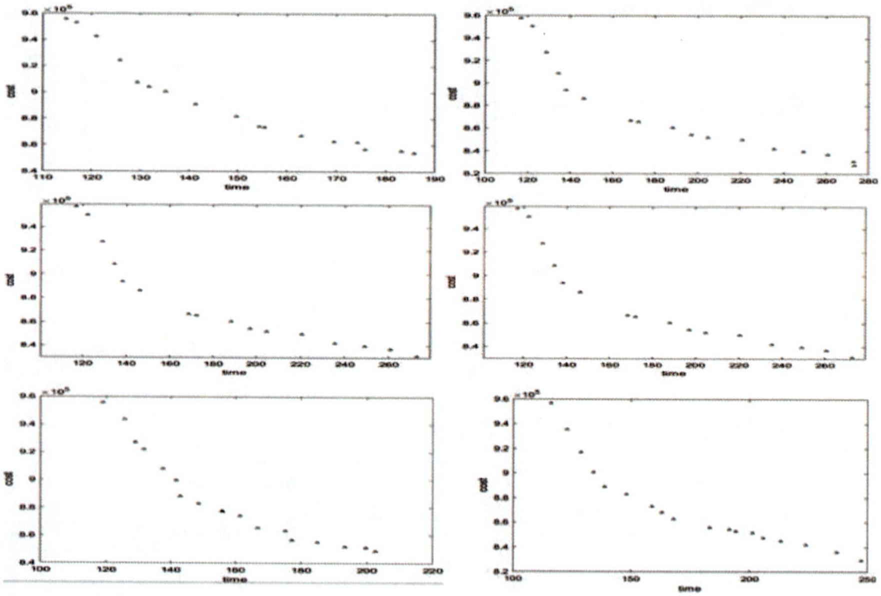

Fig. 2 Non-dominated solutions for varying combinations of GA parameters

probability (P_{cross}) and mutation probability (P_{mute}) was increased till 0.80 and 0.45. The final iteration (Fig. 2f) provides the best possible distance and spread measure, as suggested by Deb et al. [4].

The Pareto front of Fig. 2f suggested eighteen non-dominated solutions. In this case study, the selected solution has a rework 'time' of 163.2 h (27 days) at a 'Cost' of INR867963. For six working h/day, the selected solution proposed below developer allocation strategy in the rework phase;

- First quantile—1 developer for 1.2 h/day.
- The second quantile—4 developers were allocated for the full day, but the 5th developer only for 2.4 h/day.
- Third quantile—33 developers were allocated for the full day, but the 34th developer was only 3.6 h/day.
- Fourth quantile—73 developers were allocated for the full day.

The proposed solution completed the rework phase in 27 days. In the 'Control' phase, control charts were applied to monitor the success rate for completing the rework phase rework 'time' of the projects been delivered.

6 Results and Discussion

This work aimed to develop a solution to high rework 'time' at an optimum rework 'cost'. A mathematical model of the defect rework-phase was developed in Sect. 4, integrated into the Six Sigma framework. In the implementation case study of the integrated framework, the conventional phases were used to create a Decision-tree model to predict the number of defects. The predicted errors were used with the mathematical formulation to deduce the rework process as a multi-objective optimization problem between rework (i) 'time' and (ii) 'cost'. The 'gamultiobj' solver of MATLAB applied NSGA II to solve the problem. A set of non-dominated solutions were obtained with a Pareto front. The Pareto optimal front's recommended solution resulted in a rework 'cost' of INR839793 and decreased the rework 'time' to 11% of the development time.

7 Conclusion

The software SMEs seek to optimize costs while looking for solutions to their operational issues [3]. This paper proposed a mathematical model of the defect rework process, which is integrated with Six Sigma. The integrated framework is implemented in a real-time case study. In the 'Improve' phase, a Decision-tree model was created to predict the defect counts. The mathematical model was then applied in the 'Optimize' phase to formulate and solve the multi-objective problem of rework 'time' and 'cost' using a Genetic Algorithm. In the case study, the proposed solution reduced the rework 'time' by 30%.

References

1. Al-Qutaish RE, Al-Sarayreh KT (2008) Applying Six-Sigma concepts to the software engineering: myths and facts. In: Proceedings of the 7th international conference on software engineering, parallel and distributed systems (SEPADS'08), Mar 2008, pp 178–183
2. Chookittikul J, Chookittikul W (2008) Information technology strategy for Six Sigma projects in a Thai University. In: PICMET: Portland international center for management of engineering and technology, proceedings, no c, pp 917–923
3. Clarke P, O'Connor RV (2013) An empirical examination of the extent of software process improvement in software SMEs. J Softw Evol Process 25(9):981–998
4. Deb K, Pratap A, Agarwal S, Meyarivan T (2002) A fast and elitist multiobjective genetic algorithm: NSGA-II. IEEE Trans Evol Comput 6(2):182–197
5. Ghane K (2014) A model and system for applying Lean Six Sigma to agile software development using hybrid simulation. In: 2014 IEEE international technology management conference, pp 1–4
6. Grant D, Mergen AE (2009) Towards the use of Six Sigma in software development. Total Qual Manag Bus Excell 20(7):705–712
7. Gunasekaran A, Lyu J (1997) Implementation of just-in-time in a small company: a case study. Prod Plan Control 8:406–412
8. Guo M, Zhang Q, Liao X, Chen FY, Zeng DD (2020) A hybrid machine learning framework for analyzing human decision-making through learning preferences. Omega. https://doi.org/10.1016/j.omega.2020.102263
9. Langabeer J, DelliFraine J, Heineke J, Abbass I (2009) Implementation of Lean and Six Sigma quality initiatives in hospitals: a goal theoretic perspective. Oper Manag Res 2:13–27
10. Mahanti R, Antony J (2005) Confluence of Six Sigma, simulation and software development. Manag Audit J 20(7):739–762
11. Papatheocharous E, Bibi S, Stamelos I, Andreou AS (2017) An investigation of effort distribution among development phases: a four-stage progressive software cost estimation model. J Softw Evol Process 29(10):1–21
12. Pillai AKR, Pundir AK, Ganapathy L (2012) Implementing integrated Lean Six Sigma for software development: a flexibility framework for managing the continuity: change dichotomy. Glob J Flex Syst Manag 13(2):107–116
13. Pujari CG, Seetharam K (2015) An evaluation of effectiveness of the software projects developed through Six Sigma methodology. Am J Math Manag Sci 34(1):67–88
14. Raval SJ, Kant R, Shankar R (2018) Revealing research trends and themes in Lean Six Sigma: from 2000 to 2016. Int J Lean Six Sigma 9(3):399–443
15. Rodrigues NM, Lingappa AK (2014) Six Sigma and CMMI. J Comput Eng 16(4):1–5
16. Wong WY, Lee CW, Tshai KY (2012) Six Sigma in IT processes, IT services and IT products—a fact or a fad? Six Sigma beyond manufacturing in IT processes, IT services and IT products. In: Proceedings—2012 IEEE 12th international conference on computer and information technology, CIT 2012, pp 524–531

Theoretical Analysis of 1st Law and 2nd Law Efficiency of a Solar Pump for Geographical Location 28.10′N, 78.23′E

Vineet Singh, Vinod Singh Yadav, Vishal Saxena, Niraj Kumar, and Anurag Maheswari

Abstract In this paper, a solar pump is designed on the basis of the theoretical analysis on the basis of empirical relation of solar pump. The pump is designed for FET, MJP Rohilkhand University at location 28.10′N, 78.23′E, in north India. The performance of solar plate and motor pump system have been calculated on the basis of total monthly average daily radiation on three different tilt angles 18.86, 28.86 and 38.86°. On the basis of monthly average daily solar radiation maximum solar radiation with tilt angle is determined which will helpful for manual tracking. When the module temperature increases the performance of the solar pump is declined gradually for improving the performance exergy analysis is also completed.

Keywords Solar pump · Inverter · Solar plate · Discharge · Thermometer

1 Introduction

Pumping of water from earth for domestic and irrigation purpose are dependent on the conventional fuel like diesel, natural gas. These are the exhaustive, very costly and harness the environment for pumping the water for domestic, commercial, industrial and irrigation use. So, it is the need of world today to develop the alternatives of pumping the water by some other energy source. One of the energy source pv cell technology which can be used for pumping the surface, river, earth water for commercial and irrigation. The good thing with this technology is that the solar energy available in environment according the requirement of water. The water requirement in India is varied according to the variation of seasons, the maximum amount of water is required in summer season but the good thing is that huge amount of solar energy available these days. In last year's pv cell technology was costly means of producing

V. Singh (✉) · V. Saxena · N. Kumar · A. Maheswari
Department of Mechanical Engineering, F.E.T., M.J.P. Rohilkhand University, Bareilly, Uttar Pradesh, India

V. S. Yadav
Department of Mechanical Engineering, National Institute of Technology Uttarakhand, Srinagar (Garhwal), Uttarakhand 246174, India

R. Agrawal et al. (eds.), *Recent Advances in Industrial Production*, Lecture Notes in Mechanical Engineering, https://doi.org/10.1007/978-981-16-5281-3_39

power due to its high manufacturing cost and low performance efficiency and on the other hand the cost of diesel, gaseous fuel is low and efficiency is high but now due to development of easy way of manufacturing and improvement in performance pv cell technology it can full fill the demand of energy in future. So, in future pv cell may fulfil the energy demand of the world due to increase of cost and depletion of reserve of fossil fuel. The main components of the solar pump system consist of a pv cell, inverter controller, motor-pump set. Since from 1879 lot of research going on improving the performance of pv solar panel for reduction of its size and cost. Rohit et al. [1] design and developed the lab-view simulation model by putting the boost converter, LC filter in circuit for improving its performance and increasing the voltage of DC current for driving the induction motor. Kou et al. [2] developed a method for predicting the long-term performance of the solar pump system. They have used the solar panel and motor-pump manufacturer data and weather data generated by statistical equation. On the basis of these data they designed the solar pump system for future use. They have used the simulation model TRNSYS and compared the results with UW-PUMP program and finally concluded that the percentage difference of 3% in Albuquerque and 6% in Seattle. Benjamin and Richard [3] performed the theoretical and experimental analysis of solar pumps at 98 localities of Canada. Benjamin and Richard [3] performed the theoretical and experimental analysis of solar pumps at 98 localities of Canada. Belgacem [4] conducted the test on four location in Tunis at latitude 36.5°N, 10.11°E. He finally concluded that for starting the pump it will require only 60% of maximum voltage at 35Hz frequency. The overall efficiency of the system was recorded to be 3.7%. Benghanem et al. [5] developed the non linear relation between discharge and solar power for performance prediction of the solar pump system. The test were conducted by varying head (50m, 60m,70m, 80m) with 24 solar panels. Campana et al. [6] said that economic study is very much required in the case of the solar pump due to the high price of PV modules. A new optimization approach has been developed based on the initial cost of investment, revenue from the crop. It is finally concluded that the proposed model deducts the total cost to be around 18.8% in a photovoltaic water pumping system. Petela et al. [7] and [8] conducted the test on solar pump for finding the exergy of the solar pump. They found the optimum surface temperature by optimizing the exergy of the solar panels. Bhattacharjee et al. [9] performed the simulation on the solar pump at Maximum Power Point Tracking (MPPT) with Vanadium Redox Flow Battery (VRFB). Finally, it is concluded that pump characteristics parameters like the current, voltage and torque curves fluctuate in absence of VRFB when it directly connects the PV module with MPPT. Foster et al. [10] installed and studied 130 types of solar pumps in eight states of Mexico which were of few kilowatts to 2 KWp. This study provides the big data for the financial study of the solar pump technology and gives a comparison between PV technology with the traditionally used methods of irrigation.

2 Analysis of Solar Pump

Some mathematical equations and empirical relations are used to find out the theoretical performance of the solar pump on the basis of latitude, longitude, hour angle and slope of the solar plate on a particular location.

The solar radiation flux is changed by tilted the solar plate at certain angle, so total radiations are the sum of beam and diffuse radiation falling directly on the surface and reflected radiation on to the surface by surrounding. The governing equation used for calculating the total solar radiation on tilted surface is given by

$$HT/Hg = (1 - Hd/Hg) * Rb + Hd/Hg * Rd + Rr \tag{1}$$

where H_d = Total diffuse radiation.

Rb = Tilt factor find out by equation suggested by Liu and Jordan [3] for a south facing surface.

$$\begin{aligned} Rb = {} & (wst * \sin(\delta) * \sin(\varphi - \beta) + \cos(\delta) * \sin(wst) * \cos(\varphi - \beta)) \\ & /(ws * \sin(\varphi) * \sin(\delta) + \cos(\varphi) * \cos(\delta) * \sin(ws)) \end{aligned} \tag{2}$$

RD = $(1 + \cos(\beta))/2$ diffuse radiation tilt factor.

Rr = $(1 - \cos(\beta))/2$ Tilt factor for reflected radiation.

According to the specification the maximum power produced by solar panel is 335 W but due to no. of losses the actual power produced is less than the 335 W. Due to exposer of solar panels for long time in atmosphere the temperature of panel increases which increase the further heat losses in atmosphere. The actual power produced is

$$P_{max(actual)} = [(1 - L_T) * (1 - L_C) * (1 - L_B) * (1 - L_M)] * P_{max} \tag{3}$$

where L_T = Temperature loss factor, L_C = Cable loss factor, L_B = Battery loss factor, L_M = Mismatch loss factor and P_{max} = Maximum power produced by pv panel.

Temperature Loss Factor

The temperature loss factor is calculated by equation

$$L_T = (P_T)loss/P_{max} \tag{4}$$

Now

$$P_{max} = C_P * P_{max} * (T_m - T_{mref}) \tag{5}$$

$$Tm = Ta + (NOCT - 20) * (I_T/800)\,^{\circ}C \tag{6}$$

where NOCT = Nominal operating cell temperature.

$$\eta_{Cell} = V_m * I_m / Q_{in} \tag{7}$$

Now apply the 2nd law of thermodynamic for calculating the exegetic efficiency or 2nd law efficiency of the solar palate. So according to 2nd law

$$EX_{IN} - EX_{OUT} = EX_{DESTRUCTON} \tag{8}$$

The exergy supplied to the system EX_{IN} is given by [4, 5]

$$EX_{IN} = A * It * \left[1 - (4/3) * (Ta/Tsun) + (1/3) * (Ta/Tsun)^4\right] \tag{9}$$

where Ta is the atmospheric temperature, Tsun is the temperature of sun, it is the solar flux falling on the solar plate and A is the area of the solar plate.

Exergy out EX_{OUT} from the system is due to increase the temperature of the solar plate as temperature further increase more and more heat loss occur due to convection and radiation and loss in availability takes place. The exergy loss in the atmosphere due to high module temperature is given by [6].

$$EX_{OUT} = (hc + hr) * A * (Tm - Tsky) * (1 - Ta/Tm) \tag{10}$$

where hc = heat transfer coefficient $W/m^2 {*} K$ given by [7].

hc = 2.8 + 3 V, where V is the velocity of the air passes over the solar plate.

hr = radiation heat transfer coefficient given by [8].

$$hr = \varepsilon * \sigma * (Tsky + Tm) * (Tsky2 + Tm2) \tag{11}$$

where ε = Emissivity of the panels, and σ = Stefan Boltzmann constants, Tsky is the temperature of sky given by [8].

$$Tsky = (Ta - 6) \tag{12}$$

Tm = module temperature given by Eq. (9).

The electrical work is the pure form of exergy because it is 100% convertible into the work so it is recovered form of exergy.

$$EX_E = V_m * I_m \tag{13}$$

Now the exegetic efficiency

$$\Phi = \text{Exergy Recovered/Exergy Supplied}$$

$$\Phi = V_m * I_m / \left[A * It\left[1 - (4/3) * (Ta/Tsun) + (1/3) * (Ta/Tsun)^4\right]\right. \tag{14}$$

3 Result and Discussion

Figure 1 shows the variation of monthly average daily solar radiation (kWh/m^2) month wise at three different tilt angle 18.86, 28.86 and 38.86. It shows the minimum solar radiation in month of December which is considered as the worst month of the year. As per the graphical representation the maximum solar radiation is available in month of June, it occurs at the tilt angle of 18.86. The number of modules and controller size are calculated on the basis of worst month of the year in which minimum radiation is getted by sun.

Figure 2 shows the graphical representation of three important variables (Mean cell Temperature, Actual Loss and Temperature loss coefficient) with the month of the year. It can be clearly seen that the mean cell temperature (T_m) is maximum in month of June due to that the maximum energy losses occur due conduction, convection and radiation.

Figure 3 shows the graphical representation of exergy reached and the solar flux fall on the solar plate. It can be easily seen from graph that input exergy reached on the solar plate is proportional of the solar flux. When the solar flux is high the exergy available on solar plate.

Figure 4 shows the exergy lost in the atmosphere and the energetic efficiency, the graph represents the inverse relation between exergy lost in atmosphere and the energetic efficiency. It is finally concluded that maximum exergy efficiency is near

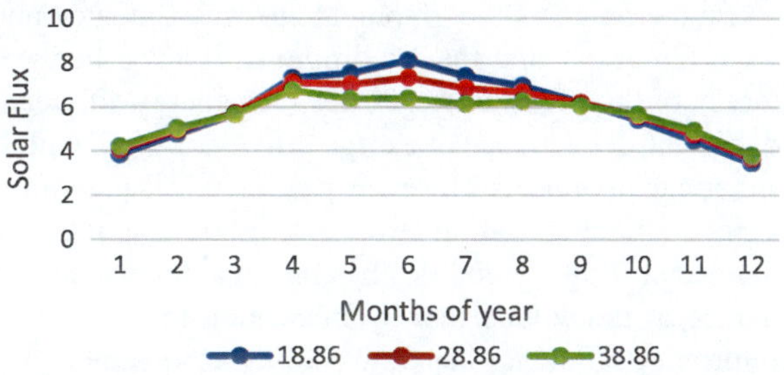

Fig. 1 Solar flux on the basis of tilt angle

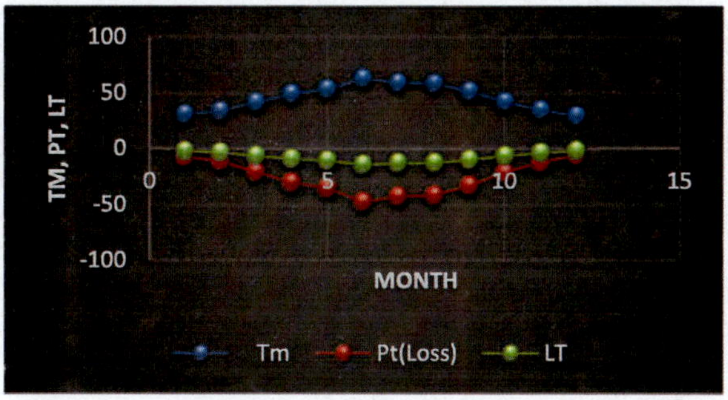

Fig. 2 Variation of cell temperature, actual loss and temperature loss coefficient with month of year

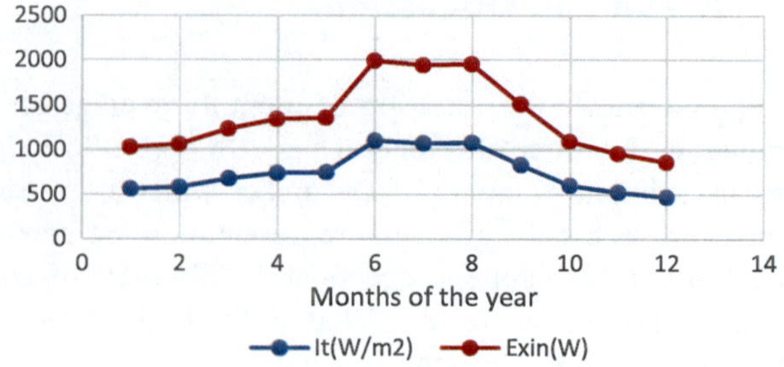

Fig. 3 Plot of graph of input exergy and solar flux fall on the solar pump

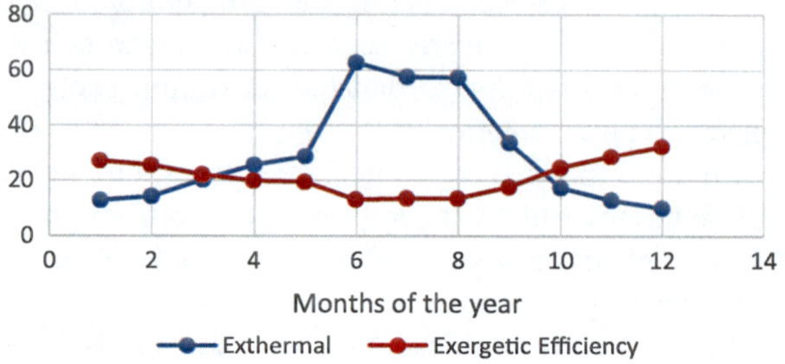

Fig. 4 Plot of graph of thermal exergy lost in atmosphere and exegetic efficiency of solar plate

to 20% in the months of spring seasons. In the spring seasons the loss is minimum as seen from Fig. 4.

Figure 5 shows the graphical energy efficiency and exergy efficiency it is noted from the graph that the maximum difference between both efficiencies is 22% in month of July. The difference between energy efficiency and exergy efficiency occurs due to the dependence of exergy efficiency on the module temperature and the atmospheric temperature. The exergy efficiency is maximum when the module is working at its nominal operating temperature but in months of May, June, and July there is intense heating of modules by sun rays. Intense heating severely affects the performance of the solar panel by increasing the loss of energy in the atmosphere. For improving efficiency separate cooling arrangement is provided either by air or water depends upon the size of the solar pump system.

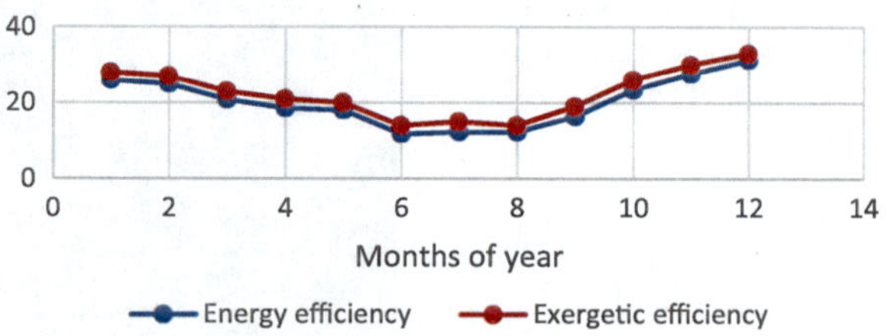

Fig. 5 Representation of its law and 2nd law efficiency

4 Conclusion

An theoretical analysis is carried out and some conclusions are made:

(i) The theoretical calculation of solar flux on inclined plate on three different tilt angle 18.86, 28.86 and 38.86° in all month of year have been completed. Which shows that in summer season 18.86° tilt angle give the maximum solar flux, in winter season 38.86° tilt angle give the maximum solar flux and in spring season 28.86° tilt angle give the maximum solar flux.

(ii) Theoretical calculations of actual power output on solar plate shows that if the cell temperature is maximum then power loss coefficient is maximum give the low efficiency.

(iii) The theoretical calculations on 1st law and exergy efficiency analysis shows that the exegetic efficiency and 1st law efficiency is minimum in month of summer and maximum in months of winter. This occur due to higher cell temperature promote the higher energy loss in summer season when highest solar energy is available. The maximum exegetic efficiency, 1st law efficiency is simultaneously 33 and 31.5% in month of December.

Acknowledgements This work was supported by the collaborative project scheme (CRS) fund under NATIONAL PROJECT IMPLEMENTATION UNIT (NPIU) (A Unit of MHRD, Govt. of India for Implementation of World Bank Assisted Projects in Technical Education) [CRS Project ID: 1-5736521897].

References

1. Rohit KB, Karve G, Khatri M (2013) Solar water pumping system. Int J Emerg Technol Adv Eng 3:225–259
2. Kou Q, Klein SA, Beckmen WA (1997) A method for estimating the long-term performance of direct-coupled PV pumping system. Pergamon
3. Liu BYH, Jordan RC (1960) The interrelationship and characteristic distribution of direct, diffuse and total solar radiation. Sol Energy 4:1
4. Belgacem BG (2012) Performance of submersible PV water pumping systems in Tunisia. Energy Sustain Dev 16:415–420
5. Benghanem M, Daffallah KO, Almohammedi A (2018) Estimation of daily flow rate of photovoltaic water pumping systems using solar radiation data. Results Phys 8:949–954
6. Campana PE, Li H, Zhang J, Zhang R, Liu J, Yan J (2015) Economic optimization of photovoltaic water pumping systems for irrigation. Energy Convers Manag 95:32–41
7. Petela R (2003) Exergy of undiluted thermal radiation. Sol Energy 74:469–488
8. Petela R (2008) An approach to the exergy analysis of photosynthesis. Sol Energy 82:311–328
9. Bathacharjee A, Mandal DK, Saha H (2016) Design of an optimized battery energy storage enabled solar pump for rural irrigation. SPECIE
10. Foster R, Cota A (2013) Solar water pumping and advance economics. Energy Procedia 57:1431–1436

Experimental Analysis of a Thermoelectric Air-Conditioning System with Desiccant Dehumidification

Anurag Maheswari, Manoj Kumar Singh, Yogesh K. Prajapati, Niraj Kumar, and Vineet Singh

Abstract Cooling system based on vapour compression refrigeration system runs on high amount of electricity and refrigerants that are responsible for green house emission. In order to protect environment and to save valuable high grade energy there is a need to replace VCRS based cooling systems with some alternate methods having less energy demands. Desiccant based system which is driven by renewable energy is one of the better substitute to replace VCRS. In the present work, a new approach for the air-condoning is adopted using the amalgam of desiccant dehumidification and thermoelectric cooler for cooling and drying of the process-air up to human comfort condition approximately 22 °C and RH 50% defied by The American Society of Heating, Refrigerating and Air-Conditioning Engineers (ASHRAE) before entering to the cooling chamber. Maximum COP of the solar power driven system is reported to be 0.861 and it as an eco-friendly alternative of VCRS based systems.

Keywords Desiccant dehumidification · Thermoelectric cooling · Coefficient of performance · Air-conditioning

1 Introduction

Approximately 40% of the energy consumed in the world is used in buildings and around 50–60% of energy is associated with HVAC systems, out of which 30% contribute to greenhouse effects [1]. Cooling system based on VCRS runs on high amount of electricity and refrigerants that are responsible for green house emission [2]. VCRS based cooling system should be replaced with some environment friendly alternative to save environment and high grade energy [3]. Desiccant wheel system reduces the air humidity with an increment in temperature and when they integrated

A. Maheswari (✉) · M. K. Singh · N. Kumar · V. Singh
Department of Mechanical Engineering, F.E.T. M.J.P. Rohilkhand University, Bareilly, U.P., India

Y. K. Prajapati
Department of Mechanical Engineering, NIT Uttrakhand, Srinagar (Garhwal), Uttarakhand 246174, India

R. Agrawal et al. (eds.), *Recent Advances in Industrial Production*, Lecture Notes in Mechanical Engineering, https://doi.org/10.1007/978-981-16-5281-3_40

with a cooling system like thermoelectric cooler results in dehumidified cooled air, which can be circulated in the preferred location to achieve desired cooling.

Desiccant wheel based air-conditioning system (DWBACS) has three main process i.e. regeneration, dehumidification and cooling [4]. Regeneration and cooling requires energy input. A solar DWBACS has been successfully used in different types of buildings [5] in which solar energy is used for regeneration. Electricity can also be used as a regeneration source for desiccant wheel that can be used in night time application.

This work presents theoretical and experimental study of DWBACS, in which regeneration occurs with an electric heater, dehumidification with desiccant wheel (zeolite material) cooling is done by thermoelectric cooler unit and the entire system is run by solar power system. Thermoelectric coolers have many advantages but they still have less application in air-conditioning as it has less efficiency compared to VCRS [6]. In this work a more efficient design of thermoelectric cooler has been proposed. Subsequent paragraphs, however, are indented.

2 Literature Review

Desiccant based dehumidification systems for air-conditioning are considered friendly for environment and cost effective in nature as compared to conventional systems like VCRS [7]. Enormous works have already been carried on desiccant cooling systems to increase the possible usage and overall effectiveness [8]. Comparison of different desiccant systems with solar regeneration show that the effect of atmosphere temperature and humidity affect the COP and exergy efficiency of the system [9]. Performance of desiccant dehumidification based evaporative coolers system investigated and concluded that maximum COP to be 0.62 [10]. As per several studies, different desiccants configurations lead to different energy consumption and cooling effect and COP. In evaporative cooled DWBACS, solar energy contributes (SF factor) 70% for regeneration and have average COP of 0.91 [11].

Solar integrated desiccant cooling system is experimentally proposed that provide required thermal comfort with SF of 60% [12]. A solar-thermal desiccant cooling system comprising hot water production has been analysed and several performance procedures used to maximize its operation based on climate problems, cooling conditions or relaxation conditions from the indoor climates and the period of use. A desiccant sub system is also analyzed to evaluate its cooling functionality, in terms of total thermal COP (ratio between cooling and regeneration capacity) [13]. Eicker et al. [14], the potency of a solar powered collector used to drive that the desiccant wheel based cooling system was evaluated, emphasizing on two different kinds of dehumidification effectiveness [14]. The performance parameters are evaluated with experimental study, like the Moisture Removal capacity, MRC, Sensible energy ratio [15].

Several researchers work on TEC (thermos-electric cooler) system and studied performance and optimization advancement of electronics system by assembling TE modules [16]. A mobile solar panel fridge designed for using rural places. It tested for assorted parameters and inside temperature has been reduced from 27 to 5 °C in approximately 44 min and COP was computed about 0.16 [17]. By controlling idling voltage electricity consumption of the refrigerator can be reduced by 32% and COP can be increased by 64% [18]. Thermoelectric modules are connected to cold and hot side heat sink so the design of both is very important. To enhance the cooling performance of TEC mini channel heat sinks are very efficient [19]. Researchers has been devoted a major amount of research to investigate desiccant wheel based air conditioning and thermoelectric cooling systems separately but their amalgam is not being studied as per author knowledge.

3 Methodology and Experimental Setup

In the proposed research, an air-conditioned system with desiccant dehumidification and thermoelectric cooling as (see Fig. 1), is developed and studied. The process starts with sucking ambient air in to the chamber of desiccant wheel through the vent. The desiccant wheel is used to absorb the moisture from the air and dehumidified it, and then this process air is transferred to thermoelectric cooler where it is cooled to the comforting temperature. Finally, this cold and dry air is entered in the cooling space. In second chamber of desiccant wheel, a small electric heater is installed to regenerate the desiccant wheel.

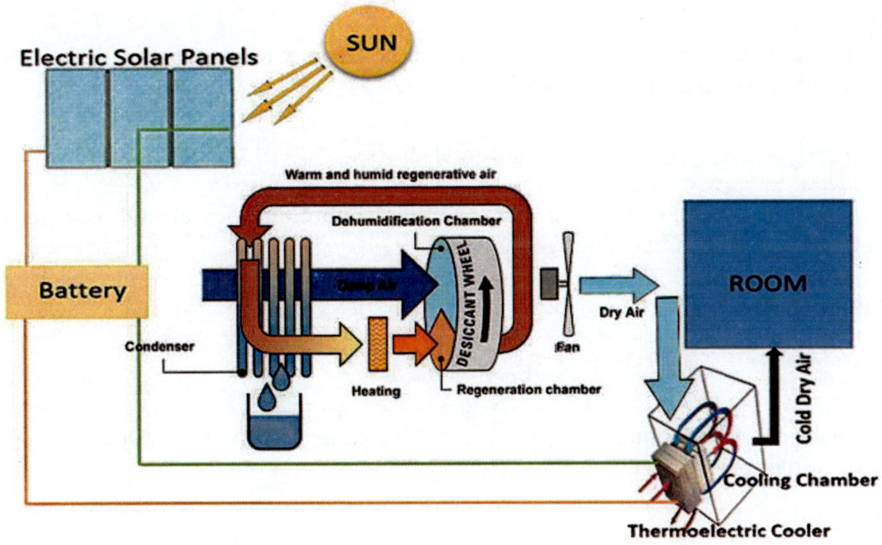

Fig. 1 Schematic of air-conditioning system

Experimental setup is consisting of four major parts i.e. Solar power system, Desiccant dehumidifier, TEC (Thermo-electric cooler) and the cooling chamber. A commercial desiccant dehumidifier is used with the dehumidification capacity of 8 L/day and for regeneration of the desiccant material, a small electric heater is incorporated within the dehumidifier. Depending upon the atmospheric condition, dehumidifier controls and maintains the relative humidity of the cooling chamber at 50%. A TEC is constructed using 28 TE module of TEC1-12706 with a frontal area of 1 ft^2. Hot side of the TE modules are connected with water cooled mini-channel heat sink and on the cold side, copper fins are used to ensure efficient cooling process air entering to the room. The exit of the dehumidifier is the entrance for the TEC, at the exit of TEC cold and dry air at required temperature is entered to the cooling area to provide the required comfort.

The walls of cooling chamber are made up of wood with thermal insulation of thermocol, the overall dimension of the cooling chamber is $4 \times 2 \times 2$ ft^3.

4 Experimental Analysis

The objective of this research is to determine the overall COP (Coefficient of performance) of the proposed system. For calculating the COP, the net cooling effect required and the total power consumed by the TEC is determined.

$$COP = \frac{Q_T}{W_{TEC}} \tag{1}$$

In the above equation, QT is the total cooling effect which means amount of heat needed to extracted from the cooling chamber.

$$Q_T = Q_s + Q_L \tag{2}$$

QS is the heat entering to the chamber after dehumidification and QL is the heat entering to the chambers through the walls.

WTEC is the net power consumed by the TEC which can be determined by using following equation:

$$W_{TEC} = n * V * I \tag{3}$$

In the above equation n is the number of TE module used, V is the electric voltage and I is the current intensity.

5 Results and Discussion

In the present work, amalgam of desiccant dehumidification and TE cooling is achieved and performance of the system is evaluated by considering two major factors COP and comfort condition inside the chamber i.e. 22 °C and RH 50% defied by ASHRAE.

The experimental investigation is carried out in the month of July at Bareilly, India. By maintaining the comfort condition inside the chamber several parameters are studied.

Figure 2 shows the variation of ambient temperature during the different time of the day and the variation the air dehumidified air entering to the TEC. The temperature of process air at the Inlet to the TEC in higher than that of ambient because of the heat added to it during the regeneration of the desiccant wheel.

Figure 3 shows the variation the cooling load requirement at the different time of the day. The major portion to the total cooling load is contributed by the heat enter in the chamber through the process air after the dehumidification. The heat entering to the chamber through the room walls is comparatively less as the room is well insulated furthermore, can be clearly seen from the trend that higher the ambient temperature higher will be the cooling load inside the chamber.

Figure 4, shows the variation of the electric voltage required by TE module with the different time of the day. It can be seen when the ambient temperature is maximum, higher electric voltage is required to run and achieve the necessary cooling and as the temperature decreases the voltage required is also decreases.

Figure 5, shows the variation of the current intensity with the time. The variation of current also follow the same trend as the electric voltage required by TE module i.e. for higher value of ambient temperature current demand also increases for the TE module.

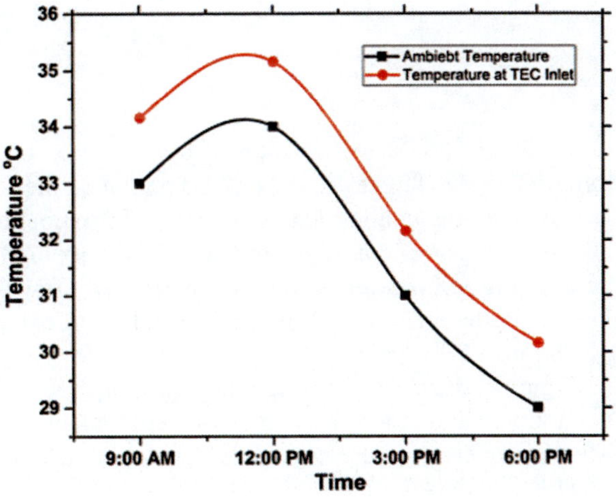

Fig. 2 Temperature variation with time of ambient and TEC inlet

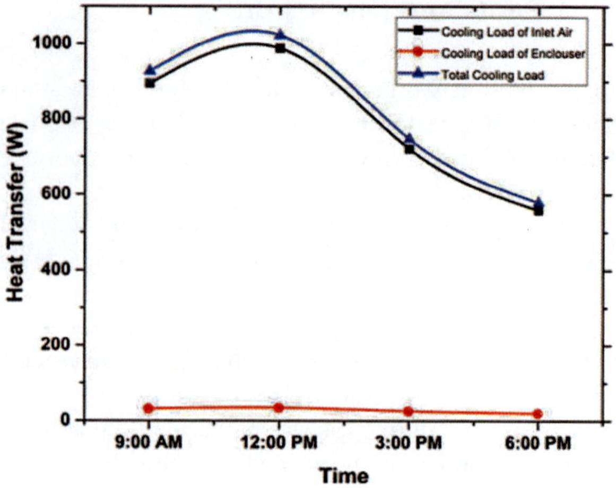

Fig. 3 Heat transfer variation with time of total cooling load, inlet air and enclosure

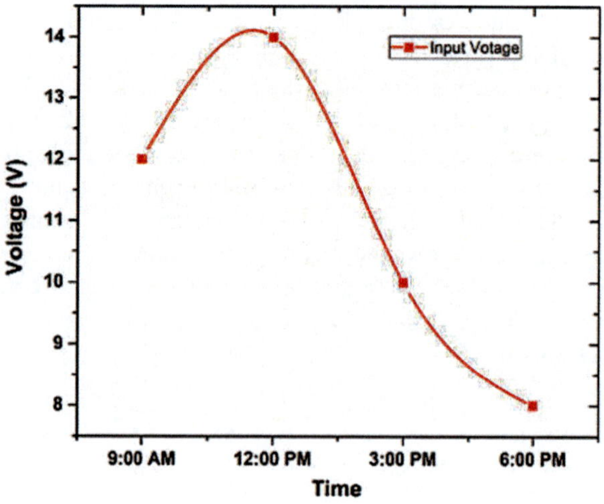

Fig. 4 Electric voltage variation with time for TE module

Figure 6, shows the variation of total cooling load and variation of the power required by the TEC to run. At 12:00 pm when the temperature of the atmosphere is maximum the cooling load is high but for extraction of that much heat from the chamber the power consumed by the TEC is very high whereas at 6:00 pm when the atmospheric temperature is very low below 30 °C the difference between the cooling load and power consumed by TEC very low in comparison with the difference at 12:00 pm.

Figure 7, shows at 12:00 pm when the temperature is maximum the value of COP is minimum and at lower value of ambient temperature the value of COP increases. This trend clearly implies that for less surrounding temperature the TEC performance of TEC is satisfactory.

Fig. 5 Current intensity variation with time for TE module

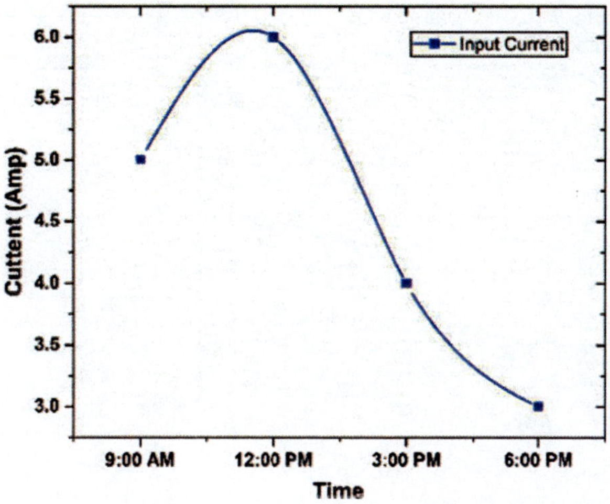

Fig. 6 Energy transfer variation with time of power consumed by of TE module and total cooling load

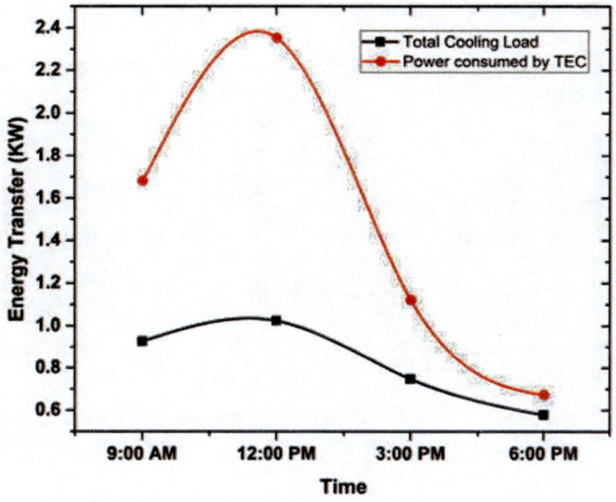

Figure 8, shows the value of the thermal resistance of the heat sink needed for the maximum cooling load and the maximum power required by the TE module is 0.431 °C/W. From the figure it is clearly seen that the thermal resistance of water cooled mini-channel used on the hot side is considerably less as per the required thermal resistance therefore the heat sink used if sufficient for effective the working of the TEC.

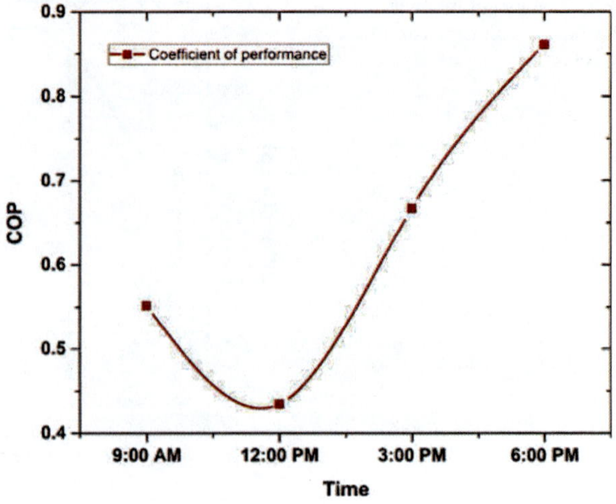

Fig. 7 COP variation with time of air-conditioning system

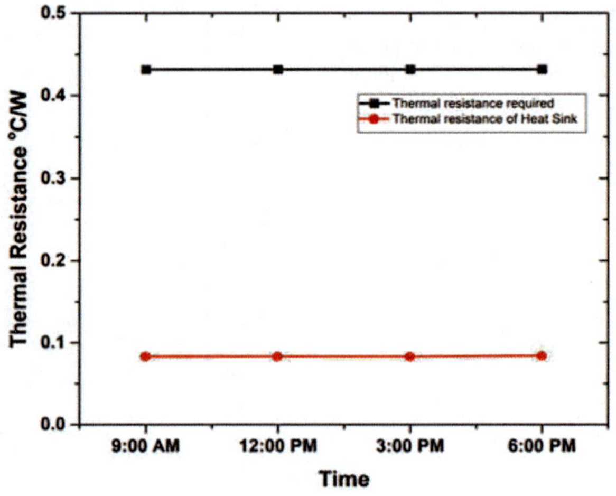

Fig. 8 Thermal resistance variation with time of heat sink and total required by TE module

6 Conclusion

In the present work a new approach for the air-condoning is adopted with using the amalgam of desiccant dehumidification and TEC for cooling and drying of the process air entering to the cooling chamber. The experimental set up is study according to the Indian weather condition in the month of July. The performance of the system is studied and based on the analysis of the results following chief findings are outlined.

- The COP of the system depend on the surrounding temperature, as the temperature rises the power consumed by the TEC increased significantly. For 14% drop of the surround temperature 90% increase in the COP is reported.

- The system is successfully able to cool and dehumidify the air with in the thermal comfort zone i.e. 22 °C and RH 50% defied by ASHRAE.

Hence it may be concluding that amalgam of Desiccant Dehumidification and TEC can provide an alternative to VCRS based system.

Acknowledgements This work was supported by the collaborative project scheme (CRS) fund under NATIONAL PROJECT IMPLEMENTATION UNIT (NPIU) (A Unit of MHRD, Govt. of India for Implementation of World Bank Assisted Projects in Technical Education) [CRS Project ID: 1-5736521412].

References

1. Heidarinejad G, Rayegan S, Pasdarshahri H (2020) Dynamic simulation of a solar desiccant cooling system combined with a ground source heat exchanger in humid climates. J Build Eng 28:101048
2. Xu F, Bian ZF, Ge TS, Dai YJ, Wang CH, Kawi S (2019) Analysis on solar energy powered cooling system based on desiccant coated heat exchanger using metal-organic framework. Energy 177:211–221
3. Berardi U, La Roche P, Almodovar JM (2017) Water-to-air-heat exchanger and indirect evaporative cooling in buildings with green roofs. Energy Build 151:406–417
4. Daou K, Wang RZ, Xia ZZ (2006) Desiccant cooling air conditioning: a review. Renew Sustain Energy Rev 10 55–77
5. Baniyounes AM, Rasul MG, Khan MMK (2013) Experimental assessment of a solar desiccant cooling system for an institutional building in subtropical Queensland, Australia. Energy Build 62:78–86
6. Patyk A (2010) Thermoelectrics: impacts on the environment and sustainability. J Electron Mater
7. Nóbrega CEL, Brum NCL (2013) Desiccant-assisted cooling: fundamentals and applications. Springer, London
8. Sultan M, El-Sharkawy II, Miyazaki T, Saha BB, Koyama S (2015) An overview of solid desiccant dehumidification and air conditioning systems. Renew Sustain Energy Rev 46:16–29
9. Abbassi Y, Baniasadi E, Ahmadikia H (2017) Comparative performance analysis of different solar desiccant dehumidification systems. Energy Build 150:37–51
10. Uçkan İ, Yılmaz T, Büyükalaca O (2017) Effect of operation conditions on the second law analysis of a desiccant cooling system. Appl Therm Eng 113:1256–1265
11. Chaudhary GQ, Ali M, Sheikh NA, ul Haq Gilani SI, Khushnood S (2018) Integration of solar assisted solid desiccant cooling system with efficient evaporative cooling technique for separate load handling. Appl Therm Eng 140:696–706
12. Comino F, González JC, Navas-Martos FJ, de Adana MR (2018) Experimental energy performance assessment of a solar desiccant cooling system in Southern Europe climates. Appl Therm Eng 165:114579
13. Enteria N, Yoshino H, Satake A, Mochida A, Takaki R, Yoshie R et al (2010) Development and construction of the novel solar thermal desiccant cooling system incorporating hot water production. Appl Energy 2010(87):478–486
14. Eicker U, Schneider D, Schumacher J, Ge T, Dai Y (2018) Operational experiences with solar air collector driven desiccant cooling systems. Appl Energy 87:3735–3747
15. Rossington D, White S, Weigand A, Sire R, Reece R, Kohlenbach P (2009) Comparison of silica gel and zeolite desiccant wheel performance. In: Proceedings of the 3rd solar air conditioning conference, Palermo (Italy), 30 Sept–2 Oct 2009, pp 230–235

16. Cai Y, Liu D, Yang JJ, Wang Y, Zhao FY (2017) Optimization of thermoelectric cooling system for application in CPU cooler. Energy Procedia 105:1644–1650
17. Abdul-Wahab SA et al (2009) Design and experimental investigation of portable solar thermoelectric refrigerator. Renew Energy 34:30–34
18. Martínez A et al (2013) reduction in the electric power consumption of a thermoelectric refrigerator by experimental optimization of the temperature controller. J Electron Mater 42:1499–1503
19. Liu X, Yu J (2016) Numerical study on performances of mini-channel heat sinks with non-uniform inlets. Appl Therm Eng 93:856–864

Fabrication of Isogrids by Conventional and Unconventional Techniques: A Comparative Review Study

K. Tripathi, K. Kukreja, and N. Gupta

Abstract In this research paper a comparison is drawn among different techniques of manufacturing grid stiffened structures, i.e., isogrid panels manufactured using non-conventional manufacturing techniques such as Fused Deposition Modeling (FDM) over existing conventional methods such as milling; by summarizing existing success analyses. These Isogrids/Grid stiffened structures are primarily building blocks used in aerospace applications such as rocket shells, satellites, space station walls and also for other structures that require additional strengthening such as armor shells, wall structures for mega buildings such as stadiums, domes, or even spider webs. A comprehensive analysis is made for different aspects of these advantages and the role of non-conventional techniques is discussed. Moreover, a hypothesis is put forward regarding design optimization and evolution in the future. The isogrids are upcoming research area which has the capacity to strengthen the aerospace, nuclear and other allied sectors.

Keywords Isogrid · Grid stiffened structures · Mass ratio · Fusion deposition modeling · Aerospace · AWJ · Laminated object manufacturing

1 Introduction

One of the industry which has gone under tremendous change since inception is the aerospace industry. The aerospace industry [1] is aiming to make effective aircraft, ranging from the passenger aero plane to rockets aiming for space exploration. Apart from advancements in the propulsion systems, the various manufacturers are now addressing a very crucial parameter called the mass ratio, which is simply the ratio of fuel mass to dry mass of the rocket. The higher the mass ratio, the more propellant is required or the dry mass of the rocket should be reduced. A higher mass ratio is required as it results in a higher delta v (a measure of impulse that is needed to perform a man oeuvre), which results in ease in maneuvering of the rocket.

K. Tripathi · K. Kukreja · N. Gupta (✉)
DTU, New Delhi, India
e-mail: Navritigupta@dce.ac.in

R. Agrawal et al. (eds.), *Recent Advances in Industrial Production*, Lecture Notes in Mechanical Engineering, https://doi.org/10.1007/978-981-16-5281-3_41

To achieve this, the reduction in the dry mass of the rocket is essential as there is a limit to increase the fuel mass. Dry mass reduction can be done by either using lightweight materials or optimizing the geometric design of the structure or even both. In the case of lightweight materials, metals like Aluminium and its alloys are being used. However the usage of Aluminium and its alloys poses some challenge. The other way is to optimize the geometric design; rocket shells are manufactured either as an isogrid or an orthogrid.

2 Isogrid Architecture

An isogrid [2] is a lattice of intersecting ribs forming an array of equilateral triangles. Their manufacturing is based on the sandwich theory, which describes the behavior of a beam, plate or shell consisting of three layers—2 face sheets and one core. This pattern, machined into solid aluminum plates, results in substantial weight savings with an acceptable reduction in structural strength. The intersections of adjacent triangles are referred to as nodes. These nodes serve as uniformly distributed attachment points for the mounting of instrumentation and other hardware. The triangular pattern is very efficient because it retains rigidity while saving material and therefore weight. Another advantage is that these structures show high stiffness and stability, especially when built from composite materials.

Figure 1 is showing a typical Isogrid. It consists of conical sub-structures. Isogrids are used wherever there is a need to increase the stiffness of thin-walled structures along with the reduction of weight. We see the use of isogrids in gas turbines engine casings, where thin-engine walls are needed to be reinforced for additional stiffness. It is seen that the hierarchical configuration of isogrids can make these structures more efficient in load-bearing capacity.

Fig. 1 A typical isogrid structure [3]

3 Manufacturing Techniques

Isogrid structures have been manufactured using different techniques. However the type of manufacturing technique is heavily dependent on the material being used for the structure itself.

The most common and heavily used material is aluminum, used for the manufacturing of rocket shells, and so on. The most common methods are/were mechanical and chemical milling processes.

In a patent application [4], many different manufacturing techniques were mentioned; one of them being chemical milling machining. One of the methods mentioned is chemical machining. In this, metal removal is achieved by reverse electroplating in which the hydroxide of the metal to be removed is produced; suspended as an emulsion in the electrolytic solution. Another method is using an NC milling machine; which focuses on manufacturing a frusta-conical structure reinforced with isogrid reinforced on its internal structure, formed from a plurality of substantially identical panels. Each panel is manufactured from the metal which will be reinforcing the final structure. The plate to be machined is positioned in an NC milling machine and the triangular pockets are formed. For a good isogrid structure, it is seen during the manufacturing process, the wall thickness of the pocket should be reduced by 1 mm. In areas where wall thickness less than 1 mm is required, chemical milling is employed. After machining, the panel is rolled or formed into the desired shape and the panels are secured together to form the final structure. The major issue with this method is the use of harmful chemicals to get the desired wall thickness of the pockets, which are hazardous to use. In another method, milling was done twice, first to remove material to the desired pocket wall thickness and second to mill around the periphery of the pocket so formed. The issue with this method is that sufficient wall thickness can't be achieved. Hence in those few pages, it was clear that a method is needed that can bring out the sufficient wall thickness without the use of chemicals.

The above-mentioned processes greatly affected commercial aspects as they are costly, require large amounts of material, and are inefficient. Milling also results in the distortion of the isogrid ribs and this technique is limited to certain isogrid geometries. Hence, a research paper [5] has mentioned the use of abrasive water jet (AWJ) technique, an unconventional method for isogrid fabrication. The potential advantages of this technique are high productivity, no residual stresses, integral machining capability and AWJ is capable of machining a wide range of isogrid geometries and for a variety of materials. More specifically, the concept of Single-angled jet in a circular tool was utilized for the study because of its versatility and ease of application. Also, before the manufacturing took place, the various parameters of AWJ like Water jet Pressure, Abrasive size, Abrasive material, Water jet flow rate, Stand-off distance etc. were determined for 2 cases; namely, linear cutting using AWJ and milling using the AWJ nozzle.

Since [6], significant progress has been made towards the optimization of isogrid designs and towards improving manufacturing techniques. A major portion of the

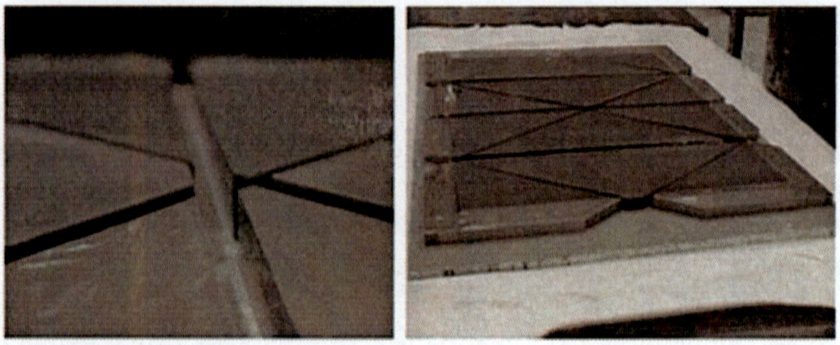

Fig. 2 Expansion block tooling [8]

efforts towards this evolution was put towards the development of composite materials in the 1970s. The use of such materials is suitable for isogrids, due to the stress being heavily distributed along the rib length as the isogrid itself is a composite structure.

Advances have been made and numerous techniques have come out which are suitable for different aspects of design and manufacturing with composite materials. Due research in mould material as well as other parameters.

In different research works, which all were aimed at reducing the drawbacks presented by conventional techniques for composite materials, by taking into focusing on changes mould design to accommodate higher nodal density, change in manufacturing technique to RFW and hybrid tooling and expansion block tooling, to achieve lower costs and higher quality through higher degrees of control of geometry [6–8] (Fig. 2).

All this evolution has led up to load specific structures called the Advanced Grid Structures focusing on:

1. Placement of fibers in a specific pattern or in a specific overlay order for each rib and so on
2. Tooling techniques suitable for different geometries
3. Mould selection to mitigate thermal expansion and other effects2 Push towards Additive Manufacturing.

4 Push Towards Additive Manufacturing

A major push has been made toward additive manufacturing techniques due to a decrease in overall material and production costs and high efficiency and accuracy while providing control over parameters like porosity. Also, the ease of prototype manufacturing for testing purposes has made a huge contribution.

In Ming Li et al. buckling tests were performed in hierarchical lattices, and tests were performed by 3D printing a plastic model using PVC engineering plastics and showed these structures have greater resistance towards local buckling and have higher global stiffness.

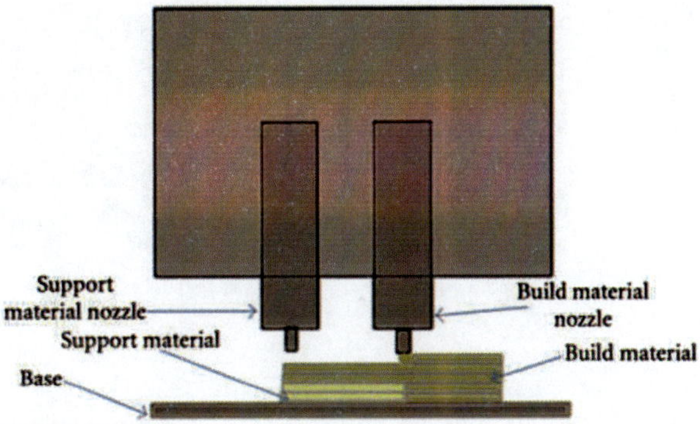

Fig. 3 The FDM process [11]

Different methods in additive manufacturing are also considered, like laminated object manufacturing (LOM) or Fused deposition Modeling (FDM) as used [9].

It is seen that there is a big role of 3D printing in the manufacturing of the isogrid structures. In a research work, a lot of focus has been given on the design of hierarchical isogrid lattice panels through additive machining [10]. In that, 3 specimen panels were taken. Each specimen consisted of 2 isogrid layers and the layers' binding together produced lattice structure with T ribs. It was seen that all these structures were 3D printed using the Laminated Object Manufacturing (LOM) technique of 3D printing, which is one of the fastest and most affordable ways to create a 3D prototype [10].

Figure 3 is depicting fusion deposition modeling technique. A lot of focus has been given on the design and characterization of an integrally stiffened structure using additive machining, certain takeaways can be obtained which highlight the role of unconventional machining, especially 3D printing [B] [9]. One major aspect highlighted was that the manufacturing of complex geometries is possible easily in 3D printing without consuming extra time as well as cost. It is also seen that 3D printing eliminates the requirement of assembly as it prints by layered manufacturing, which doesn't require assembly.

5 Role of Unconventional Techniques and Future Scope

A comprehensive problem statement was formed after the above analysis for the overall evolution of grid stiffened structures, in particular, isogrid due to its wide-ranging.

Applications in aerospace, and architecture. Figure 4 is depicting 3D printing technology being used for manufacturing isogrids. The problem statement is from an all-inclusive standpoint taking into account all parameters such as commercial aspects, and material development. It throws light on possible domains of research, namely:

Fig. 4 3D printing for manufacturing Isogrids [12]

1. Due to the emergence of the use of reusable rockets, the constraint for minimizing manufacturing duration, with the same resources, both materialistic and abstract, newer methods of production could be looked at, where focus shifts from maximum production efficiency time-wise, to methods that require a higher amount of time to execute the same operation but give a better quality of product. However, a precarious compromise must be set for each stage of this transition.
2. Due to the premise available with non-conventional manufacturing techniques, it is now possible to manufacture grid stiffened structures with lesser amounts of downsides provided (rubber mould with clearance paper reference) by conventional techniques, hence reducing imperfections and caters for design tolerances at the same time.
3. The new domain of development brought by these techniques also means that a higher degree of focus can be maintained on design optimization of the structure itself. A possible spectrum of research would be topology optimization of such grid stiffened structures for load specific purposes which may provide higher margins of cost and material saving by distributing loads in novel ways, and the manufacturing of such complex components have now been made possible by these techniques.
4. A major advantage of these manufacturing techniques is also the control over extrinsic properties due to a more local approach to the "building block" of the structure, i.e. filaments, fibers, powder chunks etc. as opposed to conventional subtractive techniques, changing the bulk properties of the overall structure and giving rise to more desired deformation behaviors on load application (insert ref).

The research in additive manufacturing technology has expanded exponentially after the development of metallic materials and alloys used in these techniques. The applications of grid stiffened structures, particularly isogrids are heavily impacted by material selection and manufacturing constraints. Lighter and more complex structures can now be produced successfully with tighter tolerances and yet lower factors of safety due to higher amounts of precise load distribution saving millions of dollars' worth production costs per rocket.

Isogrids are essentially composite structures due to multiple "layers", namely, the grid and the skin, with multiple sub-layers in hierarchical structures [10]. The structure till now has been constructed mainly out of a single material which can be viewed as a constrain. Different layers of this "composite" have different amounts of stresses and load distributions when force is applied. With the help of additive manufacturing, different metals, each with different distinctive advantages can be used in these different layers to achieve maximum overall benefit.

6 Results and Discussion

This work analyses the effect and consequences of using different manufacturing techniques for manufacturing of grid stiffened structure, from three viewpoints, namely:

1. Ease of manufacturing.
2. The reasoning for using a particular manufacturing technique and tooling methods for type and magnitude of loading on the structure.
3. Taking into account other secondary constraints as a factor, with parameters like temperature, budgetary constraints, aesthetic and applications of the structure itself.

7 Conclusion and Future Scope

It is seen that there are many areas, especially the aerospace industry which require isogrids in a large scale in order to increase the mass ratio of the aircrafts along with increase in strength. At the same time, these isogrids find a lot of use in gas turbine casings for increasing the stiffness. It is also visible that wherever hierarchical configuration of isogrids is used, load bearing capacity increases. It is also seen that a large variety of materials can be used for its manufacturing; Aluminum being the most commonly used material.

The choice of manufacturing technique will vary from material being used and at the same time, certain parameters are needed to be considered; some of them being ease of manufacturing, quality of product, cost of manufacturing along with production efficiency along with seeing whether the process is hazardous or not. From the above study, unconventional machining has taken the upper hand over the more commonly used milling on the basis of the above parameters. It was seen that AWJ technique and then additive machining has enhanced the quality of the final isogrid fabricated. Also in the future more and more use of unconventional machining will be included for the manufacturing of these structures.

References

1. Rocket Science 101: Lightweight rocket shells—Aerospace Engineering Blog. Aerospace Engineering Blog (2016). Retrieved from https://aerospaceengineeringblog.com/rocket-science-101-lightweight-rocket-shells/
2. Isogrid. En.wikipedia.org (2020) Retrieved from https://en.wikipedia.org/wiki/Isogrid
3. Huybrechts S, Hahn S, Meink T (1999) Grid stiffened structures: a survey of fabrication, analysis and design methods. In: International conference on composite materials, Paris, p 10. https://www.iccm-central.org/Proceedings/ICCM12proceedings/site/papers/pap357.pdf
4. Green R, Shore P (2005) US7631408B2 United States
5. Quest Integrated Inc. (1990) Abrasive-Waterjet machining of isogrid structures (pp. 1, 4, 6, 8). Retrieved from https://www.researchgate.net/publication/235215449_Abrasive-Waterjet_Machining_of_Isogrid_Structures
6. Kim TD (2000) Fabrication and testing of thin composite isogrid stiffened panel. Compos Struct 49(1):21–25. https://doi.org/10.1016/s0263-8223(99)00122-1
7. Sorrentino L, Marchetti M, Bellini C, Delfini A, Del Sette F (2017) Manufacture of high performance isogrid structure by Robotic Filament Winding. Compos Struct 164:43–50. https://doi.org/10.1016/j.compstruct.2016.12.061
8. Huybrechts SM, Meink TE, Wegner PM, Ganley JM (2002) Manufacturing theory for advanced grid stiffened structures. Compos A Appl Sci Manuf 33(2):155–161. https://doi.org/10.1016/s1359-835x(01)00113-0
9. Yang J (2015) Design and characterization of an innovative integrally stiffened structure using Additive Manufacturing (Post Graduate). Coventry University
10. Li M, Lai C, Zheng Q, Han B, Wu H, Fan H (2019) Design and mechanical properties of hierarchical isogrid structures validated by 3D printing technique. Mater Des 107664. https://doi.org/10.1016/j.matdes.2019.107664
11. Wong K, Wong KV, Hernandez A (2012) A review of additive manufacturing. ISRN Mech Eng 2012 (2012), Article ID 208760, 10 p. ISRN Mech Eng. 2012. https://doi.org/10.5402/2012/208760
12. Forcellese A, di Pompeo V, Simoncini M, Vita A (2020) Manufacturing of isogrid composite structures by 3D printing. Procedia Manuf 47:1096–1100. https://doi.org/10.1016/j.promfg.2020.04.123

An Integrated Lean Six Sigma Model for Enhancing the Competitive Advantage of Industries

S. K. Tiwari, R. K. Singh, and Sharad Chandra Srivastava

Abstract Overall operational excellence is one of the prime necessities for any business organization to sustain and compete in the global market. In this concern, several continuous improvement philosophies have been implemented by companies, such as lean manufacturing, six sigma, total quality management, agile manufacturing, etc. In these, lean manufacturing and six sigma shown outstanding results. Both the approaches are effective in their own, but with time the organizations implementing these approaches separately may no longer sustain improvements (Arnheiter and Maleyeff in TQM Mag 17:5–18, 2005 [1]). This is because of the various limitations associated with each of these philosophies while implementing them on the real-life problems separately. Thus, in this paper an attempt is made to combine these two popular philosophies in such a way that the weaknesses of both the manufacturing philosophies will be overcome by an integrated lean six sigma (LSS) strategy that when implemented will give better results compared to the cases when these two philosophies are implemented individually.

Keywords Lean manufacturing · Six sigma · Lean six sigma

1 Introduction

The new uprising in the manufacturing sectors has pushed great challenges for manufacturing industries. The challenges are in form of maximization of efficiency in operations, optimum utilization of plant capacity, minimization of cost of operations, quality of products and services, reduced inventories, optimum use of resources,

S. K. Tiwari (✉)
Birla Institute of Technology, Mesra—Off-campus Deoghar, Deoghar, Jharkhand 814112, India
e-mail: sanjivtiwari@bitmesra.ac.in

R. K. Singh
Birla Institute of Technology, Mesra, Ranchi, Jharkhand 835215, India
e-mail: riteshsingh@bitmesra.ac.in

S. C. Srivastava
Guru Ghasidas Central University, Bilaspur, Chhattisgarh 495009, India

R. Agrawal et al. (eds.), *Recent Advances in Industrial Production*, Lecture Notes in Mechanical Engineering, https://doi.org/10.1007/978-981-16-5281-3_42

and reduction of production wastes. In this competitive environment manufacturing enterprises are striving to develop, fast responsive and customer focused techniques that maximize the manufacturers return on all resources—capital, materials, equipment, facilities, personnel, energy and most importantly time. To meet these business objectives and to improve the competitive advantage the companies have adopted several manufacturing strategies to offer superior performance compared to their competitors. In these manufacturing strategies, lean manufacturing (LM) and six sigma (SS) have shown outstanding results; and are being considered as powerful business strategies to achieve competitive advantage.

The concept of six sigma was, initially, developed and implemented by Motorola Corporation and has gained substantial advantages. Subsequently, many more giant US companies, like GE and Allied Signal, has been applied this philosophy to remain competitive and sustain in the fast-growing market [2]. Although, the concept of lean manufacturing was first originated at Toyota Motor Corporation and hastily adopted by various companies like, Ford, Jhon Deere, Intel, Caterpillar Inc., Harley-Davidson, Nike, and many more [3]. All these large companies and many others have realized dramatic results by implementing either lean or six sigma methodologies in their organization [4].

Even though both approaches are effective in their own fashion, the organizations implementing these approaches separately may no longer able to get further improvements [1]. This means that after gaining substantial improvements by resolving the key issues, it becomes tedious to generate further improvements by implementing lean and six sigma separately [1]. Consequently, companies started to search certain alternatives to remain competitive. In this concern, the organizations that have been successfully adopted lean principles are exploring six sigma and vice-versa. This exploration originated the term 'Lean Six Sigma (LSS)', which has been used to elaborate a management system that combines these two popular philosophies [1]. The main aim of this paper is to develop an integrated lean six sigma model that combines the benefits of lean manufacturing and six sigma philosophies for improving the value of the product from the customers' viewpoint by reducing variability and wastes.

The remaining paper is organized as follows. Section 2 briefly discusses the basic of lean six sigma. In Sect. 3, a thorough review of different lean six sigma framework/model proposed by different authors is presented. Section 4 highlights the research gaps and the limitations of previous proposed lean six sigma models. In Sect. 5, a generic lean six sigma model is presented which can be implemented in any discrete industries. Finally, Sect. 6 concludes the paper.

2 Lean Six Sigma

The phrase "Lean Six Sigma (LSS)" is used to describe the integration of lean manufacturing and six sigma philosophies for improving the organizational performance [5]. In recent years, lean six sigma has been evolved as a powerful business transforming strategy and problem-solving tool for deploying continuous improvement

(CI) [6]. LSS may be defined as "a methodology that focuses on continuous improvement in order to boost the competence of an organization by reducing production costs [7], maximizing the value for shareholders by improving quality [8], and improving customer satisfaction [2]".

The origin of LSS can be traced back during 1986 in the US-based George Group [9], but was popularized during 2000 by GE [2, 8]. Subsequently, application of LSS has been drastically increased in the industries to have a competitive advantage. Some of eminent organizations that have been adopted LSS are GE, Motorola, Du Pont, Merck, Honeywell, Bank of America, Johnson & Johnson, etc. [8]. Moreover, application of LSS have not been limited to only larger organizations, but it has also been successfully deployed in SMEs (small-and-medium enterprises) [10].

LSS aims to target every type of opportunity for improvement within an organization. In fact, lean and six sigma is completing each other and there is an obvious relation between both methodologies, which makes it possible for the synergy of the two methodologies. Six sigma complements lean by providing statistical tools and knowledge for resolving the specific issues that cannot be tackled by simply implementing lean principles [4, 11]. Lean eliminates wastes and set up a standard, while six sigma reduces variation in the process. Moreover, only a few individuals of the company are associated within six sigma implementation process, however, lean involves everyone [11]. The integration of these two methodologies attempts to provide empowerment even at the higher-level process analysis stages, so that employees have true ownership of the process [4].

Further, if lean and six sigma are implemented as stand-alone philosophy, the results can never being effective and moreover they are confined by one another's requirement in the organization [11]. They may form two subcultures within the organization which are struggling for the same resources. In addition, they may limit the span of improvements. Thus, implementing six sigma in isolation cannot remove all types of waste from the process, and deploying lean principles in isolation cannot control the process statistically and remove variation from the process. This is the reason why after initial improvement they reach a plateau from where it becomes difficult to persist the continuous improvement culture [1, 11]. To overcome these limitations, the two approaches must be integrated so that lean can find a more scientific approach to tackle quality related issues and six sigma can expand its potential in tackling waste related issues effectively other than the variation by using lean principles [1, 11]. Therefore, the integration of these two approaches provides the organization more efficiency and helps to achieve superior performance and faster implementation compared to the implementation of each approach in isolation [9].

3 Lean Six Sigma Framework

The eagerness of being alive in the present cutthroat market forces the organizations to embrace different tools and techniques to remain competitive in the market. In this regard, LSS has been gained a huge popularity in the last few years. Ample of research related to LSS has been carried out in both industries as well as academia. This section highlights the contributions of various authors in framing lean manufacturing and six sigma into a single LSS integrated approach.

One of the most popular and highly cited LSS framework has been proposed by Kumar et al. [10]. In this framework lean tools were used within six sigma DMAIC framework to enhance the bottom-line results and win customer loyalty. The model has been applied in an Indian SME with the aim of reducing the percentage of defects occurring in the final product manufactured by a die-casting process. Subsequently, Su et al. [12] developed LSS methodology for improving the service quality of an IT help-desk service department in a multi-national company of Taiwan. Thomas et al. [6] developed an integrated lean six sigma (LSS) model for SMEs which has been further validated by conducting a case study of a UK based manufacturing industry. Chen and Lyu [7] proposed LSS integrated model, based on the previous LSS models proposed by different authors [10, 13], to improve the quality of the final product in touch panel manufacturing company of Taiwan.

Pepper and Spedding [4] presented a conceptual model for integrating lean principles with six sigma methodology by examining their rational approach for continuous improvement. Vinodh et al. [14] proposed a sophisticated LSS framework to overcome the deficiencies of the previous LSS framework like, lack of scientific management and insignificant lean anchorage [4], and application of lean tools only during the initial stages [10]. The framework has been tested by performing a case study in an Indian automotive valve manufacturing organization with the objective to improve the first time right (FTR) percentage. Gupta et al. [15] implemented LSS concept in a tyre manufacturing company of India to reduce the defects in radial tyres. Hilton and Sohal [5] proposed a predictive LSS model after a thorough study of different literatures related to the critical success factors (CSFs) that drive sustainable continuous improvement. Gnanaraj et al. [16] proposed a LSS model named as DOLADMAICS (Deficiency, Overcoming, Lean, Anchorage, Define, Measure, Analyze, Improve, Control, Stabilize) and implemented in an Indian SME manufacturing cylinder frames to validate the effectiveness of the proposed model at first level.

Jie et al. [17] proposed a novel LSS framework, for improving the productivity and reducing the lead time and cost of production of a label printing industry. Tenera and Pinto [18] proposed a LSS project management improvement model in the form of six sigma DMAIC cycle. The model has been tested on project management processes of a Portuguese Telecommunication Company. Garza-Reyes et al. [19] presented a refined LSS framework for reducing the ship loading commercial time of an iron ore pelletizing plant in Sultanate, Oman. Most recently, Panagopoulos et al. [20] proposed a conceptual LSS framework by combining the classical quality management tools and DMAIC lean six sigma methodology to improve the safety performance and reduce the variability of measurement system in aviation.

4 Research Gap

Literature revealed that in the two decades a noticeable work has been performed in the area of LSS. In most of the works researchers tried to present a holistic methodology for integrating lean manufacturing and six sigma. However, none of them has presented an appropriate blending of these two powerful continuous improvement philosophies. Most of the authors advocated for three different methods to integrate lean and six sigma principles into a single paradigm. In the first method, six sigma DMAIC methodology has been used as lean six sigma framework in which lean tools were used in the different phases of DMAIC methodology [7, 10, 14, 17–20]. Secondly, both the philosophies were implemented subsequently [4] and in the third method both were implemented in-parallel [12, 15, 16].

Further, there was no focus on selecting appropriate projects in the entire LSS frameworks, as selection of appropriate project is one of the vital and foremost steps in thc long term success of the organization [10, 14]. Moreover, some of the authors only presented a conceptual LSS model without validating it [4] and others confined the applicability of their LSS framework to a specific problem or industry type [7, 12, 17, 19]. All the above methods do not provide a real sense of integration these two powerful philosophies. By using these methods, it is not possible to utilize the full benefits of both the philosophies. Thus, it is necessary to develop a model that reaps the benefits of both these philosophies while eliminating the limitations.

5 Proposed Lean Six Sigma (LSS) Framework

In lean manufacturing, the value of a product is defined solely based on what the customer required and willing to pay for. Keeping in view the insight details of the manufacturing operations, various activities performed can be categorized into three main segments: (i) value-adding (VA) activities, (ii) non-value-adding (NVA) activities, and (iii) necessary but non-value-adding (NNVA) activities [21].

Therefore, the proposed model is based on the fact that total activities required to make a product (TPA) can be grouped into three main segments as discussed above. That is,

$$\text{TPA} = \text{VA} + \text{NVA} + \text{NNVA} \tag{1}$$

where TPA is total product activities, VA is value added activities, NVA is non-value-added activities, NNVA is necessary but non-value-added activities.

In this research, NNVA is considered same as the VA. Therefore, Eq. (1) can be written as:

$$\text{TPA} = \text{VA} + \text{NVA} \tag{2}$$

To make simplified calculations, let TPA = T, VA = U and NVA = V, therefore Eq. (2) can be written as:

$$\mathrm{T} = \mathrm{U} + \mathrm{V} \tag{3}$$

Let U′ and V′ be the normalized value of value-added activities, non-value-added activities, and total activities, respectively. Therefore,

$$\mathrm{U}' + \mathrm{V}' = 1 \tag{4}$$

Equation (4) represents an equation of a line in the Cartesian space in Quadrant 1 such that U′ and V′ are a set of positive rational numbers between 0 to 1 represented by the set shown below,

$$\{0 \leq (U/T, V/T) \leq 1, U \in G, V \in G, T \in G, T \neq 0\}$$

where, G is a set of positive integers. This represents the line of Lean Six Sigma as shown in Fig. 1.

Any point on the line U′ + V′ = 1 represents the total product activities T′ = 1 with its constituents U′ and V′. Now, let Z be the value of the product from a customer perspective. With the increase in non-value-added activities, the unnecessary cost associated with these activities also increases. If the same quality product is manufactured at higher cost the value of the product decreases. Let 'k' is a constant which is 'penalty'. Higher the non-value-added activities, higher will be the penalty on the value of product from the customer's perspective. The goal of LSS is to minimize the component V′, at the same time maximize the component U′ to make the total product value Z to be maximized. From Fig. 1, it can be concluded that the optimum value of this optimization model is at the coordinate point (0, 1) i.e., at maximum U′ = 1 and minimum V′ = 0. Mathematically, the model can be expressed as:

Fig. 1 Direction of lean six sigma

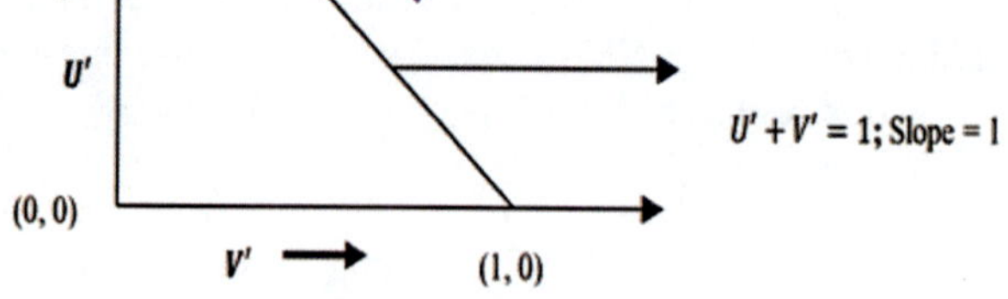

$$\text{Maximize } Z = U' + kV' = \frac{U}{U+V} + k\frac{V}{U+V}$$
$$\text{Subject to : } U' + V' \leq 1$$
$$0 \leq (U', V') \leq 1 \tag{5}$$

where, k is penalty constant which varies from case to case.

Figure 2 portrays the effect of non-value-added activities on the total product value. Therefore, it is evident that the value of the product increases with decrease in the non-value-added activities. Let us consider that U and V consist of n and m numbers of non-empty disjoint sets, respectively. Thus,

$$U = \{x_1, x_2, \ldots, x_n\}\forall x_i \; are \; VA \; activities, and$$
$$V = \{y_1, y_2, \ldots, y_m\}\forall y_i \; are \; NVA \; activities.$$

Therefore, Eq. (5) can be written as:

$$Maximize \; Z = \frac{\sum_{i=1}^{n} x_i}{\sum_{i=1}^{n} x_i + \sum_{i=1}^{m} y_i} + k\frac{\sum_{i=1}^{m} y_i}{\sum_{i=1}^{n} x_i + \sum_{i=1}^{m} y_i}$$
$$Subjected \; to : \frac{\sum_{i=1}^{n} x_i}{\sum_{i=1}^{n} x_i + \sum_{i=1}^{m} y_i} + \frac{\sum_{i=1}^{m} y_i}{\sum_{i=1}^{n} x_i + \sum_{i=1}^{m} y_i} \leq 0$$
$$0 \leq (x_i, y_j) \leq 1 \quad \forall i \; and \; \forall j \tag{6}$$

In this problem the main objective is to maximize Z which can be achieved by either by reducing V′, or by increasing U′. Moreover, the value of U′ shall always increase and V′ shall decrease. However, in real world the value of the product increases with decrease in U. This happens because with reduction of U the lead time, inventory and cost of production also reduce. However, in the equation it is just reverse i.e., ‘Z’ reduces with reduction in U. To overcome this problem, the real-world scenario is adopted in the equation. After any improvement activity U

Fig. 2 Effect of non-value-added activities on the total value of the product

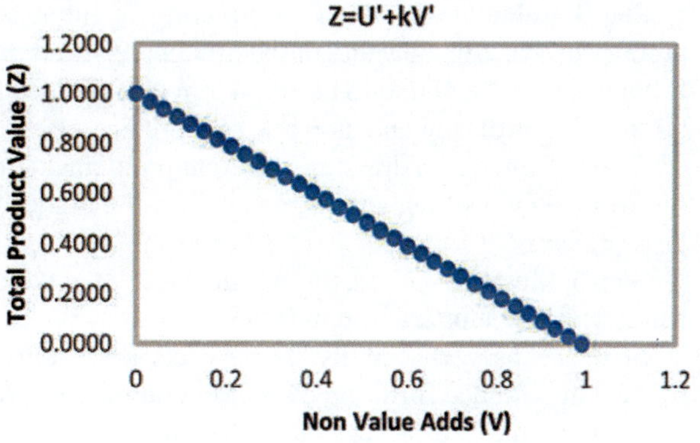

may be changed by ΔU. Thus, Eq. (5) can be written as:

$$
\begin{aligned}
&Maximize\ Z = \frac{U \pm \Delta U}{(U \pm \Delta U) + V} + k\frac{V}{(U \pm \Delta U) + V} \\
&Subjected\ to: \frac{U \pm \Delta U}{(U \pm \Delta U) + V} + \frac{V}{(U \pm \Delta U) + V} \le 0 \\
&0 \le (U \pm \Delta U, V) \le 1
\end{aligned} \tag{7}
$$

In Eq. (7), if the value of U decreases by ΔU than ΔU becomes positive (+). In other case if due to some reason U increases by ΔU than ΔU becomes negative (−). Based on the mathematical formulation LSS model is proposed in this research which is shown as a flow chart in Fig. 3. From Fig. 3, it can be realized that the proposed model is a closed loop process which ends only when V = 0 or Z becomes η (pre-defined threshold value). Moreover, the proposed model integrates the two methodologies in such way that all the NVAs (yi) present in the entire value stream can be minimized approximately up to zero. The proposed LSS model/framework is explained below.

Step 1: Draw current state map—The LSS process starts with making the current state map of the existing process using value stream mapping (VSM). In this concern, the first step is to analyze the existing process for identifying the various VAs (xi) and NVAs (yi) that make U and V, respectively. These VAs and NVAs are derived based on customers' viewpoint. Lastly, typical information like, cycle time, conversion time, percentage of defects, percentage uptime of the equipment, number of shifts the equipment works, work in-process (WIP) inventory of each stage are gathered which are required to make the current state map.

Step 2: Calculate Z—After constructing the current state map of the entire value stream, the value of the product 'Z' (in terms of time) is calculated. If the value of Z is found to be equal to η (pre-defined threshold value), then directly go to the control step and stop the LSS project, else go to the next step.

Step 3: Draw future state map—In the next step future sate map is constructed by transforming the push production system into pull production system by using the guidelines proposed by Rother and Shook [22].

Step 4: Calculate Z—After constructing the future state map by implementing pull production system, the value of the product 'Z' (in terms of time) is again calculated. If the value of Z is found to be equal to η (pre-defined threshold value), then directly go to the control step and stop the LSS project, else go to the next step.

Step 5: Define—In this step, different problems found in the current state map and future state map which are responsible for the higher V are converted into probable LSS projects.

Step 6: Measure—Next, the significance of the various LSS projects are measures using decision support system (DSS).

Step 7: Analyze—Finally the projects are analyzed and ranked in accordance to their importance form the overall operational improvement viewpoint. Thus, the most important LSS projects are selected that have the major contribution in V.

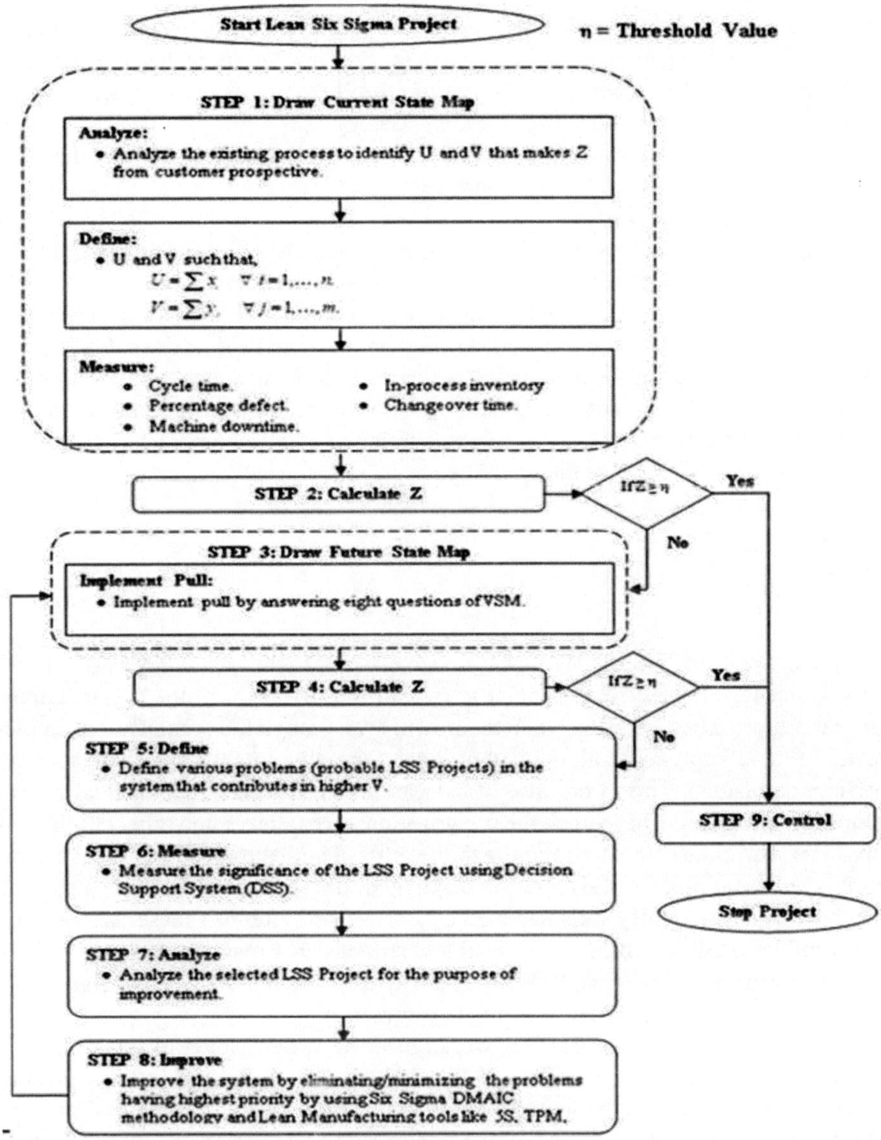

Fig. 3 Proposed lean six sigma framework

Step 8: Improve—Improve the entire process by resolving the different selected projects one-by-one in accordance with their rank. Different tools and techniques of lean manufacturing and six sigma are used for this purpose. In general, for reducing process/process variation problems DMAIC methodology of six sigma is used. And for other problems lean tools, in conjunction to other optimization techniques and statistical tools, are used. After resolving each high priority project future state map is prepared to check Z. If the value of Z reaches to η, directly move to control step, and stop the LSS improvement process or, move to the third step (draw future state map) in the search of other improvement opportunities. Due to this LSS implementation is considered as a continuous improvement strategy.

Step 9: Control—It is the last step of the LSS improvement framework. In this step, the improvements made in the previous steps should be controlled and sustained. It involves periodic check-up and analysis of the entire value stream by using different mapping and statistical tools and techniques. If any deviation is found from the threshold in the value stream, the problems are instantaneously fixed.

LSS is an iterative process which never ends and always strives for perfection. The tool set used in the above proposed LSS framework is the combination of tool sets from six sigma and lean approaches that have been widely used in both these improvement methods. The same tools set as lean manufacturing and six sigma methods can be used effectively in stages mentioned in this approach based on what tools can eventually achieve the goal of each stage mentioned.

6 Conclusion and Limitations

Lean manufacturing and six sigma have been evolved in the late 1900s as a comprehensive management system to achieve overall improvement in the organizations. Both the approaches are effective in their own, even though the organizations implementing these approaches alone may no longer are able to get improvements after attaining a plateau. This is because of the various limitations associated with each of these approaches. This forced the organizations to explore some other continuous improvement strategy that will acquire the benefits of both the approaches while overcome the limitations. This exploration originated the term 'Lean Six Sigma', which has been used to signify a management system that combines these two popular philosophies. LSS is considered as business strategy that reaps the benefits of lean manufacturing and six sigma. It focuses on delivering high performance, high value, and reliable products and services to the customers.

In the last few decades, various works have been performed to integrate lean manufacturing and six sigma [4, 6, 7, 10, 12, 14]. But still none of the work has presented an effective methodology for integrating these two philosophies. Thus, the main aim of this research is to develop an integrated LSS model that reaps the benefits of lean manufacturing and six sigma principles while discarding their limitations. The LSS framework proposed in this study will be used to improve the competitive advantage of any discrete industry. It provides a logical and well-structured LSS implementation steps. Moreover, it drives the entire organization to establish best practices to achieve competitive advantage.

The main limitation associated with this research is that only theoretical model for LSS deployment is presented. Though, empirical validation of the proposed framework is one of the prime necessities to validate its effectiveness. Thus, the future scope of this research is to validate the proposed LSS model by implementing it to solve real life problem of industries.

References

1. Arnheiter ED, Maleyeff J (2005) The integration of lean management and six sigma. TQM Mag 17(1):5–18
2. Antony J, Gijo EV, Kumar V, Ghadge A (2016) A multiple case study analysis of Six Sigma practices in Indian manufacturing companies. Int J Qual Reliab Manage 33(8):1138–1149
3. Tiwari SK, Singh RK, Srivastava SC (2020) Implementing lean paradigm in an Indian foundry facility: a case study. Int J Serv Oper Manage 36(1):20–41
4. Pepper MP, Spedding TA (2010) The evolution of lean Six Sigma. Int J Qual Reliab Manage 27(2):138–155
5. Hilton RJ, Sohal A (2012) A conceptual model for the successful deployment of Lean Six Sigma. Int J Qual Reliab Manage 29(1):54–70
6. Thomas A, Barton R, Chuke-Okafor C (2008) Applying lean six sigma in a small engineering company—a model for change. J Manuf Technol Manag 20(1):113–129
7. Chen M, Lyu J (2009) A Lean Six-Sigma approach to touch panel quality improvement. Prod Plan Control 20(5):445–454
8. Laureani A, Antony J (2011) Standards for lean six sigma certification. Int J Product Perform Manag 61(1):110–120
9. Salah S, Rahim A, Carretero JA (2010) The integration of Six Sigma and lean management. Int J Lean Six Sigma 1(3):249–274
10. Kumar M, Antony J, Singh RK, Tiwari MK, Perry D (2006) Implementing the Lean Sigma framework in an Indian SME: a case study. Prod Plan Control 17(4):407–423
11. Hess JD, Benjamin BA (2015) Applying Lean Six Sigma within the university: opportunities for process improvement and cultural change. Int J Lean Six Sigma 6(3):249–262
12. Su CT, Chiang TL, Chang CM (2006) Improving service quality by capitalizing on an integrated Lean Six Sigma methodology. Int J Six Sigma Compet Adv 2(1):1–22
13. Seth D, Gupta V (2005) Application of value stream mapping for lean operations and cycle time reduction: an Indian case study. Prod Plan Control 16(1):44–59
14. Vinodh S, Gautham SG, Ramiya RA (2011) Implementing lean sigma framework in an Indian automotive valve manufacturing organization: a case study. Prod Plan Control 22(7):708–722
15. Gupta V, Acharya P, Patwardhan M (2012) Monitoring quality goals through lean six sigma insures competitiveness. Int J Product Perform Manag 61(2):194–203
16. Gnanaraj SM, Devadasan SR, Murugesh R, Sreenivasa CG (2012) Sensitization of SMEs towards the implementation of Lean Six Sigma—an initialization in a cylinder frames manufacturing Indian SME. Prod Plan Control 23(8):599–608
17. Jie JCR, Kamaruddin S, Azid IA (2014) Implementing the Lean six sigma framework in a small medium enterprise (SME)—a case study in a printing company. In: Proceedings of the 2014 International Conference on Industrial Engineering and Operations Management Bali, Indonesia, pp 387–395
18. Tenera A, Pinto LC (2014) A Lean Six Sigma (LSS) project management improvement model. Procedia Soc Behav Sci 119:912–920
19. Garza-Reyes JA, Al-Balushi M, Antony J, Kumar V (2016) A Lean Six Sigma framework for the reduction of ship loading commercial time in the iron ore pelletizing industry. Prod Plan Control 27(13):1092–1111
20. Panagopoulos I, Atkin C, Sikora I (2017) Developing a performance indicators lean-sigma framework for measuring aviation system's safety performance. Trans Res Procedia 22:35–44
21. Monden Y (1997) The Toyota management system: linking the seven key functional areas. Productivity Press

22. Rother M, Shook J (2003) Learning to see: value stream mapping to add value and eliminate muda. Lean Enterprise Institute

Agri-fresh Supply Chain Management: A Systematic Literature Review

J. Krishna Manasvi and Rajesh Matai

Abstract The paper presents a systematic literature review of Agri-fresh supply chain management. This study emphasizes all the supply chain stages in the context of fruits and vegetables. A total of 184 relevant research publications from the year 2000 to mid-2020 were reviewed. The literature is classified based on year, across various countries, and Agri-fresh supply chain issues. The previous literature has been categorized into different issues such as post-harvest losses, technology, infrastructure, food safety, quality, logistics management, inventory management, performance management, awareness and knowledge of farmers, demand management, and pricing. Most of the research publications have concentrated on post-harvest losses and food safety aspects of the Agri-fresh supply chain. This paper provides an insight into various issues of the Agri-fresh supply chain. It uncovers the future research scope. The study includes only fresh produce such as fruits and vegetables; other perishable items like dairy, meat, flowers are excluded.

Keywords Agriculture supply chain · Agri-fresh supply chain · Horticulture supply chain · Fresh produce supply chain · Fruit supply chain · Vegetable supply chain

1 Introduction

Supply chain management is essential to match the demand and supply of any product. It provides a system's view of all the logistic activities such as production, inbound logistics, and outbound logistics. In agriculture, managing the demand and supply becomes difficult due to its perishable nature.

The agriculture supply chain is defined as: "Agriculture supply chain consists of a sequence of operations, that is concerned about the perishable nature of the produce, high fluctuations in price and demand, increasing concerns of consumer

J. Krishna Manasvi (✉) · R. Matai
Department of Management, Birla Institute of Technology and Science, Pilani, India

R. Matai
e-mail: matair@pilani.bits-pilani.ac.in

R. Agrawal et al. (eds.), *Recent Advances in Industrial Production*, Lecture Notes in Mechanical Engineering, https://doi.org/10.1007/978-981-16-5281-3_43

towards food safety and dependence on climate conditions" [1]. The fresh produce in agriculture is referred to as Agri-fresh produce. Agri-fresh supply chain is defined as "the processes from the production to consumption of fresh produce (vegetables, fruits, and flowers)" [2]. As the fresh produce is handled at various stages of the supply chain, it is necessary to maintain its quality. Agri-fresh produce contributes to various employment opportunities and has a scope of profitable ventures in farming ventures [2]. In the Agri-fresh supply chain, enhancement of the shelf-life of the product is very important. Agri-fresh supply is different from other supply chains due to its perishable nature [3].

2 Research Methodology

The Scopus database is used for scholarly peer-reviewed journals and conference proceedings. By considering a time frame from the year 2000 to mid-2020 for the systematic review. Research publications were identified using a systematic search of keywords, such as "agriculture supply chain", "Agri-fresh supply chain", "horticulture supply chain", "fresh produce supply chain", "fruit supply chain", "vegetable supply chain". Total selected papers from Scopus are 184. Most of the papers selected for the study are from high-quality peer-reviewed journals; Journal of Cleaner Production, Production and Operations Management, Supply chain management: An International Journal, Journal of Production and Economics, Food Policy, The International Journal of Logistics Management, International Journal of Operations and Production Management.

3 Descriptive Analysis of Literature

Some descriptive observations are made on the Agri-fresh supply chain literature, which is presented in the following sub-sections.

3.1 Distribution of Research Publications year-wise

This distribution represents the frequency of all the 184 research publications over the years 2000 to mid-2020. Figure 1 shows that the number of research in the Agri-fresh supply chain is increasing continuously. The majority of publications are from 2009 onwards, and this may be due to the rise of potential issues in the Agri-fresh supply chain or the interest of researchers and practitioners in this field. This analysis revealed that nearly 60 percent (113 publications) of the research publications are from recent eight years.

Fig. 1 Year-wise distribution of research publication

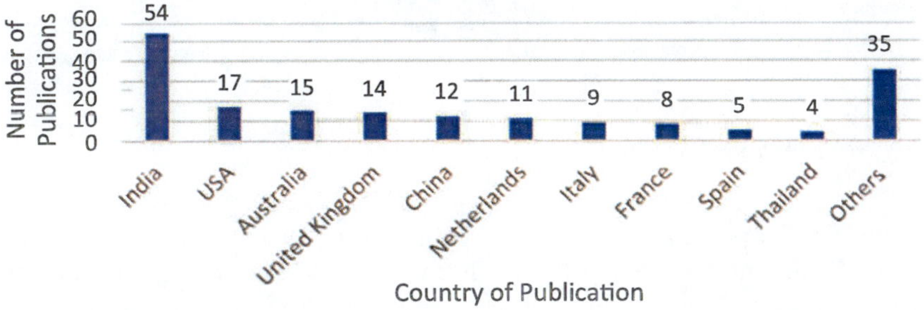

Fig. 2 Country-wise distribution of research publications

3.2 Distribution of Research Publication Country-Wise

As shown in Fig. 2, the distribution of various research publications is across developing and developed countries. The highest number of publications collected are from India, followed by the USA. Different aspects of the Agri-fresh supply chain are addressed in these countries.

Other countries include Argentina, Belgium, Brazil, Canada, Chile, Ethiopia, Fiji, Finland, Germany, Greece, Hongkong, Indonesia, Ireland, Jordon, Moscow, New Zealand, Nigeria, South Africa, Switzerland, Taiwan, Uganda, United Arab Emirates.

3.3 Distribution of Research Publication Based on Issues in Agri-Fresh Supply Chain

The previous literature has been categorized into Agri-fresh supply chain issues, such as post-harvest losses, technology, food safety, infrastructure, logistics management, awareness and knowledge of farmers, sustainability, inventory management, demand management, performance measurement, pricing, and collaboration [4]. From Fig. 3,

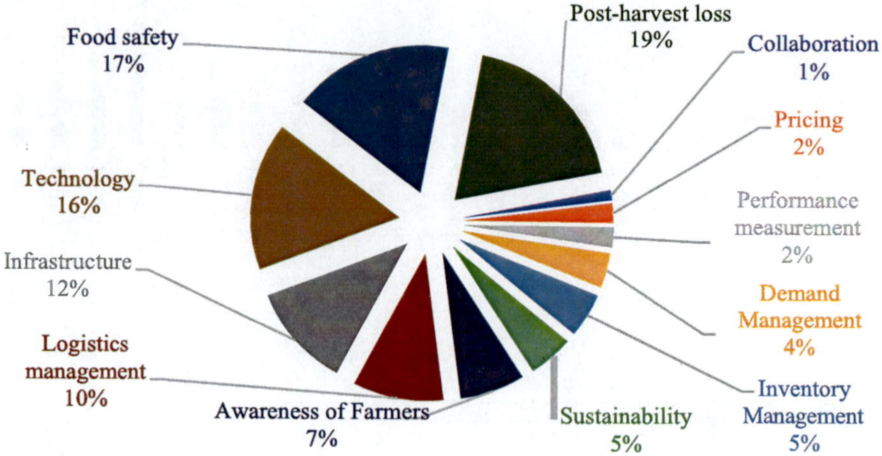

Fig. 3 Issues in agri-fresh supply chain

it is evident that post-harvest losses (35 publications, 19%), food safety (32 publications, 17%), and technology (30 publications, 16%) are the highest researched areas.

4 Various Issues of Agri-fresh Supply Chain

Significant issues identified from various publications were post-harvest losses, food safety, quality, infrastructural, technological, logistics management, awareness of farmers, sustainability, inventory management, demand management, performance measurement, pricing and collaboration [2, 5]. The following sub-sections cover the post-harvest losses, technology, food safety and quality, logistics and infrastructural issues of the Agri-fresh supply chain.

4.1 Post-harvest Loss

Post-harvest loss is a quantitative or qualitative food loss across the supply chain from harvesting to consumption level. Quantitative losses are caused due to spillage, chemical changes due to moisture content, temperature changes, and pests'. Qualitative losses are caused by poor handling and contamination by insect pests. Post-harvest losses at different stages of the supply chain, such as harvesting, food storage, processing, packaging, sales and consumption, are discussed below [6].

Losses at the harvesting level are nearly thirty percent, which is usually caused by sunburn, cracking, fruit infections, lack of adequate storage facilities [7], and lack of adequate information flow from supply chain operators and farmers [8]. Losses at the food storage level can be reduced or prevented by continually maintaining the refrigerated storage temperature of products [9], using the products before the use-by dates, and maintaining the product's adequate loaded capacity [10]. Losses at processing and packaging levels are due to changes in, firm performance (delivery performance of firms' supplier, flexibility, reliability, and responsiveness dimensions) and food supply chain uncertainties (demand-supply and price uncertainties) [11]. Losses at the consumption level are due to lack of proper training and information technology [12], overbuying products [13]. Food wastage at the harvesting level cannot be prevented in fruits and vegetable produce. Reducing or preventing wastage at other levels of farm to market is necessary [14].

4.2 Technology

Various new technologies are emerging in agriculture, such as Blockchain, the Internet of Things, Artificial Intelligence, and Big data [15]. Radiofrequency Identification (RFID) technology is being used in many industries such as packaging, shipping and transportation, manufacturing, healthcare, and medical. The application of RFID in horticulture covers tracking the inventory level across the supply chain [16]. Selection of appropriate technology is challenging; speed, reliability, and efficiency play a significant role in selection. Implementing these technologies will improve the information flow, trackability, and traceability in product loss during transportation and storage phases of agricultural produce [17]. Maintaining the flow of information in agriculture helps; advancements in information technologies are needed to know the current status of demand and supply, market price, and other information regarding agricultural produce [18]. Traceability of products is essential to maintain transparency across the supply chain. Traceability systems help in sharing information across the supply chain and also ensures the safety of Agri-fresh produce. By implementing a computerized traceability system, tracking the product across the whole supply chain becomes easy, resulting in qualitative and quantitative improvements in supply and production [19].

4.3 Food Safety and Quality

Food safety is essential for the agriculture supply chain to maintain the uniform quality of the produce [20]. With rising awareness of food safety and quality, customers prefer modern retail stores, due to the availability of high quality, food safety, security, and competitive prices offered by the retail chains [21]. To attain the market's food safety needs, collective action and public-private partnership play a vital role in farm-to-market linkage [21]. With adequate institutional support, small farmers could reach the standards of food safety [22]. Drivers of food safety risks of fruits and vegetable supply chains are climate, human behavior, and economy. Food safety risks can be avoided by attaining accurate data and taking expert opinion [23]. By improving the packaging system, the quality of the fresh produce can be maintained [20]. The freshness of the product is an essential factor of quality [24]. For the produce sensitive to freshness, it is necessary for the producer and the distributor to be coordinated to maintain the fresh produce's quality. Another factor that depends on the produce's freshness is the market demand, which can be maintained by using quality control systems from upstream suppliers to downstream customers in a supply chain. To maintain vegetable quality, factors such as storage temperature, storage time [25], reduction of system cost [26] play a significant role in the supply chain.

When fruits and vegetables are categorized according to the quality, it reduces the hazardous effects causing it. Shifting from chemicals to organic produce can also improve the quality of fruits and vegetables [27].

4.4 Logistics and Infrastructure

To attain an efficient supply chain, reduce intermediaries, increase the production volume handled through institutions and technology, implement better transportation facilities, provide accurate information to farmers regarding market prices, integrate small farmers directly with the urban market, and improve infrastructure is necessary. It allows transparency across the chain, and also, the farmer can earn more amount per kilogram sold in the market [28–32]. The logistics costs vary due to the fluctuations in demand patterns in fruits and vegetable markets; farmers need to be assisted regarding the daily volume of harvesting to overcome it [33]. The Significant challenges faced by logistic service providers are, lack of awareness regarding the maintenance of temperature during storage and transportation, channel members usually switch off the cold chain equipment in order to save electricity expenses or diesel for a specified period, which leads to a loss of perishable produce in the form of quality, color, and nutrition [34]. Furthermore, providing proper cold chain facilities across the supply chain improves the shelf life of the agricultural produce [32, 35]. By improving the efficiencies of the logistics systems, wastage of perishable products can be reduced. For a supply chain to be efficient, there should be a strong collaboration between the upstream and downstream members [36].

5 Conclusion

This paper presents a systematic literature review of AFSCM, by focusing on issues at various Agri-fresh supply chain stages. It was observed that more than 60 per cent of papers are from the recent eight years, the research in Agri-fresh supply chain has been increasing. The previous research on Agri-fresh supply chain focuses on post-harvest loss, technology, food safety and quality, infrastructure and logistics management. It is identified from the study that the most critical issues are post-harvest loss, food safety and technology.

Furthermore, there is a broader scope for research in inventory management, demand management, and pricing in Agri-fresh supply chains. Negligible studies have been conducted in the integration of various stages of Agri-fresh supply chain. Studies on implications of various technologies such as blockchain, internet of thing, artificial intelligence, big data in the Indian agriculture industry are needed.

References

1. Lemma Y, Kitaw D, Gatew G (2014) Loss in perishable food supply chain: an optimization approach literature review. Int J Sci Eng Res 5(5):302–311
2. Shukla M, Jharkharia S (2013) Agri-fresh produce supply chain management: a state-of-the-art literature review. Int J Oper Prod Manage 33(2)
3. Van Der Vorst JGAJ, Beulens AJM (2002) Identifying sources of uncertainty to generate supply chain redesign strategies. Int J Phys Distrib Logist Manag 32(6):409–430. https://doi.org/10.1108/09600030210437951
4. Siddh MM, Soni G, Jain R, Sharma MK, Yadav V (2017) Agri-fresh food supply chain quality (AFSCQ): a literature review. Ind Manag Data Syst 117(9):2015–2044. https://doi.org/10.1108/IMDS-10-2016-0427
5. Routroy S, Behera A (2017) Agriculture supply chain: A systematic review of literature and implications for future research. J Agribus Dev Emerg Econ 7(3):275–302. https://doi.org/10.1108/JADEE-06-2016-0039
6. Gardas BBB, Raut RDRD, Narkhede B (2018) Evaluating critical causal factors for post-harvest losses (PHL) in the fruit and vegetables supply chain in India using the DEMATEL approach. J Clean Prod 199:47–61. https://doi.org/10.1016/j.jclepro.2018.07.153
7. Sharma S, Shukla R (2017) Economics of post harvest losses in onion inJhunjhunu district of Rajasthan. Int J Commer Bus Manag 10(1):15–19. https://doi.org/10.15740/has/ijcbm/10.1/15-19
8. Redlingshöfer B, Coudurier B, Georget M (2017) Quantifying food loss during primary production and processing in France. J Clean Prod 164:703–714. https://doi.org/10.1016/j.jclepro.2017.06.173
9. Raut RD, Gardas BB, Narwane VS, Narkhede BE (2019) Improvement in the food losses in fruits and vegetable supply chain—a perspective of cold third-party logistics approach. Oper Res Perspect 6. https://doi.org/10.1016/j.orp.2019.100117
10. Buzby JC, Hyman J, Stewart H, Wells HF (2011) The value of retail- and consumer-level fruit and vegetable losses in the United States. J Consum Aff 45(3):492–515. https://doi.org/10.1111/j.1745-6606.2011.01214.x
11. Gokarn S, Kuthambalayan TS (2019) Creating sustainable fresh produce supply chains by managing uncertainties. J Clean Prod 207:908–919. https://doi.org/10.1016/j.jclepro.2018.10.072
12. Balaji M, Arshinder K (2016) Modeling the causes of food wastage in Indian perishable food supply chain. Resour Conserv Recycl 114:153–167. https://doi.org/10.1016/j.resconrec.2016.07.016
13. Richards TJ, Rickard B (2020) COVID-19 impact on fruit and vegetable markets. Can J Agric Econ 1–6. https://doi.org/10.1111/cjag.12231.
14. Beausang C, Hall C, Toma L (2017) Food waste and losses in primary production: qualitative insights from horticulture. Resour Conserv Recycl 126(January):177–185. https://doi.org/10.1016/j.resconrec.2017.07.042

15. Lezoche M, Panetto H, Kacprzyk J, Hernandez JE, Alemany Díaz MME (2020) Agri-food 4.0: a survey of the supply chains and technologies for the future agriculture. Comput Ind 117. https://doi.org/10.1016/j.compind.2020.103187.
16. Mapa LB, Goni F, Alam S, Aryal G (2018) Developing a radio frequency identification (RFID) as a decision support system in horticulture industry. ASEE Annu Conf Expo Conf Proc, vol 2018
17. Singh B, Sikka BK, Singh SP (2015) Enhancing global competitiveness of Indian apple: investigating the value chain perspective. Acta Hortic 1099:525–532. https://doi.org/10.17660/ActaHortic.2015.1099.64
18. Beheraa BS, Panda B, Behera RA, Nayak N, Beherae AC, Jena S (2015) Information communication technology promoting retail marketing in agriculture sector in india as a study. Procedia Comput Sci 48(C):652–659. https://doi.org/10.1016/j.procs.2015.04.148
19. Alfaro JA, Rábade LA (2009) Traceability as a strategic tool to improve inventory management: a case study in the food industry. Int J Prod Econ 118(1):104–110. https://doi.org/10.1016/j.ijpe.2008.08.030
20. Stringer R, Sang N, Croppenstedt A (2009) Producers, processors, and procurement decisions: the case of vegetable supply chains in China. World Dev 37(11):1773–1780. https://doi.org/10.1016/j.worlddev.2008.08.027
21. Singh B, Yadav I (2015) Enhancing global competitiveness for fresh produce retail shops in India: investigating consumers perceptions and opportunities. Acta Hortic 1103:267–271. https://doi.org/10.17660/ActaHortic.2015.1103.39
22. Narrod C, Roy D, Okello J, Avendaño B, Rich K, Thorat A (2009) Public-private partnerships and collective action in high value fruit and vegetable supply chains. Food Policy 34(1):8–15. https://doi.org/10.1016/j.foodpol.2008.10.005
23. Bouzembrak Y, Marvin HJP (2019) Impact of drivers of change, including climatic factors, on the occurrence of chemical food safety hazards in fruits and vegetables: a Bayesian Network approach. Food Control 97(October 2018):67–76. https://doi.org/10.1016/j.foodcont.2018.10.021
24. Cai X, Chen J, Xiao Y, Xu X (2010) Optimization and coordination of fresh product supply chains with freshness-keeping effort. Prod Oper Manag 19(3):261–278. https://doi.org/10.1111/j.1937-5956.2009.01096.x
25. Jraisat LE, Sawalha IH (2013) Quality control and supply chain management: a contextual perspective and a case study. Supply Chain Manag 18(2):194–207. https://doi.org/10.1108/13598541311318827
26. Yin HL, Wang YM (2017) An effective method for vegetable supply chain quality management. In: Chinese Control Conference on CCC, no 71362030, pp 7507–7510. https://doi.org/10.23919/ChiCC.2017.8028541
27. Gnanavel S, Manohar S, Sridhar KE, Sokkanarayanan S, Sathiyanarayanan M (2019) Quality detection of fresh fruits and vegetables to improve horticulture and agro-industries. In: Proceedings of the 4th international conference on contemporary computing and informatics, IC3I 2019, pp 268–272. https://doi.org/10.1109/IC3I46837.2019.9055558
28. Kalidas K, Nair AP, Anjana AS, Ashika KV, Athira KA (2017) Vegetable supply chain management in Kerala. Int J Commer Bus Manag 10(2):250–254. https://doi.org/10.15740/has/ijcbm/10.2/250-254
29. Dastagiri MBB et al (2012) Marketing efficiency of India's horticultural commodities under different supply chains. Outlook Agric. 41(4):271–278. https://doi.org/10.5367/oa.2012.0103
30. Negi S, Anand N (2015) Issues and challenges in the supply chain of fruits & vegetables sector in India: a review. Int J Manag Value Supply Chain 6(2):47–62. https://doi.org/10.5121/ijmvsc.2015.6205
31. Mahajan R, Garg S, Sharma PB (2011) Indian frozen peas market: a case study on FPIL. Int J Glob Small Bus 4(2):154–169. https://doi.org/10.1504/IJGSB.2011.042250
32. Sharangi AB, Datta S (2015) Value addition of horticultural crops: recent trends and future directions

33. Shukla M, Jharkharia S (2011) ARIMA models to forecast demand in fresh supply chains Manish Shukla and Sanjay Jharkharia. Int J Oper Res 11(1):2011
34. Mahajan R, Garg S, Sharma PB (2013) Frozen corn manufacturing and its supply chain: case study using SAP-LAP approach. Glob J Flex Syst Manag 14(3):167–177. https://doi.org/10.1007/s40171-013-0040-y
35. Kalpana S, Priyadarshini SR, Maria Leena M, Moses JA, Anandharamakrishnan C (2019) Intelligent packaging: trends and applications in food systems. Trends Food Sci Technol 93(September):145–157. https://doi.org/10.1016/j.tifs.2019.09.008
36. Nakandala D, Lau HCW (2019) Innovative adoption of hybrid supply chain strategies in urban local fresh food supply chain. Supply Chain Manag 24(2):241–255. https://doi.org/10.1108/SCM-09-2017-0287

Methods to Measure Residual Stresses in 3D Printed Objects: A Review

Devesh, Devender, and N. Gupta

Abstract Additive manufacturing is a rapidly growing manufacturing method used in manufacturing objects due to its accurate results and complex shapes can be made with precision. Residual stresses are the hazard to this technique which affects the mechanical properties and shape of the object that is why it is critical to measure the residual stress present in the AM part to increase the reliability of the object. These stresses can cause the strength failure and other associated failures of the object. Also if residual stresses are not removed from the object, the object is not balanced and can be fitted properly. This paper aims to put various methods in one place to measure residual stress in a 3D printed object.

Keywords 3D printing · Additive manufacturing · Neutron diffraction method · Non-destructive testing · Semi-destructive testing · X-ray diffraction method

1 Introduction

Additive manufacturing is the need of the future for accurate and complex products to be manufactured. Residual stresses are challenging the scientists and engineers by earlier failure and distortion of shapes as well as mechanical properties. It is very important to measure the residual.

Stresses present in the object to increase its reliability accordingly. There are commonly three types of methods destructive, semi-destructive, and non-destructive. Destructive in which a portion of the object is removed, and deformation is created followed by noting residual strain, then from linear elastic theory residual stresses are determined. Non-destructive methods such as neutron and X-ray diffraction methods measure crystal strain and using elastic constant stresses are determined (Fig. 1).

The most used technique is X-ray and hole drilled method (semi destructive). All these methods are flexible to use and can be repeated besides of high accuracy. In destructive methods, a part of base object is removed further, deformations are

Devesh · Devender · N. Gupta (✉)
MED, Delhi Technological University, New Delhi, India
e-mail: Navritigupta@dce.ac.in

R. Agrawal et al. (eds.), *Recent Advances in Industrial Production*, Lecture Notes in Mechanical Engineering, https://doi.org/10.1007/978-981-16-5281-3_44

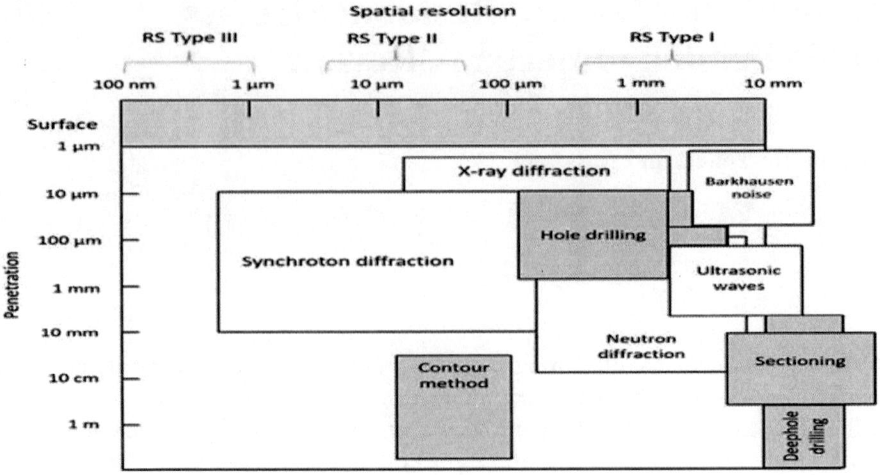

Fig. 1 Spatial resolution and penetration of RS measurement techniques [1, 2]

developed, and RS values are calculated. Semi-destructive methods are similar to destructive method only some properties are modified. Hole drilling (HD) is widely held method in which a portion is cut and then by installing strain gauges, stress profiles are measured. "In non-destructive technique (NDT), the lattice strain is estimated, and the RS is obtained using elastic constants" [1]. Some of the disadvantages of all methods include limited sample size and surface roughness and limitation in penetration power in XRD and error can occur in calculation of RS value due to inaccuracy in making hole in hole drilling.

2 Non-destructive Methods (NDT)

Non-destructive techniques are those techniques, which are aimed at testing of sample without cutting or destroying it. Such methods are radiation testing etc. Some of the nondestructive techniques are mentioned below.

2.1 Neutron Diffraction Method

Residual stress can be measured using neutron diffraction method by determining residual strains directly by quantifying the changes in lattice spacing from their stress-free condition.

A neutron beam of wavelength say λ strike the material with an interplanar spacing.

A neutron beam of wavelength say λ is allowed to strike the crystalline material having an interplanar spacing say d_{hkl}, a diffraction pattern is developed. By using Bragg relation, position of each plane (hkl) can be estimated [3].

$$2d_{hkl}\sin\theta_{hkl} = \lambda$$

The curve can be plot of angle of diffraction for each point and peak position, intensity and peak fitting errors can be estimated. The peak position get change in the specimen having residual stresses which cause strain in the strain free sample, d0, and the elastic strains ε_{hkl} are given by [4]:

$$\varepsilon_{hkl} = \frac{d_{hkl} - d_0}{d_0} = \frac{\Delta d_{hkl}}{d_0} = \frac{\sin\theta_0}{\sin\theta_{hkl}} - 1$$

where θ_0 is the angle between peak position and reference without strain [3].

Strain developed in bulk components can be measured easily using this method as high penetration depths have no effect on it unlike other strain-scanning techniques such as synchrotron radiation [3].

Type-I residual stresses can be classified using this method which "equilibrate over macroscopic dimensions" [5, 6]. If the lattice spacing is larger than reference lattice spacing then tensile stress are developed and if it is less than the reference spacing then compressive stresses will develop so comparing the measured d spacing with the stress free do spacing, the developed strain (ε) can be estimated [7]. The differences in the lattice spacing are due to the compositional irregularities and residual stresses.

The penetration power of neutron is more than X-ray which is nearly 30 mm while X-ray can merely penetrate 0.050 mm. These methods measures residual stresses at very large depth and also help in finding stress tensor.

2.2 *X-Ray Diffraction Method*

This technique is considered useful in computing the residual stresses present in the component and it very effective for materials like iron, aluminum and nickel alloys. The small depth of penetration and irradiated area helps in the measurement of residual stress distributions with depth resolution exceeding all other techniques but error could occur as it need high precision in measuring diffraction peak [8]. Diffraction peak of ranged intensity which is free from interference for any orientation of the sample surface can be produced when this technique is applied to fine grained crystalline materials to calculate stresses.

Similar to the neutron diffraction technique, Bragg relation can be used to find lattice spacing using diffraction angle 2θ and X-ray wavelength. σ_{11} and σ_{22} are applied stresses having zero stress orthogonally or perpendicular. The stress σ_φ which required determining is at angle φ to the σ_{11}.

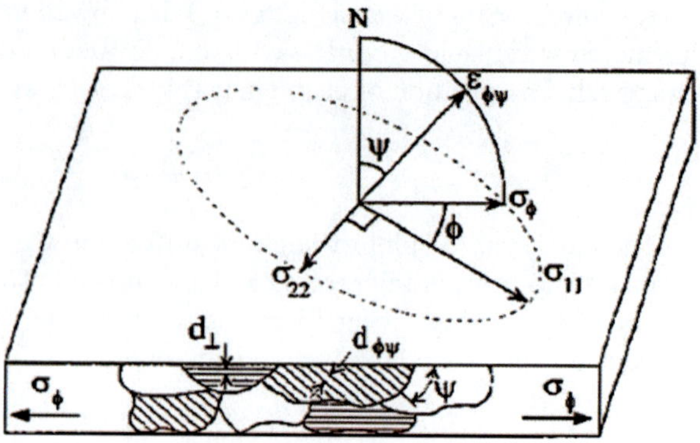

Fig. 2 Change in lattice spacing with tilt ψ shown by plane stress developed at free surface for a uniaxial stress σ_φ parallel to one edge [8]

In Fig. 2, the strain vector, $\varepsilon_{\varphi\psi}$ is at an angle ψ.

$$\varepsilon_{\phi\psi} = \left(\frac{1+v}{E}\right)\sigma_\phi \sin^2\psi - \left(\frac{v}{E}\right)(\sigma_{11} + \sigma_{22}) \tag{1}$$

strain $\varepsilon_{\varphi\psi}$ is calculated using the equation which includes surface stress σ_φ in any particular direction of the angle φ within the direction (φ, ψ) and the principal stresses present in the surface [8]. The strain in the crystal lattice can be determined by using the formula [8]:

$$\varepsilon_{\phi\psi} = \frac{\Delta d}{d_0} = \frac{d_{\phi\psi} - d_0}{d_0} \tag{2}$$

Here

$d_{\varphi\psi}$ Gap/space in between the planes of direction of φ and ψ
d_o stress-free lattice spacing

On putting the values in Eq. 1 and solving for $d_{\varphi\psi}$ gives

$$d_{\phi\psi} = \left[\left(\frac{I+v}{E}\right)_{(hkl)} \sigma_\phi d_0\right] \sin^2\psi - \left(\frac{v}{E}\right)_{(hkl)} d_0(\sigma_{11} + \sigma_{22}) + d_0 \tag{3}$$

The disparity may occur in the elastic constants because of anisotropy, which is about 40% from mechanical testing value [6, 7]. Equation (3) describes the variation of lattice spacing at any direction with $\sin^2\psi$ as linear and association between lattice spacing and the biaxial stresses. On putting $\sin^2\psi = 0$, the intercept is the stress free lattice (d_0) subtracting the ration of passion's contraction:

$$\frac{\partial d_{\phi\psi}}{\partial \sin^2\psi} = \left(\frac{1+v}{E}\right)_{(hkl)} \sigma_\phi d_0 \tag{4}$$

On solving for stress σ_{φ};

$$\sigma_{\phi} = \left(\frac{E}{I + v}\right)_{(hkl)} \frac{l}{d_0}\left(\frac{\partial d_{\phi\psi}}{\partial \sin^2 \psi}\right) \tag{5}$$

Though none of the techniques measure the true surface of a material, by using X-ray diffraction method we can determine strain which is nearest to the surface. Analytical, instrumental and sample dependent errors are some of the important category of errors which are to be taken care of while using this technique.

3 Semi Destructive Method

The semi destructive method of testing includes methods in which a little portion of whole work piece is destroyed during the testing process for example whole drilling etc.

3.1 Hole Drilling

It is a semi destructive method; in this technique strain caused by drilled hole is measured directly and RS value is determined by measured strain [9]. This method helps in calculating very accurate value of surface residual stress at 1–2 mm depth. In this strain gage (rosette strain gauge) are glued on the specimen surface at the place where hole has to be produced [10]. Rosette usually consists three strain gages which are positioned at 0°, 90°, and 135° [11]. The strain gauge is to calculate the micro displacement caused due to released stresses by the incremental in the depth of the hole, the gauge can measure the displacement because of change in resistor length. Data loggers are connected to each gauge which can measure strain in each step of the drilling process [10]. Equations (9) and (10) can be used to convert the strain to stresses [10]. Temperature can be increased due to the rpm of the drill speed for that environment temperature should be kept constant for accurate results [12].

$$A_n = \left(\frac{\varepsilon_n^1 + \varepsilon_n^2}{2\Delta h_n(\sigma_{1n} + \sigma_{2n})}\right) \tag{6}$$

$$B_n = \left(\frac{\varepsilon_n^1 - \varepsilon_n^3}{2\Delta h_n(\sigma_{1n} - \sigma_{2n}) \cos 2\theta_n}\right) \tag{7}$$

$$\theta_n = \left(\frac{1}{2}\right) \tan^{-1}\left[\frac{\varepsilon_n^1 - 2\varepsilon_n^2 + \varepsilon_n^3}{\varepsilon_n^3 - \varepsilon_n^1}\right] \tag{8}$$

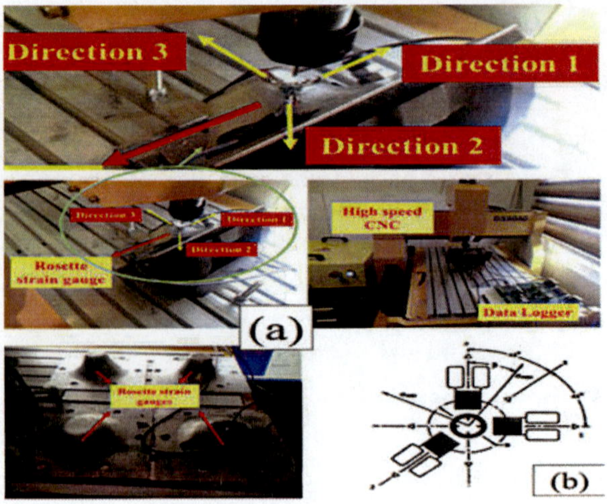

Fig. 3 Showing **a** setup for hole drill method **b** rosette strain gauges [10]

$$\sigma_{min} = \frac{\varepsilon_n^2(A_n + B_n \cos 2\theta_n) - \varepsilon_n^1(A_n - B_n \sin 2\theta_n)}{2B_n A_n(\sin 2\theta_n + \cos 2\theta_n)\Delta h_n} \quad (9)$$

$$\sigma_{max} = \frac{\varepsilon_n^1(A_n + B_n \sin 2\theta_n) - \varepsilon_n^2(A_n - B_n \cos 2\theta_n)}{2B_n A_n(\sin 2\theta_n + \cos 2\theta_n)\Delta h_n} \quad (10)$$

$$\tau_{\max} = \frac{(\sigma_{\max} - \sigma_{\min})}{2} \quad (11)$$

where n is the number of increments and A_n and B_n are the calibration coefficients, $\sigma\sigma_{1n}$ and $\sigma\sigma_{2n}$ are applied stresses. ε_n^1 and ε_n^2 are the strain in the direction 1 and 2 respectively [10].

There are few limitations in applications such as the hole need to be drilled with decent positional accuracy. The calculation of residual stress are affected by position of hole [9].

Figure 3 is showing the semi destructive testing technique, which is aided by usage of strain gauges.

3.2 *Hole Drilling with DIC*

The conventional DIC procedure for residual stresses start with the displacement measurement followed by numerical manipulation for strain, finally stress calculation using theoretical or numerical FEM model [9]. In this method the images are recorded and digitalized of the surface before drilling the hole and after the drilling hole. For better accuracy, the camera must have an excellent focal length and pixels. RS values are determined with the help of displacement measured value after hole drilling.

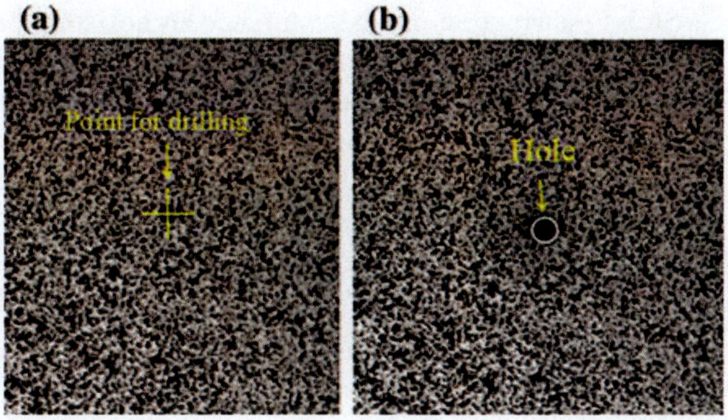

Fig. 4 Images **a** showing mark of point for drill hole or reference image and **b** showing drilled hole [12]

Procedure for hole drilling with DIC:

First to select the area where hole has to drill and them polish it and clean it. Then calibrate the relative position of cameras. Mark the point where the hole has to drill and take picture of the sample surface before drilling the hole, this will be the reference image for the correlation calculation Fig. 2a With the help of a drill machine, make a hole at the marked point and an image of the specimen surface after the drilling hole should be taken Fig. 2b. With the help of 3D DIC software, the displacement can be calculated [12]. The RS value can be calculated mathematically [12] (Fig. 4).

The above figure is elaborating the hole drilling method (semi destructive testing).

4 Destructive Method

Destructive method of testing is that method in which the work-piece is damaged during the testing procedure; however the testing has to be done.

4.1 Contour Method

It is the destructive method which destructs the body part. It is based totally on the Buckner's elastic superposition principle. It uses the apparatus which is generally used in the workspaces, it can produce 2D mapping of RS and different component of stress tensor can also be produced with multiple cuts. After the cutting of components, metrology data has to be processed using some software (such as MATLAB, Abacus and pyCM) [4], it requires wire electric discharge machine (EDM) and coordinates measuring machine or non-profilometer such as confocal, differential intensity or interferometry based instrument.

Stress distribution can be estimated in horizontal direction and cut the material along the desire plane, deformation is produced at the cut faces because of stress relaxation, then measure the deformation using profilometer and finally virtual stresses are required to deform the body to original shape.

Specimen cutting, surface profile measurement and data analysis are three basic phases (Fig. 5).

For the cutting of specimen wire electric discharge machining (WEDM) is mostly used in this method. It is preferred because of roughness produced in the finishing of WEDM.

Define the length scale for minimum residual; deformation can be measure by profiling instrument.

The data recording by profilometer required to process and measured data sets need to be aligned, this can be done by translating and rotating the one data set in a plane to overlap the other. Then interpolate two sets of data on the same grid [13]. Extrapolate the data to the cuts parts' perimeter, then averaging the measured surface data and then removing the measurements which could produce error. Flattening and smoothing the data sets and finally estimating the data for the FE model [13] meshing the model and putting the elastic material values and properties. Residual stresses can be calculated.

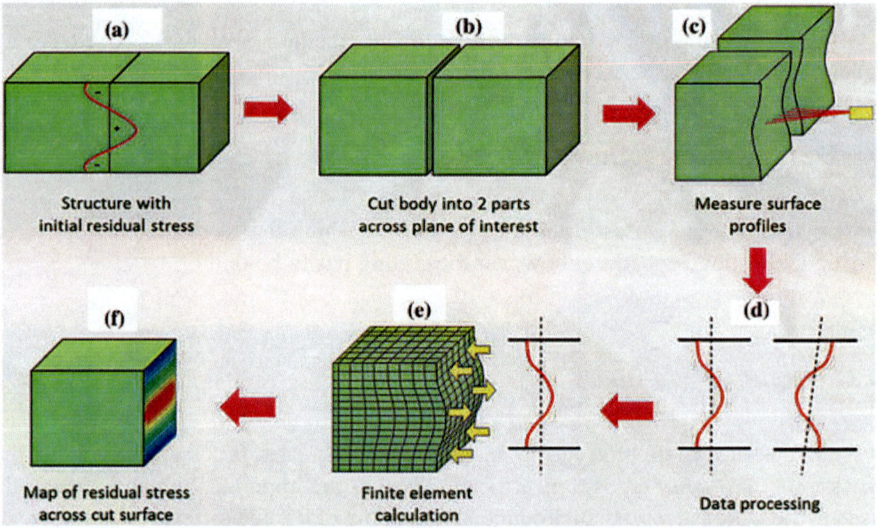

Fig. 5 Image showing the working of contour method [14]

5 Conclusion

The goal of the research was to present various methods to determine residual stresses in 3D printed objects and they are discussed in brief. Also, the limitations of techniques are discussed but their solution is out of the scope of this review. To further investigate the same, future work can be carried out focusing on the problems faced with each technique and their best alternative solution.

References

1. Acevedo R, Sedlak P, Kolman R, Fredel M (2020) Residual stress analysis of additive manufacturing of metallic parts using ultrasonic waves: state of the art review. J Mater Res Technol 9(4):9457–9477. https://doi.org/10.1016/j.jmrt.2020.05.092
2. Rossini NS, Dassisti M, Benyounis KY, Olabi AG (2012) Methods of measuring residual stresses in components. Mater Des 35. https://doi.org/10.1016/j.matdes.2011.08.022
3. Sinclair-Adamson R, Luzin V, Duguid A, Kannoorpatti K, Murray R (2020) Residual stress distributions in cold-sprayed copper 3D-printed parts. J Therm Spray Technol 29(6):1525–1537. https://doi.org/10.1007/s11666-020-01040-7
4. Roy MJ, Stoyanov N, Moat RJ, Withers PJ (2020) pyCM: An open-source computational framework for residual stress analysis employing the Contour Method. SoftwareX 11:100458. https://doi.org/10.1016/j.softx.2020.100458
5. Withers PJJ, Bhadeshia HKDHKDH (2001) Residual stress. Part 1—measurement techniques. Mater Sci Technol 17(4):366–375. https://doi.org/10.1179/026708301101510087
6. Withers PJ, Bhadeshia HKDH (2016) Editorial Board. Environ Int 97:IFC. https://doi.org/10.1016/s0160-4120(16)30771-1
7. Goel S et al (2020) Residual stress determination by neutron diffraction in powder bed fusion-built Alloy 718: influence of process parameters and post-treatment. Mater Des 195:109045. https://doi.org/10.1016/j.matdes.2020.109045
8. Prevéy PS (1996) Current applications of XRD diffraction residual stress measurement. Dev Mater Charact Technol ASM Int 513:103–110
9. Gao J, Shang H (2009) Deformation-pattern-based digital image correlation method and its application to residual stress measurement. Appl Opt 48(7):1371–1381. https://doi.org/10.1364/AO.48.001371
10. Alinaghian M, Alinaghian I, Honarpisheh M (2019) Residual stress measurement of single point incremental formed Al/Cu bimetal using incremental hole-drilling method. Int J Light Mater Manuf 2(2):131–139. https://doi.org/10.1016/j.ijlmm.2019.04.003
11. Walker D (2001) Residual stress measurement techniques. Adv Mater Process 159(8):30–33
12. Peng Y, Zhao J, shu Chen L, Dong J (2021) Residual stress measurement combining blind-hole drilling and digital image correlation approach. J Constr Steel Res 176:106346. https://doi.org/10.1016/j.jcsr.2020.106346
13. Hosseinzadeh F, Kowal J, Bouchard PJ (2014) Towards good practice guidelines for the contour method of residual stress measurement. J Eng 2014(8):453–468. https://doi.org/10.1049/joe.2014.0134
14. https://www.stressmap.co.uk/contour-method/

Corrosion Performance in Grain Structure of C22 in Acidic Environment

Aezeden Mohamed, Kamalakanta Muduli, Devendra K. Yadav, and Pankaj Jena

Abstract This paper aims to describe the impact of the acid attack on the internal grains system of HASETTOY C22. This includes a discussion of different phases available and grain boundary attack precipitation. Microscopic observations revealed that chromium depletion zone formation along the interface between grains led to localized corrosive attack and found that this attack increased with exposure temperature.

Keywords Chromium depletion zone · Grain boundary · Hastelloy C22 · Severe corrosive HCl

1 Introduction

HASTELLOY C22 is a polycrystalline material formed of many grains with different orientations. Like other alloys, interfaces of grains of HASTELLOY C22 are usually the significant sites for cracks and corrosion attack preference. Often impurities and precipitates localized at the interfaces of the grains. Many alloys such as HASTELLOY C22 are subject to intergranular attack because of the phases and particles present at the interface between grains [1]. Alloys such as stainless-steel alloys and nickel-based chromium alloys experience chromium depletion free zone surrounding the interfaces of the grains formed due to thermal treatment and corrosion attack. The region surrounding the grains is characterized as a less noble region that produces different compositions observed as distinct color contrast. The corrosion attack not only attacks the grain zones' interfaces and moves towards the grain

A. Mohamed · K. Muduli
Department of Mechanical Engineering, University of Technology, Lae, Papua New Guinea

D. K. Yadav (✉)
Department of Mechanical Engineering, National Institute of Technology Calicut, Kozhikode, Kerala, India

P. Jena
Department of Production Engineering, Veer Surendra Sai University of Technology, Burla, India

R. Agrawal et al. (eds.), *Recent Advances in Industrial Production*, Lecture Notes in Mechanical Engineering, https://doi.org/10.1007/978-981-16-5281-3_45

center, leading to grain area reduction and as alloy exposed to acid attack as more grain areas are exposed to further damage and finally break down the entire grains.

Selection of heat treatment process and acid attack can lead to deleterious interface between boundary structure due to chromium depletion adjacent to the interfaces between grains produce carbides precipitation at the interface of grains, making alloys subject to interfaces attack [1, 2].

HASTELLOY C22 is a chromium-based alloy often used extensively in an application with high corrosive acids, including chemical processing, petroleum refiners' plants, and nuclear waste containers because of their resistance to corrosive environments. Despite the fact this alloy with a mixture of alloying elements include; chromium, iron, molybdenum, and titanium, titanium together with carbon can avoid chromium carbide precipitation, yet the problem of chromium carbides can't be avoided and is more favorable than titanium carbides [3–6].

This paper aims to describe the effects of acid attack on the interfaces of grains in HASTELLOY C22. This includes different temperature, morphology of interface and grains structure and possibly other phases available.

2 Experimental Procedure

2.1 Materials

The investigated material is HASTELLOY C22 with the chemical composition, as shown in Table 1. HASTELLOY C22 solution treated annealed at 1000 °C for 1.5 h in a sealed tube furnace filled with Ar gas followed by cooling in a furnace. Prior alloy samples immersion into high concentrated hydrochloric acid for electrochemical reaction test. Samples are grinded with SiC sandpapers from coarser to finer grit numbers series from 320 to 600 followed by polishing with alumina slurries and washed into an organic solvent reagent acetone and ethyl alcohol (C_3H_6O and C_2H_6O) followed by air dryer to remove vapor water from surface samples.

Each sample was visually examined after each test and then taken for optical micrograph evaluation to detect any chromium depletion zones and interface grains corrosion. Samples etched in a mixture of nitric acid and ethyl alcohol (4% Nital) before the corrosion test and microstructure observations, as shown in Fig. 1. Conventional techniques used to characterize and compare the metallurgical microstructures. All alloy samples prior (baseline) and after electrochemical tests were evaluated.

Table 1 Chemical composition wt. (%) of HASTELLOY C22

Cr	Fe	Mo	Al	Ti	C	Si	Mn	Cu	S	P	Ni
22	16.9	13.7	1.6	0.4	0.3	0.3	0.2	0.06	0.001	0.001	Bal

Fig. 1 Image of HASTELLOY C22 electrolytically etched in 4% Nital revealed linear grain and annealed twins

2.2 *Electrochemical Behavior*

Samples were then picked up for the electrochemical test, as shown in Fig. 2 were measured by performing electrochemical polarization tests using a commercial apparatus made by Princeton Applied Research Corrosion Measurement Instrument as described in details by the author of this paper published elsewhere [7, 8]. The cell was a 500 ml Erlenmeyer flask containing the sample (working electrode), two 0.25 inches diameter graphite counter electrodes, and a saturated calomel reference electrode (SCE) [7, 8]. The corrosive medium concentrated hydrochloric acid solution. It is worth noting that HASTELLOY C22 samples in acid solution corrode with the production of more stable metals as oxides and hydroxides. However, this may not affect the present comparative study since alloy samples of a specific thermal

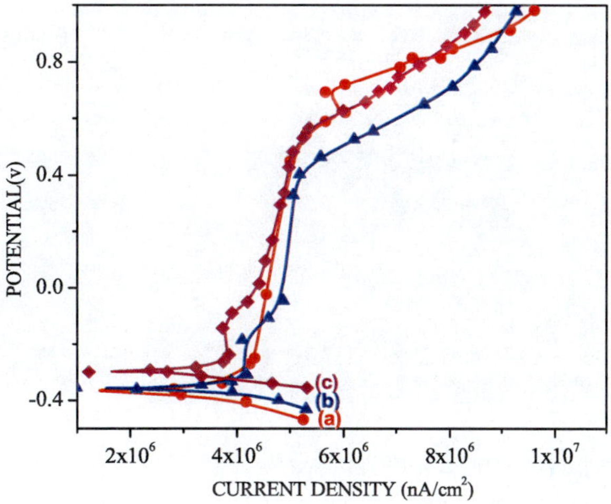

Fig. 2 Potentiodynamic polarization plots HASTELLOY C22: **a** at 50 °C; **b** at 60 °C; **c** at 70 °C

treatment corroded under different temperature electrochemical conditions. Samples surface morphology was observed before and after corrosion immersions using image analyzer microscope.

3 Results

Figure 2 shows the electrochemical test results of HASTELLOY C22 at 50, 60, and 70 °C. Samples after test were directly examined with an image analyzer microscope.

Under an optical microscope, two phases were visible as the temperature test increased. The test sample at 50 °C (Fig. 3a) shows interfaces between grains, not progressive stages representing interfaces of grains and adjacent area interfaces. As the temperature increased, we could see more the acid solution further attack and developed interface and area. Chromium depletion zones develop and expand with a time of exposure at 60 °C (Fig. 3b).

Investigation revealed evidence that microstructure is composed of grains area and interface area region (Fig. 3). From the surface observation, the sample indicated a transition to developing a straightforward interface and surrounding area. This possible evidence of chromium depletion zone developed and extended starting from a temperature of 50 °C and higher, as shown in (Fig. 4). As the test temperature increases, as the chromium depletion zone grows, as can be observed from Figs. 3 and 4, the contrast differences of grain boundary morphology between the depletion zone and remaining grains. The test sample revealed the morphology of the oxide layers along the grains' interface is evident that commonly found to be the current study, indicating HASTELLOY C22 samples producing chromium free region. HASTELLOY C22 sample (Fig. 3) after the corrosion test developed two major areas observed oxides layers along interfaces presents chromium carbides

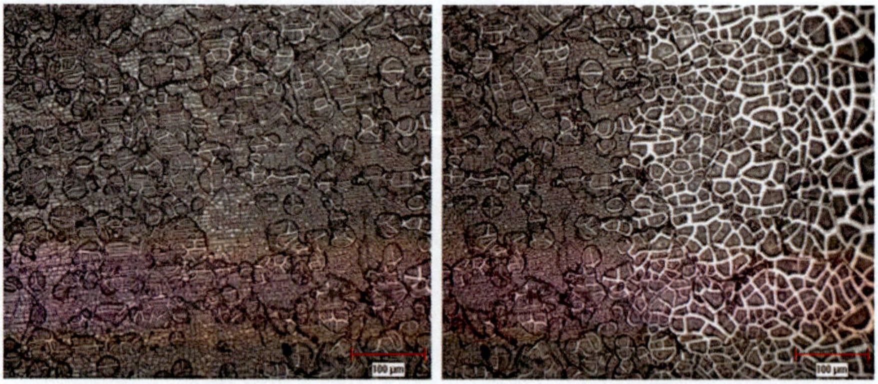

(a) (b)

Fig. 3 Images of HASTELLOY C22 after the electrochemical test at a temperature of **a** at 50 °C from the center of the sample; **b** at 60 °C from the sample's edges

Fig. 4 Images of HASTELLOY C22 after the electrochemical test at a temperature of 70 °C

surrounding the grains, forming different morphologies. While the sample micrograph prior exposure to elevated temperature and then too aggressive corrosion acid solution (Fig. 3), all samples presented an area free zone alongside the grains' interfaces (see Fig. 3). According to these analyses, the surface sample observations gradually change with test temperature, producing a ditch zone alongside the interface between grains, as shown clearly in Fig. 4.

HASTELLOY C22 samples tested at a temperature of 50, 60, and 70 °C show irregular grain boundaries thickness morphologies (Figs. 3a, b and 4), that if compared with the alloy samples tested at a temperature of 50, 60, and 70 °C showing that interfaces between grains and zone alongside interfaces have regular shape morphologies as shown in Figs. 3 and 4 respectively.

Fine line located within the center of grain boundary zones is visible in Fig. 5 at high magnifications. Grain boundaries represent this fine line (2). With exposure, the temperature increased, evidence of chromium depletion zones or attack zones become thicker zone (1). These chromium depletion zones, possibly due to localized attack, dissolve grain boundary zones and result in severe falling out of entire grains (3), which eventually leads to final fracture. The initial reaction within and adjacent grain boundaries zones extend deep in the grain center's interior with increasing exposure temperature to the concentrated acid solution and causes severe damage to the grain structure itself, which may eventually lead to drop out, as shown in Fig. 5.

4 Discussion

Chromium depletion and dissolution zones are a localized type of corrosion adjacent to the interfaces between grains, with little or no corrosion observed at the grain area

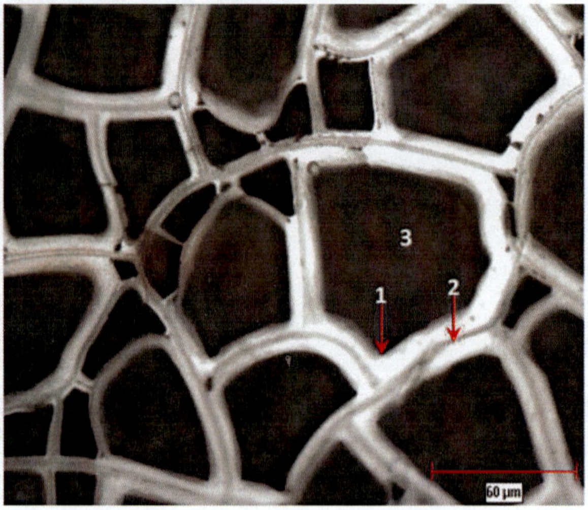

Fig. 5 Image at higher magnification indicated: (1) Oxide layer; (2) grain boundary; (3) grain

center. This results in differences in conductivity and resistivity alloy microstructure system. Furthermore, making a small interface area act as an active zone called an anodic area remains unstable. In contrast, the large grain area becomes a passive site called the cathodic area remains stable. As a result, there is a large cathode to a small anode area, which prompts an expansion in the grain boundary zones' consumption rate. Also, to localized attack is the phases present in the grain boundaries that behave as anodes or cathodes to the grain area's remainder. During heat treatment, chromium reacts with carbon to form chromium carbides such as $Cr_{23}C_6$ and Cr_6C. The chromium carbide content is an unstable phase. Controlling thermal treatment at a temperature equal to or less than 800 °C, the thermal treatment process of the current work is designed at 950 °C, which is sufficient to dissolve and form chromium carbides that produce an unstable phase. Required time to develop the nucleation and growth required for second phases such as β-phase and gamma-phase. Both nucleation sites of the chromium carbides and second phases grow at the grains interface. Thus, it is essential to design a desire and a proper heat treatment process to avoid producing precipitates free zone adjacent to the interface between grains. The optimum heat treatment process is that heat treatment can distribute chromium alloying elements homogeneous in the grain matrix. Chromium usually diffuses by slow substitutional in grains and faster interstitial in the grain boundaries. In comparison, carbon diffused by interstitial in grains and grain boundaries. Furthermore, to avoid an intergranular attack, carbon is a critical element. It is imperative to minimize carbon content diffuses faster by interstitial in grains and grain boundaries and combine with chromium producing chromium carbides: $23Cr + 6C \rightarrow Cr_{23}C_6$. Three different phases present in grain boundaries won't show grain limit carbides after air cooling from arrangement strengthening. Whenever wanted, grain boundary carbide precipitation can be effortlessly smothered during the thermal process along these lines. The sort of carbide shaped during heat treatment relies on the temperature as shown in Fig. 5. At higher temperatures, roughly from 850 to 1050 °C, the carbides are both NbC as thin interface films between grains, and MC, where M is principally Ni, Cr, and Mo. At temperatures in the 700–900 °C range, the grain boundary carbides are primarily M_3C_6, where M is almost entirely Cr. After intermediate temperature heat treatments, all these three types of carbides usually can be found, as shown in Fig. 5. The M6C and $M_{23}C_6$ carbides generally have blocky, unpredictable shapes and structure a progression of isolated, discrete grain boundary particles. In a practical sense, the outcomes give a decent representation of how a minor change in chemistry can nearly bother the stage change behavior [9]. As a result, grains and grain interfaces are indicated with increased test temperature applied, as can be seen in Figs. 3, 4, and 5, respectively. From Fig. 5, it is essential to select the proper heat treatment temperature. For example, the second phase presents mostly gamma and possible M_6C annealing temperature of 1000 °C. Below 1000 °C, other phases can be formed at the interface area (see Fig. 5) when tested in corrosive solution (50% H_2SO_4 + 42 g/l $Fe_2(SO_4)_3$. On the other hand, alloy C22 alloy is an experience or subject to interface attack temperature after just 1 h for a temperature of between 900 and 1000 °C [10]. Alloy C22 consists of multigrain and grain boundaries. Unlike the grains, the grains' interfaces are heavily attacked by the acidic solution because of the absence of the element's resistance to acid attack at the grain interface. These particles formed a region at the interface and reduced attack resistance alongside the interface [10]. Chromium dissolutions along with the interface between grains Particular alloying element such as chromium is dissolved leafing behind particle-free zone and weakening that are known for particles Dissolutions that formed along with the interface between grains and consequently results in oxide layers along with these areas, resulting, and a narrow channel along the interface become anodic as compared to the rest of the area of grain structure [10]. The corrosion attack accelerated at the little track along with the interface between grains. As the episode continuously progresses towards the grain center as consequences, more and more grain areas are attacked with time on under a severe condition attack will force out remaining grains [10]. The alloy's corrosion behavior has a direct effect on properties, e.g., strength, ductility, and toughness. Alloy C22 prone to interface attack due to chromium phase present known as anodic region side, whereas the remaining grain area is known as the catholic region [10]. This alloy is observed to have a massive attack alongside the interface producing oxides and developing gradually and slowly towards the center of the grain and eventually attack the entire grain structure [10].

5 Conclusions

Controlling and designing thermal treatment is the key to producing the desired homogenous particle distribution in grains, and it is the interface system of HASTELLOY C22. Microscopic observations revealed distinct intergranular corrosion products initiated, built up, and growing adjacent at the grains' interfaces and growing with exposure acid attack. This behavior becomes more evident for smaller grain sizes. It produces more grain boundaries and has more corrosion and oxidization layer products, leading to more small grains.

Annealed alloy C22 produced distinct morphologies of intergranular corrosion product. Some phases formed, particularly MC and gamma. For example, others, such as grain boundary carbides and gamma, can be useful for erosion opposition or quality purposes.

References

1. Fontana MG, Staehle RW (eds) (1972) Advances in corrosion science and technology, 2 Plenum Press, New York
2. Evans UR (1960) The corrosion and oxidation of metals: scientific principles and practical applications. Edward Arnold (Publishers) Ltd., London
3. Palumbo G, Aust KT, Lehockey EM, Erb U, Lin P (1998) Influence of grain boundary character distribution on sensitization and intergranular corrosion of alloy 600. Scripta Mater 38:1685–1695
4. Bennett BW, Pickering HW (1987) Effect of grain boundary structure on sensitization and corrosion of stainless steel. Metall Trans 18A:1117–1120
5. Pan Y, Adams BL, Olson T, Panayotou N (1996) Grain-boundary structure effects on intergranular stress corrosion cracking of alloy X-750. Acta Mater 44(12):4685–4695
6. Floreen S, Fuchs GE, Yang WJ (1994) The metallurgy of alloy 625, Knolls Atomic Power Laboratory, P. 0. Box 1072, Schenectady, New York, pp 12301–1072
7. Aezeden M (2016) On the capability of in-situ exposure in scanning and auger electron spectroscopy for investigating corrosion property of engineering alloys. Innov Corros Mater Sci 6(2):140–145
8. Aezeden M, Jack C, William C (2012) Anodic polarization behavior of Nickel-based alloys in neutral and very acidic solutions. J Corros Sci Eng 15:1–30
9. Stephen FG, Fuchs E, Walter JY, The metallurgy of alloy 625, Knolls Atomic Power Laboratory, P. 0. Box 1072, Schenectady, New York 12301-1072
10. NACE, International to protect people, assets, and the environment from corrosion. ISO 9001-2008

An Evolutionary Tomographic Reconstruction Procedure for Defect Identification Using Time-of-Flight of Ultrasound

Shyam Prasad Kodali and Boggarapu Nageswara Rao

Abstract Structural integrity of engineering materials is influenced by external as well as internal defects. Tomographic reconstruction is useful for detection of internal defects in a nondestructive way. A novel tomographic reconstruction procedure is described and shown to have the potential to identify any defects present in a materials cross-section. The algorithm is designed in line with the principles of real coded genetic algorithms (RCGA). The proposed algorithm does not require the user to input exact characteristic property of material defects assumed to be present in the material cross-section being examined. The algorithm works its way with some finite range of the characteristic property as input. Results of several numerical studies demonstrate the effectiveness of proposed RCGA based reconstruction procedure.

Keywords Non-destructive testing · Tomographic reconstruction · Real coded genetic algorithms · Time-of-flight of ultrasound

1 Introduction

Several industries including the aerospace, ship building, petrochemical, automotive, steel, railway, defense industries rely on non-destructive evaluation (NDE) for the identification of internal defects without causing any permanent damage to the components being tested. During the past couple of decades, computed tomography (CT) has developed as an effective NDE procedure suitable in characterizing defects in engineering materials [1, 2]. Principally CT is a two-step process; the object being investigated is first irradiated with some form of energy like X-rays or Ultrasound and either the transmitted or reflected energy is recorded which is referred to as projections. In the next step called reconstruction, the projections are fed to a reconstruction procedure that can reveal the inside of the object. Different reconstruction algorithms and their variations are reported in the literature with each of them having some merits and demerits. Transform methods and series expansion

S. P. Kodali · B. N. Rao (✉)
Department of Mechanical Engineering, Koneru Lakshmaiah Education Foundation, Vaddeswaram, Guntur, Andhra Pradesh 522502, India

R. Agrawal et al. (eds.), *Recent Advances in Industrial Production*, Lecture Notes in Mechanical Engineering, https://doi.org/10.1007/978-981-16-5281-3_46

methods are the oldest and still in use. Widely used transform methods like filtered back projection and the convolution back projection methods though quicker, require many projections. On the other hand, popular series expansion methods like algebraic reconstruction technique, and multiplicative algebraic reconstruction technique take longer time but produce better reconstructions [3, 4]. In recent times reconstruction methods developed with the application of genetic algorithms are being reported as good alternatives particularly when only limited projection data is available.

In this work, a novel reconstruction procedure using real coded genetic algorithms is designed in contrast to earlier reported reconstruction procedures using binary coded genetic algorithms [5–9]. Also, with the present algorithm the user does not have to input exact characteristic property of material defects assumed to be present in the material cross-section being examined, it only requires some finite range of the characteristic property. The projection data is numerically simulated ultrasound time-of-flight [10]. Results of several numerical studies validate the effectiveness of proposed reconstruction procedure.

2 Proposed Reconstruction Procedure

One needs a reconstruction algorithm or procedure to obtain the cross-section of object concealed in the acquired projection data. The reconstruction procedure proposed here is designed on the concepts of real coded genetic algorithms which work directly with variable values without any coding of the variables [11]. The flowchart in Fig. 1 illustrates the significant features of the proposed reconstruction procedure.

Genetic algorithms work on the evolution of a set of randomly initialized solutions referred to as a population. The population is iteratively improved with the repeated application of the selection, crossover, and mutation operators until convergence is reached [12]. The algorithm is terminated when either a specified maximum number of generations (iterations) are completed or when the fitness of the best population member reaches a specified minimum value. The fitness of a population member as defined by Eq. (1), is the means of evaluating the goodness of a population member. The genetic operator selection creates a mating pool of the relatively better solutions in the current population. With crossover operation new and optimistically better population of solutions (children) is generated from the working on the solutions in mating pool (parents). With mutation operation, random alterations are introduced to a very small fraction of the children that helps maintain diversity in the population and reduce the chances of the algorithm converge to a local optimum. The tournament selection operator [13], the simulated binary crossover operator [14, 15], and the polynomial mutation operator [16] are used in our work. The proposed algorithm is elitist meaning, a small fraction of parents is placed in the children population without subjecting them to crossover and mutation operators. This ensures one does not lose all the better or improved solutions even after crossover or mutation operations [17]. Each population member corresponds to the discretized cross-section

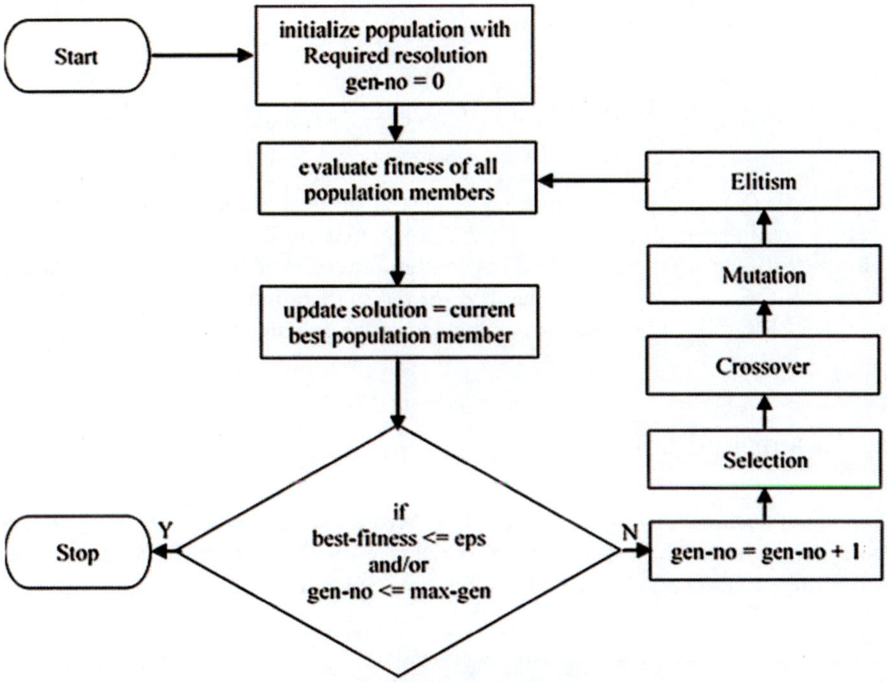

Fig. 1 Real coded genetic algorithm reconstruction procedure

of the test object and is represented as a matrix of real numbers mathematically as illustrated in Fig. 2a. The computer visualization of same is displayed as shown in Fig. 2b. The matrix of real values corresponds to the material characteristic properties to be reconstructed. The required projection data or the input to reconstruction algorithm, which is ultrasound time-of-flight in the study is acquired following numerical

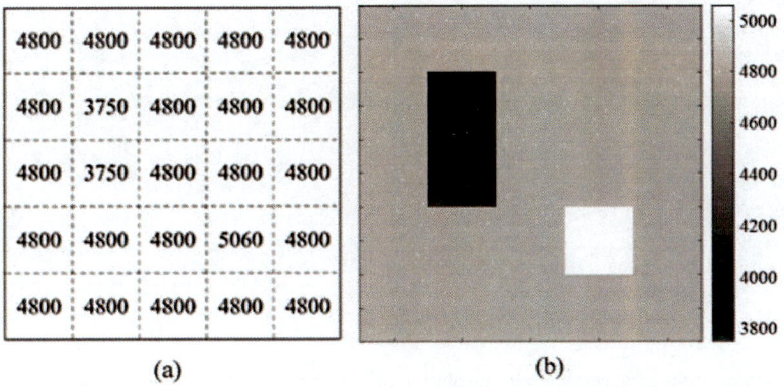

Fig. 2 Representation of discretized test cross-section and population member

simulation procedure detailed in [10].

$$\varphi(i) = \sum_{c=1}^{C}\sum_{t=1}^{T}\sum_{r=1}^{R}\left\{TOF_{Pop}^{(i)}(c,t,r) - TOF_{Specimen}(c,t,r)\right\}^2 \quad (1)$$

Here, C is the total configurations, T is the transmitters in each configuration, R is the receivers in each configuration, $TOF_{Pop}^{(i)}(c,t,r)$ is the TOF of the ray from the tth transmitter to the rth receiver in the cth configuration considering the ith population member, $TOF_{Specimen}(c,t,r)$ is the TOF of the ray from the tth transmitter to the rth receiver in the cth configuration considering the specimen to be reconstructed.

3 Results and Discussions

Numerical studies performed on synthetic test cross-section having one central defect are reported here. The reconstruction is achieved with three grid resolutions and with different parent and material combinations shown in Figs. 3, 4, and 5. Before

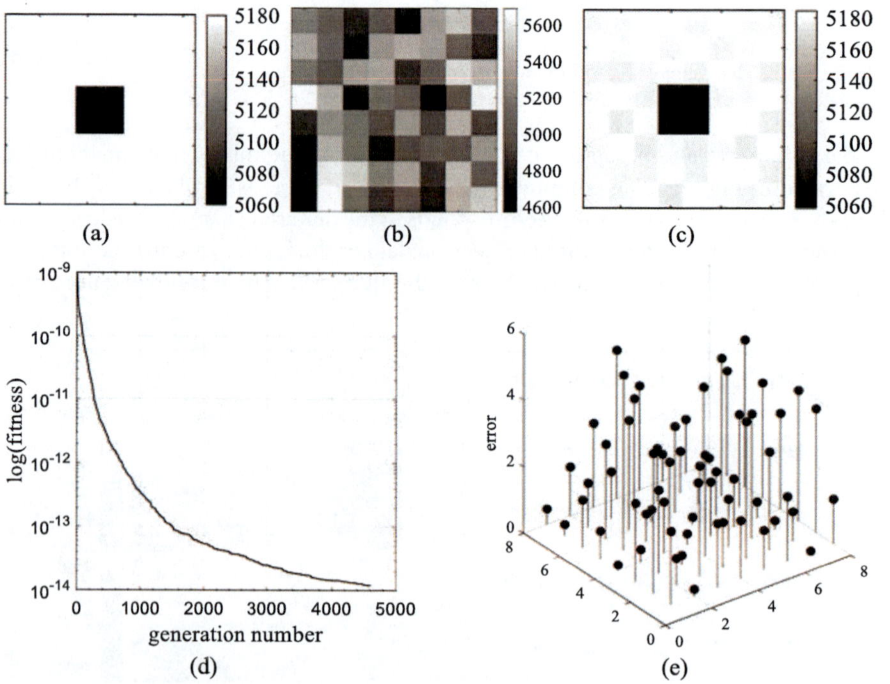

Fig. 3 Results of reconstruction with resolution of 8 × 8: **a** Test cross-section to be reconstructed, **b** initial random solution, **c** reconstructed solution, **d** fitness history plot, **e** error in each pixel of the reconstructed cross-section

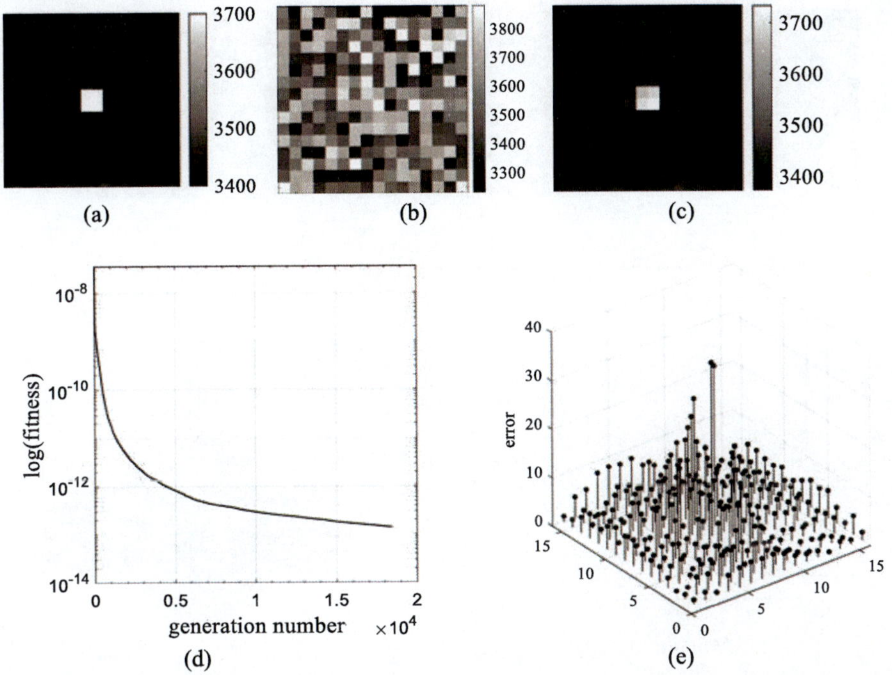

Fig. 4 Results of reconstruction with resolution of 16 × 16: **a** Test cross-section to be reconstructed, **b** initial random solution, **c** reconstructed solution, **d** fitness history plot, **e** error in each pixel of the reconstructed cross-section

applying the proposed reconstruction procedure, its sensitivity to the real coded genetic algorithm parameters is studied. Several trial reconstructions indicate that the algorithm performance depends on population size, crossover probability, and probability of mutation. The required GA population size is found to be three times the grid resolution. Crossover probability between 0.7 and 0.95 is found suitable in all reconstructions. It is observed that the algorithm results are very sensitive to probability of mutation and trial reconstructions showed mutation rate of approximately the inverse of the square of grid resolution is found to give stable results. Figure 3 shows the results obtained for reconstruction with grid resolution of 8 × 8. The characteristic property of parent and defect material in the test cross-section shown in Fig. 3a can be read as 5060 and 5180 m/s using property scale placed to the right of the cross-sectional image. Figure 3b is one of the thirty-two randomly initialized starting solutions (population member) using which the reconstruction algorithm arrived at the final solution shown in Fig. 3c. Notice the range of characteristic property values, from 4554 to 5699 m/s, with which each discretized cell of the initial guess solution shown in Fig. 3b is randomly assigned.

A quick comparison between Fig. 3a, c demonstrates the algorithms' ability to successfully reconstruct the unknown test cross-section in terms of locating the

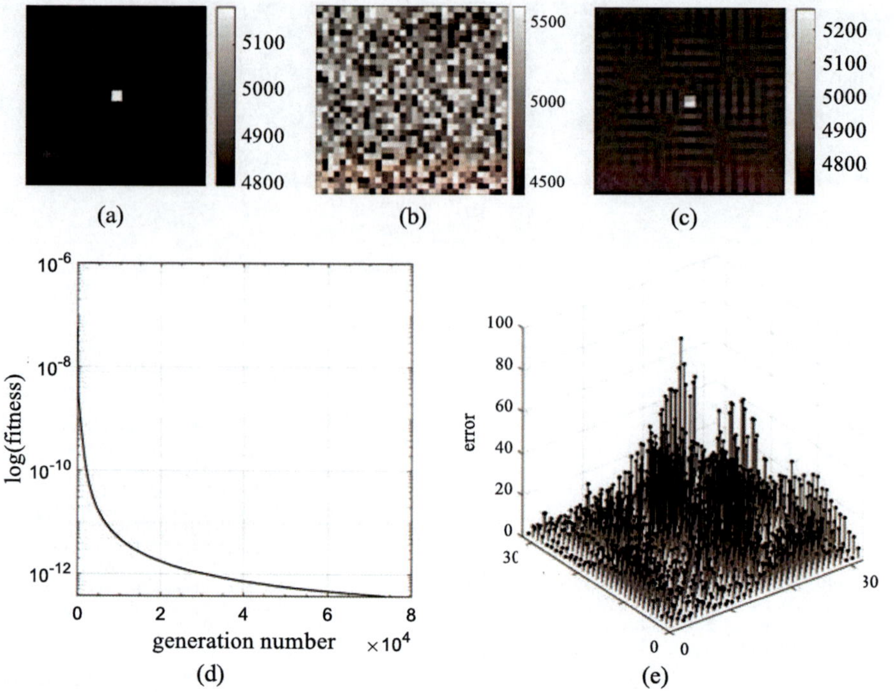

Fig. 5 Results of reconstruction with resolution of 32 × 32: **a** Test cross-section to be reconstructed, **b** initial random solution, **c** reconstructed solution, **d** fitness history plot, **e** error in each pixel of the reconstructed cross-section

defect. Post processing reveals certain deviation or error in the reconstructed characteristic property in each of the discretized pixels as shown in Fig. 3e. The minimum and maximum error is found to be 0.05 and 5.27 units, respectively. Realizing that it is cumbersome to comment on the overall quality of reconstructions based on deviations in individual pixels, the root mean square error (RMSE) is used as a single quantifiable metric. The RMSE is calculated taking square root of the summation of the squares of deviation of the reconstructed property from the actual value in each individual cells. The RMSE is found to be 2.36 units for the case discussed. The improvement in the fitness of the best population member with number of iterations or generations is illustrated as a semi-log plot in Fig. 3d. Notice that the fitness improvement curve is steeper during the first few generations and gradually flattens out, which is typical characteristic of genetic algorithms.

Results of reconstruction with a grid resolution of 16 × 16 are illustrated in Fig. 4. Observe from Fig. 4a that the characteristic properties of parent and defect material in the test cross-section are 3400 units and 3700 units, respectively. The proposed algorithm could capture the cross-section material distribution with an RMSE of 6.61 units. The deviation from actual property value in each of the pixels is shown in Fig. 4e. Figure 4d illustrates the fitness improvement trend which shows that the reconstruction took more generations compared to reconstruction with 8 × 8 resolution.

The results of reconstruction with a grid resolution of 32 × 32 are shown in Fig. 5. The test cross-section shown in Fig. 5a has two materials of represented by characteristic property values of 4800 units and 5180 units. The results are also like those obtained in earlier cases. Figure 5e shows the error in reconstructed characteristic property values of individual cells and the RMSE in reconstruction is found to be approximately 18 units. Also notice from the fitness history plot shown in Fig. 5d that the reconstruction required even higher number of generations when compared to reconstruction with 16 × 16 resolution.

4 Conclusions

A novel reconstruction methodology modeled on the principles of real coded genetic algorithms is proposed for the estimation of the number of defects, their location, geometry, and characteristic properties. The proposed algorithm relieves the user from having to input the exact characteristic property values which is the case with earlier reported reconstruction procedure which is built using principles of binary coded genetic algorithms. Several numerical studies conducted on different synthetic test cross-sections indicate that the proposed methodology can handle a range of tomographic reconstructions with maximum error in reconstructed characteristic properties being less than three percent. Further, the overall reconstruction quality metric, the root mean square error in reconstruction is shown to be below one percent of the characteristic properties.

References

1. De Chiffre L, Carmignato S, Kruth JP, Schmitt R, Weckenmann A (2014) Industrial applications of computed tomography. CIRP Ann 63(2):655–677
2. Ponikiewski T, Katzer J (2016) X-ray computed tomography of fibre reinforced self-compacting concrete as a tool of assessing its flexural behavior. Mater Struct 49:2131–2140
3. Kak AC (1998) Malcolm slaney: principles of computerized tomographic imaging. Society for Industrial and Applied Mathematics, Philadelphia
4. Geyer LL, Schoepf UJ, Meinel FG, Nance JW, Bastarrika G Jr, Leipsic JA, Paul NS, Rengo M, Laghi A, De Cecco CN (2015) State of the art-iterative CT reconstruction techniques. Radiology 276(2):339–357
5. Delsanto PP, Romano A, Scalerandi M, Moldoveanu F (1998) Application of genetic algorithms to ultrasonic tomography. J Acoust Soc Am 104:1374–1381
6. Kodali SP, Bandaru S, Deb K, Munshi P, Kishore NN (2008) Applicability of genetic algorithms to reconstruction of projected data from ultrasonic tomography. In: Keijzer M (ed) GECCO 2008, Proceedings of the 10th annual conference on genetic and evolutionary computation. ACM, New York, pp 1705–1706

7. Kodali SP, Deb K, Bandaru S, Munshi P, Kishore NN (2009) Simulation studies on a genetic algorithm based tomographic reconstruction using time-of-flight data from ultrasound transmission tomography. In: Kolehmainen M, Toivanen P, Beliczynski B (eds) Adaptive and natural computing algorithms ICANNGA 2009, LNCS, vol 5495. Springer, Heidelberg, pp 253–262
8. Kodali SP, Deb K, Munshi P, Kishore NN (2009) Comparing GA with MART to tomographic reconstruction of ultrasound images with and without noisy input data. IEEE Congress on Evolutionary Computation 2009. IEEE, pp 2963–2970
9. Kishore NN, Munshi P, Ranamale MA, Ramakrishna VV, Arnold W (2011) Tomographic reconstruction of defects in composite plates using genetic algorithms with cluster analysis. Res Nondestr Eval 22(1):31–60
10. Kodali SP, Nageswara Rao B (2020) Tomographic reconstruction of isotropic materials using genetic algorithms with ultrasound time-of-flight projection data. J Comput Appl Res Mech Eng (in Press)
11. Wright AH (1991) Genetic algorithms for real parameter optimization. Found Genet Algorithms 1:205–218
12. Deb K (1999) An introduction to genetic algorithms. Sadhana 24(4–5):293–315
13. Miller BL, Goldberg DE (1995) Genetic algorithms, selection schemes and the varying effects of noise. IlliGAL report no. 95009, University of Illinois at Urbana-Champaign
14. Deb K, Kumar A (1995) Real-coded genetic algorithms with simulated binary crossover-Studies on multi-modal and multi-objective problems. Complex Syst 9(6):431–454
15. Deb K, Agrawal RB (1995) Simulated binary crossover for continuous search space. Complex Syst 9(2):115–148
16. Deb K, Deb D (2014) Analyzing mutation schemes for real-parameter genetic algorithms. Int J Artif Intell Soft Comput 4(1):1–28
17. Deb K, Pratap A, Agarwal S, Meyarivan T (2002) A fast and elitist multi-objective genetic algorithm-NSGA-II. IEEE Trans Evol Comput 6(2):0181–0197

Manufacturing Is Not as Usual: Lessons Learnt from COVID-19 Pandemic

Anbesh Jamwal, Rajeev Agrawal, and Monica Sharma

Abstract Manufacturing organizations need reconfigurability in their manufacturing systems to complete the unpredictable and volatile demands of the market. Reconfigurable manufacturing systems require fewer implementation efforts as compared to other types of manufacturing systems. COVID-19 pandemic has changed the manufacturing scenario and also twisted the customer requirements. At present, manufacturing industries need to meet customer expectations as well as maintain the market value of their organization with low efforts. Industries are now reconfiguring their manufacturing plants to satisfy the surge demands of specific products from the customer. To practitioners, reconfigurable manufacturing systems require less reconfiguration effort during the dynamic changes in the market and help to complete the market demands in minimum time. In early 2020, when the world is hit by a novel coronavirus (COVID-19) there was a sudden increase in the demand for sanitizer gel, PPE kits, ventilators and masks many industries in India reconfigured their manufacturing systems with low reconfiguration effort. This study analyzes the reaction upheld by the Indian manufacturing sector to the COVID-19 pandemic, to drive practical insights that how manufacturing industries responded to the market demands with reconfiguration efforts. Additionally, we have provided an insight into how the linkage between dynamic capability theory with reconfigurable theory will be helpful for industries.

Keywords COVID-19 · Manufacturing · Supply chain · Reconfigurable manufacturing system · India

A. Jamwal · R. Agrawal (✉) · M. Sharma
Department of Mechanical Engineering, Malaviya National Institute of Technology, J.L.N. Marg, Jaipur, Rajasthan 302017, India
e-mail: ragrawal.mech@mnit.ac.in

M. Sharma
Department of Management Studies, Malaviya National Institute of Technology, J.L.N. Marg, Jaipur, Rajasthan 302017, India

R. Agrawal et al. (eds.), *Recent Advances in Industrial Production*, Lecture Notes in Mechanical Engineering, https://doi.org/10.1007/978-981-16-5281-3_47

1 Introduction

COVID-19 outbreak has changed the manufacturing scenario in the world. COVID-19 has disrupted the manufacturing firms and supply chains over the globe. Manufacturing organizations in both emerged and emerging economies reported the disruption in their manufacturing activities due to the COVID-19 outbreak. Around 35% of manufacturing industries reported that the COVID-19 outbreak has disturbed their manufacturing practices due to the shorten of raw material supplies [1]. COVID-19 outbreak was started from Wuhan city of China which is known as the largest transportation hub of China [2, 3]. Around 48.7 million cases of n-CoV have been reported till November 2020 which resulted in lockdowns and manufacturing units shut down in many countries due to extensive human causalities. China contributes a total of 19% of the Global GDP and also known as the traditional hub of manufacturing [4]. China is also a major manufacturer of bio-medical products, pharmaceuticals, modern manufacturing (Iron, Automobile and Steel components) [5]. The outbreak has disrupted all the supplies from China due to the stoppage of air-flights and cargo ships. Furthermore, the sudden rise in demand for medical equipment and pharmaceuticals has forced manufacturing organizations to revisit their manufacturing systems [6]. The World Health Organization has also declared an emergency for this outbreak. For this reason, many countries have in the world has imposed partial or total lockdowns in their country [7]. For the precaution reasons for COVID-19 people are wearing masks and using sanitizers for sanitization purposes. In this landscape, there is a sudden rise in demand for both sanitizers and masks. Also, the numbers of cases are increasing at a rapid rate. Emerging economies in which health infrastructure is not strong has imposed the total lockdowns in their country. Also, the demand for ventilators has increased due to the global pandemic.

Industries now need the manufacturing systems which can capable to handle such type of situations in future outbreaks. Reconfigurable manufacturing systems (RMS) are capable to handle such type of situations in future. RMS was firstly introduced in 1999 which can be considered as a suitable choice of manufacturing systems capable to handle volatile market conditions and demands in a short time [8]. Unlike other manufacturing systems, e.g. Flexible manufacturing systems (FMS) RMS are designed with a specific level of flexibility that is capable to handle the predefined product family. The overall structure of RMS consists of machines that are reconfigurable and can be easily removed, added or reconfigured to meet the market expectations and demands [9].

Indeed, in this emergency and global outbreak manufacturing industries have to meet customer expectations and demands. Also, manufacturing organizations have to keep their reconfigurability effort as low as possible. In this situation, it is difficult to achieve reconfigurability in manufacturing. Therefore, based on this research question can be addressed for the present study:

RQ1: How reconfigurable manufacturing systems can help to handle the fluctuating demands in the present outbreak and future outbreaks?

RQ2: What reconfigurable manufacturing-related insights are manufacturing organizations reacting to n-CoV outbreak providing to both the business and academia community?

This research work captures the insights from the reaction—March 2020 when the lockdown started in India—to the outbreak when Indian industries started serving the Indian markets.

1.1 Reconfigurable Manufacturing Systems

With the development of Industrialization and transportation at the end of the twentieth century and the beginning of the 21st-century competition among industries has become widespread and globally. At present dedicated manufacturing lines which were initially designed to manufacture products in larger quantities at lower cost has become obsolete [10–13]. Now the market conditions are unpredictable and dynamic as occurs during the COVID-19 pandemic which resulted in the introduction of new products with higher demand fluctuation [12]. To cope up with such type of situations FMS were introduced which generally consists of computer numerical control machines (CNCs) and capable to produce a high variety of products. These manufacturing systems are designed with generalized flexibility. When compared to DMLs, FMS has a lower production rate and requires some initial cost for technology advancement for flexibility. Many industries in emerging economies are facing financial issues so FMS were not a suitable choice as the manufacturing systems for many industry sectors [9]. By considering the disadvantages and advantages of FMS and DML, RMS was introduced which is capable to cope up with fluctuating market demands. As discussed by Korean et al. (1999) a manufacturing system can be considered as "reconfigurable manufacturing systems" if manufacturing systems having the following characteristics [8] :

I. Modularity: In reconfigurable manufacturing systems modularity is defined as a modular structure that consists of both the software and hardware.
II. Integrability: In reconfigurable manufacturing systems modularity is defined as the common interface where the machines and modules can be integrated.
III. Customization: In reconfigurable manufacturing systems modularity is defined as the flexibility that is required for a product family.
IV. Convertibility: In reconfigurable manufacturing systems modularity is defined as the ability of a manufacturing system to evolve functionality over a while.
V. Diagnosability: In reconfigurable manufacturing systems modularity is defined as the ability of the manufacturing system to diagnose rapidly and effectively.
VI. Scalability: In reconfigurable manufacturing systems modularity is defined as the ability of a manufacturing system to adapt the production capacity to cope up with the volatile demands from the customer and the market.

The above-discussed characteristics of manufacturing systems reduce the reconfiguration efforts in manufacturing organizations. This makes reconfigurability of the in-manufacturing systems a suitable choice for post-COVID-19 manufacturing. The above discussed six characteristics related to reconfigurability "Convertibility" and "Scalability" are closely related to the manufacturing systems. Both these characteristics directly help in achieving reconfigurability in manufacturing systems and provide exact functionality and capacity in manufacturing systems when required. In the past studies, several authors have worked on reconfigurable manufacturing systems in which enablers related to convertibility and scalability are reported.

2 Research Design

In this study, we have considered the Indian manufacturing organization which were serving and reacting during the n-CoV outbreak. The different industry sizes were considered for the analysis. India has reported more than 84 million cases of n-CoV till November 2020. The Indian government has imposed a total lockdown in the country in April 2020. Due to lack of supplies and lack of labour most of the manufacturing organizations were shut down. Only industries that produce essential goods e.g. food products, pharmaceutical and medical equipment were operating with n-CoV precautions guidelines. The situation of the n-CoV is still not completely in control. Indian Government has banned large gatherings e.g. social gatherings and corporate gatherings. It is not possible to directly visit industries and take interviews in this dramatic situation. Consequently, information was collected with industry experts with video conferencing and telephone calls. The other information sources were the company's website and Business magazines published in India during this period. For the more reliability of the study, we have followed the triangulation empirical study in which we have also considered the public documents, research articles, industry reports which report "India" in their content. The collected information from these databases was contextualized into the available theory on reconfigurable manufacturing systems. In this study theoretical sampling is used for sample selection. Table 1 shows the sample of investigation considered in the present study.

3 Cases and Results from Indian Manufacturing Industries

In the manufacturing sector it is found that the majority of industries except INDc1-Healthcare and pharmaceutical industries operating in the Northern region of India and INDd1, INDd2 would have had their manufacturing units shut down or not in the operating stages. The main reason behind this is due to (1) Lack of demand from the customer end due to COVID-19 transmission fear. Also, the lack of support from labour due to transmission fear was the main reason. COVID-19 has affected the Indian market and business. Also, the Government of India has instructed to shut down the non-critical production of manufacturing goods temporarily to minimize the COVID-19 transmission.

Table 1 Industry sector and their production stages

Industry ID	Industry sector	Product	Production stages
INDa1	Rubber manufacturing	Large scale	All stages of production
INDa2	Plastic manufacturing	Large scale	Manufacturing of one product
INDa3	Consultancy	Small scale	Manufacturing of one product
INDb1	Textile manufacturing	Large scale	All stages of production
INDb2	Textile manufacturing	Medium-scale	All stages of production
INDb3	Textile manufacturing	Large scale	All stages of production
INDb4	Textile manufacturing	Large scale	All stages of production
INDb5	Textile manufacturing	Large scale	All stages of production
INDb6	Textile manufacturing	Small scale	All stages of production
INDb7	Textile manufacturing	Small scale	Fabrication only
INDb8	Textile manufacturing	Small scale	All stages of production
INDb9	Textile manufacturing	Small scale	All stages of production
INDb10	Textile manufacturing	Large scale	All stages of production
INDb11	Packaging	Small scale	All stages of production
INDc1	Healthcare and pharmaceutical	Large scale	All stages of production
INDc2	Chemical manufacturing	Large scale	All stages of production
INDc3	Chemical manufacturing	Medium-scale	All stages of production
INDd1	Medical equipment manufacturing	Medium-scale	All stages of production
INDe1	Medical equipment manufacturing	Small scale	All stages of production
INDe2	Automobile products	Large scale	Only limited stages of production
INDe3	Automobile products	Large scale	Only limited stages of production

3.1 Analysis of Cases from Manufacturing Industries

We have analyzed and interpret the results from 21 manufacturing industries from India. The majority of industries were selected from the Northern region of India. Based on industry type and their operations we have sorted all the 21 industries into five groups discussed in Table 1.

In the first group, only two industries were taken from the different industry sectors which were cooperated with a manufacturing consultancy industry (INDa3) which was operating in the field of mechanical and industrial engineering and helped to improve the essential requirements for COVID-19 i.e., ventilators and beds. Also, the consultancy has supported some startups based on smart hand sanitizer machine which helps to minimize the transmission of COVID-19 later directly or indirectly. This industry helps to provide consultancy to industries for designing low-cost bed and other essential equipment. Consequently, the INDa1 which were a rubber manufacturing industry started manufacturing masks in a very short period during the pandemic. The manufacturing consultancy industry provides consultancy to other industries to produce healthcare devices with its 3D printing technology.

In the second group total of 11 industries were included in which mostly textile industries were included (except INDb11 which was a packaging industry). These industries helped to ensure the meet the market demands of masks and maintain the supply to hospitals and COVID care centers. INDb1 has quickly changed its manufacturing system configuration and introduced one manufacturing line dedicated to mask production (more than 20 workers involved). Moreover, manufacturing consultancy has provided the consultancy to INDb1 for the distribution of masks and other essential products during the pandemic all over India. INDb2 has also changed its manufacturing line and set up an online dedicated for the N95 mask production. INDb4 setup special machines with more production efficiency for the mask and gloves production. Most machines were automated so that there will be minimal contact between the workers and machines. INDb8 and INDb10 converted their production plants to a gloves production unit which were capable to meet higher demand from the market side.

In the third group, we have included three chemical industries that produced the sanitizer gel in the pandemic time and completed the heavy demands from the market. For example, INDc1 dedicated one of its production units to sanitizer production only. In the meantime, the rumours from the news and media demand for alcohol-based sanitizer was increased. INDc1 make sure to maintain the quality of sanitizer as per standards so that transmission of COVID-19 can be minimized.

In the fourth group, only one industry was considered INDd1 which was responsible for the production of oxygen cylinders. These oxygen cylinders were highly required for patients with serious conditions and admitted to Intense care units.

In the last group of industries mostly medical equipment industries were included in which INDe1 produced the medical devices during the pandemic time and 2 industries which were initially produced the automobile products produced the ventilators and stretcher during the pandemic time while other industries INDe2 and INDe3 who were responsible for the production of some components and also helped in the assembly operations.

4 Results and Discussion

Overall the reconfigurability of any industry can be synthesized into the four main points which have been explained Fig. 1.

1. As already stated, reconfigurability in manufacturing systems allows industries to change their manufacturing systems quickly and adapt to volatile demands

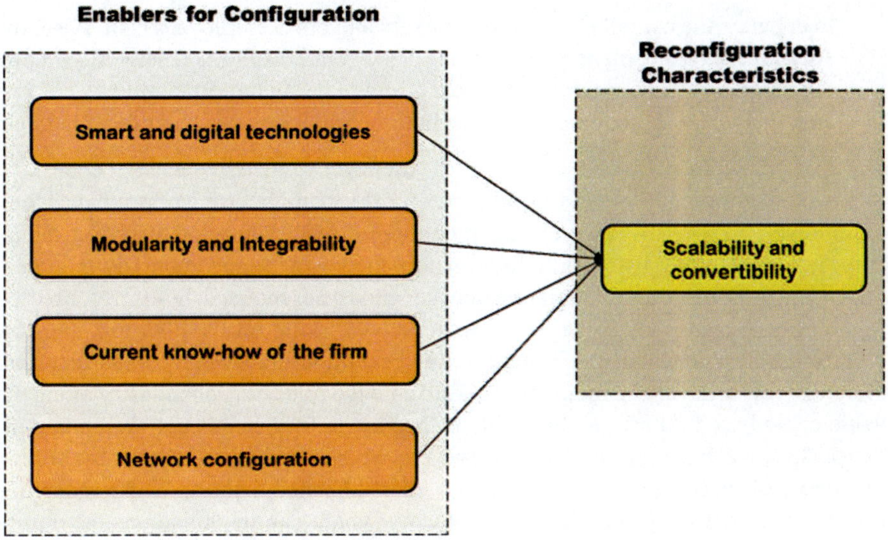

Fig. 1 Reconfigurability characteristics and enablers

from the customer end in which different part families can be produced in minimal time. The COVID-19 epidemic has shown that there is an immediate requirement for reconfigurable manufacturing systems in industries by which they can quickly change their configuration and meet the dynamic market changes. For example, the sudden rise in the demand for sanitizers and masks. Also, in developing nations like India, the automobile industries are known as reconfigurable industries as they can quickly change their configuration. In this pandemic, some automobile industries have quickly changed their configurations and produced other essential items. Generally, automobile industries provide an opportunity to produce a large number of products with different product families.

2. Configuration of the network of manufacturing industries involved in the end product value chain activities. There are many examples of successful collaboration between the industries which shows the importance of linkage between industries to create the value chain.
3. Modularity in end products of any industry enables the integrability and modularity characteristics of manufacturing processes in any manufacturing industry. In our study INDa1 industry which clearly defined their processes which needs the reconfiguration effort, i.e., manufacturing of a mold during the pandemic which help to produce the valve instead of using the traditional valve which were needed by conventional snorkeling masks. Also, INDa2 allowed by its modularity characteristics identify its role in the value chain activities.
4. The role of smart and digital technologies is very important. These digital technologies help to prevent the employees from the virus spread. With the combination of AR and VR technologies (Augmented reality and Virtual reality) it helps

to enhance the overall skills of the workers and other employees [10]. Also, the social media platform helps a lot to build new relationships and networks among the other industries. Additive manufacturing plays an important role by conversion of specific functionalities in the automobile industries and other chemical industries.

The collaboration between the industries for the manufacturing of products along the product value chain helped in the reduction of reconfiguration effort for industries. In a developing nation like India where the business and economy is largely dependent on SMEs the collaboration between industries reduced the effort for reconfigurability. Because SMEs in India are already working on the high mix and low volume production which can be achieved by reconfigurability in the manufacturing systems. This allowed the industries to distribute the roles of each industry along the value chain based on availability. This is the reason further this research broadens the perspective from an industry level to the whole supply chain level by linking the theory of reconfigurability and other well-established theories. Some industries also worked on the dynamic capability theory which can be defined as the ability of a particular industry to "*integrate, build and reconfigure its systems by rapidly changing its environment for the dynamic changes in the market*". However, some studies in the past reported that dynamic theories help to describe that how business is responding to the dynamic changes in the market but fails to describe the capabilities which need to operationalized during the dynamic changes. By linking both the reconfigurability theory and dynamic theory it might be possible for the SMEs for developing nations to bridge the gaps between the effective resource utilization during the dynamic changes in the market with the different manufacturing level in an industry.

5 Conclusion

In the COVID-19 due to the dynamic changes in the market by the sudden increase in demand for sanitizer gel and masks many industries including the MSMEs and SMEs reconfigured their manufacturing systems by reconfigurable theory. These industries were already working on the reconfigurable manufacturing systems concept, with the low reconfiguration effort these industries were able to complete the higher demand for the sanitizer gel and other essential items. However, some industries after the pandemic worked on PPE kit production which helped India to become a PPE kit manufacturer. Based on the analysis of some selected industries and their contribution during the pandemic to ensure the regular supply for essential goods we have provided insights to some enablers for convertibility and scalability characteristics and possible ways by industries can achieve reconfigurability in their manufacturing systems with the low reconfiguration effort. The main limitation of this study is due to the COVID-19 spread, rules and regulations for no physical meetings by local laws it was not possible to conduct the in-depth analysis of presented cases. In the future, this study can be extended by focusing on the cases of industries considered in this study which will help to provide a detailed analysis of manufacturing systems, enablers and challenges faced by industries for the reconfiguration during the pandemic. This will also help to analyze the managerial challenges of reconfiguration in manufacturing systems. This study focused on the analysis of multiple cases from the manufacturing industries which will help to provide insights to practitioners and managers about reconfiguration theory. Also, this study shows how can industries can take benefits by linking dynamic capability theory with reconfigurability theory.

References

1. Agarwal S, Jamwal A, Gupta S (2020) Effect of COVID-19 on the Indian economy and supply chain. Preprints 2020050148. https://doi.org/10.20944/preprints202005.0148.v1
2. Jamwal A, Bhatnagar S, Sharma P (2020) Coronavirus disease 2019 (COVID-19): current literature and status in India
3. Zuniga JM, Cortes A (2020) The role of additive manufacturing and antimicrobial polymers in the COVID-19 pandemic. Expert Rev Med Dev 17(6):477–481
4. Zimmerling A, Chen X (2021) Innovation and possible long-term impact driven by COVID-19: manufacturing, personal protective equipment and digital technologies. Technol Soc 65: 101541
5. Rubio-Romero JC, Pardo-Ferreira MC, Torrecilla-García JA, Calero-Castro S (2020) Disposable masks: disinfection and sterilization for reuse, and non-certified manufacturing, in the face of shortages during the COVID-19 pandemic. Saf Sci 129: 104830
6. Tietze F, Vimalnath P, Aristodemou L, Molloy J (2020) Crisis-critical intellectual property: findings from the COVID-19 pandemic. IEEE Trans Eng Manage
7. Shokrani A, Loukaides EG, Elias E, Lunt AJ (2020) Exploration of alternative supply chains and distributed manufacturing in response to COVID-19; a case study of medical face shields. Mater Des 192: 108749
8. Koren Y, Heisel U, Jovane F, Moriwaki T, Pritschow G, Ulsoy G, Van Brussel H (1999) Reconfigurable manufacturing systems. CIRP Annals 48(2): 527–540
9. Mehrabi MG, Ulsoy AG, Koren Y (2000) Reconfigurable manufacturing systems: key to future manufacturing. J Intell Manuf 11(4):403–419
10. Jamwal A, Agrawal R, Sharma M, Giallanza A (2021) Industry 4.0 technologies for manufacturing sustainability: a systematic review and future research directions. Appl Sci 11(12):5725
11. Jamwal A, Agrawal R, Sharma M, Kumar V, Kumar S (2021) Developing a sustainability framework for Industry 4.0. Procedia CIRP 98:430–435
12. Li X, Wang B, Liu C, Freiheit T, Epureanu BI (2020) Intelligent manufacturing systems in COVID-19 pandemic and beyond: framework and impact assessment. Chinese J Mech Eng 33(1):1–5
13. Okorie O, Subramoniam R, Charnley F, Patsavellas J, Widdifield D, Salonitis K (2020) Manufacturing in the time of COVID-19: an assessment of barriers and enablers. IEEE Eng Manag Rev 48(3):167–175

Machine Learning in CAD/CAM: What We Think We Know So Far and What We Don't

Smriti Upmanyu, Anil Upmanyu, Anbesh Jamwal, and Rajeev Agrawal

Abstract Development in manufacturing activities has shifted the industries towards the adoption of Industry 4.0 practices. Now the industries are adopting key enabling technologies of Industry 4.0 such as machine learning and artificial intelligence in the business practices and manufacturing activities to reduce the human effort and complete the customer demands in the minimum time. The adoption of CAD/CAM with the machine learning techniques in the design engineering has changed the market scenario as well as the business trends all over the world. Still the research is limited in this area. Industries are not aware about the current research progress in this area and there is no study which has mapped the research progress of machine learning in CAD/CAM and the discussed how industries can take benefit from this technology. This study aims at the study mapping of CAD/CAM research progress with machine learning in which we have discussed what has been done and what are some areas which still needs further research work. This study uses the Scopus database for the bibliographic data collection and bibliometrix R package for the analysis of data.

Keywords Computer aided manufacturing · Computer aided design · Machine learning · Industry 4.0 · Bibliometrics · Bibliometric analysis · Scopus

S. Upmanyu
Department of Computer Applications, Rabindranath Tagore University, Bhopal, Madhya Pradesh, India

A. Upmanyu
Prema Software Solutions, Bhopal, Madhya Pradesh, India

A. Jamwal · R. Agrawal (✉)
Department of Mechanical Engineering, Malaviya National Institute of Technology, J.L.N. Marg, Jaipur, Rajasthan 302017, India
e-mail: ragrawal.mech@mnit.ac.in

R. Agrawal et al. (eds.), *Recent Advances in Industrial Production*, Lecture Notes in Mechanical Engineering, https://doi.org/10.1007/978-981-16-5281-3_48

1 Introduction

At present industries are moving towards the adoption of Industry 4.0 which is also known as fourth industrial revolution [1]. Industry 4.0 term was coined in Germany in 2011 and was the high technology plan for the German industries. Industry 4.0 can be defined as "current trends in data exchange and automation in manufacturing activities" [2]. The key enabling technologies for Industry 4.0 are block-chain technology, artificial intelligence, internet of things and additive manufacturing [3]. Machine learning and deep learning are known as the subset for the artificial intelligence. Now the application of machine learning algorithms in Computer aided design (CAD) and computer aided manufacturing (CAM) is gradually becoming a research hotspot for many researchers.

Artificial Intelligence (AI) is commonly used in CAD and due to the increase in the power of computing CPU/GPU, the use of Machine Learning has also started becoming common [4]. This paper discusses some implementations of Machine Learning in engineering software (CAD/CAM).

Research has been conducted in the CAD/CAM, but most of these have not been adopted by the industry [5–7]. This paper lists out the current scenarios in which ML have been used in CAD/CAM by the industry. These methods include the adaptive Icon and searching for parts based geometry.

The manufacturing sector is changing in present time to meet the volatile customer demands [8]. Now the industries are focusing to execute faster and cheaper projects [9]. AI can be used in the industries to reduce the repetitive functions as well as CAM users [10]. In the next we have discussed about the uses of machine learning in the part search.

1.1 Uses in Part Search

Searching is ancient and used by everyone who uses a computer. Types of search used to search parts are listed below (engineers can use any combination of searches given below).

Text-Based Search
Text-based search is simple and is the search used by everyone daily. It searches using the file name, Part Number, file properties, dimension or callout [7].

Feature Base Search
This searches using simple features like cylinder, cone, hole, slot. or complex features like circular pattern or linear pattern. It can be augmented using Machine Learning [11].

Profile Search
This searches using profiles of parts with help of Machine Learning. A constrain profile is given as input and outputs parts with similar profiles. In this method, a mashed Image can also be used as input for the profile [12].

Similar Part Search
A 3D geometry part is taken as input and similar parts will be searched for on Internet/Cloud or local server, and results are displayed on the screen [13].

Partial Part Search
In this case, the user can input the portion or surface of the part or a combination of those. The search will then search for parts with those surfaces/portion on cloud/Internet or local server automatically [14].

Inverse Part Search
This is mostly the same as partial part search except that it searches for the male/female inverse of the surface. e.g. if you have plastic part then both the halves of mold can be searched using inverse part search [15].

In the present study we have mapped out the research work done in the CAD/CAM area with machine learning techniques. The research questions for the present study are:

1. What is the publication trend, keywords and authors working in this area?
2. How implementation in GUI can be done?

The next sections of paper discuss about the research methodology adopted and bibliographic results obtained from the analysis.

2 Research Methodology

In the present study we have collected the data from the largest scientific database Scopus. Scopus is one of the largest databases for scientific database which includes the peer reviewed research articles, book chapters and conferences papers [16]. In the past few studies on the systematic literature review various authors have reported the advantages of using Scopus database over other databases [5, 6]. The initial search for the article collected started with the keywords in this study we have used the specific keywords for the article collection which are:

1 "CAD" OR "CAM" OR "computer aided manufacturing" OR "computer aided design" AND "Machine learning"

These are the above-mentioned keywords used as the search string for the article collection. The article which were not related to the machine learning or doesn't related the use of these keywords were excluded from the database. Documents other than research articles, conference articles and review articles were excluded from the final database. Articles published in other language except English were also excluded from the database. The final set of articles 1254 documents were considered for the study from the time span of 1990–2021. In which we found that before the 2011 articles were majorly focused on the artificial intelligence concept than the machine learning. It is true that machine learning is the subset of the artificial intelligence but to maintain the reliability and validity of study we have considered the literature of machine learning in this article. Further Table 1 shows the main information about the publication in CAD/CAM area with the machine learning techniques.

Table 1 Main information about publications in CAD/CAM with machine learning

Description	Results
MAIN INFORMATION ABOUT DATA	
Timespan	1990:2021
Sources (Journals, Books, etc.)	669
Documents	1254
Average years from publication	3.78
Average citations per documents	14.73
Average citations per year per doc.	2.792
References	44,878
DOCUMENT TYPES	
Article	652
book chapter	32
conference paper	454
Review	73
DOCUMENT CONTENTS	
Keywords plus (ID)	7160
Author's keywords (DE)	2922
AUTHORS	
Authors	4637
Author appearances	6148
Authors of single-authored documents	42
Authors of multi-authored documents	4595
AUTHORS COLLABORATION	
Single-authored documents	81
Documents per author	0.27
Authors per document	3.7
Co-authors per documents	4.9
Collaboration index	3.92

2.1 Results and Discussion

In this section we have presented the results obtained from bibliometric based review from the articles collected from Scopus database. The analysis is done with the Bibliometrix R package by R studio on Mac OS Big Sur version 11.2.2. The raw bibliographic data is exported from database in BibTex format and this format is used for the analysis in R studio by using the command.

```
>library (bibliomertix)
> detach("package:bibliometrix", unload = TRUE)
> biblioshiny()
```

The results from analysis is shown in next sections.

Annual Scientific Production in Articles

The annual scientific production of CAD/CAM articles is found by the R package in which we have observed the annual growth rate of 14.87% in the article publication. Publication in the area was very less till 2004 as this is the time when artificial intelligence (AI) in the CAD/CAM was in the developing stages very few authors were known about the capabilities and advantages of using AI in machine learning. The concept of Industry 4.0 came in 2011 from Germany which promote the area of AI and other technologies which plays an important role in the development of Industry 4.0 as the key enabling technologies. From the Fig. 1 we can see that there is increase in the number of publications after the 2012. Now around the world more

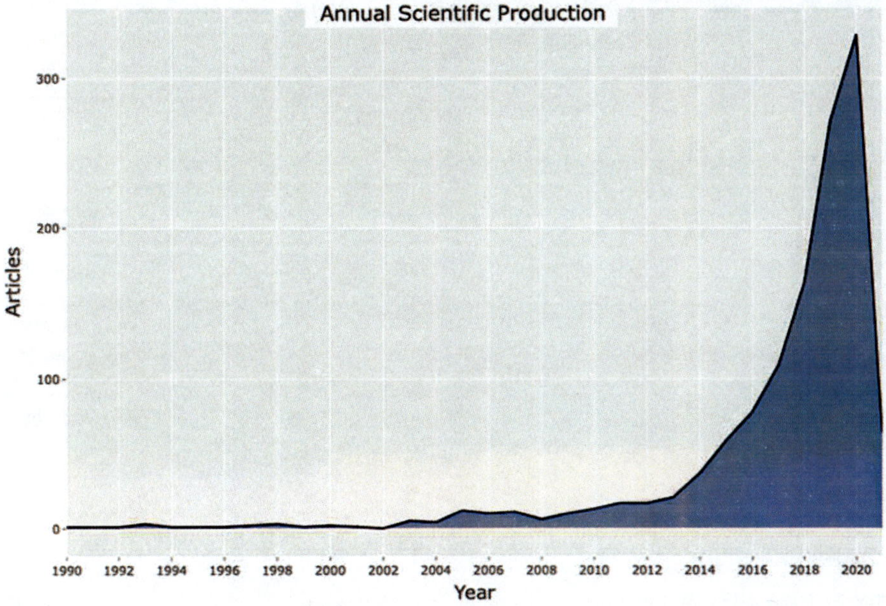

Fig. 1 Annual scientific production in the publications

researchers are working in this area which has recently promote this area and have found out new research scopes in related to designing and manufacturing.

Three Field Plots

Three field plot help to visualize the main items of bibliographic data in the three fields through the Sankey diagram. These three fields can be authors, keywords, countries, journals, abstracts, titles and references. In our case we have considered the three main items as: countries, authors and keywords which help to visualize which countries are working on CAD/CAM and who are the authors from these countries are working in this research area? What kind of keywords they have used in their research? Fig. 2 shows the three field plot for the bibliographic data in which USA and China are working on the CAD/CAM in machine learning area. Total of six most productive authors are associated with these countries and they are working on the keywords like machine learning, deep learning, CAD, feature selection and artificial intelligence. The source code for three field plot is shown below:

```
threeFieldsPlot(
M,
fields = c ( "AU", CO", "DE")
n= c (10, 10, 10)
width = 1200
height = 600
)
```

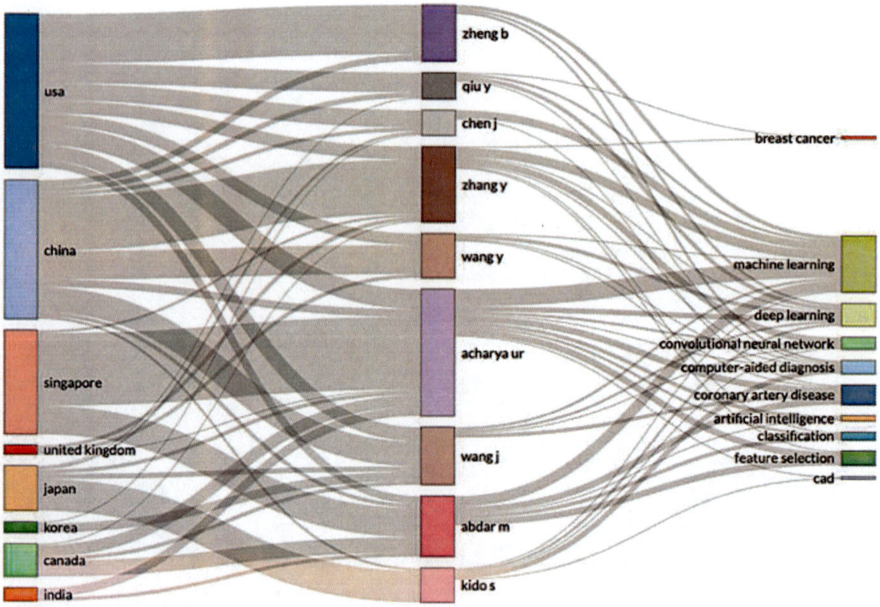

Fig. 2 Three field plot between countries, authors and keywords

Table 2 Source clustering through Bradford's law

Source output	Rank	Freq.	Cum freq.
Progress In biomedical optics and imaging—Proceedings of SPIE	1	49	49
Lecture notes in computer science	2	36	85
Computer methods and programs in biomedicine	3	27	112
Proceedings of SPIE—The International Society for Optical Engineering	4	23	135
Advances in intelligent systems and computing	5	18	153
Computers in biology and medicine	6	18	171
ACM international conference proceeding series	7	15	186
Medical physics	8	15	201
IEEE access	9	14	215
IEEE transactions on medical imaging	10	14	229

Source Clustering Using Bradford's Law

The Bradford law's helps in the source clustering by finding the relationship between the top productive journals *(n)* and their cumulative yield *R(n)*. Generally, the quantitative value of the journal is measure by its literature yield which can be calculated as:

$$R(n) = h\log(\frac{n}{u} + 1) + R(0) \quad for \;\; n \geq 0$$

Here *h, u and R (0)* are constant.

Table 1 represents the source clustering using the Bradford's law in which ranking of top 10 journals is done with the Bradford's law. It is found that *Progress in Biomedical optics and Imaging—Proceedings of SPIE* is at most productive source with the frequency of 49 and cumulative frequency 49. Similarly, *Lecture notes in Computer Science* is at second rank with the cumulative frequency of 85. Top 10 journals with ranks, frequencies and cumulative frequencies is shown in Table 2.

Author Productivity Measurement Using Lotka's Law

Lotka law's helps to calculate the frequency of publications done by authors in the particular research area. Lotka law's states that number of authors from same research area contributing *x* contributions in the given period of time is the fraction of number of single contributions done by author by following the Formula $1/x^{\alpha}$.

Here α nearly always equals to two.

The general formula is:

$$X^n Y = C$$

Table 3 Lotka's law author productivity

Documents written	No. of authors	Proportion of authors
1	3829	0.826
2	524	0.113
3	134	0.029
4	64	0.014
5	40	0.009
6	9	0.002
7	12	0.003
8	12	0.003
9	4	0.001
10	4	0.001

Or

$$Y = C/X^n$$

Here X represents the number of publications and Y represents the relative frequency of authors associated with X publications and C and$n$ are the constants.

In the analysis we have found out that 3829 authors have written only 1 article and 524 authors have written 2 articles. Only 4 authors have published more than 10 articles which is 0.001 of total proportion of authors. The authors productivity based on Lotka's law is presented in Table 3.

Local and Global Citation Structure

Local citations calculates how many time a particular document or an author included in this collection have cited by the documents (articles, conference articles) included in that collection. The local citation score for the articles is calculated by:

```
localCitations(M, fast.search = FALSE, sep = ";", verbose = FALSE)
```

Global citation calculates how many time a particular document or an author included in one collection have cited by the documents (articles or conference articles) included in other collection. The global citation score for the articles is calculated by:

```
globalCitations(M, fast.search = FALSE, sep = ";", verbose = FALSE)
```

The local and global citation of top cited documents is shown Table 4.

Co-occurrence Network Analysis

Co-occurrence network analysis helps to visualize graphical relationship between the keywords in bibliographic data. In this study we have visualize the co-occurrence network analysis between 2922 keywords. In this we have used association normalization approach with Louvain clustering algorithm and considered total of 50 nodes.

Table 4 Local and global citation structure

Document	LC	GC	LC/GC ratio (%)	NLC	NGC
SHIN HC, 2016, IEEE TRANS MED IMAGING	27	1950	1.38	16.45	33.88
ALIZADEHSANI R, 2018, COMPUT METHODS PROGRAMS BIOMED	20	36	55.56	21.45	1.87
KOOI T, 2017, MED IMAGE ANAL	18	383	4.70	21.56	16.68
ABDAR M, 2019, COMPUT METHODS PROGRAMS BIOMED	16	47	34.04	41.69	8.50
SETIO AAA, 2016, IEEE TRANS MED IMAGING	14	523	2.68	8.53	9.09
RIBLI D, 2018, SCI REP	12	166	7.23	12.87	8.63
XIE W, 2016, NEUROCOMPUTING	12	84	14.29	7.31	1.46
AL-MASNI MA, 2018, COMPUT METHODS PROGRAMS BIOMED	11	95	11.58	11.80	4.94
EL-BIALY R, 2015, PROCEDIA COMPUT SCI	11	55	20.00	9.14	1.60
ALIZADEHSANI R, 2019, COMPUT BIOL MED	10	26	38.46	26.06	4.70

The graphical parameters considered are 0.7 opacity with 50 number of labels and 6 label size. Here in the visualization we can see that machine learning, deep learning, neural networks and learning algorithms are in the same clusters. We can also see that keywords like CAD, computer aided design and CAM shows the good linkage with another cluster keywords. The Co-occurrence network for keywords analysis is shown in the Fig. 3.

Conceptual Structure Map by Multiple Corresponding Analysis

The conceptual structure function in bibliometrix package helps to produce three map structure in which conceptual structure map is the factorial map of the research articles with the highest number of contributions and most cited research articles. Figure 4 represent the different clusters for CAD/CAM research in machine learning and corresponding variables.

In the graph the vertical and horizontal dimensions for multiple comparison analysis is used for the analysis of conceptual map. For the horizontal category the left side of graph has a strong research area for machine learning and deep learning algorithms for CAD/CAM research while right side having the weak research work on machine learning algorithms. In the graph the research is shown in the two different clusters which shows the research in CAD/CAM.

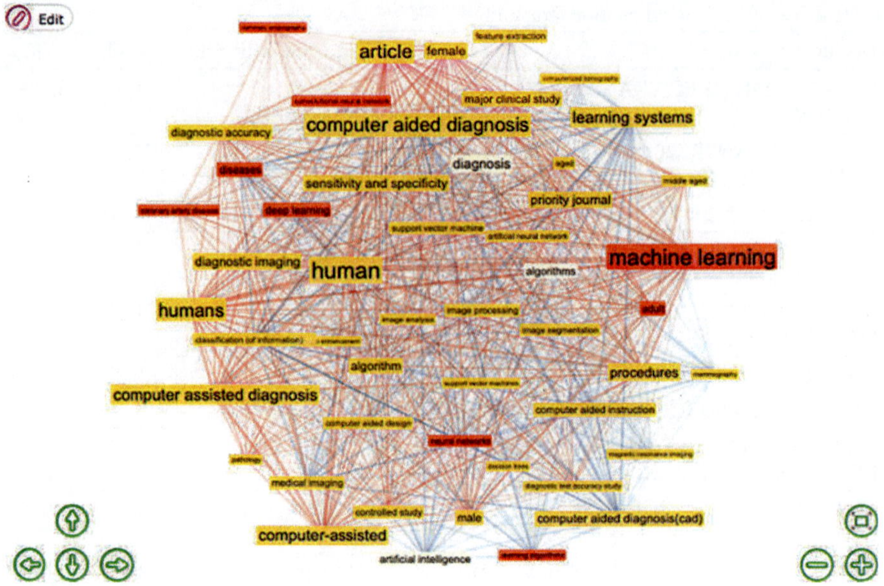

Fig. 3 Co-occurrence network for keywords

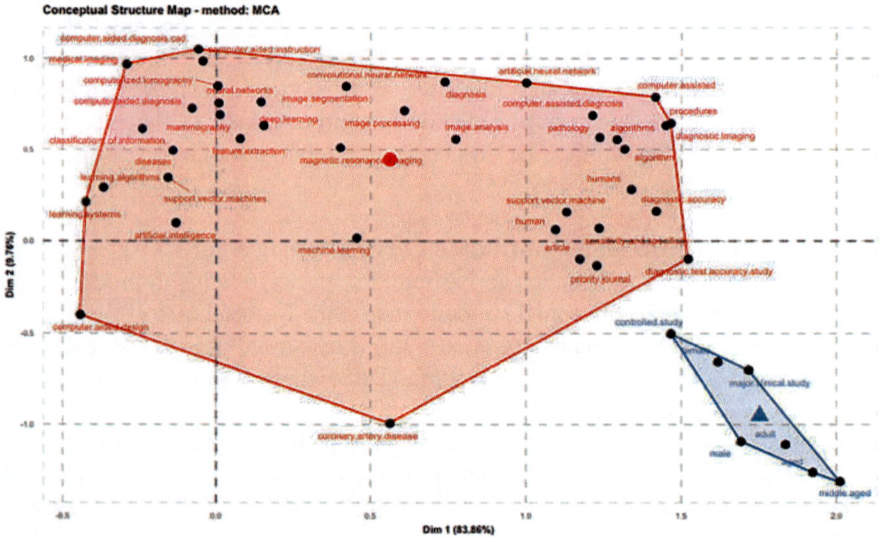

Fig. 4 Conceptual structure map by multiple corresponding analysis

3 Implementation in GUI

To give ease to user CAD/CAM companies are making time to time changes in the GUI which are as fallows.

3.1 Last Use on Top

The user interface will show the last used command will be displayed on top in a stack.

3.2 Adaptive User Interface

This feature heavily uses Machine Learning. The CAD Software track all icon which the user has used. Depending on the most used commands. Product software automatically creates a set of commands in adaptive user interface. This also suggests the next command depending on the previous command.

In some CAD/CAM software, there is a further facility to copy the GUI of an experienced user and give it to a new user.

To give ease to user CAD/CAM companies are making time to time changes in the GUI which are as fallows.

3.3 Last Use on Top

The user interface will show the last used command will be displayed on top in a stack.

3.4 Adaptive User Interface

This feature heavily uses Machine Learning. The CAD Software track all icon which the user has used. Depending on the most used commands. Product software automatically creates a set of commands in adaptive user interface. This also suggests the next command depending on the previous command.

In some CAD/CAM software, there is a further facility to copy the GUI of an experienced user and give it to a new user.

4 Conclusion

In the present study we have utilized the Scopus database for the collection of articles related to CAD/CAM and machine learning. We found that AI in CAD/CAM is in its initial stages of development. Very few industries are adopting this technique due to complexity and requirement of skilled operators. It is true that industries can take benefits from the adoption of machine learning techniques in CAD/CAM but each of technique has their advantages and limitations and it requires higher knowledge about digitalization and machine learning. In this paper we have discussed how implementation can be done in GUI. Moreover, we have discussed about the research progress in this area with help of Scopus database by use of R studio. Further, this study can be extended by considering the Web of Science and other relevant databases for study mapping.

References

1. Lasi H, Fettke P, Kemper HG, Feld T, Hoffmann M (2014) Industry 4.0. Bus Inf Syst Eng 6(4): 239–242
2. Lasi H, Fettke P, Kemper HG, Feld T, Hoffmann M (2014). Industry 4.0. Bus Inf Syst Eng 6(4): 239–242
3. Jamwal A, Agrawal R, Sharma M, Kumar A, Kumar V, Garza-Reyes JAA (2021) Machine learning applications for sustainable manufacturing: a bibliometric-based review for future research. J Enterp Inf Manage
4. Yamaguchi S, Lee C, Karaer O, Ban S, Mine A, Imazato S (2019) Predicting the debonding of CAD/CAM composite resin crowns with AI. J Dent Res 98(11):1234–1238
5. Papadiochou S, Pissiotis AL (2018) Marginal adaptation and CAD-CAM technology: a systematic review of restorative material and fabrication techniques. J Prosthet Dent 119(4):545–551
6. Janeva NM, Kovacevska G, Elencevski S, Panchevska S, Mijoska A, Lazarevska B (2018) Advantages of CAD/CAM versus conventional complete dentures—a review. Open Access Maced J Med Sci 6(8):1498
7. Pereira ALC, Medeiros AKB, Santos SK, Almeida ÉO, Barbosa GAS, Carreiro ADFP (2020) Accuracy of CAD-CAM systems for removable partial denture framework fabrication: a systematic review. J Prosthet Dent
8. Jamwal A, Agrawal R, Sharma M, Giallanza A (2021) Industry 4.0 technologies for manufacturing sustainability: a systematic review and future research directions. Appl Sci 11(12):5725
9. Jamwal A, Agrawal R, Sharma M, Kumar V (2021) Review on multi-criteria decision analysis in sustainable manufacturing decision making. Int J Sustain Eng 1–24
10. Thalji G, Jia-mahasap W (2017) CAD/CAM removable dental prostheses: a review of digital impression techniques for edentulous arches and advancements on design and manufacturing systems. Curr Oral Health Rep 4(2):151–157
11. Goujat A, Abouelleil H, Colon P, Jeannin C, Pradelle N, Seux D, Grosgogeat B (2019) Marginal and internal fit of CAD-CAM inlay/onlay restorations: a systematic review of in vitro studies. J Prosthet Dent 121(4):590–597
12. Amesti-Garaizabal A, Agustín-Panadero R, Verdejo-Solá B, Fons-Font A, Fernández-Estevan L, Solá-Ruíz MF (2019) Fracture resistance of partial indirect restorations made with CAD/CAM technology: a systematic review and meta-analysis. J Clin Med 8(11): 1932

13. Patil M, Kambale S, Patil A, Mujawar K (2018) Digitalization in dentistry: CAD/CAM—a review. Acta Sci Dental Sci 2(1):12–16
14. Abdulla MA, Ali H, Jamel RS (2020) CAD-CAM technology: a literature review. Al-Rafidain Dental J 20(1):95–113
15. Carvalho IFA, Marques S, Araújo FM, Azevedo LF, Donato H, Correia A (2018) Clinical performance of CAD/CAM tooth-supported ceramic restorations: a systematic review
16. Jamwal A, Agrawal R, Sharma M, Kumar V, Kumar S (2021) Developing a sustainability framework for Industry 4.0. Procedia CIRP 98:430–435

Assesment of Traditional and Hybrid Controller for Controlling Robotic Manipulator System

Aditi Saxena, Jitendra Kumar, Vinay Kumar Deolia, and Debanik Roy

Abstract Controlling of robot has always became an exigent work for engineers these controllers were came into existence since 1950, when a PID controller was used to control complex systems the structure of this controller was very easy to design and has always remain a most popular controller in the industry because of its lucidity and low cost. it is sufficient enough to handle linear system but was unfit for nonlinear system, researchers and scientist has tried to resolve the drawbacks of PID by fuzzy logic controller which was effective for nonlinear system further as the year passes by new controlling techniques has arrived which were more powerful in terms of controlling in comparison to the previous ones. neuro logic controller came to handle those systems which are complex, highly nonlinear and poorly dynamically analyzed and further this controller was combine with the Traditional controllers and give rise to neuro PID controller and neuro Fuzzy PID controller. This paper presents a detailed review on various controlling techniques which are being used in the controlling of robotic manipulator starting from the development of the first controller till now.

Keywords Controlling · Nonlinear system · Robotic manipulator first section · Traditional controller · Hybrid controller

1 Introduction

As we all know that today's world is moving towards robotization and automation, there is high necessity of robot in today's world with the help of robot we can produce high quality of products in lesser period of time, skilled labor shortage issue can be fulfilled, pressure to increase the production rates to compete the market would be handled smoothly, best quality of products can be developed in shorter period of time and also productivity can be increased. The word robot is being originated from a Czech word robot a which means slaved or force labor but in order to manipulate the

A. Saxena (✉) · J. Kumar · V. K. Deolia · D. Roy
IET Department of Electronics and Communication Engineering, GLA University, Mathura 281406, India

R. Agrawal et al. (eds.), *Recent Advances in Industrial Production*, Lecture Notes in Mechanical Engineering, https://doi.org/10.1007/978-981-16-5281-3_49

robot we highly need a controller as without a controller robot could itself became a disaster, in terms of robotics robot arm is known as robotic manipulator. There are various controllers and techniques which are used to control and manipulate the manipulator. Controlling of robot has always became an exigent work for engineers, parameters of controller are needed to be tuned very finely accurately and precisely in order to control a robotic manipulator and also to achieve high trajectory tracking, hence in the upcoming year it has become a very tedious and challenging task to develop such control system which is able to handle complexities and non-linearity's [1–3]. To cope up with today's demand we need to design an intelligent control system which could be adaptive and robust. To design such control system various controllers and control techniques are used which also took us towards interdisciplinary fields, for example suppose we are talking about to control a robotic manipulator so the first thing which we need is the physical model of the robot which falls under mechanical engineering next step involve designing of controller for the plant (manipulator) so this controller designing would fall under electronics and communication field and in order to make the controller more intelligent adaptive and autonomous we may can used artificial intelligence, neural network which falls under computer science and engineering so in short we can say that this research topic is itself a combination of various interdisciplinary branch [4–6], this interdisciplinary research has very widely increase the area of research by adding fuzzy control, neural network [7–10]. Fuzzy controller is a remarkable combination of artificial intelligence and control engineering [11, 12] this intelligent controller has made a successfully mark in the commercial applications [13] as fuzzy controller has many advantages that it can easily handle all the non-linearity's. It has less rise time and maximum overshoot, fuzzy logic provides certain sort of flexibility to the research engineering's as there parameters are not just confined between true and false fuzzy logic provides a solution to a problem who somewhere or the other lies between true and false here we can take an example of controlling an autonomous car so for break system fuzzy logic provides flexibility such that we can handle the break if suddenly some pedestrian or some vehicle comes in our way [14–17].

2 Brief Analysis of Fuzzy Controller

In 1788 the first controller was experienced by James watt, the feedback control loop technique has been used for controlling fly-ball governor, which was acting similar like proportional controller and the main work of this controller was to control and regulate the speed of the engine [18–21]. After this several years pass by and generous work has been done but the whole PID controller came into existence in 1911 where this controller combination has been used for controlling the automatic sheep steering. The control law that we now used commonly arise in 1922 further minor-sky in detail discussed and analyzed the properties of the three controller and also published a paper on it [22–24]. Pneumatic controller came into existence since 1933 its name was model 56R fulscope controller this model was introduced by

Taylor instrument company this model feature was that it was completely proportional controller however this controller alone cannot control a process variable as this controller has some drawbacks too, in this controller we cannot remove error permanently, according to theories steady state error is inversely proportional to proportional controller so each time there is some residual error or offset which remains present in it [25–27]. So this has become a tedious task for control engineering's, To remove this permanent residual error hence various research has been carried out in this direction and finally in 1930 they came up with a solution that set point can be reset either to a higher or to a lower value this set point will be kept on resetting until the error became zero basically what is happening in this method was that the error is being integrated again and again whenever we will reset the set point and the output from here is been merged with the proportional controller and which has given rise to the term proportional integral controller (PI) this PI controller is being first introduced by Foxboro in 1934–1935, But this PI controller alone could not give optimum result as there were some drawbacks in it like if the integral action is in more power and more and more integration has been performed the accuracy was affected and the accuracy of achieving perfect feedback control is not up to the mark in PI controller, also a problem or situation could occur which is known as hunting where when the controller tries to overcorrect the error, at the same time a new error is formed in the opposite direction whose magnitude can also be greater than the previous one and when this happens the controller start navigating its output and swap the output between fully on and fully off [28]. To remove these difficulties and to handle these type of phenomenon in 1935 a new controller arrived with the advance version of the above discussed controller in which there is an action present in the controller called pre-act which helps in handling the reset and proportional controller this controller was introduce by Taylor instrument company after this in the same year Foxboro instrument company added an hyper reset controller, the derivative of the error signal are being provided by the action due to this control action by adding these two action. These two controller gives an control action to the integral of error signal and hence both work as PID controller [25]. The PID controller is a combination of all the three controller and the block diagram of classical PID controller is shown in the figure, the advantage of derivative controller present here is that it provides a spike to the controller output whenever due to setpoint an abrupt change occur this controller helps in handling the change [26] but the biggest issue which arrive in this PID controller was to find the suitable parameters of PID controllers (Fig. 1).

This issue was resolved in the year 1942–1943 by the Taylor engineers Ziegler and Nichols when they developed a method of tunning which is still very popular in 20th century and the name to this tunning method was given on the basis of their name [29, 30] hey also published their two research paper on this tunning method the first research paper was on open loop test on the plant and the second research paper was on closed loop test on plant further 2–3 years research has been carried out in the tunning of the parameter and then in 1950 cohen and coon [25] give the flexibility in tunning by providing alternative choice of choosing parameter for certain plant

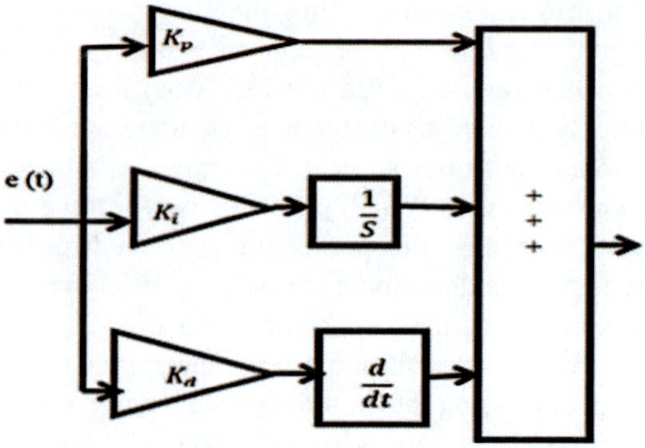

Fig. 1 PID control system

3 Brief Analysis of Fuzzy Controller

Intelligent techniques were arrive in 1990s which has overcame the drawbacks of classical and traditional PID controller and the combination of PID controller with fuzzy logic completely resolve all the drawbacks of PID and Fuzzy controller and also increase the advantages as a whole and as a result it has been noticed that by combining these two techniques a robust and much better control system can be achieved [26]. Control system which are based on fuzzy logic are in detail analyzed by Wang and Kwok [29] and then they themselves suggest the combination of fuzzy pd and fuzzy I controller, further by making a small improvement in the fuzzy PI controller and Gatland [31] proposed a fuzzy three term controller with small change in fuzzy PI controller in which a normal and simple two dimensional rule base is being used Vandoren VJ. [30] focused on the design concepts of controller related to the structures of fuzzy and pid and the combinational structure of fuzzy pid controller after analyzing the structure they also did a comparative study on fuzzy PD and fuzzy pi controller. The best advantage of these type of controller are that due to fuzzy logic the designing and development of this controller would now be became much more cheaper in comparison to the traditional controllers this combination also eliminates the individual drawbacks of fuzzy and PID controller fuzzy logic rule base and works are much more similar to human for computation of this logic very easy mathematics is being used for non-linear and complex systems when we use this controller for controlling a robotic manipulator the result which we achieved are very precise whether we can talk in terms of trajectory tracking or to handle uncertainty further more research have been done and many researchers have used this controller (Fig. 2).

This combination [32] of pid and fuzzy logic has been used for controlling photovoltaic inverters which are connected in grid this combination can now easily control the nonlinear objects like electricity grid also it automatically changes the value of gain whenever there is a change occur in the load and the desired power. A fuzzy self-tuning PID [33] semi-global regulator for controlling the robotic manipulator it contains a proof of semi-global asymptotic stability and has used this controller on

Fig. 2 Explaining process of fuzzy controller

two link robotic manipulator. A comparison of [34] conventional and fuzzy based sliding mode PID controller for robotic manipulator, and the simulation has been done on MATLAB SIMULINK [35].

4 Brief Analysis of Neural Network Controller

If the plant would be non-linear than the function will be nonlinear. To drive the plant which is linear it is very tough to choose and provide the correct and appropriate input vector which can drive the plant from the initial or present state to the desired one that is from the past year the approach which we are using to solve this issue is around a number of operating points we linearize the plant for non-linear plant this approach requires high design effort by the researchers and the engineers and the computation is also bit tough in this approach.

Basically, for training a controller through neural network the main aim is to drive the plant to the desired state for which it is highly required to produce a correct signal here the value of will decide the present state of the time for training of the neural network various diverse methods have been used like reinforcement learning inverse control and optimal controller. Various different approaches for training have been used and proposed by Stearns Widrow and Jordan. The above figure shows how the training process starts in an neural network firstly the plant is in initial state whether it is linear or non-linear after this randomly input of the plants started generated at time the current state of the plant and the input of the neural net are set to be equal, back-propagations method is used to trained the network by this method we can predict the next state of the plant in back-propagation the plant identification is been done automatically by the neural network [36, 37].

5 Conclusion

The review of various controlling techniques has been presented in this paper, starting from the development of the traditional PID controller and its eccentricity are being presented in an consecutive order, further the advancement of the traditional PID controller using fuzzy logic has been discussed as fuzzy logic is based on heuristic approach due to which it plays a remarkable role in the improvement of PID controller, For making the controller more intelligent Fuzzy logic has been combine with PID controller and give rise to Fuzzy PID controller this combination results in reduced rise time and overshoot but simultaneously runtime has been increased and this controller was not robust for the implementation of learning strategy as the capability to receive the feedback was very low in fuzzy PID controller, after doing a lot of research neuro controller has make it existence in the control field its numerical strength has helped to perform more than one work at a time and further this neuro controller has been combine with PID Controller and fuzzy controller prominent results were obtained by the combination of neuro PID controller in comparison to neuro fuzzy controller while neuro fuzzy works well for controlling and for making the controller adaptive by using adaptive Neuro fuzzy interference system, further we can apply various optimization techniques like genetic algorithm and particle swan optimization for performing optimal tunning of the gains. The main purpose of this study is to give an overview to the researchers and control engineer who are keenly interested to work in this area so that if they will once go through the paper, they will be knowing various techniques merits and demerits of each technique so that while implementing new controller they keep all this overview in mind.

References

1. Bequette BW (1991) Nonlinear control of chemical process: a review. Ind Eng Chem Res 30(7):1391–1398
2. Seborg DE (1994) A perspective on advanced strategies for process control. MIC—Model Identif Control 15(3):179–189
3. Stephanopoulos G, Han C (1996) Intelligent systems in process engineering: a review. Comput Chem Eng 20(6–7): 743–791
4. Passino KM (1995) Intelligent control for autonomous systems. IEEE Spectr 32(6):55–62
5. Passino KM (1996) Intelligent control. In: Levine W (ed) The control handbook. CRC Press, Boca Raton, pp 999–1001
6. Antsaklis P, Passino KM (1993) An introduction to intelligent and autonomous control. Kluwer Academic Publishers, Norwell
7. Chen CT, Peng S-T (1999) Intelligent process control using neural fuzzy techniques. J Process Control 9(6): 493–503
8. Zumberge J, Passino KM (1998) A case study in intelligent versus conventional control for a process control experiment. J Control Eng Pract 6(9):1055–1075
9. Zumberge J, Passino KM (1996) A case study in intelligent versus conventional control for a process control experiment. In: Proceedings of the 1996 IEEE international symposium on intelligent control, Dearborn, MI, USA, pp 37–42
10. Kuswadi S (2001) Review on intelligent control: its historical perspective and future development. IECI J Ser 3(2):38–46
11. Bernard JA (1988) Use of a rule-based system for process control. IEEE Control Syst Mag 8(5):3–13
12. Reznik L, Ghanayem O, Bourmistrov A (2000) PID plus fuzzy controller structures as a design base for industrial applications. Eng Appl Artif Intell 13(4):419–430

13. Chiu S (1997) Developing commercial applications of intelligent control. IEEE Control Syst Mag 17(2):94–100
14. Driankov D, Hellendoorn H, Reinfrank M (1993) An introduction to fuzzy control. Springer-Verlag, N Y
15. Passino KM, Yurkovich S (1998) Fuzzy control. Menlo Park, Addison Wesley Longman, CA
16. Chen G, Pham TT (2001) Introduction to fuzzy sets, fuzzy logic and fuzzy control systems. CRC press
17. Chen G (2006) Introduction to fuzzy systems. Chapman and Hall/CRC press
18. Zhang H, Liu D (2006) Fuzzy modeling and fuzzy control. Birkhäuser, Boston
19. Jantzen J (2007) Foundations of fuzzy control. John Wiley Sons
20. Yen J, Langari R (2003) Fuzzy logic: intelligence, control, and information. Pearson Education, India
21. Lee KH (2004) First course on fuzzy theory and application. Springer
22. Yager RR, Filev DP (2002) Essentials of fuzzy modeling and control. Wiley-India
23. Ross TJ (2005) Fuzzy logic with engineering applications, 2nd edn. Wiley-India
24. Coales JF (1956) Historical and scientific background of automation. Engineering 182:363–370
25. Bennett S (1996) A brief history of automatic control. IEEE Control Syst Mag 16(3):17–25
26. Bennett S (1993) Development of the PID controller. IEEE Control Syst Mag 13(6):58–65
27. Bennett S (1979) A history of control engineering 1800–1930. Peter Peregrinus, Steven Age
28. Minorsky N (1922) Directional stability of automatically steered bodies. J Am Soc Naval Eng 34(2):290–309
29. Bennett S (2001) The past of PID controllers. Annu Rev Control 25:43–53
30. VanDoren VJ (2003) PID: still the one. In: Control engineering, pp 28–29
31. Bennett S (2000) The past of PID controllers. In: Proceedings of IFAC workshop on digital control: past, present and future of PID control
32. Liptak BG (2003) Process control and optimization, 4th edn. Instrument Engineer's Hand Book, CRC press, London
33. Shinskey FG (1994) Feedback controllers for the process industries. McGraw-Hill, New York
34. Author F, Author S (2016) Title of a proceedings paper. In: Editor F, Editor S (eds) Conference 2016, LNCS, vol 9999. Springer, Heidelberg, pp 1–13
35. Author F, Author S, Author T (1999) Book title, 2nd edn. Publisher, Location
36. VanDoren VJ (2000) Understanding PID control. Control Eng 53–56
37. Author F Contribution title. In: 9th International proceedings on proceedings. Publisher, Location, pp 1–2